Methods of Inverse Problems in Physics

Author

Dilip N. Ghosh Roy
Technical Research Associates
Salt Lake City, Utah

CRC Press
Boca Raton Ann Arbor Boston

Library of Congress Cataloging-in-Publication Data

Ghosh Roy, Dilip N., 1939-
Methods of inverse problems in physics / Dilip N. Ghosh Roy.
p. cm.
Includes bibliographical references.
ISBN 0-8493-6258-X
1. Inverse problems (Differential equations) 2. Mathematical physics. I. Title.
QC20.7.D5G48 1991
535'.01'515353—dc20 90-1499
CIP

Direct all inquiries to CRC Press, Inc., 2000 Corporate Blvd., N.W., Boca Raton, Florida, 33431.

International Standard Book Number 0-8493-6258-X

Library of Congress Card Number 90-1499
Printed in the United States

PREFACE

Inverse problems occur in numerous disciplines of physical sciences and constitute a significant portion of current research in physics and mathematics. The objective of this volume is to discuss the essence and the nature of inverse problems, their various settings and perspectives, and the diverse methods of their solutions as they arise in physical, mathematical, and engineering sciences. Considering the enormous importance of inverse problems, it is natural that a number of beautiful and elegant books and review articles exist on the subject. However, the focus is mostly on treating these problems in the context of specific settings such as quantum scattering, integral geometry, or imaging. Moreover, the more elegant works on the topic are also usually very specialized and are beyond the reach of anyone but the properly initiated. This book seeks to fill in these gaps.

The book is aimed at a broad audience that includes graduate students and researchers in inverse problems, as well as those who are not directly involved in this work, but are sufficiently interested in acquiring a reasonably deep knowledge of the field.

AUTHOR

Dilip N. Ghosh Roy is a graduate of Northwestern University, Evanston, Illinois. His doctoral dissertation was on the inversion of spectral intensities emitted by a plasma source. Since then, he has been involved in laser physics, nonlinear optics, and imaging-related inverse problems both in the U. S. and Europe. His current research is in the applications of inverse scattering transform to problems in optics, especially stimulated light scattering and optical phase conjugation.

TABLE OF CONTENTS

Chapter 10

Chapter 11

SYMBOLS

A boldface lower-case letter, such as **x**, usually denotes a vector quantity and a boldface upper-case letter, such as **L**, designates either a matrix or an operator.

L_1	space of absolutely integrable functions
L_2	space of square integrable functions
L_∞	space of continuous functions complete under uniform norm
C	space of continuous functions
C_∞	space of continuously differentiable functions
C_m	space of m-th differentiable functions
$C_{0\infty}$	space of continuously differentiable functions with compact support
$L_m^{\ 1}$	space of functions f(x) such that $\int_R(1+\lvert x\rvert^m)f(x)\,dx < \infty$
$\mathcal{L}$	Schwartz space
$\overline{\mathcal{L}}$	dual of Schwartz space
$\mathcal{D}$	space of the distributions
$\overline{\mathcal{D}}$	dual of the space of the distributions
P^N	space of all hyperplanes in the Euclidean space R^N
$\mathcal{L}_H, \mathcal{D}_H$	space $\mathcal{L}$, $\mathcal{D}$ on P_N and satisfying Lemma 12.1
H^1, H^2	Hardy spaces
Δ	the Laplacian operator
$\subset$	belongs to
$\cap$	intersection of spaces or sets
O,o	order symbols
$\cdot$	usually $\partial/\partial t$ (also $\partial/\partial k$ in Chapters 7 and 8)
$'$	space derivative unless otherwise specified
$\lVert\ \rVert_1$	L_1-norm
$\lVert\ \rVert_2$	L_2-norm
$\lVert\ \rVert_\infty$	L_∞-norm
$\sim$	Laplace transform unless otherwise specifed
$⨍$	Cauchy principal value integral
$\oint$	contour integral
$\int_\Omega$	integral over the hypersurface of a generalized unit sphere
$\wedge$	direct Fourier transform unless otherwise specified
$\vee$	inverse Fourier transform unless otherwise specified
$\check{f}$	Radon transform of f
H	Heaviside's function
$\underline{\mathbf{J}}$	a Jacobi matrix
$\mathbf{J}$	a persymmetric Jacobi matrix
$\mathbf{S}$	permutation matrix
$\mathbf{T}(x,y)$	generalized displacement operator
$\mathbf{T}(x,\lambda)$	transmission line matrix operator
$(\ )^T$	transpose operation
$\mathfrak{R}$	Radon transform operator
$\mathfrak{R}^*$	adjoint of the Radon transform operator
$\mathbf{A}^+$	generalized inverse
$\hat{H}$	Hilbert transform
ρ,σ	spectral functions
Y_{lm}	spherical harmonic
$C_l^{\,v}$	Gegenbauer polynomial
S_ϕ	spherical mean
Ω_n	total surface measure of a generalized unit sphere in an n-dimensional Euclidean space
x^λ, $x_+^{\ \lambda}$, $\lvert x\rvert^\lambda$, $x_-^{\ \lambda}$, r^λ	various generalized functions
δ	Dirac delta function
q_N	monic polynomial of degree N

$\xi_+ = x' - x$, $\xi_- = x - x'$

$F_\pm = (1/2ik)[\exp(2ik\xi_\pm) - 1]$

$\int_+ = \int_x^\infty$, $\int_- = \int_{-\infty}^x$

$\int_{+z} = \int_z^\infty$, $\int_{-z} = \int_{-\infty}^z$

Chapter 1

INTRODUCTION

It was perhaps John Tyndall[1] who observed more than a century ago the first scientific scattering of light by particles smaller in size than the wavelength of the incident radiation. Since Tyndall's time the science of scattering of waves and particles has undergone a long development. Rayleigh's[2] analysis of the bluish hue in the scattered light observed by Tyndall, Mie's[3] theory of scattering of electromagnetic waves by spherical particles, Einstein's[4] and Smolouchowski's[5] theory of light scattering by dielectric fluctuations caused by thermal agitation, Compton's[6] scattering of X-rays by atoms, Thomson's[7] scattering of plane electromagnetic waves by free electrons, well known phenomenon of neutron scattering,[8-10] scattering of waves and particles by quantum mechanical potentials,[11-15] Mandelshtam-Brillouin[16-18] scattering of light by acoustic phonons, and finally, Raman[19] scattering of photons by molecular vibrations that results in the frequency shift related to the internal dynamics of the molecules are some of the landmarks in this development. Sophisticated theories have been developed by Landau-Placzek,[20] Rytov,[21] and Komarov and Fisher[22] to explain the structures of the Mandelshtam-Brillouin doublet and of the nonpropagating entropy fluctuations in the Rayleigh lines. With the introduction of laser sources, the classical, spontaneous scattering of light has undergone dramatic advances both theoretically and experimentally, resulting in a number of powerful spectroscopic and other diagnostic techniques. The literature in this area is too extensive to mention in any detail here. However, an exhaustive account of these developments can be found in References 23 to 27. With the development of the Q-switched, pulsed, high-power laser pumps, stimulated scattering of light came into being. Almost every spontaneous scattering phenomenon was found to have its stimulated counterpart such as stimulated Brillouin scattering,[28-30] stimulated Raman scattering,[31-33] stimulated Rayleigh scattering,[34-35] and stimulated polariton scattering.[36] For details of these nonlinear optical phenomena, the reader is referred to Bloembergen,[37-38] Shen,[39] and a review article by Kaiser and Maier.[40] Scattering of acoustic waves has also undergone a lengthy development. The literature on acoustical scattering is extensive and reference is made to Morse and Ingard[41] and Beyer.[42]

Scattering is a very important tool in studying the structure and dynamics of matter and radiation, and there is hardly any important discipline of science and engineering that does not use scattering as an investigative probe. Whether in a highly sophisticated quantum mechanical problem of potential scattering or in the mundane task of locating a crack in an automobile crankshaft, method of scattering finds universal application. Fields of research as diverse as astrophysics and turbulence, plasma physics and flaw evaluation, quantum physics and medical imaging, lasers and seismology, nuclear physics and aerosol theory — the list can be extended easily — all use scattering phenomenon in one form or another in order to extract information about the system concerned. "Direct" or "forward" scattering involves determining the scattered fields or particles when the scattering potential or the distribution of the scatterers is known. The "inverse" or "backward" problem, on the other hand, consists of finding the distribution of the scatterers or the nature of the potential from the information on the scattered fields or particles. It is this backward scattering problem that has received by far the most attention in recent years and has been applied in almost all major branches of science and engineering.

I. STRUCTURE OF AN INVERSE PROBLEM

Consider the equation

$$y''(x, k) + \left[k^2 - q(x)\right] y(x, k) = 0 \tag{1.1}$$

x is the space variable that may belong to either R^+, the positive half-line, or to R, the entire real axis. k may be real or complex. Appropriate boundary conditions must, of course, be prescribed. Equation 1.1 describes the scattering of plane waves by a "potential" q(x), the dependent variable y(x,k) constituting the "scattered" field. The direct problem associated with Equation 1.1 can be stated as follows. Given the function q(x), find the solution y(x,k) satisfying certain prescribed boundary conditions. The inverse problem, on the other hand, is the reverse: given a certain "data" associated with the solution y(x,k) satisfying certain boundary conditions, determine the function q(x). The direct problem consists of finding the solution of Equation 1.1 under prescribed conditions. On the other hand, the inverse problem is to reconstruct the differential equation *cum* boundary conditions from the data.

The formulation of a direct problem involves setting up a certain number of functions. One part of these functions determines the differential equation, for example, through the determination of the coefficients in the differential operator, while the other part involves determining the initial and boundary conditions of the problem. The end result of a direct problem is a new function, the solution, determined by this given set of functions. In an inverse problem, some of these functions may be unknown and are to be determined. This requires additional information on the solution of the direct problem in place of these unknown functions. The additional information may consist of the solution itself, defined on a certain set of independent variables, or some functionals of the solution.

The structure of an inverse problem may be described as follows. We follow Newton[43] here. A linear differential equation or a system of such equations is given. The system contains a function or functions such as q(x) in Equation 1.1 which are usually real valued. We will call these functions "potentials". Newton calls them "forces". The equation or a system thereof may be the Schrödinger equation with an atomic potential or may represent an electromagnetic problem, a transmission line, a problem of sound propagation in a medium, or a problem in elasticity. The corresponding "potential" may then be an actual atomic potential or an index of refraction, the transmission line impedance, the acoustic speed, the impedance of the elastic medium, or the elastic modulus. The solutions of the differential equation or the system of equations subject to certain prescribed boundary conditions lead to the determination of a set of quantities which can be observed either directly or indirectly such as the reflection or the transmission amplitudes of scattering. Such "observables" will be called the "scattering data" or the "excitation response". The data are usually complex valued and are not functions in the space of the differential equation. Instead, the data may depend on parameters such as frequency, energy or angles, the parameters that occur either in the differential equation itself or in the boundary conditions. This point will be discussed in Chapter 9.

The direct problem can be considered to be a map on **M** from the potential q to the scattering data S, i.e., **M**: q → S. The major thrust of the direct problem is to investigate the functional dependence of the scattering data on the potential. That is, the direct problem is to determine the map **M** and investigate its properties. The direct problem is also important in determining how the data depend on the relevant parameters. This information is crucial if the corresponding inverse problem is to be solved. The direct problem yields valuable information about the condition or the conditions that may have to be imposed on the potentials if a solution is to exist. Thus the potential q(x) in Equation 1.1 may have to be restricted to a certain class of functions if a physically meaningful solution is to be obtained. Determining such conditions on the potential may prove to be very complex, as will be seen in Chapter 7.

The essence of the inverse problem is to create the substance from the shadow and is necessarily more complex. It has to deal with the following questions.

1. Is it possible to recreate the substance at all? That is, does there exist an inverse solution? This is the question of existence. (See note at the end of the chapter.)
2. Assuming that a solution does exist, what are the necessary and sufficient conditions on the scattering data in order that the reconstruction is unique? In mathematical terms, if **M** is a map from the potential to the scattering data, i.e., $\mathbf{M}$: q $\rightarrow$ S, is the inverse map $\mathbf{M}^{-1}$: q $\leftarrow$ S one to one? If not, the inverse problem is ill-posed and the reconstruction is not unique. This step is the most important in solving an inverse problem. Physically, this poses the question: what information is needed in order to recover the "potentials" uniquely? If more than one q(x) corresponds to the same data S, then the transform S must be enlarged, if possible, until uniqueness is obtained. This point is illustrated in Example 7 of Chapter 2.
3. What necessary and sufficient conditions on the data ensure that they would correspond to a realistic system? Suppose that the direct problem has restricted the "admissible" potentials to a class C and has ensured that the map **M** on this class C leads to a class S^C of "admissible" data. The question is, how does one recognize if the data given $\subset$ S^C? $\subset$ is a mathematical symbol meaning "belongs to". Newton[40] shows that sometimes it is convenient to enlarge the class $C \rightarrow C'$, such that $C \subset C'$. Then the inverse problem for $\mathbf{M}$: $C \rightarrow S^C$ will be imbedded in a larger problem $\mathbf{M}'$: $C' \rightarrow S^{C'}$, $S^C \subset S^{C'}$. If the map $\mathbf{M}'$ is unique, then the data were in S^C. This is possible, provided it is easier to characterize C′. (More details are given in the note at the end of the chapter.)
4. A methodology is required, that is, an actual procedure for reconstructing the potential must be designed. This is a must for a physicist, but may be optional for a mathematician.
5. Finally, one needs to know how stable the reconstruction is with respect to any perturbation in the data. In other words, does the solution depend continuously on the data? The stability of the solution will be discussed later in this chapter.

Motivated by numerous applications, physicists, mathematicians, engineers, geophysicists, and workers in many other disciplines have studied a wide range of inverse scattering problems. There is a common thread underlying these problems even though they occur in seemingly different contexts. The inverse spectral problems for Schrödinger operators and vibrating strings, algorithms for transmission-line synthesis, determination of the shape of the vocal tract in speech research, identification of acoustic impedance or conductivity profiles in layered-earth models, the design of digital filters in cascade form, and the derivation of fast lattice-form linear least-squares prediction algorithms all share a common mathematical foundation. They all require a procedure that identifies from the scattering data the depth-dependent information about the system. The data are the "excitation-response" pair either in frequency or in time domain.

II. TWO DIFFERENT SETTINGS OF INVERSE PROBLEMS

Inverse problems seem to have arisen in two entirely different settings. The first, and the one to command the most attention, is in the context of scattering of plane waves by a potential, as represented by Equation 1.1. These are the inverse problems in scattering. Their importance derives primarily from the Schrödinger problem in quantum mechanical potential scattering.[11-15] In the context of scattering in a dielectric or an elastic medium, Helmholtz's equation[44] is used, namely,

$$\Delta f(\mathbf{x}) + n(\mathbf{x})\, f(\mathbf{x}) = 0 \tag{1.2}$$

where Δ is the Laplacian and $n(\mathbf{x})$ represents the inhomogeneity of the medium (refractive index for an electromagnetic and impedance for an elastic medium) to be reconstructed. The solutions of these inverse scattering problems are usually expressed as integral equations of the second kind. It may be recalled that an integral equation of the second kind is of the form

$$f(x) = g(x) + \lambda \int_a^b K(x, x') f(x') \, dx' \tag{1.3}$$

where f(x) is the unknown function to be determined, g(x) is an inhomogeneous term which is the solution of the scattering free problem, i.e., with no potential, and $K(x,x')$ is the "kernel". If the integration limits are fixed, then the equation is of the Fredholm type. If the upper limit is the parameter x itself such that $K(x,x') = 0$, $x' > x$, then the kernel is said to be "triangular" and the equation is of the Volterra type. λ is called the "characteristic value" of the equation.[45] The eigenfunctions of the Schrödinger equation (1.1) admit of such representations. This will be discussed in Chapters 4, 5, and 10. All well-known inverse solutions of scattering problems such as the Lippmann-Schwinger equation,[46,47] the Gel'fand-Levitan equation on the semi-axis,[48,49] and the Marcenko equation[50,51] on the entire real axis are all Fredholm's integral equations of the second kind. The kernel of the Lippmann-Schwinger equation is the Green's propagator in free space. The Gel'fand-Levitan equation has the Povzner-Levitan[52,53] transform as the kernel, whereas Marcenko's integral equation contains the Levin transforms[54,55] of the Jost functions or the Jost solutions[12,13,55] in the integral operator. The integral equations of this type are usually characteristics of the boundary value problems. In quantum scattering, for example, it is the asymptotic data, i.e., data at infinity which are often used in the reconstruction of the potential.

The fundamental equation for scattering in one space dimension x, $x \subset R$ or R^+, is given by Equation 1.1, a singular Sturm-Liouville equation that arises in both quantum and classical physics. In quantum scattering, the equation arises as follows. For a spherically symmetric potential q(r), the time-independent Schrödinger equation is

$$-\left(\frac{\hbar^2}{2m}\right)\left[r^{-2}\left[\frac{\partial}{\partial r}\left(r^2\frac{\partial}{\partial r}\right)\right] - \frac{\mathbf{L}^2}{\hbar^2 r^2}\right]y + q(r)y = Ey \tag{1.4}$$

y is the wavefunction, m the mass, E the energy and the operator $\mathbf{L}$ is given by:

$$\mathbf{L}^2 = -\hbar^2\left[\left(\frac{1}{\sin\theta}\right)\frac{\partial}{\partial\theta}\left(\sin\theta\frac{\partial}{\partial\theta}\right) + \left(\frac{1}{\sin^2\theta}\right)\left(\frac{\partial^2}{\partial\phi^2}\right)\right]$$

where $\mathbf{r} = (r,\theta,\phi)$ is the position vector in spherical polar coordinates. $\mathbf{L}^2$ satisfies the following eigenvalue problem

$$\mathbf{L}^2 \, Y_{lm}(\theta,\phi) = l(l+1)\,\hbar^2 Y_{lm}(\theta,\phi) \tag{1.5}$$

$l(l+1)\hbar^2$ being the eigenvalue and $Y_{lm}(\theta,\phi)$ the spherical harmonic. The wavefunction y is expanded in terms of the partial waves[6]

$$y(k,r) = \sum_{l=0}^{\infty} \sum_{m=-l}^{l} C_{lm}(k) \, R_{lm}(k,r) \, Y_{lm}(\theta,\phi) \tag{1.6}$$

where $C_{lm}(k)$ is the expansion coefficient on the wave number k, $R_{lm}(k,r)$ is the radial part of the wavefunction, and $k^{-1} = \hbar/(2mE)^{1/2}$. Replacing Equation 1.6 in Equation 1.4, and using the eigenvalue problem (1.5), the radial equation is obtained, namely

$$-\left(\frac{\hbar^2}{2m}\right)\left[r^{-2}\frac{d}{dr}\left(r^2\frac{d}{dr}\right)-\frac{l(l+1)}{r^2}\right]R_l(k,r)+q(r)\,R_l(k,r)=E\,R_l(k,r) \qquad (1.7)$$

Substituting $u_l(k,r) = rR_l(k,r)$ and $U(r) = 2mq(r)/\hbar^2$ in Equation 1.7, the radial equation is obtained in its familiar form given by

$$\left[\frac{d^2}{dr^2}+k^2-\frac{l(l+1)}{r^2}-U(r)\right]u_l(k,r)=0 \qquad (1.8)$$

For $l = 0$, Equation 1.8 reduces to the scattering equation (1.1). In quantum theory, the $l = 0$ condition is referred to as the "S-wave" scattering.

The scattering equation is not just a quantum mechanical equation, but is widespread in classical physics also. As an example, consider the vibration of a one-dimensional elastic medium of variable density ρ(x) and velocity v(x,t), the elastic modulus K being a function of x. The dynamical equation for this system is

$$\rho(x)\,\dot{v}(x,t)=S' \qquad (1.9)$$

"·" and "′" are differentiations with respect to t and x, respectively, and S(x,t) is the stress, which is related to K and the displacement u(x,t) as

$$S(x,t)=K(x)\,u'(x,t) \qquad (1.10)$$

where, $v(x,t) = \dot{u}(x,t)$.

Taking the Fourier transform of Equations 1.9 and 1.10 with respect to time:

$$i\omega\rho(x)\,v\hat{}(x,\omega)=S'\hat{}(x,\omega) \qquad (1.11)$$

and

$$i\omega S\hat{}(x,\omega)=K(x)\,V'(x,\omega) \qquad (1.12)$$

The symbol ^ denotes the Fourier-transformed variable, and the Fourier transform of a function f(t) is defined as

$$(2\pi)^{\frac{1}{2}}\,f\hat{}\,(\omega)=\int_{-\infty}^{\infty}f(t)\exp(-i\omega t)\,dt$$

Eliminating v^(x,ω) between Equations 1.11 and 1.12 yields

$$\left\{\left[\frac{1}{\rho(x)}\right]S'\hat{}(x,\omega)\right\}'+\left\{\frac{\omega^2}{K(x)}\right\}S\hat{}(x,\omega)=0 \qquad (1.13)$$

Let us introduce two new variables

$$z = \int_0^x \left[\frac{\rho(z')}{K(z')}\right]^{-\frac{1}{2}} dz'$$

and

$$Z\hat{}(z,\omega) = S\hat{}(x,\omega)\left[K(x)\ \rho(x)\right]^{-\frac{1}{4}}$$

In terms of z and Z^(z,ω), Equation 1.13 reduces to:

$$\left[\left(\frac{d^2}{dz^2}\right) + \omega^2 + Q(z)\right] Z\hat{}(z,\omega) = 0 \tag{1.14}$$

where $Q(z) = -\frac{3}{10}\left[\left(\frac{p'}{p}\right)^2 + \left(\frac{r'}{r}\right)^2\right] - \frac{1}{8}\left(\frac{p'r'}{pr}\right) - \frac{1}{4}\left(\frac{p''}{p} + \frac{r''}{r}\right),$

$$p = \frac{1}{\rho(x)} \quad \text{and} \quad r = \frac{1}{K(x)}$$

Equation 1.14 is again the same scattering equation (1.1), but in a classical situation.

The second type of inverse problems arises in a nonboundary value context and is concerned with the reconstruction of geometrical or configurational distributions of some parameter or parameters from a set of "integrated" data, i.e., data which contain simultaneous contributions from many different elements of the distribution. The word "distribution" simply means a physical distribution and not in the sense of distribution in the theory of generalized functions. Following a terminology introduced by Gel'fand,[56] these will be referred to as inverse problems in "integral geometry". An integral-geometric problem is, therefore, a problem in which a function is to be determined from its integrals over a family of hyperplanes or hypersurfaces, depending on the dimension of the space involved. The parameters to be reconstructed in these problems can be refractive index, density, temperature, emission/attenuation coefficients, velocity, so forth and so on. Problems of this nature include inversions of astronomical data, of line-integrated spectral intensities in spectroscopy, determination of plasma parameters, imaging of human anatomy by X-ray transform, and a host of others. Figure 1 illustrates a physical situation typical of the problems in this category. The information emanating from the source S reaches the detector D and in the transit is integrated by the intervening hyperplanes of the object. The hyperplanes reduce to lines and planes for two-dimensional and three-dimensional problems, respectively.

The well-known solution of this problem is credited to Johannes Radon.[57] Extensive discussions of the transform bearing his name are given in References 58 to 60. The well-known Abel's inversion[43] is a special case of Radon's inversion. The integral equations associated with the inverse problems of the second type are generally of the first kind, i.e., of the form

$$F(x) = \lambda \int_a^b K(x,x')\ f(x')\ dx'$$

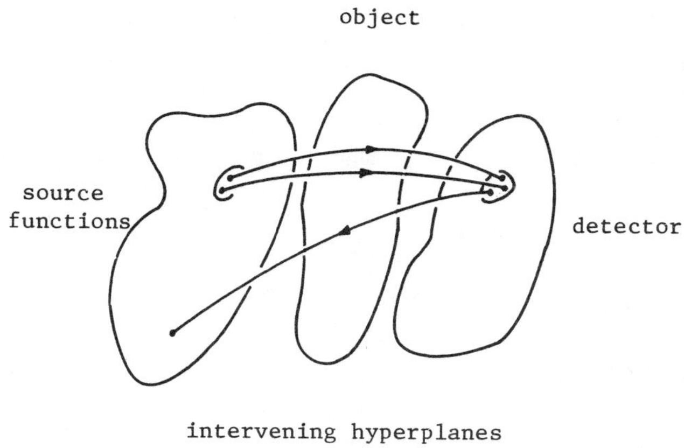

FIGURE 1. A typical integral-geometric inverse problem. The data received by the detector is the source information after being integrated by the intervening hyperplanes between the source and the detector. The problem is to determine the function from its integrals over the family of the hyperplanes or hypersurfaces.

F is the integrated data, that is, the integral of the function f(x) over a family of hyperplanes or surfaces. From these data, the distribution f(x) is to be reconstructed. The limits are almost always fixed and the equations are, therefore, of the Fredholm type.

The inversion problems in integral geometry are mostly ill-posed.[61] Qualitatively, this means that the inverse solutions of these integral equations either do not exist or are infinitely many. Even if a unique analytical solution does exist, the solution is not stable, i.e., depends discontinuously on the data. A small disturbance in the data can cause severe perturbation of the solution. This is unlike a well-posed problem where the discontinuity is no more than a small perturbation. The reason lies in the very nature of the integral equations involved. Physically, ill-posedness indicates an irrecoverable loss of information caused by the smearing of the data by the kernel. A generalization of this classical definition of the ill-posedness is given by Tikhonov and Arsenin[62] and will be discussed later in this chapter.

III. QUANTUM SCATTERING

The problem of inverse scattering has received by far the most attention in quantum mechanics. The fundamental problem here is to reconstruct the potential from the scattering data which reduces to reconstructing the Schrödinger equation from the asymptotic information on its solutions. For scattering on the entire real axis, the scattering data consist of the "spectral transform" of the potential which contains information of the continuous spectrum (via the reflection coefficient), of the bound states (through the point spectrum) and the bound state normalization constants. For scattering on the half-axis (radial problem), it is the information of the partial-wave phase shifts and the bound states that constitute the scattering data. The physical picture here is that of a rectilinearly propagating energy E directed parallel to a vector k and incident upon a central field of force V(r). In quantum mechanics, such a propagation is described in terms of the momentum eigenfunction exp(ik·r), which is a plane wave with k as its wave vector. The scattering is, therefore, that of a plane wave from the spherically symmetric potential V(r). The process is described by a stationary scattering wavefunction which is expanded in terms of the spherical harmonics. The components of this expansion depend upon the angular momentum quantum number l (in addition to the momentum k) and are called "partial waves" of angular momentum l. Expanding the momentum eigenfunction exp(ik·r) in terms of the spherical harmonics and comparing with the partial wave expansion of the scattering wavefunction shows that the effect of the central potential is to introduce phase shifts (relative to the plane wave) in the outgoing scattered waves. The analysis of the scattering data in terms of these partial wave-phase shifts dates back to Faxen and Holtsmark.[63] The questions of whether the phase shifts alone can determine the potential uniquely, whether phase shifts at all angular momentum at one energy or at one angular momentum at all energies are required, or whether the scattering matrix itself is sufficient to determine the potential uniquely when the bound states are present were investigated by Levinson,[64,65] Bargmann,[66,67] Marcenko,[68] and Borg.[69] A brief and insightful account of these pioneering attempts, including that of Ambartsumian[70] in a slightly different context, has been given by Levitan and Gasymov.[49] Although of central importance to the inverse solution of the scattering potential, these uniqueness proofs do not lead to the actual determination of the potential.

The first attempts at the actual reconstruction of the potential were by Fröberg[71] and Hyelleraas.[72] But Bargmann[66,67] showed by actually constructing contrary cases that the work of Fröberg and Hylleraas was inconclusive since only the scattering matrix was considered as the data, even if the "redundant zeros" discovered by Ma[73] were ignored. The work of Fröberg and Hyelleraas was extended by Holmberg[74] in the light of the investigations on the uniqueness problem.

The research reached a new height when Gel'fand and Levitan[48] published their now celebrated, classic paper on the determination of the Schrödinger equation on the half-line from the spectral function. In this famous formalism, the wave function is represented as a linear Volterra integral equation of the second kind in terms of the "free" wave function exp(ikx), i.e., the solutions of Equation 1.1 with the potential q(x) set to zero. Specifically,

$$y(x,k) = \exp(ikx) + \int_{x}^{\infty} K(x,x')\exp(ikx')\,dx'$$

The kernel K(x,x′) is triangular, independent of k, and known as the "Povzner-Levitan transform".[52,53] This is similar to the Lippmann-Schwinger equation in the Born approximation. The kernel, however, is different. The inverse solution is given by

$$q(x) = 2\{K(x,x')\}' \quad \text{at } x = x'$$

K(x,x′) is obtained either as the solution of the Gel'fand-Levitan integral equation (see Chapters 5 and 6) or as the solution of a Cauchy problem of Goursat type.[75,76] The potential is then obtained via the value of the Goursat solution along a characteristic. Extensions of the Gel'fand-Levitan scheme to scattering in three space dimensions where the kernel ceases to be triangular are primarily due to Fadeev[77] and Newton.[78,79] The Gel'fand-Levitan theory was applied immediately by Jost and Kohn[80] and by Levinson[81] to the inverse scattering problem. The theory was discussed extensively by Kay and Moses, who also introduced a more general approach in the context of the inversion of an electromagnetic reflection problem and the application of the full three-dimensional Schrödinger equation. The significant contribution of these authors can be found in a collection of their papers.[82] The problem of determining the potential from the scattering at one angular momentum (the S-scattering specifically) was discussed by Bargmann, whereas the contrary case of determining the potential from all phase shifts at one energy was approached by Wheeler,[83] Regge,[84] and Newton,[85] and Sabatier,[86] Redmond,[87] and Martin and Targonski[88] for superposed Yukawa potentials.

One of the well-known approaches to the problem of inversely reconstructing the potential is due to Jost and Kohn.[89] The Jost-Kohn method involves the construction at the origin of what is now known as the Jost function from the phase shift data. On the basis of this function, an auxiliary potential is calculated. The actual potential to be recovered is expanded in a series, each term of which is determined in terms of the auxiliary potential. The authors proved the convergence of the series even in the presence of a bound state and also formulated clearly the scattering data required for the unique determination of the potential. The importance of the Jost-Kohn formalism from the viewpoint of numerical calculations was pointed out by Prosser.[90] The Jost-Kohn or the Gel'fand-Levitan formalism has been extended to various scattering contexts such as coupled equations,[91-93] spin 1/2 particles,[94] the Klein-Gordon equation,[95] the Dirac equation,[96,97] and to separable potentials.[98]

Bargmann's work[66,67] deserves special mention for reasons to be discussed below. Bargmann considered the radial Schrödinger equation for the Jost solutions. However, instead of inverse solving for the potential, he assumed a certain form of the solution, namely, the Jost solutions as a polynomial in k plus a function of the radial coordinate only. Using the radial equation, Bargmann then constructed the potentials corresponding to the given Jost solutions. A far-reaching consequence of Bargmann's construction is that it is possible for the potential to undergo continuous deformation in shape while the eigenvalues remain unchanged. This is the so-called "iso-spectral" case of the Schrödinger equation. That this can happen is the crucial fact in the formulation of the "inverse scattering transform". The construction of Bargmann's potentials is directly related to the multisoliton solutions of Korteweg and deVries' equation. Bargmann's work, therefore, is a precursor to the development of the inverse scattering techniques, the significance and importance of which will be explained below. A good account of the importance of Bargmann's work in the development of the more complete inverse scattering techniques can be found in a text by Lamb.[99]

IV. EVOLUTION EQUATIONS AND NONLINEAR OPTICS

Perhaps the most spectacular advances in inverse scattering have taken place in solving (via inverse scattering transform) a class of nonlinear "evolution" equations,[100-132] the most well known being those of Korteweg and de Vries or the K-dV equation,[133-157] the nonlinear Schrödinger equation,[158-162] and the Sine-Gordon equation.[163-173] The origin of this line of research is usually associated with the famous Fermi-Pasta-Ulam problem[174] of resonant coupling between oscillators investigated extensively by Zabusky and Kruskal.[175] The pioneering work is due to Gardner, Greene, Kruskal, and Miura[135] and Lax.[100] These equations govern an astoundingly large number of problems in a wide variety of disciplines such as interactions between light pulses and resonant atoms, wave propagation in plasmas, lattice dislocations, motion of a vortex filament, shallow-water waves, and relativistic field theory. The remarkable feature is an interconnection between a linear system such as that of Sturm and Liouville with a parameter-dependent potential and a nonlinear system such as the K-dV equation. The fundamental "soliton" solutions of the latter obey a superposition principle analogously known in a linear system. The soliton solutions are then related to the potential. Details will be discussed in Chapters 9 and 10.

Coherent interactions between resonant atoms and optical pulses form one of the most important branches of modern physics and have extensive applications in high-resolution spectroscopy of atoms and molecules, laser physics, high-precision electromagnetic measurements and coherent pulse propagation. These are the counterparts in the optical frequency region of nuclear magnetic resonance which falls in the radio-frequency domain. Within the framework of semiclassical theory interactions of two- and three-level atoms with a resonant radiation are governed by the optical Maxwell-Bloch equations and give rise to a wide variety of physical phenomena that include photon echo, self-induced transparency, two-photon resonance, free induction decay, optical mutation, and superfluorescence.[176-181] Although of recent origin and of relatively short history, the derivation of analytical results in these areas via inverse scattering techniques is already an independent field of study and is gaining momentum rapidly. The application of the inverse scattering transform shows that the resonant interaction between an electrical field and an assemblage of two-level atoms yield solitons of self-induced transparency and the solitons of the nonlinear Schrödinger equation. The application of the inverse scattering techniques to optical Bloch equations (including inhomogeneous broadening and nonlinear propagation effects) has given rise to some important results in the analysis of coherent processes during the interaction between ultrashort light pulses and resonant media. The inverse scattering transform has also been applied in finding an exact solution to two-component scattering of an initially unchirped resonant hyperbolic secant pulse. References 182 to 198 should be consulted for further details of this important and fascinating area of research.

The theoretical solution of the three-wave interaction problem involving three parametrically coupled resonant or near-resonant waves is a longstanding problem in physics. This occurs in various contexts such as nonlinear optics, fluid dynamics, and plasma physics. Application of inverse scattering techniques to these problems in a homogeneous medium has led to the now famous Zakharov-Manakov[199-201] solution. The formalism has been extended to interactions of wave packets in time varying and inhomogeneous medium with quadratic nonlinearity and rapidly oscillating initial conditions. In the context of stimulated light scattering involving propagating modes such as stimulated Raman and Brillouin scattering, Zakharov-Manakov formalism has been applied successfully. A complete solution of the three-wave problem is also of direct interest from the viewpoint of optical pulse compression,[202-205] which is receiving considerable attention because of its relevance to fusion research. A very high intensity "pump" pulse (usually produced by a focused high-power, Q-switched laser) is sent through an optically nonlinear medium in which is also present an "idler" wave due to spontaneous scattering. The intense pump beam induces coherent fluctuations (acoustical phonons if the medium is Brillouin active and optical phonons or molecular vibrational excitations, if the medium is Raman active, or both). These induced fluctuations, which are intense because of the high laser intensity and high nonlinearity of the medium, then parametrically couple the pump and the idler. As a consequence, the latter and also the fluctuations grow rapidly at the expense of the pump. The scattered light beam (it is the backscattering which is usually considered) eventually reaches an enormous magnitude (several times that of the peak pump intensity) in a pulse width which could be orders of magnitude less than the pump width. The success of the inverse scattering technique to this problem can be understood since the problem is closely related to the nonlinear Schrödinger equation. As another variant of this class of problems, a long wave-short wave interaction in fluids and plasmas has been solved via inverse scattering transform by reducing the governing equations to a pair of linear integral equations of the Marcenko type. Literature in this field is vast; References 206 to 220 provide an overview. Recently, inverse scattering transform was applied to the quasi-steady-state stimulated Raman and Brillouin scattering.[220a] A recent important development in nonlinear optics is that of "optical phase conjugation".[221-223] In optical phase conjugation, a light wave is made to traverse a medium which introduces distortions in the wave fronts. On reflection from a "phase-conjugated mirror", and upon a second traversal through the same distorting medium, the wave fronts are restored to their original states. There are a variety of ways for creating a phase conjugated mirror, the well-known techniques being those of stimulated Raman or Brillouin scattering and four-wave mixing.[224-227] Insofar as the inverse problems are concerned, this phenomenon has important applications in optical image reconstructions.[228-232]

Somewhat related are the problems of optical self-focusing and filamentation of a laser pulse propagating in neutral dielectrics and plasmas.[233-238] Such self-focusing and filamentation of an intense laser pulse lead to the destruction of the medium and are well known[237] to reduce the threshold for stimulated Raman scattering in dielectric fluids. Similar problems occur in nonlinear pulse propagation in optical fibers. Intense picosecond pulses in self-focusing fibers are responsible for self-constriction of these pulses. Techniques of inverse scattering have been applied in studying the evolution of the width, amplitudes, and nonlinear frequency chirping of pulses during their propagation in the fibers. Analytical solutions have also been obtained for asymmetric evolution of the picosecond pulses in optical fibers due to relaxations of the nonlinearity of both self-constricting and self-spreading of such pulses. These phenomena are governed by the nonlinear Schrödinger equation or one of its generalizations and, hence, the application of inverse scattering techniques can be appreciated. References 239 to 244 provide an overview of this area of research.

V. PLASMA PHYSICS

Another area where inverse scattering methods are applied widely is plasma physics. The plasma medium is well known to sustain an enormous variety of waves, and the major thrust of plasma research is on investigating the nature of these waves and their propagation characteristics.[245-248] It is, therefore, no surprise that a rich portion of plasma phenomena is governed by nonlinear evolution equations, leading to the fact that plasma physics is a fertile ground for the application of inverse scattering methods. For example, the evolution of the nonstationary Langmuir solitons in the presence of driven, electron-damped ion sound is governed by the nonlinear Schrödinger equation,[249] the evolution of the electron-acoustic waves in a current carrying collisionless magnetized plasma follows a modified K-dV equation,[250] and modulational instability of long Alfvén waves parallel to the magnetic field is analyzed in terms of inverse scattering solution of the modified nonlinear Schrödinger equation.[251] Application of inverse scattering techniques to these problems has led to such famous formalism as that of Zakharov and Shabat.[159] Zakharov-Shabat analysis is the asymptotic form of the previously mentioned Zakharov-Manakov[199-201] theory. A powerful perturbation analysis based on the inverse scattering methods known as the "Karpman-Maslov" problem[252] has been developed to study the stability of the perturbed propagating soliton envelopes. Using these potent methods of analysis based on inverse scattering techniques, a host of plasma phenomena has been analyzed such as the Langmuir solitons, modulational instabilities of Alfvén waves and broad Langmuir turbulence, electron-acoustic solitons in a current carrying magnetized plasma, nonlinear Landau damping and many others. The literature on plasma physics is enormous; References 250 to 269 provide an exposition of the application of the inverse scattering methods in plasma physics.

VI. ELECTRON DENSITY PROFILES

Inverse scattering theory is used to reconstruct electron density profiles from the analytic representation of the reflection coefficients. The most well-known work in this area is that of Kay and Kay and Moses.[82] The complex reflection coefficient is represented as a rational function of the frequency variable k. The Gel'fand-Levitan integral equation of one-dimensional inverse scattering theory is then solved to obtain the electron-density profile function. The method has important applications in ionospheric studies. Free electron densities of the ionosphere have been determined from the vertical radio-sounding data in the form of ionograms by using the one-dimensional inverse scattering theory of Gel'fand and Levitan. Thus, the ionospheric electron densities in the valley between two peaks that do not allow Abel inversion has been determined using Gel'fand-Levitan formulation. This formalism has also been applied in determining ionospheric structural profiles from the reflection coefficients. References 270 to 275 discuss these problems.

VII. PHYSICAL OPTICS

Another field of investigation in which scattering and its inverse problems are applied widely is physical optics. The problems in this classical area of physics are too numerous and varied to analyze in such a short review. We mention only briefly some recent inverse scattering problems in this area, namely, reconstruction of the spectral spatial-coherence function of a quasimonochromatic planar source from the angular distribution of the far-field scattered data; inversion of the characteristic function of scatterers with spatial and temporal dependence from phase and range-normalized far-field backscattering; determining surface microroughness from scattered radiation; profile inversion of the refractive index in diffusion-fabricated waveguides from the reflection data; identification of objects whose inverse scattering can be approximated by a finite collection of distinct scattering centers through combined physical optics and the geometrical theory of diffraction; retrieval of a perfectly conducting diffraction grating from the efficiency curve; recovery of the refractive index layer backed by a perfectly conducting plane; identifying conducting cones from measured transient response data for only a few viewing angles; recovering target information from band and aspect-limited scattering data; application of a T-matrix to reconstruct the shape of a smooth reflecting object when the fields on the surface obey Dirichlet or Neumann boundary conditions; reconstruction of refractive index profile of an inhomogeneous planar object from the reflection coefficient; application of Radon's transform to far-field inverse scattering problems, especially those concerning unresolved questions related to band- and aspect-direction-limited sparse-data cases; and, finally, holographic and tomographic image reconstruction of diverse sorts. It is interesting to note that the inverse scattering method has also been applied to quantum optical problems, such as understanding the quantum process of photodetection in terms of the microscopic model of the fields through inversions for counting probabilities, and inverse detection and scattering problems for a class of distributions which characterize the fluctuations in scattered intensity for a wide class of experimental situations. An exhaustive account of various problems of this category is given by Colton and Cress.[276] An overview of inverse problems in classical optics can be found in References 277 to 299.

VIII. INTEGRAL GEOMETRY

The integral-geometric inverse problems range over a variety of scientific disciplines. Their most widespread use is associated with the inversion problems in astrophysics and gravitational dynamics. Radon's original work was in this context. Some of the important problems are the determination of the mass distribution in a galactic disc, inversion of the Bremsstrahlung continuum spectra, and interpretation of the electron spectra from the synchrotron spectra. Instead of referring to these problems individually, we refer to a recent monograph by Craig and Brown,[300] which contains further references regarding problems in astrophysics. In a completely different context, a very popular application of the integral-geometric inversion problems is in computed tomography. The basic problem here is to extract depth-dependent information of an object from projections of source functions at various angles. A vast literature has developed on this subject, primarily because of its useful application in medicine. For an exhaustive account and associated bibliography of this area, reference is made to a book by Deans.[301] Reconstruction of the indices of refraction in a dielectric medium is another aspect of the integral-geometric problems. Inversion of line integrated spectral intensities for obtaining electron and ion temperature distributions and distributions of absorption and emission coefficients in an ionized plasma medium is another fertile ground for the integral-geometric inverse problems. For optically thin, axisymmetric plasmas, Abel's inversion[302-306] is the standard procedure. However, when the plasma is absorbing and asymmetric, Abel's inversion fails and other methods of integral-geometric inversion must be applied.[307-309]

The most interesting and sophisticated developments in the integral-geometric problems are due to the Russian mathematicians in connection with potential theory, gravitation, Cauchy problems, problems of telegraph equation, linearized wave equations, and inverse kinematic problems of geophysics. Developments in this area are too involved to discuss at any length in this introduction. Instead, we refer to the important work of Romanov.[310] One of the most interesting aspects of the research by these mathematicians in this area is the recovery of the coefficients of the lower-order derivatives in a linear hyperbolic differential equation through integral-geometric inversions. We present a very brief and terse description of this fascinating research referring to Reference 310 for details.

Consider the Cauchy problem for the function u(**x**,**X**,t),

$$\ddot{u} = \mathbf{L}u + 4\pi\delta(\mathbf{x} - \mathbf{X},\ t),\quad \mathbf{x},\ \mathbf{X} \subset R^3 \quad u \equiv 0, \quad t < 0 \tag{1.15}$$

where **X** is a fixed point and a parameter in the problem. **L** is an elliptic operator written as

$$\mathbf{L}u = \sum_{i,j=1}^{3} a_{ij}(\mathbf{x})\, u_{x_i x_j} + \sum_{i=1}^{3} b_i u_{x_i} + c(\mathbf{x})u$$

$$\sum_{i,\ j=1}^{3} a_{ij}(x)\, \alpha_i \alpha_j \geq \mu \sum_{i=1}^{3} \alpha_i^2, \quad \mu > 0 \tag{1.16}$$

$a_{ij} \subset C^3$, $b_i \subset C^2$, and $c \subset C$. C^n is the space of continuous functions possessing n derivatives. α is any vector.

Let D be a region in R^3 bounded by surface S (not necessarily closed). Let the coefficients a_{ij} (i, j = 1, 2, 3) of the operator **L** be known. Let **X** run over a point set P of the surface S. The solution u(**x**,**X**,t) of Equation 1.15 is considered to be known for **x**, $\mathbf{X} \subset$ P at times t lying in the neighborhood of the value $t = \tau(\mathbf{x},\mathbf{X})$, where τ is the time of propagation between x and X. The problem is to find in D the coefficients of Equation 1.16 for the lower-order derivatives, that is, to determine the b_i's and c.

The above problem can be reduced to an integral-geometric problem. One defines a function:

$$\sigma(\mathbf{x},\mathbf{X}) = \underset{t\to\tau+0}{\text{limit}} \int_0^t u(\mathbf{x},\mathbf{X},t)\, dt$$

By the condition of the problem, σ(x,X) is a known function of the points x and X of the set P. Based on a procedure developed by Sobolev,[311,312] Romanov proves that for any pair of points **x**, $\mathbf{X} \subset$ P, the following integral relation hold

$$\int_{\Gamma(\mathbf{x},\mathbf{X})} \left[\sum_{i=1}^{3} \tilde{b}_i(\mathbf{x}) \left(\frac{dx_i}{ds} \right) \right] ds = f(\mathbf{x},\mathbf{X}) \tag{1.17}$$

where,

$$\tilde{b}_i(\mathbf{x}) = \sum_{k=1}^{3} b_k(\mathbf{x}) b_{ki}(\mathbf{x})$$

The $b_{ki}(x)$s are the elements of the matrix inverse to $\mathbf{A} = (a_{ij})$. The integral runs over the geodesic $\Gamma(\mathbf{x},\mathbf{X})$ joining $\mathbf{x}$ and $\mathbf{X}$ in a metric in which the differential distance ds is written as

$$ds = \left[\sum_{i,j=1}^{3} b_{ij} dx_i dy_j \right]^{\frac{1}{2}}$$

The existence and regularity of the geodesic inside the region D may impose restrictions on the coefficients a_{ij}. The function $f(\mathbf{x},\mathbf{X})$ can be computed in terms of the function $\sigma(\mathbf{x},\mathbf{X})$ and the coefficients a_{ij} (i, j = 1, 2, 3). The contour of the integral of Equation 1.17 contains only the coefficients of the higher derivatives and is thus known. This reduces the problem of determining the b_i's from known integrals of a linear combination of these functions along the geodesics represented by $\Gamma(\mathbf{x},\mathbf{X})$ joining the points on the surface S.

The problem of determining the coefficient $c(\mathbf{x})$ also reduces to a problem in integral geometry. Since the problem for the coefficients b_i (i = 1, 2, 3) has been solved in which the coefficient $c(\mathbf{x})$ is not involved, then in deriving an equation for $c(\mathbf{x})$, the coefficients b_i's (i = 1, 2, 3) can be considered to be known. The coefficient $c(\mathbf{x})$ is given by the relation

$$\int_{\Gamma(\mathbf{x},\mathbf{X})} R(\mathbf{x},\mathbf{X},\mathbf{y})\, c(\mathbf{y})\, d\tau = V\big(\mathbf{x},\mathbf{X},\tau(\mathbf{x},\mathbf{X})\big) - \int_{\Gamma(\mathbf{x},\mathbf{X})} Q(\mathbf{x},\mathbf{X},\mathbf{y})\, d\tau$$

R, V and Q are functions containing σ and involving the conjugate operator $\mathbf{L}^*$ and hence are known on the point set on the surface S. Thus, the determination of coefficients b_is and c is reduced to the solution of integral-geometric problems.

Generally speaking, integral-geometric problems are ill-posed in the classical sense. The question of ill- or well-posedness is a very important one and merits elaboration. The problem can be best explained within the framework of functional analysis of differential equations. A differential equation defines a whole set of solutions which usually depend upon a certain number of arbitrary constants or functions. Appropriate boundary and initial conditions are imposed in order to single out the solution describing the given physical process corresponding to the equation. A differential equation thus involves a pair of function spaces (U,V), U being the solution space and V the space to which the conditions of the equation belong. A differential equation is thus an operator equation

$$Au = v, \quad u \subset U \text{ and } v \subset V$$

The operator $\mathbf{A}$ determined by the differential equation and the additional conditions then relates an element u of the solution space U to an element v of the function space V containing the additional conditions. The classical notion of well-posedness of the differential equation consists of three criteria: (1) the problem solution must exist for *any* $v \subset V$ and must belong to the function space U, (2) the solution must be unique in the space U, and (3) the solution must be stable for *any* $v \subset V$. Mathematically speaking, the last condition is as follows. A solution is stable to small perturbations in $v \subset V$, if for any $\delta > 0$, there exists an $\varepsilon > 0$, such that for any element $v^{\sim} \subset V$ for which

$$\left\| v - v^{\sim} \right\|_V < \delta$$

the inequality

$$\left\| u - u^{\sim} \right\|_U < \varepsilon, \quad Au = v \text{ and } Au^{\sim} = v^{\sim}$$

holds.

In essence, well-posedness in the classical sense implies existence, uniqueness, and stability to small perturbations in the data. The reason for italicizing the word *any* above will become clear shortly. For a linear operator with the pair (U,V) as Banach spaces, the classical well-posedness has the existence of the inverse operator $\mathbf{A}^{-1}$ as the necessary and sufficient condition.

The classical definition of well-posedness is restrictive in many practical situations. There are many instances of boundary value problems describing real physical processes whose solutions will not exist if subjected to this classical criteria. An often cited example is that of Hadamard for Laplace's equation on a finite interval.[313] Moreover, the data to be prescribed must be obtained through measurements which usually alter the function space V that may forbid the solution to exist in the space U, although the process is physical and no doubt exists. Tihonov[62] generalized the classical definition of well-posedness for a broad class of inverse problems. Tihonov's generalization consists of replacing the underlined *any* of the classical definition by less restrictive conditions. The second criterion of uniqueness is left undisturbed, whereas conditions 1 and 3 are modified. The condition of *any* $u \subset U$ is replaced by those $u \subset$ a subset M of the space U and the solution is assumed to belong to M *a priori*. This is simple and easily justifiable on physical grounds. The condition *any* $v \subset V$ of the classical notion is replaced by $v \subset R = \mathbf{A}M$. The stability criterion is replaced by the fact that in Tihonov's generalized definition, $v, v^{\sim} \subset R = \mathbf{A}M$ and $u, u^{\sim} \subset M$. The continuous dependence of the solution on the data is retained, but only insofar as it does not take the solution out of the space M. In summary, the new definition requires that a solution be unique in a class of solutions belonging to a given subset M of a function space. The existence of a solution in this set is assumed *a priori* for some set of data. The classical requirement of continuous dependence of the solution on the data is retained, but it is interpreted differently. It is required that solutions depend continuously only on those data which do not take the solutions out of M. Problems which are ill-posed classically become well-posed in light of Tihonov's generalization. As an example,[314] consider the problem of determining a function $u(\mathbf{x},y) \subset C$, the space of the continuous functions, the function being even in y and $\mathbf{x} \subset R^n$, from its integrals over arbitrary spheres whose centers lie on the hyperplane $y = 0$. This problem is equivalent to solving the Darboux equation[315,316] for the spherical mean $U(\mathbf{x},y,r)$ of $u(\mathbf{x},y)$ at the point $(\mathbf{x},y)$ over a sphere of radius r. The Darboux equation for U is given by

$$U'' + (n-1)\frac{U'}{r} - \Delta_{x,y} U = 0$$

subject to the conditions

$$U(\mathbf{x},0,r) = f(\mathbf{x},r) \text{ and } U_y(\mathbf{x},0,r) = 0$$

The primes represent differentiation with respect to r and the Laplacian Δ is in $\mathbf{x}$ and y. This problem is known to be ill-posed in the classical sense, that is, a small change in the data $f(\mathbf{x},r)$ can result in finite changes in the solution $U(\mathbf{x},y,r)$. Courant[314] has shown that the solution is unique in that compactum C where the functions can be determined uniquely if the data $f(\mathbf{x},r)$ is restricted to the cylinder $\{|\mathbf{x}| \le \varepsilon, 0 \le r < \infty\}$, ε being a positive number. Thus, the classical definition of continuous dependence on the data is retained, but in any compactum belonging to the function space C. A lucid and insightful discussion of classical and Tihonov's generalization of well- and ill-posedness has been given by Romanov.[313]

One of the traditional and fertile grounds for inverse problems is in geophysical research. The principal task here is to determine the internal structure of the earth from surface and subsurface measurements. Since direct geophysical explorations are confined to a narrow subsurface layer (the deepest geophysical borehole is only about 11 km deep),[313] the demand for inverse solutions is hardly surprising. Broadly speaking, there are basically two types of problems in geophysical inversions. The first, and the older of the two, can be formulated as follows. Let $D \subset R^3$ be a domain with a boundary surface S. Let M be a certain set $\subset D \cup S = \overline{D}$, $\mathbf{x} \subset S$, and $\mathbf{X} \subset \overline{D}$. Let $v(\mathbf{x})$ be the speed of signal propagation and $t(\mathbf{x},\mathbf{X})$ the time required for the signal to travel from $\mathbf{X}$ to $\mathbf{x}$. The first inverse problem of geophysics runs as follows: determine $v(\mathbf{x})$, $\mathbf{x} \subset D$, given $t(\mathbf{x},\mathbf{X})$ for $\mathbf{x} \subset S$ and $\mathbf{X} \subset \overline{D}$. The function $t(\mathbf{x},\mathbf{X})$ satisfies the boundary value problem

$$\left|\nabla_t(\mathbf{x},\mathbf{X})\right| = \frac{1}{v(\mathbf{x})}$$

and

$$t(\mathbf{x},\mathbf{X}) \sim 0\left(\left|\mathbf{x}-\mathbf{X}\right|\right), \quad \mathbf{x} \to \mathbf{X}$$

The function $t(\mathbf{x},\mathbf{X})$ defines the wavefront emanating from the source $\mathbf{X}$ at the instant of time t. Following Romanov,[313] we will designate these as "inverse kinematic problems in seismology". The original formulation of the inverse kinematic problems are credited to the German geophysicists Herglotz[317] and Wiehert and Zoeppritz[318] on the assumption of the spherical symmetry of the earth and for the set M = S.

The second set of inversion problems in geophysics depends not on the geodesic $t(\mathbf{x},\mathbf{X})$, but on the seismic records of the vector of displacements $\mathbf{u}(\mathbf{x},t,\mathbf{X})$ of the surface points as a function of time t. The problem is to infer the density, the Lamé parameters of elasticity of the earth, and the vector source function $f(t,\mathbf{X})$ from the knowledge of the vector displacement $\mathbf{u}(\mathbf{x},t,\mathbf{X})$. Mathematically, the problem is to determine the density ρ, the Lamé parameters μ, λ, and the function $f(t,\mathbf{X})$, given $\mathbf{u}(\mathbf{x},t,\mathbf{X})$, $\mathbf{x} \subset S$, $\mathbf{X} \subset M \subset \overline{D}$, and $t \ge 0$. Romanov[313] calls these problems "inverse dynamic seismic surveying".

IX. GEOPHYSICS

The inverse problems in geophysics take various forms. Some of the well-studied problems in this area are determinations of density, compressibility, velocity, bulk, and shear modulus profiles from reflection seismic data in one, two, and three dimensions, deconvolution problems for geophysical time series, and various inversion problems associated with stratified elastic media in general. Well-known numerical procedures such as that of Backus and Gilbert,[319] dynamic or downstream deconvolution,[320-325] and layer-peeling algorithms which have been recently used in connection with nonuniform transmission line problems[326,327] were originally formulated in geophysical research. Some of these techniques will be discussed in Chapter 11. The important works in geophysical inversions in the West are listed in References 328 to 351. The Soviet literature on this subject is extremely important, widely known, and vast. It is not possible to do justice to the deep Soviet literature on this subject in this short introduction. Extensive references on Soviet geophysical research activities are contained in a recent publication by Romanov.[313]

X. ACOUSTICS

Another traditional and widely studied area of inverse problems is acoustics. The problem here is principally the inversion of acoustical scattering data in order to determine the acoustical impedance profiles of elastic media of various sorts. The problems in this important field of research are, therefore, closely allied to those in geophysics, and a number of references given above in connection with geophysical inversion also apply to the inversions of acoustical impedance profiles. In addition to those mentioned above, the reader is directed to References 352 to 358, which describe the inversion of acoustical impedance profiles. However, unlike their geophysical counterparts, the inversion problems in acoustics have found wide applications in imaging sciences, medical applications in particular. Acoustical holography[359-363] and diffraction tomography[364,365] are well known. A closely related problem of this nature is in speech synthesis. The most well-known work in this regard is that of Gopinath and Sondhi[366] and Sondhi and Resnick.[367]

XI. POTENTIAL THEORY

One more inverse problem is associated with the theory of potential, gravitational potential in particular. The general formulation of this problem is as follows. Consider a domain D and its boundary S. A potential induced by a body (masses, an electromagnetic source, etc.) inside S is known outside the domain D. The problem is to determine the shape and density of the body. As an example, let us consider the Newtonian potential. Let a body with density $f(\mathbf{x})$ be located inside a limited surface S in the space R^3. Let D be the domain of a function $g(\mathbf{x})$, that is, the set of points where $g(\mathbf{x}) \neq 0$. Let $\mathbf{x}(D)$ be a characteristic function of the point set D. Then Newton's potential $q(\mathbf{x})$ obeys Poisson's equation

$$\Delta q = -4\pi\, f(\mathbf{x})\, \mathbf{x}(D)$$

The inverse problem is to determine the functions $f(x)$ and $x(D)$ from the known solution to the above Poisson's equation outside the surface S. Rigorous discussions of this problem can be found in References 368 to 377. An interesting discussion of the determination of the potential from the knowledge of the parameters of its dependence and from observations of the motion of a particle in this potential field at various times is given by Kagiwada[377] in the context of a discrete multipoint boundary value problem using quasi-linearization technique.

XII. ANALYTICAL MECHANICS

Let us mention another class of inverse problems in the context of theoretical mechanics. The essence of these inverse problems is this. Consider Newton's equations of motion

$$m_k \ddot{r}_k - \mathbf{F}_k(t, r, \dot{r}) = 0, \quad k = 1,\ 2\ \ldots.. N$$

where, m_k is the mass of the k-th particle, r_k is the position vector, and F_k contains the force terms. The forces may be conservative and/or dissipative. A force is conservative if it is derivable from a potential,[378] say, U(r, t), i.e.,

$$\mathbf{f}_k(t,r,\dot{r}) = -\frac{\partial U}{\partial r_k} + \frac{d}{dt}\left(\frac{\partial U}{\partial \dot{r}_k}\right)$$

where $\mathbf{f}_k$ designates a conservative force. The dissipative forces, designated $F_k(t,r,\dot{r})$ are those that are not derivable from a potential. Hence, one can write Newton's equations as:

$$m_k\ddot{r}_k - \mathbf{f}_k(t,r,\dot{r}) - F_k(t,r,\dot{r}) = 0$$

to include both conservative and dissipative forces.[378]

An alternative formulation of mechanics is in terms of a "Lagrangian" L or in terms of a "Hamiltonian" H. The Lagrangian L is given by

$$L = T(r) - V(t,r,\dot{r})$$

where, T and V are the kinetic and the potential energy, respectively. The Hamiltonian is related to the Lagrangian

$$H = p_k\dot{r}_k - L$$

where $p_k = \partial L/\partial \dot{r}_k$ is the momentum. In terms of the Lagrangian L, the dynamical equations[378] are written as:

$$F_k(t,r,\dot{r}) = -\frac{\partial L}{\partial r_k} + \frac{d}{dt}\left(\frac{\partial L}{\partial \dot{r}_k}\right) \tag{1.18}$$

Equation 1.18 is called the "Euler-Lagrange" equation or simply the Lagrangian equation of motion.

In the "direct" problem, the Lagrangian is assigned and the equations of motion are then computed through Lagrange's equation. The "inverse" problem, on the other hand, consists of assigning the equations of motion and then computing the Lagrangian. In other words, one considers a given system of Newton's equations of motion, then studies the conditions under which a Lagrangian capable of representing such a system exists, and finally constructs the Lagrangian. Thus, a system of Newton's equation of motion is represented by an Euler-Lagrange equation for a given Lagrangian such that the solutions of the two systems coincide. Let

$$P_k(t,q,\dot{q},\ddot{q}) = 0$$

represent a system of Newton's equation of motion in configuration space with

$$\dot{q} = \frac{dq}{dt},\quad \ddot{q} = \frac{d^2q}{dt^2},\quad q = \{q_k(t)\},\quad k = 1,\ 2\ \ldots\ldots N$$

where q is a "generalized coordinate". If $L(t,q,\dot{q})$ is a Lagrangian in terms of the generalized coordinate q, then the inverse problem consists of studying the conditions under which

$$P_k(t,q,\dot{q},\ddot{q}) = \frac{\partial L}{\partial q_k} - \frac{d}{dt}\left(\frac{\partial L}{\partial \dot{q}_k}\right)$$

and finally constructing such a Lagrangian.

The significance of solving an inverse problem of this type in mechanics is twofold. First, constructing a Lagrangian allows one to construct the Hamiltonian which represents the phenomenology of either an atomic or a mechanical system. Second, and of primary significance, is the fact that the forces in a Newtonian system are not in reality derivable from a potential. The inverse problem allows one to study systems with arbitrary nonconservative forces. This is of utmost importance in studying systems with dissipation. These inverse problems in mechanics are detailed exhaustively in a monograph by Santili,[379] where the author reveals the difficulties in solving problems (classical and quantum) in the presence of dissipation and shows how such problems can be treated in inverse solutions.

XIII. DISCRETE SYSTEMS

The problems mentioned so far were mostly continuous problems. The systems, the reconstructions of which constituted the central theme of the methodologies presented above, were continuous systems that could be described by differential equations. The corresponding inverse solutions were also continuous, as exemplified by the linear integral or differential equations expressing the reconstructions. In contrast to these continuous systems, there exist a vast number of problems where the systems considered are in discrete forms. Some of these were cited in References 320 to 329. This is motivated by numerous applications and practical computational considerations. A thorough and exhaustive account of general discrete systems is the well-known treatise by Atkinson.[380] The discrete systems are usually characterized by linear matrix differential equations instead of ordinary or partial continuous differential equations. The inverse solutions involve matrix inversions as opposed to inverting linear integral equations. Well-known discrete formalisms of the Gel'fand-Levitan equation are due to Case and Kac[381] and Case.[382] Discrete versions also have been formulated for scattering in two-space dimensions.[383,384]

It appears that considerations of inverse problems involving discretized systems have proceeded along two distinct lines characterized by two different matrices to be inverted. One involves the inversion of a Jacobian matrix, and the associated inverse problems are those of the mechanical systems in vibration such as spring-mass systems, torsional and transverse beam vibrations, molecular chains, and so forth. The basic matrix equation governing these systems is given by

$$\mathbf{A}\ddot{\mathbf{u}} + \mathbf{C}\mathbf{u} = \mathbf{F}$$

where **C** is a tridiagonal matrix called the "stiffness" matrix, **A** a diagonal "inertia" or "mass" matrix, and **F** a vector representing various forces acting on the system. **A** contains the "masses" involved and **C** the "spring" constants. For free vibration, this equation can be Fourier transformed to the eigenvalue problem

$$(\mathbf{B} - \lambda\mathbf{I})\mathbf{V} = 0$$

where **B** is another symmetric, tridiagonal, and positive definite or semidefinite matrix if **C** shares these properties and **I** is the identity matrix. **B** is termed a "Jacobian" matrix. The inverse solution of reconstructing the masses and the spring constants is obtained via inverting the Jacobian matrix **B**. The inverse solutions of discrete systems began in the early 1930s in the U.S.S.R. with the research of Krein and Gantmacher.[385-390] A recent, detailed account of the inverse problems in discrete mechanical vibrational systems can be found in a monograph by Gladwell.[391]

The second type of inverse problems in discrete systems involves inverting Toeplitz and/or Hankel matrices. The physical problems associated with the inversion of these matrices occur in geophysics, Schrödinger scattering, vibrating strings, systems identification, and vocal tract reconstruction and electrical engineering problems. The model of such discrete medium inversions is that of a lossless, nonuniform transmission line. In essence, the line is discretized into differential segments over which the line characteristic, namely, the impedance, is assumed to be uniform. An input current or voltage is applied at the boundary and the "excitation-response" pairs of time functions are also measured at the boundary. This information is then related to the depth-dependent parameters that are to be reconstructed. Wilcox[392] and Brown[393] give a mathematical exposition of the synthesis of the nonuniform transmission line. Another important contribution is due to Caflisch[394] and Youla.[395] However, one of the pioneering works in this important branch of inverse problems is due to Gopinath and Sondhi.[396] References 326 to 327 should be consulted for recent reviews of the problems in nonuniform transmission line synthesis. Inversion problems in discrete systems will be considered in Chapter 11.

XIV. MISCELLANEOUS

Finally, we close our review with a brief mention of some miscellaneous applications of inverse scattering. Noteworthy among these are acoustic spectroscopy or identifying an elastic target from its resonant eigenvibrations in the echoes of the reflected acoustic waves, combustion diagnostics by multiangular spectrographic absorption, determination of the refractive index profile of a diffusion-fabricated waveguide, recovering surface microroughness from light scattering, inverse scattering with a stochastic initial condition, inverse scattering in photon counting, application of inverse scattering problems for blood vessels, and inverse scattering analysis of pressure transients in a petroleum reservoir.[397-405]

Another beautiful, yet rarely mentioned inverse problem occurs in glaciology, the study of the structure and dynamics of glaciers. An in-depth discussion of the glacier theory has been written by Lliboutry.[406] For temperate mountain glaciers, the classical study of glacier dynamics involves determining the mass balance and the surface velocity at a set of locations or stakes along its surface. The most important parameter to know is the sliding velocity. This, of course, requires that the velocity distributions at various depths be known. This gives rise to the inverse problem in glacier dynamics, namely, the determination of the velocity distributions at various depths in the glacier ice from the knowledge of the surface data, which consists of the surface velocity and its gradient. The rheological law of permanent steady creep of glacier ice has been investigated by Duval,[407] the knowledge of which is indispensable for solving the inverse problem.

Mathematically, the inverse problem consists of determining the velocity distribution v(y,z) from the Navier-Stokes equation

$$(\eta v')' + (\eta \dot{v})_{\dot{z}} + \rho g \sin\alpha = 0$$

subject to the surface data given by

$$v(y, z = 0) = v_{sur}(y)$$

and

$$v(\dot{z}=0)=0$$

The glacier is considered to be cylindrical with its surface the x-y plane which is at an angle of inclination of α to the z-axis pointing downwards. In other words, the problem as enumerated above is two-dimensional. The rheological law giving the viscosity η is that of Glen[408] and is written as

$$\eta = B^{-1/n}\left(v'' + \ddot{v}\right)^{(1-n)/2n}$$

B and n being constants. $'$ denotes $\partial/\partial y$, and $\cdot$ as well as the subscript z stand for $\partial/\partial z$.

The direct problem consists of solving the Navier-Stokes equation with the sliding velocity given.[409-410] If the bedrock depth is known from seismological or other measurements, then the solution of the inverse problem yields the sliding velocity. An extensive discussion of this inverse problem, including its stability and instability for downward numerical extrapolation, and the longstanding inverse solution by Somigliana[411] for Newtonian viscosity, i.e.,

$$\eta\left(v'' + \ddot{v}\right) + \rho g \sin\alpha = 0$$

are presented by Hantz and Lliboutry.[412]

It is highly interesting that an inverse problem very similar to that of glacier flow also arises in a dramatically different physical situation, namely, in an inclined-angle centrifuge.[413-414] In an inclined-angle centrifuge, the ordinary centrifuge action is enhanced by inclining the walls of the centrifuge,[415] the so-called "Boycott effect".[416] However, a detailed analysis of this practical, useful and life-saving problem is not known to the author.

We conclude our discussion of miscellaneous inverse problems with a highly interesting inverse problem of three-dimensional, potential compressible fluid flow around a body of arbitrary shape. The direct problem involves determining the flow characteristics due to the presence of known physical boundaries or objects. The inverse problem or the design problem is concerned with the opposite effect, namely, that of the shapes and configurations of the bodies or boundaries corresponding to the given flow characteristics which for potential flow consist of the "potential" (ϕ) and the "stream" functions (ψ). For example, the velocity for pressure distributions could be specified in terms of these functions and the inverse solution would consist of expressing the Cartesian coordinates (x,y,z) of the flow in terms of the specified potential and stream functions. The problem is too complex to discuss in any detail here and we refer the reader to the report published by Davis and Jeppson.[417]

NOTE

The existence problem may or may not arise depending upon the situation. If it is known a priori that there is an underlying potential corresponding to the data in S^C, then the problem is merely that of reconstructing the potential, the so-called "reconstruction" problem in inverse theory. On the other hand, if the existence of such an underlying potential is in question, the so-called "construction" problem, then the problem of existence not only does arise, but can be quite severe.

The construction problem is intricately associated with that of "characterization" and "consistency". There may be a large set of functions S that may be considered as legitimate data in so far as the form and general properties are concerned. However, due to the fact that the potentials are restricted to the class C whether because of the inherent nature of the

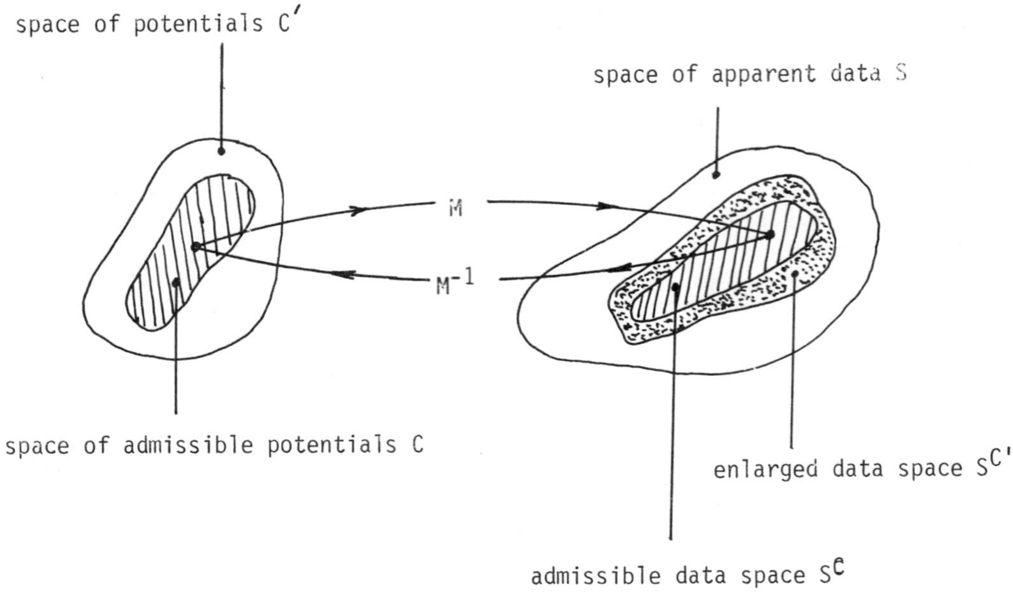

FIGURE 2. A schematic illustrating the problem of characterization.

problem or because of the present state of our technical knowledge in handling the inverse problems, the data set must be restricted to $S^C \subset S$. This is shown pictorially in Figure 2. However, the map **M**: C - -> S^C is only a definition and does not imply anything more. In other words, the map does not imply recognizability of the admissible data $\subset S^C$. The problem of recognizability is known as the "characterization" problem and is an embedding problem as described in the text. It is important to note that contrary to our intuitive thinking, any contraction of the set C is detrimental to the characterization problem. It is interesting that characterization via embedding **M′**: C′ - -> $S^{C'}$, $C \subset C'$, $S^C \subset S^{C'}$, is also a means of rendering an ill-posed problem (almost always the case with inverse construction problems) well-posed.

The characterization problem is particularly severe when the data set overdetermines the potential. In other words, when there is redundancy in the data set. The characterization problem is more difficult under this condition since one is now faced with the additional problem of determining exactly which subset of the data set is admissible. This is also a "consistency" problem since the remainder of the data set may be incompatible with this restricted subset. A well known example is inverse scattering in three-dimensions, where the potential is recovered on the basis of the scattering amplitude $A(k,\theta,\theta')$ (which is a complex-valued function in $R \times S^2 \times S^2$ - - -> C, S^2 being a sphere in three-dimensions) in the Born or the high energy limit k - - -> ∞. The scattering amplitude is a function of five variables, energy and two angles each for $\theta,\theta' \subset S^2$, whereas, the underlying potential is a function in the three-dimensional Euclidean space. The data contained in $A(k,\theta,\theta')$ is highly redundant and finding admissibility usually requires that severe restrictions be imposed on $A(k,\theta,\theta')$. Viewed from another angle, since the large $|k|$ limit is used, the data is insensitive to values of $A(k,\theta,\theta')$ for finite k's making it difficult to recognize admissibility of the functions $A(k,\theta,\theta')$.

REFERENCES

1. **J. Tyndall:** On the blue colour of the sky, the polarization of skylight, and on the polarization of light by cloudy matter generally, *Philos. Mag.*, 37, 1869. (See also, Note on the formation and phenomena of clouds, *Philos. Mag.*, 38, 1869., On the colour of water and the scattering of light in water and in air, *R. Inst. Proc. VI*, 1872.)
2. **Lord Rayleigh:** *The Scientific Papers of Lord Raleigh*, Vol. 1, Dover, New York, 1964.
3. **G. Mie:** A contribution to the optics of turbid media, especially colloidal metallic suspensions, *Ann. Phys.*, 25, 1908.
4. **A. Einstein:** Theorie des Opaleszenz von homogenen Flüssigkeiten und Flüssigkeitsgenuschen in der Nahe des kristische Zustsande, *Ann. Phys.*, 33, 1910.
5. **M. Smolouchosky:** Molekular-Kinetische Theorie des Opaleszenz von Gazen im kritischen Zustande, sowie einiger verwandter Enscheinungen, *Ann. Phys.*, 25, 1908.
6. **A. H. Shankland, Ed.:** *The Scientific Papers of Arthur Holy Compton*, University of Chicago Press, 1973.
7. **J. D. Jackson:** *Classical Electrodynamics*, John Wiley & Sons, New York, 1962.
8. **L. van Hove:** Correlations in space and time and Born approximation scattering in systems of interacting particles, *Phys. Rev.*, 95, 249,
9. **R. J. Glauber:** The scattering of neutons by systems of nuclei, *Phys. Rev.*, 87, 1952 (abstract).
10. **J. Faber, Jr. Ed.:** *Neutron Scattering*, AIP Conference Proceedings No. 89, Argonne National Laboratory, 1981.
11. **D. B. Pearson:** *Quantum Scattering and Spectral Theory*, Academic Press, London, 1988.
12. **J. R. Taylor:** *Scattering Theory*, John Wiley & Sons, New York, 1972.
13. **C. J. Joachain:** *Quantum Collision Theory*, North-Holland, Amsterdam, 1983.
14. **T. Regge:** *Potential Scattering*, North-Holland, Amsterdam, 1965.
15. **W. O. Amerin, J. M. Jauch, and K. B. Sinha:** *Scattering Theory in Quantum Mechanics: Physical Principles and Mathematical Methods*, W. A. Benjamin, Reading, MA, 1977.
16. **L. Brillourin:** Diffusion de la lumiere et des rayons par un corps transparent homogene, *Ann. Phys. (Paris)*, 17, 1922.
17. **I. L. Fabelinskii:** *Molecular Scattering of Light*, Plenum Press, New York, 1968.
18. **E. Gross:** Change of wavelength of light due to elastic heat waves at scattering in liquids, *Nature*, 126, 1930.
19. **C. V. Raman:** *The Scattering of Light*, Indian Academy of Sciences, 1978.
20. **L. D. Landau and G. Placzek:** *Phys. Z. Sovietun.*, 5, 1934.
21. **S. M. Rytov:** Correlation theory for Rayleigh scattering of light. I, *Sov. Phys. JETP*, 6, 1958.
22. **L. I. Komarov and I. Z. Fisher:** Theory of Rayleigh scattering of light in liquids, *Sov. Phys. JETP*, 16, 1963.
23. **H. Z. Cummins and H. L. Sweeney:** Light beating spectroscopy, in *Progress in Optics*, Vol. 8, E. Wolf, Ed., North-Holland, Amsterdam, 1970.
24. **B. Chu:** *Laser Light Scattering*, Academic Press, New York, 1974.
25. **P. A. Fleury and J. P. Boon:** Laser light scattering in fluid systems, in *Advances in Chemical Physics*, Vol. 24, I. Prigogine and S. A. Rice, Eds., John Wiley & Sons, New York, 1973.
26. **W. M. Gelbert:** Depolarized light scattering by simpe fluids, in *Advances in Chemical Physics*, Vol. 26, I. Prigogine and S. A. Rice, Eds., John Wiley & Sons, New York, 1973.
27. **H. Z. Cummins and E. R. Pike:** *Photon Correlation and Light Beating Spectroscopy*, Proc. NATO Advanced Study Institute, Plenum Press, New York, 1973.
28. **R. Y. Chiao, C. H. Townes, and B. P. Stoicheff:** Stimulated Brillouin scattering and coherent generation of intense hypersonic waves, *Phys. Rev. Lett.*, 12, 1964.
29. **N. M. Kroll:** Excitation of hypersonic vibrations by means of photoelastic coupling of high-intensity light waves to elastic waves, *J. Appl. Phys.*, 36, 1965.
30. **C. L. Tang:** *J. Appl. Phys.*, 37, 1966.
31. **R. W. Hellwarth:** Theory of stimulated Raman scattering, *Phys. Rev.*, 130, 1963.
32. **A. Penzkofer, A. Laubereau, and W. Kaiser:** High intensity Raman interactions, in *Progress in Quantum Electronics*, Vol. 6, T. S. Moss and S. Stenholm, Eds., 1979.
33. **Y. R. Shen and N. Bloembergen:** Theory of stimulated Brillouin and Raman scattering, *Phys. Rev.*, 137, 1965.
34. **V. S. Starunov:** *Phys. Lett.*, 26A, 1968. (See also, *Sov. Phys. JETP*, 30, 1970.)
35. **R. M. Herman and M. A. Gray:** Theoretical prediction of the stimulated thermal Rayleigh scattering in liquids, *Phys. Rev. Lett.*, 19, 1967.
36. **B. A. Akanaev, S. A. Akhmanov and Yu. G. Khronopulo:** *Sov. Phys. JETP*, 28, 1969.
37. **J. A. Armstrong, N. Bloembergen, J. Ducuing and P. S. Pershan:** Interactions between light waves in nonlinear media, *Phys. Rev.*, 127, 1962.

38. **N. Bloembergen:** *Nonlinear Optics*, W. A. Benjamin, New York, 1965.
39. **Y. R. Shen:** *The Principles of Nonlinear Optics*, John Wiley & Sons, New York, 1984.
40. **W. Kaiser and M. Maier:** Stimulated Rayleigh, Brillouin and Raman Spectroscopy, in *Laser Handbook*, F. T. Arechhi and E. O. Schulz-Dubois, Eds., North-Holland, Amsterdam, 1972.
41. **P. M. Morse and K. U. Ingard:** *Theoretical Acoustics*, McGraw-Hill, New York, 1968.
42. **R. T. Beyer:** *Nonlinear Acoustics*, U.S. Government Printing Office, Washington, D.C., 1974.
43. **R. Newton:** The Marchenko and Gel'fan-Levitan methods in the inverse scattering problem in one and three dimensions, in *Conference on Inverse Scattering Theory and Application,* J. B. Bednar, R. Redner, E. Robinson and A. Weglein, Eds.: Society for Industrial and Applied Mathematics, Philadelphia, 1983. (Regarding the general exposition of inverse problems, see also **J. B. Keller:** Inverse problems, *Am. Math. Mon.*, 83, 1967; **R. L. Parker:** Understanding inverse theory, *Annu. Rev. Earth Planet. Sci.*, 5, 1977; **P. C. Sabatier:** Inverse problems — an introduction, *Inverse Probl.*, 1, 1985; **H. Hochstadt:** *Integral Equations,* Wiley-Interscience, New York, 1973; **I. D. Kulagin, L. H. Sorokin, and E. A. Dubrovskaya:** Evaluation of some numerical methods for solving Abel's integral equation, *Opt. Spectros. (USSR),* 32, 1972; **S. I. Herlitz:** A method for computing the emission distribution in cylindrical light sources, *Ark. Fys.*, 23, 1963.)
44. **I. Tolstoy:** *Wave Propagation*, McGraw-Hill, New York, 1973.
45. **S. G. Mikhlin:** *Integral Equations*, Pergamon Press, Oxford, 1957.
46. **B. A. Lippmann and J. Schwinger:** Variational principles for scattering processes. I, *Phys. Rev.*, 79, 1950.
47. **I. Schiff:** *Quantum Mechanics*, McGraw-Hill, New York, 1968.
48. **I. M. Gel'fand and B. M. Levitan:** On the determination of a differential equation from its spectral function, *Am. Math. Soc. Transl.*, 1, 1955.
49. **B. M. Levitan and M. Gasymov:** Determination of a differential equation by two of its spectra, *Russ. Math. Surv.*, 1964.
50. **V. A. Marcenko:** *Sturm-Louisville Operators and Applications*, Birkhanser-Verlag, Boston, 1986.
51. **A. K. Agranovich and V. A. Marcenko:** *The Inverse Problem of Scattering Theory*, Gordon and Breach, London, 1963.
52. **A. Ya. Povzner:** On differential equations of Sturm-Lionville type on a half-axis, *Am. Math. Soc. Transl.*, 5, 1950.
53. **B. M. Levitan:** Application of generalized displacement operators to second order differential equations, *Am. Math. Soc. Transl.*, 59, 1951.
54. **P. Deift and E. Trubowitz:** Inverse scattering on the line, *Commun. Pure Appl. Math.*, 32, 1979.
55. **K. Chadan and P. C. Sabatier:** *Inverse Problems in Quantum Scattering Theory*, Springer-Verlag, New York, 1977. A revised and much expanded second edition has just been published.
56. **I. M. Gel'fand:** Integral geometry and its connection with representation theory, *Usp. Mat. Nauk*, 15, 1960.
57. **J. Radon:** Uber die Berstimmung von Funktionen durch ihre Integralwerte langs gewisser Mannigfaltigkeiten, Berichte Sachsische Akademie der Wissenschaften, Leipzig, *Math. Phys. Kl.*, 69, 1917.
58. **I. M. Gel'fand, M. I. Graev and N. Ya. Vilenkin:** *Generalized Functions*, Vol. 5, Academic Press, New York, 1966.
59. **S. Helgason:** *The Radon Transform*, Birkhanser, Boston, 1980.
60. **D. Ludwig:** The Radon transform on Euclidean space, *Commun. Pure Appl. Math.*, 19, 1966.
61. **M. M. Labrentiev, V. G. Romanov and S. P. Shishtskii:** *Ill-posed Problems of Mathematical Physics and Analysis, American Mathematical Society*, 1986.
62. **A. N. Tikhonov and V. Ya. Arsenin:** *Methods of Solving Incorrect Problems*, Halsted Press, Washington, D.C., 1977.
63. **F. Faxen and J. Holtsmark:** *Z. Phys.*, 45, 1927.
64. **N. Levinson:** Determination of the potential from the asymptotic phase, *Phys. Rev.*, 75, 1949.
65. **N. Levinson:** On the uniqueness of the potential in a Schrodinger equation for a given asymptotic phase, *Dan. Vid Selsk. Mat. Fys. Medd.*, 25, 1949.
66. **V. Bargmann:** On the connection between phase shifts and scattering potential, *Rev. Mod. Phys.*, 21, 1949.
67. **V. Bargmann:** Remarks on the determination of a central field of force from the elastic scattering potential, *Phys. Rev.*, 75, 1949.
68. **V. A. Marchenko:** Concerning the theory of a differential operator of the second order, *Dokl. Akad. Nauk*, 72, 1950.
69. **G. Borg:** Uniqueness theorems in the spectral theory of $y' + (\lambda - q(x))\, y = 0$, in Proc. 11th Congr. Scand. Mathematicians, J. Crundt, Ed., Trondheim, Tanums Furlag, 1949.
70. **V. Ambarzumian:** Über eine Frage der Eigenwerttheorie, *Z. Phys.*, 53, 1929.

71. **C. Fröberg:** Calculation of the interaction between two particles from the asymtotic phase, *Phys. Rev.*, 72, 1948.
72. **E. A. Hyellaraas:** Calculation of a perturbing central field of force from the elastic scattering phase shift, *Phys. Rev.*, 74, 1948.
73. **S. T. Ma:** Redundant zeros in the discrete energy spectra in Heisenberg's theory of characteristic matrix, *Phys. Rev.*, 69, 1946.
74. **B. Holmberg:** A remark on the uniqueness of the potential determined from the asymptotic phase, *Nuovo Cimento*, 9, 1952.
75. **R. Courant and K. O. Friedrichs:** *Methods of Mathematical Physics*, Vol. 2, New York University, 1943.
76. **P. Garabadien:** *Partial Differential Equations*, John Wiley & Sons, New York, 1964.
77. **L. D. Fadeev:** Increasing solutions of the Schroedinger equation, *Sov. Phys. Dokl.*, 11, 1966. (Also see Reference 78.)
78. **R. G. Newton:** The Gel'fan-Levitan method in the inverse scattering problem, in *Scattering Theory in the Inverse Problem*, J. A. La Vita and J. P. Marchand, Eds., D. Reidel, Dordrecht, Holland, 1974.
79. **R. G. Newton:** Inverse scattering II. Three dimensions, *J. Math. Phys.*, 21, 1980; **R. G. Newton:** Inverse scattering. III. Three dimensions, continued, *J. Math. Phys.*, 22, 1981; **R. G. Newton:** Inverse scattering. IV. Three dimensions: generalized Marchenko construction with bound states, and generalized Gel'fand Levitan equations, *J. Math. Phys.*, 23, 1982. A complete account of this subject is given by the author's most recent book *Inverse Schrödinger Scattering in Three-Dimensional,* Springer-Verlag, 1988. See also chapter XIV of the second edition of Reference 53. Both of these appeared after the present monograph was completed.
80. **R. Jost and W. Kohn:** Equivalent potentials, *Phys. Rev.*, 88, 1952.
81. **N. Levinson:** Certain explicit relationships between phase shift and scattering potential, *Phys. Rev.*, 89, 1953.
82. **I. W. Kay:** *Inverse Scattering Papers (1955-1962)*, Math. Science Press, Brooklyn, MA, 1982.
83. **J. A. Wheeler:** Scattering and potential, *Phys. Rev.*, 99, 1955.
84. **T. Regge:** Introduction to complex orbital momenta, *Nuovo Cimento*, 14, 1959.
85. **R. G. Newton:** Construction of potentials from the phase shift at fixed energies, *J. Math. Phys.*, 3, 1962.
86. **P. C. Sabatier:** Asymptotic properties of the potentials in the inverse scattering problem at fixed energy, *J. Math. Phys.*, 1966.
87. **P. J. Redmond:** Some remarks concerning a pathological matrix in the inverse-scattering problem, *J. Math. Phys.*, 5, 1964.
88. **A. Martin and Gy. Targonski:** On the uniqueness of a potential fitting a scattering amplitude at a given energy, *Nuovo Cimento*, 20, 1961.
89. **R. Jost and W. Kohn:** Construction of a potential from a phase shift, *Phys. Rev.*, 87, 1952.
90. **R. T. Prosser:** Formal solutions of inverse scattering problems, *J. Math. Phys.*, 10, 1969.
91. **R. G. Newton and R. Jost:** The construction of potentials from the S-matrix for systems of differential equations, *Nuovo Cimento*, 1, 1955.
92. **M. G. Krein:** On the theory of accelerants and S-matrices of canonical differential systems, *Math. Rev.*, 19, 1956.
93. **R. G. Newton:** Relation of a potential to S matrix, *Phys. Rev.*, 100, 1955.
94. **R. G. Newton:** Analytical properties of radial wavefunctions, J. Math. Phys., 1, 1960.
95. **E. Corinaldisi:** Construction of potentials from phase shift and binding energies of relativistic equations, *Nuovo Cimento*, 11, 1954.
96. **F. Prats and J. S. Toll:** Construction of Dirac equation central potential from phase shifts and bound states, *Phys. Rev.*, 113, 1959.
97. **M. Verde:** The inversion problem in wave mechanics and dispersion relations, *Nucl. Phys.*, 9, 1958-1959.
98. **M. Bolsterli and J. Mackenzie:** Determinaton of separable potential from phase shift, *Physics*, 2, 1965.
99. **G. L. Lamb, Jr.:** *Elements of Soliton Theory*, John Wiley & Sons, New York, 1980.
100. **P. D. Lax:** Integrals of nonlinear equations of evolution and solitary waves, *Commun. Pure Appl. Math.*, 21, 1968.
101. **M. J. Ablowitz and H. Segur:** *Solitons and Inverse Scattering Transform*, Society for Industrial and Applied Mathematics, Philadelphia, 1981.
102. **A. C. Newell:** *Solitons in Mathematics and Physics*, Society for Industrial and Applied Mathematics, Philadelphia, 1985.
103. **S. Novikov, S. V. Manakov, L. P. Pitaevskii and V. E. Zakharov:** *Theory of Solitons: The Inverse Scattering Method*, Consultants Bureau, New York, 1984.
104. **M. G. Crandall:** *Nonlinear Evolution Equations*, Academic Press, London, 1978.

105. **M. Boiti, F. Pempinelli and G. Soliani, Eds.:** *Nonlinear Evolution Equations and Dynamical Systems*, Lecture notes in Physics, Vol. 120, Springer-Verlag, Berlin, 1980.
106. **R. M. Miura, Ed.:** *Backlund Transformations, the Inverse Scattering Method, Solitons, and Their Applications*, Lecture Notes in Mathematics, Vol. 515, Springer-Verlag, Berlin, 1976.
107. **A. C. Newell:** The general structure of integrable evolution equations, *Proc. R. Soc.*, A365, 1979.
108. **V. E. Zakharov and A. B. Shabat:** Integration of nonlinear equations of mathematical physics by the method of inverse scattering. II, *Funct. Anal. Appl.*, 13, 1980.
109. **R. Hirota:** Direct methods of finding exact solutions of nonlinear evolution equations, in *Backlund Transformations*, R. M. Miura Ed., Springer-Verlog, Berlin, 1976.
110. **T. Kawata and H. Inoue:** Inverse scattering method for the nonlinear evolution equations under nonvanishing conditions, *J. Phys. Soc. Jpn.*, 44, 1978.
111. **W. Eckhaus and A. Van Harten:** *The Inverse Scattering Transformation and the Theory of Solitons*, North-Holland, Amsterdam, 1981.
112. **T. R. Taha and M. J. Ablowitz:** On analytic and numerical aspects of certain nonlinear evolution equations. I-III, Clarkson College preprints, IFNS No. 14 to 16, 1982.
113. **J. M. Ball:** Remarks on blow-up and non-existence theorems for nonlinear evolution equations, *Q. J. Math. (Oxford)*, 28, 1977.
114. **V. G. Makhankov:** Dynamics of classical solitons (in non-integrable systems), *Phys. Rep.*, 35, 1978.
115. **V. G. Makhankov, G. Kummer and A. B. Shvachka:** Many dimensional U(1) solitons, their interactions, resonances and bound states, *Phys. Scr.*, 20, 1979.
116. **R. K. Bullough and P. J. Caudrey, Eds.:** *Solitons, Topics in Current Physics 17*, Springer-Verlag, Berlin, 1980.
117. **R. K. Dodd, J. C. Eilbeck, J. D. Gibbon and H. C. Moris:** *Solitons and Nonlinear Waves*, Academic Press, New York, 1982.
118. **F. Calogero and A. Degasperis:** *Spectral Transform and Solitons I*, North-Holland, Amsterdam, 1982.
119. **G. Eilenberger:** *Solitons*, Springer-Verlag, Berlin, 1980.
120. **J. L. Bona, W. C. Pritchard and L. R. Scott:** Solitary-wave interaction, *Phys. Fluids*, 23, 1980.
121. **A. R. Bishop and T. Schneider:** Solitons and condensed matter physics, *Springer Ser. Solid State Sci.*, 8, 1978.
122. **P. G. Drazin:** *Solitons*, Cambridge University Press, Cambridge, 1983.
123. **A. S. Davydov:** *Theory of Molecular Excitons*, Plenum Press., New York, 1971.
124. **A. S. Davydov and N. I. Kislukha:** Solitons in one-dimensional molecular chains, *Sov. Phys. JETP*, 44, 1976.
125. **A. S. Davydov:** Solitons in molecular systems, *Phys. Scr.*, 20, 1979.
126. **K. Longrenn and A. Scott:** *Solitons in Action*, Academic Press, New York, 1978.
127. **A. C. Scott, F. Y. F. Chu and D. W. McLaughlin:** The soliton: a new concept in applied science, *Proc. IEEE*, 61, 1973.
128. **N. J. Zabusky:** A synergetic approach to problems of nonlinear dispersive wave propagation and interaction, in *Nonlinear Partial Differential Equations*, W. Ames, Ed., Academic Press, New York, 1967.
129. **N. J. Zabusky:** Solitons and energy transport in nonlinear lattices, *Comput. Phys. Commun.*, 5, 1973.
130. **P. L. Bhatnagar:** *Nonlinear Waves in One-Dimensional Dispersive Systems*, Clarendon Press, Oxford, 1979.
131. **G. B. Whitham:** *Linear and Nonlinear Waves*, John Wiley & Sons, New York, 1974.
132. **A. Jefferey and T. Kawahara:** *Asymptotic Methods in Nonlinear Wave Theory*, Academic Press, New York, 1964.
133. **D. J. Korteweg and G. de Vries:** On the change of form of long waves advancing in a rectangular canal, and on a new type of long stationary waves, *Philos. Mag.*, 39, 1895.
134. **P. D. Lax:** Almost periodic solutions of the KdV equation, *SIAM Rev.*, 18, 1976.
135. **C. S. Gardner, J. M. Greene, M. D. Kruskal, and R. M. Miura:** Method for solving the Korteweg-de Vries equation, *Phys. Rev. Lett.*, 19, 1967.
136. **R. M. Miura:** Korteweg-de Vries equation and generalizations. I. A remarkable explicit nonlinear transformation, *J. Math. Phys.*, 9, 1968.
137. **R. M. Miura, C. S. Gardner and M. D. Kruskal:** Korteweg-de Vries equation and generalizations. II. Existence of conservation laws and constants of the motion, *J. Math. Phys.*, 9, 1968.
138. **C. S. Su and C. S. Gardner:** The KdV equation and generalizations. III. Derivation of the Korteweg-de Vries equation and Burgers' equation, *J. Math. Phys.*, 10, 1969.

139. **C. S. Gardner:** Korteweg-de Vries equation and generalizations. IV. The Korteweg-de Vries equation as a Hamiltonian system, *J. Math. Phys.*, 12, 1971.

140. **M. D. Kruskal, R. M. Miura and C. S. Gardner:** Korteweg-de Vries equation and generalizations. V. Uniqueness and nonexistence of polynomial conservation laws, *J. Math. Phys.*, 11, 1970.

141. **C. S. Gardner, J. M. Greene, M. D. Kruskal and R. M. Miura:** Korteweg-deVries equation and generalizations. VI. Methods of exact solutions, *Commun. Pure Appl. Math.*, 27, 1974.

142. **M. D. Kruskal:** The Korteweg-de Vries equation and related evolution equations, in *Nonlinear Wave Motion*, A. C. Newell, Ed., Lectures in Applied Mathematics 15, American Mathematics Society, Providence, RI, 1974.

143. **R. Hirota:** Exact solution of the Korteweg-de Vries equation for multiple collisions of solitons, *Phys. Rev. Lett.*, 27, 1971.

144. **R. M. Miura:** The KdV equation: a model equation for nonlinear dispersive waves, in *Nonlinear Waves*, S. Leibovitch and A. Seebass, Eds., Cornell University Press, Ithaca, NY, 1974.

145. **R. M. Miura:** The Korteweg-de Vries equation: a survey of results, *SIAM Rev.*, 18, 1976.

146. **S. Tanaka:** KdV equation: construction of solutions in terms of scattering data, *Osaka J. Math.*, 11, 1974.

147. **M. Wadati and M. Toda:** The exact n-soliton solution of the Korteweg-de Vries equation, *J. Phys. Soc. Jpn.*, 32, 1972.

148. **M. Wadati:** The exact solution of the modified Korteweg-de Vries equation, *J. Phys. Soc. Jpn.*, 32, 1972.

149. **V. E. Zakharov and L. D. Faddeev:** KdV equation: a completely integrable Hamiltonian system, *Funct. Anal. Appl.*, 5, 1972.

150. **P. D. Lax:** Periodic solutions of the KdV equation, *Commun. Pure Appl Math.*, 28, 1975.

151. **B. Dubrovin and S. P. Novikov:** Periodic and conditionally periodic analogs for the many-soliton solutions of the KdV equation, *Sov. Phys. JETP*, 40, 1975.

152. **S. P. Novikov:** A periodic problem for the KdV equation. I, *Funct. Anal. Appl.*, 8, 1974.

153. **J. L. Bona and R. Smith:** The initial value problem for the KdV equation, *Philos. Trans. R. Soc.*, A278, 1975.

154. **A. Cohen:** Existence and regularity of solutions of the KdV equation, *Arch. Ration. Mech. Anal.*, 79, 1979.

155. **N. Hershkowitz, T. Romesser and D. Montgomery:** Multiple soliton production and the KdV equation, *Phys. Rev. Lett.*, 29, 1972.

156. **J. W. Miles:** The Korteweg-de Vries equation: a historical essay, *J. Fluid Mech.*, 106, 1981.

157. **F. D Tappert:** Numerical solutions of the KdV equation and its generalisations by the split-step Fourier method, *Lect. Appl. Math. Am. Math. Soc.*, 15, 1974.

158. **V. E. Zakharov and S. V. Manakov:** On the complete integrability of the nonlinear Schrodinger equation, *Theor. Math. Phys.*, 19, 1974.

159. **V. E. Zakharov and A. B. Shabat:** Exact theory of two-dimensional self-focusing and one-dimensional self-modulation of waves in nonlinear media, *Sov. Phys. JETP*, 34, 1972.

160. **A. C. Newell:** The inverse scattering transform, in *Solitons, Topics in Current Physics 17*, R. K. Bullough and P. J. Caudrey, Eds., Springer-Verlog, Berlin, 1980.

161. **D. R. Nicholson and M. V. Goldman:** Damped nonlinear Schrodinger equation, *Phys. Fluids*, 19, 1976.

162. **T. Kawata, J. Sakai and N. Kobayashi:** Inverse method for the mixed nonlinear Schrodinger equation and nonlinear wave equations, *J. Phys. Soc. Jpn.*, 48, 1980.

163. **A. Barone, F. Exposito, C. J. Magee and A. C. Scott:** Theory and applications of the Sine-Gordon equation, *Riv. Nuovo Cimento*, 1, 1971.

164. **J. Rubenstein:** Sine-Gordon equation, *J. Math. Phys.*, 11, 1970.

165. **V. E. Zakharov, L. A. Tahtadjan and L. D. Fadeen:** A complete description of the solutions of the Sine-Gordon equation, *Sov. Phys. Dokl.*, 19, 1975 .

166. **V. E. Zakharov, L. A. Takhtadzhyan and L. D. Faddeev:** Complete description of the solution of the sine-Gordon equation, *Sov. Phys. Dokl.*, 19, 1975.

167. **R. Hirota:** Exact solution of the sine-Gordon equation for multiple collisions of solitons, *J. Phys. Soc. Jpn.*, 33, 1972.

168. **R. Hirota:** Nonlinear partial difference equations. II. Discrete sine-Gordon equation, *J. Phys. Soc. Jpn.*, 43, 1977.

169. **G. Leibbrandt, R. Morf and S. Wang:** Solutions of the sine-Gordon equation in higher dimensions, *J. Math. Phys.*, 21, 1980.

170. **G. Constabile, R. Parmentier, B. Savo, D. W. McLaughlin and A. C. Scott:** Exact solutions of the sine-Gordon equation describing oscillations in a long (but finite) Josephson junction, *Appl. Phys. Lett.*, 32, 1978.

171. **J. F. Currie:** Aspects of exact dynamics for general solutions of the SG equation with applications to domain walls, *Phys. Rev.*, A16, 1977.
172. **P. J. Caudrey, J. C. Eilbeck and J. D. Gibbon:** The sine-Gordon equation as a model classical field theory, *Nuovo Cimento*, 25B, 1975.
173. **J. D. Gibbon, I. N. James and I. M. Moroz:** The sine-Gordon equation as a model for a rapidly rotating baroclinic fluid, *Phys. Scr.*, 20, 1979.
174. **E. Fermi, J. R. Pasta and S. M. Ulam:** in *Collected Works of E. Fermi*, Vol. 2, University of Chicago Press, Chicago, 1965.
175. **N. J. Zabusky and M. D. Kruskal:** Interaction of solitons in a collisionless plasma and the recurrence of initial states, *Phys. Rev. Lett.*, 15, 1965.
176. **J. D. Abella, N. A. Kurnit, and S. R. Hartmann:** Photon echoes, *Phys. Rev.*, 141, 1966.
177. **L. Allen and J. H. Eberly:** *Optical Resonance and Two-level Atoms*, John Wiley & Sons, New York, 1975.
178. **I. M. Beterov and V. P. Chebotayev:** Three level systems and their interaction with radiation, in *Prog. Quantum Electron.*, 3, 1974.
179. **V. S. Letokhov and V. P. Chebotayev:** *Nonlinear Laser Spectroscopy*, Springer Series in Optical Sciences, Vol. 4, D. L. MacAdam, Ed., Springer-Verlag, Berlin, 1977.
180. **R. Balian, S. Haroche, and S. Liebermann Eds.:** *Frontiers in Laser Spectroscopy*, North-Holland, Amsterdam, 1977.
181. **W. Demtroder:** *Laser Spectroscopy*, Springer-Verlog, Heidelberg, 1982.
182. **R. E. Slusher:** Self-induced transparency, *Progress in Optics*, Vol. 12, E. Wolf Ed., North-Holland, Amsterdam, 1973.
183. **J. D. Gibbon, P. J. Caudrey, R. K. Bullough and J. C. Eilbeck:** An N-Soliton solution of a nonlinear optics equation derived by a general inverse method, *Lett. Nuovo Cimento*, 8, 1973.
184. **M. J. Ablowitz, D. J. Kaup, A. C. Newell and H. Segur:** Coherent pulse propagation, a dispersive, irreversible phenonmenon, *J. Math. Phys.*, 15, 1974.
185. **D. J. Kaup:** Coherent pulse propagation: a comparison of the complete solution with the McCall-Hahn theory and others, *Phys. Rev.*, A16, 1977.
186. **J. C. Eilbeck, J. D. Gibbon, P. J. Caudrey and R. K. Bullough:** Solitons in nonlinear optics. I. A more accurate description of the 2 pulse in self-induced transparency, *J. Phys.*, A6, 1973.
187. **P. J. Caudrey, J. D. Gibbon, J. C. Eilbeck and R. K. Bullough:** Exact multisoliton solutions of the self-induced transparency and sine-Gordon equations, *Phys. Rev. Lett.*, 30, 1973.
188. **P. J. Caudrey, J. C. Eilbeck, J. D. Gibbon and R. K. Bullough:** Exact multisoliton solution of the inhomogeneously broadened self-induced transparency equations, *J. Phys.*, A6, 1973.
189. **P. J. Caudrey, J. C. Eilbeck, J. D. Gibbon and R. K. Bullough:** Multiple soliton and bisoliton bound state solutions of the sine-Gordon equation and related equations in nonlinear optics, *J. Phys.*, A6, 1973.
190. **P. J. Caudrey, J. C. Eilbeck and J. D. Gibbon:** An N-soliton solution of the reduced Maxwell-Bloch equations, *JIMA*, 14, 1974.
191. **R. K. Bullough, P. M. Jack, P. W. Kitchenside and R. Saunders:** Solitons in laser physics, *Phys. Scr.*, 20, 1979.
192. **J. C. Eilbeck and R. K. Bullough:** The method of characteristics in the theory of resonant or nonresonant nonlinear optics, *J. Phys.*, A5, 1972.
193. **G. L. Lamb:** Higher conservation laws in ultrashort optical pulse propagation, *Phys. Lett.*, 32A, 1970.
194. **G. L. Lamb:** Analytic descriptions of ultrashort optical pulse propagation in a resonant medium, *Rev. Mod. Phys.*, 43, 1971.
195. **G. L. Lamb:** Phase variation in coherent optical pulse propagation, *Phys. Rev. Lett.*, 31, 1973.
196. **G. L. Lamb:** On the connection between lossless propagation and pulse profile, *Physica*, 66, 1973.
197. **E. A. Manykin, S. O. Elyutin, S. M. Zakharov, V. N. Likhachev and A. I. Maimistov:** Coherent phenomena during the interaction between light pulses and resonance media, *Izv. Akad. Nauk SSSR Ser. Fiz.*, 46, 1982.
198. **R. Michalska-Trautman:** Exact solution of the initial eigenvalue problem for coherent pulse propagation, *Phys. Rev.*, A23, 1981.
199. **V. E. Zakharov and S. V. Manakov:** Resonant interaction of wave packets in nonlinear media, *Sov. Phys. JETP Lett.*, 18, 1973.
200. **V. E. Zakharov and S. V. Manakov:** Asymptotic behaviour of non-linear wave systems integrated by the inverse scattering method, *Sov. Phys. JETP*, 44, 1976.
201. **V. E. Zakharov and S. V. Manakov:** The theory of resonance interaction of wave packets in nonlinear media, *Sov. Phys. JETP*, 42, 1976.
202. **D. T. Hon:** Pulse compression by stimulated Brillouin scattering, *Opt. Lett.*, 5, 1980.

203. **J. J. Ewing, R. A. Haas, J. C. Swingle, E. V. George and W. P. Krupke:** Optical pulse compressor systems for laser fusion, *IEEE J. Quantum Electron.*, QE15, 1979.

204. **J. R. Murray, J. Goldhar, D. Eimerl and A. Szoke:** Raman pulse compressor for excimer lasers for applications to laser fusion, *IEEE J. Quantum Electron.*, QE15, 1979.

205. **D. N. Ghosh Roy, D. V. G. L. N. Rao and H. Bronk:** Optical pulse compression in a cholesteric liquid crystal, *Appl. Phys. Lett.*, 39, 1981.

206. **D. J. Kaup:** The three wave interaction — a nondispersive phenomenon, *Stud. Appl. Math.*, 55, 1976.

207. **D. J. Kaup:** Simple harmonic generation: an exact method of solution, *Stud. Appl. Math.*, 59, 1978.

208. **D. J. Kaup, A. H. Reiman and A. Bers:** Space-time evolution of nonlinear three-wave interactions. I, *Rev. Mod. Phys.*, 51, 1979.

209. **A. H. Reiman:** Space-time evolution of nonlinear three wave interactions. II, *Rev. Mod. Phys.*, 51, 1979.

210. **D. J. Benney and A. C. Newell:** The propagation of nonlinear wave envelopes, *J. Math. Phys.*, 46, 1967.

211. **D. J. Benney:** Significant interactions between long and short waves, *Stud. Appl. Math.*, 55, 1976.

212. **Y. Ma:** The complete solution of the long-wave-short-wave resonance equations, *Stud. Appl. Math.*, 59, 1978.

213. **S. S. Rangnekar and R. H. Enns:** Application of the inverse scattering transform method to stimulated Brillouin backscattering in a generator setup: the Zakharov-Manakov solution, *Can. J. Phys.*, 59, 1981.

214. **R. H. Enns, D. B. Guenther, and S. S. Rangnekar:** An application of the inverse scattering method to the 3-wave interaction in nonlinear optics, *Can. J. Phys.*, 58, 1980.

215. **R. H. Enns and S. S. Rangnekar:** The three-wave interaction in nonlinear optics, *Phys. Status Solidi*, 94, 1979.

216. **A. C. Newell and L. G. Redekopp:** Breakdown of Zakharov-Shabat theory and soliton creation, *Phys. Rev. Lett.*, 38, 1977.

217. **V. E. Zakharov and A. B. Shabat:** A scheme for integrating the nonlinear equations of mathematical physics by the method of the inverse scattering problem. I, *Funct. Anal. Appl.*, 8, 1974.

218. **C. Lange and A. C. Newell:** The post buckling problem for thin elastic shells, *SIAM J. Appl. Math.*, 21, 1971.

219. **J. C. Lewis and J. A. Tjon:** Resonant production of solitons in the RLW equation, *Phys. Lett.*, 73A, 1979.

220. **A. A. Andreev:** Decay interaction of wave packets in an inhomogeneous nonlinear medium and in a time-varying medium, *Sov. Phys. Tech. Phys. Lett.*, 5, 1979.

220a. **D. N. Ghosh Roy and D. V. G. L. N. Rao:** Stimulated light scattering and inverse scattering transform: in the forth coming volume honoring N. Bloembergen, edited by M. D. Levensen, Word Science Publishing, New York, 1990.

221. **N. G. Basov, Ed.:** *Phase Conjugation of Laser Emission*, Nova Science, Commack, 1988.

222. **B. Ya. Zel'dovich, N. F. Pilipetsky and V. V. Shknunov:** *Principles of Phase Conjugation*, Springer-Verlag, Berlin, 1985.

223. **David M. Pepper:** Nonlinear optical phase conjugation, in *Laser Handbook*, M. I. Stitch and M. Bass, Eds., Elsevier Science, 1985.

224. **R. W. Hellwarth:** Generation of time-reversed wavefront by nonlinear reflection, *J. Opt. Soc. Am.*, 67, 1977.

225. **A. Yariv and D. M. Pepper:** Amplified reflection, phase conjugation, and oscillation in degenerate four-wave mixing, *Opt. Lett.*, 1, 1977.

226. **B. Ya. Zel'dovich, V. I. Popovichev, V. V. Ragulsky, and F. S. Faizullov:** On the relationship between wavefronts of reflected and exciting light in stimulated Brillouin scattering, *Pisma Zh. Eksp. Teor. Fiz.*, 15, 1972.

227. **V. N. Blaschuk, B. Ya. Zel'dovich, A. V. Mamaev, N. F. Pilipetsky, V. V. Shkunov:** Complete phase conjugation of depolarized radiation by degenerate four-photon interaction (theory and experiment), *Sov. J. Quantum Elect.on.*, 10, 1980.

228. **G. Martin, L. K. Lam and R. W. Hellwarth:** Generation of a time-reversed replica of a nonuniformly polarized image-bearing optical beam, *Opt. Lett.*, 5, 1980.

229. **A. Yariv:** Introduction to Optical Electronics, Holt, Reinhart & Winston, New York, 1976.

230. **B. Fische and S. Sternklar:** Image transmission and interferometry with multimode fibres using self-pumped phase conjugation, *Appl. Phys. Lett.*, 46, 1985.

231. **O. Ikeda, M. Takehara and T. Sato:** High-performance image-transmission through a turbulent medium using multiple reflectors and adoptive focusing combined with four-wave mixing, *J. Opt. Soc. Am.*, Al, 1984.
232. **J. W. Goodman, W. H. Huntley, Jr., D. W. Jackson and M. Lehmann:** Wavefront-reconstruction imaging through random media, *Appl. Phys. Lett.*, 8, 1966.
233. **P. L. Kelley:** Self-focusing of optical beams, *Phys. Rev. Lett.*, 15, 1965.
234. **P. Lallemand and N. Bloembergen:** Self-focusing of laser beams and stimulated Raman gains in liquids, *Phys. Rev. Lett.*, 15, 1965.
235. **J. E. Bjorkholm and A. Ashkin:** CW self-focusing and self-trapping of light in sodium vapor, *Phys. Rev.*, 32, 1974.
236. **S. A. Akhmanov, R. V. Kokhlov and A. P. Sukhorukov:** Self-focusing, self-defocusing and self-modulation of laser beams, in *Laser Handbook*, F. T. Arecchi and E. O. Schultz-Dubois, Eds., North-Holland, Amsterdam, 1972.
237. **M. Maier, W. Kaiser and J. A. Giordmaine:** Backward stimulated Raman scattering, *Phys. Rev.*, 177, 1969.
238. **K. Konno and H. Suzuki:** Self-focusing of laser beam in nonlinear media, *Phys. Scr.*, 20, 1979.
239. **H. Nakatsuka, D. Grischkowsky and A. C. Balant:** Nonlinear picosecond-pulse propagation through optical fibres with positive group velocity dispersion, *Phys. Rev. Lett.*, 47, 1981.
240. **L. F. Mollenauer, R. H. Stolen and J. P. Gordon:** Experimental observation of picosecond pulse narrowing and solitons in optical fibres, *Phys. Rev. Lett.*, 45, 1983.
241. **A. Hasegawa and Y. Kodama:** Amplification and reshaping of optical solitons in glass fibres. I and II, *Opt. Lett.*, 7, 1982.
242. **A. Hasegawa and F. Tappert:** Transmission of stationary nonlinear pulses in dispersive dielectric fibres, *Appl. Phys. Lett.*, 23, 1973.
243. **D. Anderson:** Variational approach to nonlinear pulse propagation in optical fibers, *Phys. Rev.*, A27, 1983.
244. **A. B. Shvartsburg:** The controlled non-linear evolution of TE and TH ps-pulses in self-focusing fibre, *Opt. Quantum Electron.*, 14, 1982.
245. **N. A. Krall and A. W. Trivelpiece:** *Principles of Plasma Physics*, McGraw-Hill, New York, 1973.
246. **L. Spitzer Jr.:** *Physics of Fully Ionized Gases*, Interscience, New York, 1956.
247. **R. C. Davidson:** *Methods in Nonlinear Plasma Theory*, Academic Press, New York, 1972.
248. **J. Weiland and H. Wilhelmsson:** *Coherent Nonlinear Interaction of Waves in Plasmas*, Pergamon Press, New York, 1977.
249. **P. Krocek:** The dynamics of Langmuir solitons in the presence of ion-sound damping, *Plasma Phys.*, 23, 1981.
250. **M. Mohan and B. Buti:** Electron-acoustic solitons in current-carrying magnetized plasmas, *Plasma Phys.*, 22, 1980.
251. **E. Mjolhus:** A note on the modulational instability of long Alfven waves parallel to the magnetic field, *J. Plasma Phys.*, 19, 1978.
252. **V. I. Karpman and E. M. Maslov:** Perturbation theory for solitons, *Sov. Phys. JETP*, 46, 1978.
253. **V. I. Karpman:** *Nonlinear Waves in Dispersive Media*, Pitman, London, 1975.
254. **Y. A. Berezin and V. I. Karpman:** Nonlinear evolution of disturbances in plasmas and other dispersive media, *Sov. Phys. JETP*, 24, 1967.
255. **A. Yu. Berezin and V. I. Karpman:** Theory of nonstationary finite amplitude waves in a low-density plasma, *Sov. Phys. JETP*, 19, 1964.
256. **J. Weiland, Y. H. Ichikawa and H. Wilhelmsson:** A perturbation expansion for the NLS equation with application to the influence of nonlinear Landau damping, *Phys. Scri.*, 17, 1978.
257. **Y. H. Ichikawa, M. Wadati, K. Konno and T. Shimizu:** On A New Series of Integrable Evolution Equations, Report No. IPPJ-484, Nagoya University, Japan, 1980.
258. **Y. H. Ichikawa:** Topics on solitons in plasmas, *Phys. Scr.*, 20, 1979.
259. **B. Fried and Y. H. Ichikawa:** On the nonlinear Schrodinger equation for Langmuir waves, *J. Phys. Soc. Jpn.*, 34, 1973.
260. **H. Ikezi:** Experiments on ion-acoustic solitary waves, *Phys. Fluids*, 16, 1973.
261. **M. Ikezi, R. Taylor and R. D. Baker:** Formation and interaction of ion acoustic solitary waves, *Phys. Rev. Lett.*, 25, 1970.
262. **A. Hasegawa:** *Plasma Instabilities and Nonlinear Effects*, Springer-Verlag, Berlin, 1975.
263. **M. Q. Tran:** Ion acoustic solitons in a plasma: a review of their experimental properties and related theories, *Phys. Scr.*, 20, 1979.
264. **G. N. Balanis:** The plasma inverse problem, *J. Math. Phys.*, 13, 1972.

265. **N. Yajima and M. Oikawa:** Formation and interaction of sonic-Langmuir solitons — inverse scattering method, *Prog. Theor. Phys.*, 56, 1976.

266. **M. Kako and G. Rowlands:** Two dimensional stability of ion acoustic solitons, *Plasma Phys.*, 18, 1976.

267. **G. P. Leclert, C. F. F. Karney, A. Bers and D. J. Kaup:** Two-dimensional self-modulation of lower hybrid waves in inhomogeneous plasmas, *Phys. Fluids*, 22, 1979.

268. **N. R. Pereira and F. Y. F. Chu:** Damped double solitons in the nonlinear schrodinger equation, *Phys. Fluids*, 22, 1979.

269. **B. N. Breizman and V. M. Malkin:** Dynamics of modulation instability of a broad langmuir wave spectrum, *Sov. Phys. JETP*, 1980.

270. **K. R. Pechenick and J. M. Cohen:** Exact solutions to the valley problem in inverse scattering, *J. Math. Phys.*, 24, 1983.

271. **K. R. Pechenick and J. M Cohen:** Inverse scattering — exact solution of the Gel'fand-Levitan equation, *J. Math. Phys.*, 22, 1981.

272. **I. W. Kay:** A review of the one dimensional inverse scattering problem for stratified inhomogeneous media, *Int. Symp. Dig. Ant. Prop*, 221(4), 1979. (See also Reference 80.)

273. **M. H. Reily and A. K. Jordan:** The applicability of an inverse method for reconstruction of electron-density profiles, *IEEE Trans. Ant. Prop.*, AP29, 1981.

274. **P. K. Shukla and K. H. Spatschek:** Profile modifications at critical density of laser-produced plasmas, *J. Plasma Phys.*, 19, 1978.

275. **S. Ahn and A. K. Jordan:** Profile inversion of simple plasmas and nonuniform regions: three-pole reflection coefficient, *IEEE Trans. Ant. Prop.*, AP24, 1976.

276. **D. Colton and R. Kress:** Integral equation methods, in *Scattering Theory*, John Wiley & Sons, New York, 1983.

277. **H. P. Baltes, Ed.:** *Inverse Scattering Problems in Optics*, Springer-Verlag, 1980.

278. **H. P. Baltes:** Retrieval and synthesis problems in optics — an overview, Top. Meet. Sig. Recov. Synth., Optical Society of America, Washington, D.C., 1983.

279. **H. P. Baltes and B. Steinle:** Radiometry with fields of large coherence area, *Nuovo Cimento*, 41B, 1977.

280. **B. Steinle and H. P. Baltes:** Radiant intensity and spatial coherence for finite planar sources, *J. Opt. Soc. Am.*, 67, 1977.

281. **B. Steinle and H. P. Baltes:** Evanescent waves, finite source area and the inverse scattering problem, *Helv. Phys. Acta*, 49, 1976.

282. **G. Antes, H. P. Baltes and B. Steinle:** Spatial coherence of planar sources specified from the radiant intensity, *Helv. Phys. Acta*, 49, 1976.

283. **R. P. Porter:** Inverse scattering for smooth reflecting objects, *Proc. SPIE Int. Soc. Opt. Eng.*, 358, 1982.

284. **N. N. Bojarski:** A survey of the physical optics inverse scattering identity, *IEEE Trans. Ant. Prop.*, AP30, 1982.

285. **H. J. Meckelburg:** Scatterer reconstruction from multistatic far-field data, *Electron. Lett.*, 18, 1982.

286. **N. H. Earhat:** Holography and inverse scattering, *Isr. J. Technol.*, 18, 1980.

287. **M. A. Fiddy, T. J. Hall, M. Nieto-Vesperinas, G. Ross and J. Wood:** Encoding of information — optical measurements for inverse scattering problems, *Optik*, 59, 1981.

288. **F. B. Gross and J. D. Young:** Physical optics imaging with limited aperture data, *IEEE Trans. Ant. Prop.*, AP29, 1981.

289. **N. H. Farhat:** Holography, wave-length diversity and inverse scattering, Conf. Optics in Four Dimensions, Ensanada, Mexico, 1980.

290. **A. Roger and D. Maystre:** Inverse scattering method in electromagnetic optics: application to diffraction gratings, *J. Opt. Soc. Am.*, 70, 1980.

291. **A. Roger and D. Maystre:** The inverse scattering problems in optics, Proc. 11th Cong. Int. Comm. Optics, ICO-11, Instituto de Optica, Madrid, Spain, 1978.

292. **A. J. Devaney:** Inverse source and scattering problems in optics, Conf. Optics in Four Dimensions, Ensanada, Mexico, 1980.

293. **C.-M. Ho and W. M. Boerner:** Extension of physical optics inverse scattering using polarization utilization, Int. Symp. Dig. Ant. Prop. 672-5, Part II, Quebec, Canada, 1980

294. **G. A. Dike, E. C. Burt and R. F. Wallenberg:** An application of geometrical theory of diffraction (GTD) and physical optics to object identification through inverse scattering analysis, Int. Symp. Dig. Ant. Prop. 668-71, Part II, Quebec, Canada, 1980.

295. **G. Ross, M. A. Fiddy and H. Moezzi:** The solution to the inverse scattering problem based on fast zero location from two measurements, *Opt. Acta*, 27, 1980.

296. **R. D. Mager and N. Bleistein:** An examination of the limited aperture problem of physical optics inverse scattering, *IEEE Trans. Ant. Prop.*, AP26, 1978.
297. **S. K. Chaudhuri and W. M. Boerner:** Polarization utilization in profile inversion of a perfectly conducting prolate spheroid, *IEEE Trans. Ant. Prop.*, AP25, 1977.
298. **P. L. E. Uslenghi:** Inverse scattering methods, *Alta Freq.*, 45, 1976.
299. **S. K. Chaudhuri and W. M. Boerner:** A monostatic inverse scattering model based on polarization utilization, *Appl. Phys.*, 11, 1976.
300. **I. J. D. Craig and J. C. Brown:** *Inverse Problems in Astronomy*, Adam Hilger, 1986.
301. **S. R. Deans:** *The Radon Transform and Some of its Applications*, John Wiley & Sons, New York 1983.
302. **R. W. Porter:** Numerical solution for local emission coefficients in axi-symmetric self-absorbed sources, *SIAM Rev.*, 6, 1964.
303. **O. H. Nestor and H. N. Olsen:** Numerical methods for reducing line and surface probe data, *SIAM Rev.*, 2, 1960.
304. **M. P. Freeman and S. Katz:** Determination of the radial distribution of brightness in a cylindrical luminous medium with self-absorption, *J. Opt. Soc. Am.*, 50, 1960.
305. **J. Richter:** *Plasma Diagnostics*, North-Holland, Amsterdam, 1968.
306. **D. N. Ghosh Roy and R. S. Tankin:** Measurements of continuum radiation from an argon plasma, *J. Quant. Spectrosc. Radiat. Transfer*, 12, 1972.
307. **D. N. Ghosh Roy:** Radon inversion of spectral intensities, *J. Quant. Spectrosc. Radiat. Transfer*, 31, 1984.
308. **C. D. Maldonado and H. N. Olsen:** New methods for obtaining emission coefficients from emitted spectral intensities, *J. Opt. Soc. Am.*, 56, 1966.
309. **D. N. Ghosh Roy:** Steepest descent inversion of plasma spectral intensities, *J. Quant. Spectrosc. Radiat. Transfer*, 37, 1987.
310. **V. G. Romanov:** *Integral Geometry and Inverse Problems for Hyperbolic Equations*, Springer-Verlag, Berlin, 1974.
311. **S. L. Sobolev:** The wave equation for inhomogeneous media, *Tr. Seismol. Inst.*, 6, 1930.
312. **S. L. Sobolev:** On the integration of the wave equation in inhomogeneous media, *Tr. Seismol. Inst.*, 42, 1934.
313. **V. G. Romanov:** *Inverse Problems of Mathematical Physics*, VNU Science Press, Utrecht, The Netherlands, 1987.
314. **R. Courant and D. Hilbert:** *Methods of Mathematical Physics II*, Interscience, New York, 1962.
315. **E. L. Ince:** *Ordinary Differential Equations*, Dover, New York, 1956.
316. **C. R. Darboux:** *Bull. Sc. Math.*, 4, 1873.
317. **G. Herglotz:** Uber die Elastizitat der Erde bei Berucksichtigung ihrer variablen Dichte, *Z. Math. Phys.*, 52, 1905.
318. **E. Wiechert and K. Zoeppritz:** Uber Erdbedenwellen, *Nachr. Konigl. Ges. Wiss. Goettingen*, 4, 1907.
319. **G. Backus and F. Gilbert:** Uniqueness in the inversion of inaccurate gross earth data, *Philos. Trans. R. Soc. London Ser. A*, 266, 1978.
320. **C. L. Pekeris:** Direct method of interpretation in resistivity prospecting, *Geophysics*, 5, 1940.
321. **E. A. Robinson:** Dynamic predictive deconvolution, *Geophys. Prosp.*, 23, 1975.
322. **V. Bardan:** Comments on dynamic predictive deconvolution, *Geophys. Prospect.*, 25, 1977.
323. **E. A. Robinson and S. Treitel:** *Geophysical Signal Processing*, Prentice-Hall, Englewood Cliffs, NJ, 1980.
324. **E. A. Robinson:** Spectral approach to geophysical inversion by Lorentz, Fourier and Radon transforms, *Proc. IEEE*, 70, 1982.
325. **J. G. Berryman and R. R. Greene:** Discrete inverse methods for elastic waves in layered media, *Geophysics*, 45, 1980.
326. **A. M. Bruckstein and T. Kailath:** Inverse scattering for discrete transmission-line models, *SIAM Rev.*, 29, 1987.
327. **A. M. Bruckstein, B. C. Levy and T. Kailath:** Differential methods in inverse scattering, *SIAM J. Appl. Math.*, 45, 1985.
328. **A. M. Bruckstein and T. Kailath:** Spatio-Temporal Scattering and Inverse Problems, ISL Report, Stanford University, Stanford, CA, 1983.
329. **J. A. Ware and K. Aki:** Continuous and discrete inverse scattering problems in a stratified elastic medium. I. Plane waves at normal incidence, *J. Acoust. Soc. Am.*, 45, 1969.
330. **K. P. Bube and R. Burridge:** The one-dimensional inverse problem of reflection seismology, *SIAM Rev.*, 25, 1983.

331. **R. Burridge:** The Gel'fand-Levitan, the Marchenko and the Gopinath-Sondhi integral equations of inverse scattering theory regarded in the context of inverse impulse response problems, *Wave Motion*, 2, 1980.
332. **R. Carrol and F. Santosa:** Scattering techniques for a one-dimensional inverse problem in geophysics, *Math. Methods Appl. Sci.*, 3, 1981.
333. **B. Gjevick, A. Nilsen and J. Hojen:** An attempt at the inversion of reflection data, *Geophys. Prospect.*, 24, 1976.
334. **R. G. Newton:** Inversion of reflection data for layered media: a review of exact methods, *Geophys. J. R. Astron. Soc.*, 65, 1981.
335. **W. Symes:** Stable solution of the inverse reflection problem for a smoothly stratified medium, *SIAM J. Math. Anal.*, 12, 1981.
336. **J. P. Corones, M. E. Davison and R. J. Krueger:** Direct and inverse scattering in the time domain via invariant imbedding equations, *J. Acoust. Soc. Am.*, 74, 1983.
337. **J. F. Claerbout:** Syntheses of a layered medium from its acoustic transmission response, *Geophysics*, 33, 1968.
338. **J. F. Claerbout:** *Fundamentals of Geophysical Data Processing*, McGraw-Hill., New York, 1976.
339. **J. K. Cohen and N. Bleistein:** An inverse method for determining small variations in propagation speed, *SIAM J. Appl. Math.*, 32, 1977.
340. **P. L. Goupillaud:** An approach to inverse filtering of near-surface layer effects from seismic records, *Geophysics*, 26, 1961.
341. **N. Bleistein:** *Mathematical Methods for Wave Phenomenon*, Academic Press, Orlando, 1984.
342. **J. K. Cohen and N. Bleisten:** Velocity inversion procedure for acoustic waves, *Geophysics*, 44, 1979.
343. **A. J. Berkhout:** Seismic migration: imaging of acoustic energy by wave field extrapolation, in *Developments in Solid Earth Geophysics*, Vol. 12, Elsevier, New York, 1980.
344. **W. A. Schneider:** Integral formulation for migration in two and three dimensions, *Geophysics*, 43, 1978.
345. **R. H. Stolt:** Migration by Fourier transform, *Geophysics*, 43, 1978.
346. **P. Sabatier:** On geophysical inverse problems and constraints, *J. Geophys.*, 43, 1977.
347. **S. Coen:** The inverse problem of spherically symmetric fluid earth, *J. Geophys. Res.*, 88, 1983.
348. **S. Coen:** The inverse problem of the shear modulus and density profiles of a layered earth — torsional vibration data, *J. Math. Phys.*, 10, 1981.
349. **D. C. Stickler and P. A. Deift:** Inverse problem for a stratified ocean and bottom, *J. Acoust. Soc. Am.*, 70, 1981.
350. **A. B. Weglein, W. E. Boyse and J. E. Anderson:** Obtaining three-dimensional velocity information directly from reflection seismic data: an inverse scattering formalism, *Geophysics*, 48, 1981.
351. **P. Hubral:** On getting reflection coefficients from waves, *Geophys. Prospect.*, 26, 1978.
352. **J. M. Blackledge:** Reconstruction of the three-dimensional acoustic impedance profile, *J. Phys. D*, 17, 1984.
353. **F. Santosa:** Numerical scheme for the inversion of acoustical impedance profile based on the Gelfand-Levitan method, *Geophys. J. R. Astron. Soc.*, 70, 1982.
354. **F. Santosa and H. Schwetlick:** The inversion of acoustical impedance profile by methods of characteristics, *Wave Motion*, 4, 1982.
355. **W. Symes:** Impedance profile inversion via the first transport equation, *J. Math Anal. Appl.*, 94, 1983.
356. **D. Colton and R. Kress:** The construction of solutions to acoustic scattering problems in a spherically stratified medium. I, *Mech. Appl. Math.*, 31, 1978; **D. Colton and R. Kress:** The construction of solutions to acoustic scattering problems in a spherically stratified medium. II, *Q. J. Mech. Appl. Math.*, 32, 1979.
357. **A. B. Weglein and M. T. Sylvia:** Scattering theory approach to the identification of the Helmholtz equation: a near-field solution, *J. Acoust. Soc. Am.*, 69, 1981.
358. **S. Coen:** On the elastic profiles of a layered medium from reflection data. I. Plane wave sources, *J. Acoust. Soc. Am.*, 70, 1981.
359. **B. P. Hildebrand:** Acoustic holography, in *Methods of Experimental Physics*, Vol. 19, P. D. Edmonds, Ed.,
360. **B. P. Hildebrand and B. B. Brenden:** *An Introduction to Acoustical Holography*, Plenum Press, New York, 1972.
361. **G. W. Stroke:** *Ultrasonic Imaging and Holography*, Plenum Press, New York, 1974.
362. **N. Booth, Ed.:** *Acoustical Holography*, Vol. 6, Plenum Press, New York, 1975.
363. **L. Kessler, Ed.:** *Acoustical Holography*, Vol. 7, Plenum Press, New York, 1975.
364. **A. Metherell, Ed.:** *Acoustical Imaging*, Vol. 8, Plenum Press, New York, 1978.

365. **J. F. Greenleaf:** Computerized emission tomography, in *Methods of Experimental Physics*, Vol. 19, P. D. Edmonds, Ed.,
366. **B. Gopinath and M. M. Sondhi:** Determination of the shape of the human vocal tract from acoustical measurements, *Bell Syst. Tech. J.*, 196, 1970.
367. **M. M. Sondhi and J. R. Resnick:** The inverse problem for the vocal tract: numerical methods, acoustical experiments and speech synthesis, *J. Acoust. Soc. Am.*, 73, 1983.
368. **P. S. Novikov:** On uniqueness of the inverse problem of the theory of potential, *Dokl. Akad. Nauk SSSR*, 18, 1938.
369. **V. K. Ivanov:** The inverse potential problem for a body close to a given one, *Izv. Akad. Nauk SSSR Ser. Mat.*, 20, 1956.
370. **V. K. Ivanov:** On the existence of a solution to the inverse logarithmic potential problem in finite form, *Dokl. Akad. Nauk SSSR*, 106, 1956.
371. **A. I. Prilepko:** On the Theory of Inverse Problems for Geralized Potentials, Ph.D. thesis, Novosibirsk: Mathematical institute, 1968.
372. **I. M. Rapoport:** On the two dimensional inverse problem in potential theory, *Dokl. Akad. Nauk SSSR*, 28, 1940.
373. **I. M. Rapoport:** On stability in the inverse potential problem, *Dokl. Akad. Nauk SSSR*, 31, 1941.
374. **P. S. Novikov:** On uniqueness for the inverse potential problem, *Dokl. Akad. Nauk SSSR*, 18, 1938.
375. **L. N. Sretenskii:** *Newtonian Potential Theory*, Moscow-Gostehizdat, Leningrad, 1964.
376. **L. N. Sretenskii:** On the uniqueness of shape of an attracting body determined from the values of its exterior potential, *Dokl. Akad. Nauk SSSR*, 99, 1954.
377. **H. H. Kagiwada:** *System Identification: Methods and Applications*, Addison-Wesley, London, 1974.
378. **H. Goldstein:** *Classical Mechanics*, Addison Wesley, Reading, MA, 1950.
379. **R. M. Santilli:** *Foundation of Theoretical Mechanics*, I, Springer-Verlag, New York, 1978.
380. **F. V. Atkinsen:** *Discrete and Continuous Boundary Problems*, Academic Press, New York, 1964.
381. **K. M. Case and M. Kac:** A discrete version of the inverse scattering problem, *J. Math. Phys.*, 14, 1973.
382. **K. M. Case:** The discrete inverse scattering problem on one dimension, *J. Math. Phys.*, 15, 1974.
383. **M. L. Oristaglio:** Discrete approach to inverse scattering in two dimensions, *Proc. SPIE Int. Soc. Opt. Eng.*, 358, 1982.
384. **S. Coen, M. Cheney and A. Weglein:** Inversion of the 2.5-D acoustic equation, in *Conference on Inverse Scattering Theory and Application*, J. B. Bednar, R. Redner, E. Robinson and A. Weglein, Eds.: Society for Industrial and Applied Mathematics, Philadelphia, 1983.
385. **M. G. Krein:** On the spectrum of a Jacobian matrix, in connection with the torsional oscillations of shafts, *Mat. Sb.*, 40, 1933.
386. **M. G. Krein:** On nodes of harmonic oscillations of mechanical systems of a certain special type, *Mat. Sb.*, 41, 1934.
387. **M. G. Krein:** Determination of the density of a non-homogeneous cord from its frequency spectrum, *Dokl. Akad. Nauk SSSR*, 76, 1951.
388. **M. G. Krein:** On the inverse problem for a nonhomogeneous cord, *Dokl. Akad. Nauk SSSR*, 82, 1951.
389. **M. G. Krein:** Some new problems in the theory of Sturm systems, *Prikl. Mat. Mekh.*, 16, 1952.
390. **F. P. Gantmacher and M. G. Krein:** *Oscillation Matrices and Kernels and Small Vibrations of Mechanical Systems*, State Publishing House of Technical-Theoretical Literature Translation, Moscow-Leningrad, 1961.
391. **G. M. L. Gladwell:** *Inverse Problems in Vibration*, Martinus Nijhoff, The Hague, 1988.
392. **C. H. Wilcox:** Electric wave propagation on non-umiform coupled transmission lines, *SIAM Rev.*, 6, 1964.
393. **G. L. Brown:** *The Inverse Reflection Problem For Electric Waves on Non-Uniform Transmission Lines*, Ph.D. thesis, University of Wisconsin, Madison, 1965.
394. **R. E. Caflisch:** An inverse problem for Toeplitz matrices and the synthesis of discrete transmission lines, *Linear Alg. Appl.*, 38, 1981.
395. **D. C. Youla:** Analysis and synthesis of arbitrarily terminated lossless transmission lines, *IEEE Trans. Circuit Theory*, CT13, 1964.
396. **B. Gopinath and M. M. Sondhi:** Inversion of the telegraph equation and the synthesis of nonuniform lines, *Proc. IEEE*, 59, 1971.
397. **D. Brill, G. C. Gaunaurd and H. Uberall:** Mechanical eigenfrequencies of axisymmetric fluid objects: acoustic spectroscopy, *Acustica*, 53, 1983.

398. **B. Duchene and W. Tabbara:** A geometric optics method for assessing an inverse scattering problem for blood vessels. I. A multistatic single frequency approach, *IEEE Trans. Sonics Ultrason.*, SU30, 1983; **B. Duchene and W. Tabbara:** A geometric optics method for assessing an inverse scattering problem for blood vessels. II. A monostatic multifrequency approach, *IEEE Trans. Sonics Ultrason.*, SU30, 1983.

399. **R. Goulard and P. J. Emmerman:** Combustion diagnostics by multiangular absorption, *J. Plasma Phys.*, 19, 1978.

400. **A. Selloni:** Microscopic models of photodetection, *J. Plasma Phys.*, 19, 1978.

401. **N. Bleistein and J. K. Cohen:** Progress on a mathematical inversion technique for non-destructive evaluation, *Wave Motion*, 2, 1980.

402. **S. S. Abdullaev and F. Kh. Abdullaev:** Determination of the profile of diffused waveguides by inverse scattering theory, *Sov. Tech. Phys. Lett.*, 4, 1978.

403. **J. N. Elgin and D. J. Kaup:** Inverse scattering theory with stochastic initial potentials, *Opt. Commun.*, 43, 1982.

404. **A. J. Devaney:** Inverse scattering as a form of computed tomography, *Proc. SPIE Int. Soc. Opt. Eng.*, 358, 1982.

405. **W. J. Wilson:** An inverse scattering approach to the pressure transient analysis of petroleum reservoirs, in *Conference on Inverse Scattering Theory and Application*, J. B. Bednar, R. Redner, E. Robinson and A. Weglein, Eds.: Society for Industrial and Applied Mathematics, Philadelphia, 1983.

406. **L. A. Lliboutry:** The glacier theory, *Adv. Hydrosci.,* 7, 1971.

407. **P. Duval:** Creep and fabrics of polycrystalline ice under shear and compression, *J. Glaciol.*, 27, 1981.

408. **J. W. Glen:** The creep of polycrystalline ice, *Proc. R. Soc. London Ser. A,* 228, 1955.

409. **L. Reynaud:** Flow of a velley glacier with a solid friction law, *J. Glaciol.*, 12, 1973.

410. **J. F. Nye:** The flow of a glacier in a channel of rectangular elliptic or parabolic cross-section, *J. Glaciol.*, 5, 1965.

411. **C. Somigliana:** Sul coefficiente di attrito interno del ghiaccio a la determinazione della profondita ghiacciai, *Boll. Com. Glaciol. Ital.*, 6, 1925.

412. **D. Hantz and L. Lliboutry:** The inverse problem for velley glacier flow, *J. Glaciol.*, 27, 1981.

413. **A. Acrivos and E. Herbolzheimer:** Enhanced sedimentation in settling tanks with inclined walls, *J. Fluid Mech.*, 92, 1979.

414. **W. D. Hill, R. R. Rothfus, and K. Li:** Boundary-enhanced sedimentation due to settling convection, *Int. J. Multiphase Flow*, 3, 1979.

415. **H. P. Greenstein and M. Ungarish:** On the enhancement of centrifugal separation, *J. Fluid Mech.*, 157, 1985.

416. **A. E. Boycott:** Sedimentation of blood corpuscles, *Nature*, 104, 1920.

417. **A. L. Davis and R. W. Jeppson:** Solving three-dimensional potential flow problems by means of an inverse formulation and finite differences, Utah State Research Laboratory/Utah State University Research Report PRWG-96-2,1973.

Chapter 2

SOME EXAMPLES OF INVERSE PROBLEMS

I. INTRODUCTION

Let us consider a few examples of inverse problems from various disciplines of the physical sciences. The objective is to illustrate the nature of inverse solutions and the underlying methodologies. Obtaining inverse solutions may involve solving the corresponding direct problems. However, since the primary goal is to illustrate the inverse solutions, the direct problems will not be solved in any great detail. Instead, their solutions will be considered to be known whenever possible.

II. EXAMPLES

A. EXAMPLE 1: HEAT CONDUCTION

The first example, which is due to Cannon,[1] involves heat propagation in finite homogeneous conductors. The problem is as follows: determine the unknown heat diffusivity of a finite homogeneous heat conductor, given the rate of heat flow into a boundary at a single instant of time and the temperatures at the boundaries. This example will illustrate the salient features of an inverse solution as discussed in Chapter 1, namely, the existence, uniqueness, boundedness, and stability of the solution.

Consider the following problem:

$$\begin{aligned} &\ddot{u} = ku'', && 0 < x < 1, \;\; 0 < t \le T \\ &u(0,t) = f(t), && 0 \le t \le T, \\ &u(0,0) = 0, && \\ &u(1,t) = g(t), && 0 \le t \le T, \\ &u(1,0) = 0, && \\ &u(x,0) = 0, && 0 \le x \le 1 \end{aligned} \tag{2.1}$$

and

$$-\rho c k u'(0,t_0) = h, \qquad 0 < t_0 \le T$$

In Problem 2.1, k is the heat diffusivity, ρ the mass density, c the specific heat, h a constant, and f(t) and g(t) are functions of time. Differentiation of u(x,t) with respect to t and x are denoted by $\dot{u}$ and u', respectively. The mass density and the specific heat are given, and h is the heat flow rate at the boundary x = 0, which is measured at a single time t_0. The functions f(t) and g(t) are also known. The only unknown is the heat diffusivity k, which is to be determined. It is further assumed that f and g $\subset C_\infty$, $0 \le t \le T$, where C_∞ is the space of continuously differentiable functions.

The direct solution of the heat conduction problem (2.1) involves a theta function,[2] M(z,s), defined by:

$$M(z,s) = \left(\frac{1}{\pi s}\right)^{\frac{1}{2}} \sum_{n=-\infty}^{+\infty} \exp\left[-\frac{(z+2n)^2}{4s}\right], \qquad s > 0 \tag{2.2}$$

In terms of M, the solution of the direct problem is as follows.[3]

$$u(x,t) = -\int_0^t M'[x, k(t-t')]\, f(t')\, k\, dt' + \int_0^t M'[x-1,\, k(t-t')] g(t')\, k\, dt' \tag{2.3}$$

Taking the derivative of Equation 2.3 with respect to x yields

$$u'(x,t) = -\int_0^t M''[x,\, k(t-t')]\, f(t')\, k\, dt' + \int_0^t M''[x-1,\, k(t-t')]\, g(t')\, k\, dt' \tag{2.4}$$

Using the heat conduction equation itself yields

$$M[x-z,\, k(t-t')] = -k\, M''[x-z,\, k(t-t')], \quad x \neq z, \quad t \neq t' \tag{2.5}$$

$\dot{M}$ in Equation 2.5 is the derivative of M with respect to the variable t′. Using Equation 2.5 in Equation 2.4, we obtain

$$u'(x,t) = \int_0^t \dot{M}[x,\, k(t-t')]\, f(t')\, dt' - \int_0^t \dot{M}[x-1,\, k(t-t')]\, g(t')\, dt' \tag{2.6}$$

As $t' \to t$, i.e., $s \to 0$, the function $(\pi s)^{-1/2} \exp(-\xi/4s) \to \delta(\xi)$ and, hence, vanishes if $\xi \neq 0$. Then for $0 < x < 1$, it follows that

$$\lim_{t' \to t} M[x-z,\; k(t-t')] = 0, \qquad x \neq z$$

Using this limit and recalling that f(0) = g(0) = 0 according to the hypotheses of the problem, we integrate Equation 2.6 by parts and obtain

$$u'(x,t) = -\int_0^t M[x,\, k(t-t')]\, f(t')\, dt' + \int_0^t M[x-1,\, k(t-t')]\, g(t')\, dt' \tag{2.7}$$

Evaluating Equation 2.7 in the limit $x \to 0$ yields

$$\lim_{x \to 0} u'(x,t) = -\int_0^t M[0,\, k(t-t')]\, f(t')\, dt' + \int_0^t M[-1,\, k(t-t')]\, g(t')\, dt' \tag{2.8}$$

This limiting procedure is justified as follows. The kernel M can be considered as a sequence of functions in x which are integrable over the interval (0,t). Also, the limit of M exists as $x \to 0$. Hence, Lebesgues' theorem of dominated convergence[4] allows us to perform the limiting procedure. It may be recalled that according to the theorem of dominated convergence

$$\lim_{n \to \infty} \int_0^t x_n(t)dt = \int_0^t x(t)dt$$

where, $x_n(t)$ is a sequence of functions integrable over an interval (0,t) which approaches a limit x(t), except possibly over a set of measure zero. x(t) is integrable, that is, there exists a function f(t) such that $|x_n(t)| < f(t)$ for sufficiently large n.

Let

$$u'(0,t)_{x\to 0} = G(t,k)$$

and

$$-\rho ckG(t_0,k) = F(k)$$

It follows from Problem 2.1 and Equation 2.8 that

$$h = F(k) \tag{2.9}$$

The solution of the inverse problem of determining k from the data h is then obtained as follows. Using different values of k in Equation 2.9, we construct the function F(k). The value of k for which the identity (Equation 2.9) is satisfied is then the desired solution. This yields

$$k = -h\left[\rho cG(t_0,\ k)\right]^{-1} \tag{2.10}$$

The function F(k) could also be constructed by solving Problem 2.1 numerically for different values of k and then applying the identity (Equation 2.9). Therefore, if a solution exists for the system (2.1), then k must satisfy the identity (2.9) and a solution for k is given by Equation 2.10. Moreover, k must be a positive quantity.

Suppose that a positive solution for k does exist which satisfies condition (2.9). Is the solution unique? If it can be shown that the function F(k) is monotonically increasing for k > 0, then Equation 2.9 must have only one positive solution if a solution exists. Now F(k) contains the product terms kM. From the functional form of M given by Equation 2.2 it is clear that kM[0,k(t – t′)] and kM[(–1,k(t – t′)] belong to C_∞. Since by the hypotheses of the problem, f and g also belong to C_∞, we conclude that the function F(k) is also a C_∞ function. Moreover, if h > 0, f′(t) ≥ 0, and g′(t) ≥ 0 for $0 \le t \le t_0$, and if it is assumed that f(t) ≠ 0 identically, then it follows from Equation 2.8 that F′(k) > 0, k > 0, proving that F(k) is indeed a strictly increasing function for k > 0.

It is also necessary to show that F(k) -→ 0 as k → 0 and that F(k) → ∞ as k → ∞ if F(k) is to represent h. This follows from the functional form of h, namely,

$$h = -\rho cku'(t_0,0)$$

Using the fact that exp (–x) < (1/x), x > 0, Equation 2.2 implies that

$$M \le \text{constant.}\ k^{1/2}$$

Hence,

$$F(k) \le \text{constant.}\ k^{3/2}$$

Therefore, F(k) → 0 as k → 0.

From Equations 2.2 and 2.8, one obtains

$$F(k) = (\rho ck/\pi)^{1/2} \int_0^{t_0} (t-t')^{-1/2}\left[Af'(t') + Bg'(t')\right]dt'$$

where

$$A = \sum_{n=-\infty}^{\infty} \exp\left[-n^2/k(t-t')\right]$$

and

$$B = \sum_{-\infty}^{\infty} \exp\left[-(2n-1)^2/4k(t-t')\right]$$

Hence, it follows that:

$$F(k) \geq \left(\rho c k^{\frac{1}{2}}\right)\left\{\int_0^{t_0}\left[f'(t')\Big/\{\pi(t_0-t')\}^{\frac{1}{2}}\right]dt'\right\} \tag{2.11}$$

This inequality shows that F(k) $\to \infty$ as k $\to \infty$. Collecting all these, we have

$$F'(k) > 0, \quad k > 0$$

$$F(k) \to 0, \; k \to 0$$

and

$$F(k) \to \infty, \quad k \to \infty$$

Hence, F(k) is a strictly increasing positive function, proving thereby the uniqueness of the solution. These results can be summarized into the theorem below.

Theorem 2.1

Let f(t) and g(t) $\subset C_\infty$ for $0 \leq t \leq T$, with $f(t) \not\equiv 0$. Moreover, let $f'(t) \geq 0$, $g'(t) \leq 0$, $0 \leq t \leq t_0$, and $h > 0$. Then Equation 2.9 has exactly one positive solution k.

From the uniqueness of the solution of Equation 2.9 follows the uniqueness of the solution to the system (2.1) itself. Otherwise, suppose that the solution to 2.1 is unique, whereas Equation 2.9 admits of nonunique solutions. Let k_1 and k_2, $k_1 \neq k_2$, be two solutions of Equation 2.9. Then, the pair (k_i, u_i), i = 1, 2, will form two distinct solutions to the system (2.1), which is contrary to the hypothesis. This leads to the following theorem.

Theorem 2.2

If f(t) and g(t) $\subset C_\infty$, $0 \leq t \leq T$, then Problem 2.1 possesses a unique solution if and only if the solution to Equation 2.9 is unique.

Is the solution obtained stable? That is, does the solution depend continuously on the data? From Equations 2.10 and 2.8 we can write

$$k = k(f', g', h)$$

that is, k depends on the functions f′, g′, and h. To prove the stability of the solution, it suffices to show that k is Lipschitz continuous[5] in its arguments. In other words, if k_i corresponds to h_i, f_i, and g_i, i = 1, 2, and Theorem 2.1 is satisfied, then Lipschitz continuity implies that

$$\|k_1 - k_2\| \leq \sum_{j=1}^{3} N_j \|P_{j1} - P_{j2}\|$$

where the subscript j, j = 1, 2, and 3, corresponds to h, f′, and g′, respectively, and ||•|| represents an appropriate norm. The N_js are positive constants not dependent upon the k_is.

From Equations 2.10 and 2.11 and from the fact that k > 0, we have

$$k \le h^2 \left\{ \rho c \int_0^{t_0} \left[f'(t') \Big/ \{\pi(t-t')\}^{\frac{1}{2}} \right] dt' \right\}^2 = a \tag{2.12}$$

Since we are dealing with functions in the space C_∞, the corresponding norm is the supremum norm,[6] that is,

$$\|f\|_{t_0} = \sup_{0 \le t \le t_0} |f(t)|$$

over the interval $0 \le t \le t_0$.

Consider

$$h_i = F_i(k_i), \quad i = 1, 2$$

where the functions f_i and g_i satisfy the hypotheses of Theorem 2.1. Now,

$$h_1 - h_2 = F_1(k_1) - F_2(k_2)$$

$$= F_1(k_1) - F_1(k_2) + F_1(k_2) - F_2(k_2)$$

From this, we can write

$$|F_1(k_1) - F_1(k_2)| \le |h_1 - h_2| + |F_1(k_2) - F_2(k_2)| \tag{2.13}$$

From the definition of F(k), we have

$$|F_1(k_2) - F_2(k_2)| = \rho c k_2 |G_1(k_2) - G_2(k_2)|$$

The argument t_0 has been suppressed. From Equation 2.8

$$|G_1(k_2) - G_2(k_2)| \le \|f_1' - f_2'\|_{t_0} \int_0^{t_0} |M(0, k(t_0 - \tau))| \, d\tau +$$

$$+ \|g_1' - g_2'\|_{t_0} \int_0^{t_0} |M(-1, k(t_0 - \tau))| \, d\tau \tag{2.14}$$

Using Equation 2.2 in Equation 2.14, performing the integrations, and using Equation 2.12 yields the following inequality

$$|F_1(k_2) - F_2(k_2)| \le C_2 \|f_1' - f_2'\|_{t_0} + C_3 \|g_1' - g_2'\|_{t_0}$$

where,

$$C_2 = 2\rho c(at_0/\pi)^{1/2}[1 + 4at_0/3]$$

and

$$C_3 = 32 / 3\rho c(a^3 t_0^3 / \pi)^{1/2}$$

Since F(k) is continuously differentiable, we have

$$|F_1(k_1) - F_1(k_2)| \geq \min|\dot{F}_1(k_2) - \dot{F}_2(k_2)||k_1 - k_2|$$

$\dot{F}$ is the derivative of F(k) with respect to k. From Equation 2.11 and the bound for k given by Equation 2.12, it follows that

$$|F_1(k_1) - F_1(k_2)| \geq C_1|k_1 - k_2|$$

where

$$C_1 = \{(\rho c/2)\} \int_0^{t_0} (p/q)\, dt'$$

$$p = \min[f_1'(t'),\ f_2'(t')]$$

and

$$q = [\pi(t_0 - t')\ \max(a_1,\ a_2)]$$

Collecting these results together into the inequality (2.14) yields

$$|k_1 - k_2| \leq C_1|h_1 - h_2| + C_2\|f_1' - f_2'\|_{t_0} + C_3\|g_1' - g_2'\|_{t_0}$$

This implies that k = k(f′,g′,h) is Lipschitz continuous. The inverse solution is, therefore, stable and varies continuously with the data.

B. EXAMPLE 2: DETERMINATION OF THE DIFFUSION COEFFICIENT

This example, which is discussed after Romanov,[7] illustrates the inversion of a parabolic partial differential equation for obtaining the coefficient of diffusion. This is an example of how the transforms of Fourier and Laplace can be used in solving certain inverse problems. Let us consider the diffusion equation

$$a(x, y)\, \dot{u}(x, y, t) = \Delta u(x, y, t) + \delta(x - \mathbf{X},\ y,\ t) \tag{2.15}$$

The initial and the boundary conditions are

$$\begin{aligned} u &= 0, & t &< 0 \\ u' + hu &= 0, & y &= 0 \end{aligned} \tag{2.16}$$

Here, u′ is the derivative of u with respect to the variable y, a is a parameter which is positive and bounded and is directly associated with the diffusion coefficient D(x,y), and Δ is the Laplacian in **x** and y. In the problem (2.15 - 2.16), $\mathbf{x} \subset R^n$ and $y \geq 0$. That is, the problem of diffusion is being considered in the domain $y \geq 0$ of the space of variables, **x**, y, $\mathbf{x} \subset R^n$. $\mathbf{X} \subset R^n$ is an arbitrary point and functions as a parameter. This simply defines the location of the diffusive source in the space R^n. The constant h is a given finite number.

Define the Laplace transform pair as

$$f^{\sim}(s) = \int_0^{\infty} f(t)\exp(-st)\,dt, \qquad f(t) = (1/2\pi i)\int_{\sigma-i\infty}^{\sigma+i\infty} f^{\sim}(s)\exp(st)\,ds$$

where s is the Laplace transform variable and the contour in the inverse Laplace transform is to the right of all singularities of the integrand $f^{\sim}(s)$. The symbol ~ denotes Laplace transform. From Laplace transforming problem (2.15 - 2.16), we obtain

$$\Delta u^{\sim} + sau^{\sim} + \delta(x - X,\ y) = 0 \tag{2.17}$$

and

$$u^{\sim\prime} + hu^{\sim} = 0 \qquad u^{\sim} = u^{\sim}(\mathbf{x}, \mathbf{X}, y, s) \tag{2.18}$$

The data consists of the information

$$\left[\partial u^{\sim}/\partial s\right] = f(x, X), \qquad y = s = 0$$

Given this information, the problem is to recover the coefficient a(**x**,y).

Let us expand the solution $u^{\sim}(\mathbf{x},\mathbf{X},y,s)$ of the Laplace-transformed problem (2.17 - 2.18) in powers of s and write

$$u^{\sim}(x, X, y, s) = \sum_{n=0}^{\infty} s^n\, u_n^{\sim}(x, X, y) \tag{2.19}$$

Then

$$\left[\partial u^{\sim}/\partial s\right]\Big|_{s=0} = u_1^{\sim}(\mathbf{x}, \mathbf{X}, y)$$

and

$$f(x, X) = u_1^{\sim}(x, X, 0) \tag{2.20}$$

Substituting the expansion (2.19) into Equation 2.17 and 2.18 and equating the coefficients of the zero-th and the first power of s, we obtain:

$$\Delta u_0^{\sim} + \delta(x - X,\ y) = 0 \tag{2.21}$$

$$\Delta u_0^{\sim\prime} + hu_0^{\sim} = 0, \qquad y = 0 \tag{2.21a}$$

$$\Delta u_1^{\sim} + au_0^{\sim} = 0 \tag{2.22}$$

$$u_1^{\sim\prime} + hu_1^{\sim} = 0, \quad y = 0 \tag{2.22a}$$

Next, we take the Fourier transform of Equation 2.21 and 2.21a with respect to the variable $(\mathbf{x} - \mathbf{X})$, and of Equations 2.22 and 2.22a with respect to both $\mathbf{x}$ and $(\mathbf{x} - \mathbf{X})$. This is prompted by the fact that the solution of Equation 2.21 is a function of $(\mathbf{x} - \mathbf{X})$, whereas $u^{\sim}_1$ depends on $\mathbf{x}$,y and $\mathbf{X}$. Thus

$$u_0^{\sim} = u_0^{\sim}(\mathbf{x} - \mathbf{X},\ y) = u_0^{\sim}(\mathbf{l}, y)$$

and

$$u_1^{\sim} = u_1^{\sim}(\mathbf{x}, \mathbf{X}, y) = u_1^{\sim}(\mathbf{X} + \mathbf{l}, \mathbf{X}, y)$$

where,

$$\mathbf{l} = \mathbf{x} - \mathbf{X}$$

and the argument s, the Laplace transform variable, has been suppressed. This results in the following equations and boundary conditions

$$\hat{u}_0^{\sim\prime\prime} - |\mathbf{k}|^2 \hat{u}_0^{\sim} + \delta(y) = 0 \tag{2.23}$$

$$\hat{u}_1^{\sim\prime\prime} - |\mathbf{k}|^2 \hat{u}_1^{\sim} + \hat{a}(\mathbf{p}, y)\, \hat{u}_0^{\sim}(\mathbf{k} - \mathbf{p},\ y) = 0 \tag{2.24}$$

$$\hat{u}_0^{\sim\prime} + h\hat{u}_0^{\sim} = 0, \quad y = 0 \tag{2.25}$$

$$\hat{u}_1^{\sim\prime} + h\hat{u}_1^{\sim} = 0, \quad y = 0$$

$$\hat{u}_0^{\sim} = \hat{u}_0^{\sim}(\mathbf{k}, y) \text{ and } \hat{u}_1^{\sim} = \hat{u}_1^{\sim}(\mathbf{k}, \mathbf{p}, y) \tag{2.26}$$

where ^ denotes the Fourier transform, and $\mathbf{k}$ and $\mathbf{p}$ are the Fourier transform variables conjugate to the space variables $\mathbf{l}$ and $\mathbf{X}$, respectively. The Fourier transform of $u^{\sim}_0(\mathbf{l},y)$ is written as

$$\hat{u}_0^{\sim}(\mathbf{k}, y) = \int_{R^n} u_0^{\sim}(\mathbf{l}, y) \exp(i\mathbf{k} \cdot \mathbf{l})\, d\mathbf{l}$$

and the transform of $u^{\sim}_1(\mathbf{x},\mathbf{X},y)$ is

$$\hat{u}_1^{\sim}(\mathbf{p}, \mathbf{k}, y) = \int_{R^{2n}} u_1^{\sim}(\mathbf{l}, \mathbf{X}, y) \exp(i\mathbf{k} \cdot \mathbf{l} + i\mathbf{p} \cdot \mathbf{X})\, d\mathbf{l}\, d\mathbf{X}$$

It is, of course, assumed that the solutions u($\mathbf{x}$,y,t) decreases as $|\mathbf{x}| + y \to \infty$ in such a manner that the Fourier transforms exist.

The Green's function for Equation (2.23) with the boundary condition (2.25) satisfies the following problem

$$G'' - \lambda^2 G = \delta(z - z')$$

$$G' + dG = 0, \quad z = 0$$

where $\lambda = |\mathbf{k}|$ and $G = G(z,z',\lambda)$. The Green's function is given by[8]

$$G(z,z',\lambda) = F\begin{cases}\left\{\cosh\lambda z - \left(\dfrac{d}{\lambda}\right)\sinh\lambda z\right\}\exp(-\lambda z'), & 0 \le z \le z' \\ \left\{\cosh\lambda z' - \left(\dfrac{d}{\lambda}\right)\sinh\lambda z'\right\}\exp(-\lambda z), & z' \le z\end{cases} \tag{2.27}$$

where

$$F = 1/(d - \lambda)$$

In terms of this function, the solution of Equation 2.23 can be written as

$$\hat{u}_0^{\sim}(\mathbf{k}, y) = -G(y, 0, |\mathbf{k}|) = \left[1/(|\mathbf{k}|) - h\right]\exp(-|\mathbf{k}|y)$$

The solution of Equation 2.24 subject to the boundary condition (2.26) is given by

$$\hat{u}_1^{\sim}(\mathbf{k}, \mathbf{p}, y) = \int_0^{\infty} G(y, z, |\mathbf{k}|)\, G(z, 0, |\mathbf{k} - \mathbf{p}|)\, \hat{a}(\mathbf{p}, z)\, dz$$

Setting y to zero and using the Green's function given by Equation 2.27, we have

$$\hat{u}_1^{\sim}(\mathbf{k}, \mathbf{p}, 0) = \frac{1}{(h - |\mathbf{k}|)(h - |\mathbf{k} - \mathbf{p}|)}\int_0^{\infty} \exp\left[-(|\mathbf{k}| + |\mathbf{k} - \mathbf{p}|)z\right]\hat{a}(\mathbf{p}, z)\, dz \tag{2.28}$$

Now, Equation (2.28) can be rewritten in terms of the data $f^{\sim}(\mathbf{x},\mathbf{X})$, namely

$$\hat{u}_1^{\sim}(\mathbf{k}, \mathbf{p}, 0) = F \cdot T \cdot \left[u_1^{\sim}(\mathbf{x}, \mathbf{X}, 0)\right] =$$

$$\int_{R^{2n}} f(\mathbf{1} + \mathbf{X},\ \mathbf{X})\exp\left[i\{(\mathbf{k}\cdot\mathbf{l}) + (\mathbf{p}\cdot\mathbf{X})\}\right]\, \mathbf{dl}\, \mathbf{dx}$$

$$= \hat{f}(\mathbf{k}, \mathbf{p}) \tag{2.29}$$

From Equations 2.28 and 2.29, the function $\hat{a}(\mathbf{p},z)$ to be determined is represented by the following integral equation

$$\hat{b}(\mathbf{k}, \mathbf{p}) = \int_0^{\infty} \exp\left[-(|\mathbf{k}| + |\mathbf{k} - \mathbf{p}|)z\right]\hat{a}(\mathbf{p}, z)\, dz \tag{2.30}$$

where,

$$\hat{b}(\mathbf{k},\mathbf{p}) = \hat{f}(\mathbf{k},p)\ (h-|\mathbf{k}|)(h-|\mathbf{k}-\mathbf{p}|)$$

The Fourier transform $\hat{a}(\mathbf{p},z)$ of the unknown function $a(\mathbf{x},y)$ is thus expressed through the Fourier transform of the data $f(\mathbf{x},\mathbf{X})$.

Equation 2.30 is the Laplace transform of $\hat{a}(\mathbf{p},z)$ in z with respect to the Laplace transform variable

$$s = |\mathbf{k}| + |\mathbf{k}-\mathbf{p}|$$

The transform can be inverted uniquely and the inversion is given by the expression[9]

$$\hat{a}(\mathbf{p},z) = \operatorname*{limit}_{n\to\infty}\ \left[(-1)^n n^{-1} s^{n+1} \hat{b}^{(n)}(\mathbf{p},s)\right]\Big|_{s=(n/z)}$$

where $\hat{b}^{(n)}$ means the n-th derivative of $\hat{b}$ with respect to s. The final reconstruction is then obtained by taking the inverse Fourier transform of $\hat{a}(\mathbf{p},z)$. In order to make the problem more general, a positive constant term can be added to the coefficient function, that is, we consider functions of the more general form

$$a(\mathbf{x},y) = A + a(\mathbf{x},y)$$

In this case, according to Equation 2.30, the function $\hat{b}$ will contain a singular term $(A/2|\mathbf{k}|)$ which will define the constant A.

Summarizing, we have the following theorem.

Theorem 2.3

Let $a(\mathbf{x},y)$ in Equation 2.15 be expressed as:

$$a(\mathbf{x},y) = A + a_1(\mathbf{x},y), \qquad y \geq 0$$

where A is a positive constant. For every fixed y, let $a_1(\mathbf{x},y)$ have a Fourier transform depending continuously on y. Then $a(\mathbf{x},y)$ can be determined uniquely from the data specified at the boundary $y = 0$ for all x and t.

C. EXAMPLE 3: PLASMA PROFILE INVERSION

This example is due to Baldwin and Kaufmann[10] and illustrates a technique for recovering the depth-dependent electron density distribution in a "cold" plasma from measurements of the frequency-dependent impedance. A plasma is a "quasineutral" gas of charged and possibly neutral particles exhibiting "collective" behavior. The word quasineutrality implies that except in the immediate neighborhood of a charged particle, the medium as a whole is charge neutral, that is, the electrons and the ions have equal number densities in any macroscopic volume element. For a thermal plasma at a finite temperature, quasineutrality is characterized by a parameter called the "Debye length" λ_D, which is a measure of the distance over which a charged particle is shielded from the leakage of the potential due to the presence of another neighboring charged particle. A plasma is quasineutral if the Debye shielding distance is small compared to a characteristic dimension of the plasma medium, that is, $\lambda_D \ll L$. Unlike a neutral medium where the particles move undisturbed until collisions take place, a particle in a plasma medium is continuously affected by the presence of other particles because of the long-range Coulomb forces. This constitutes the collective behavior of the charged particles in a plasma medium. The criterion for exhibiting collective behavior is that the number of particles N_D in a Debye sphere, that is, a sphere of radius λ_D, be much larger than unity or $N_D >> 1$. As a consequence of this collective behavior, the electrons, if disturbed, oscillate around their equilibrium positions. This is known as “Langmuir” or “plasma” oscillations. The frequency ω_p of these oscillations, called the “electron plasma frequency”, is given by the equation

$$\omega_p(x) = \left[4\pi n_e(x)e^2/m_e\right]^{1/2}$$

where n_e is the electron number density, and e and m_e are the electron charge and mass, respectively. The number density may depend on the depth x inside the plasma medium. The plasma frequency characterizes a threshold, that is, no electromagnetic wave can propagate inside the plasma if the frequency of the wave is below the plasma frequency.

Let us consider a plasma medium with the following characteristics

1. The electron number density $n_e(x)$ is an even function of a single coordinate x.
2. The plasma is “cold”, that is, there is no thermal motion.

Let us restrict our attention to a plasma slab of length L which may be subject to a uniform magnetic field in a plane transverse to the coordinate x. The uniformity of the magnetic field, if any, is in its magnitude and not in its direction. It will be further assumed that the slab acts as a capacitor, the frequency-dependent impedance of which can be measured without disturbing the plasma.

For a dielectric medium with a dielectric constant ε, Poisson’s equation reads

$$\nabla \cdot \mathbf{D} = 4\pi\rho$$

where ∇ is the gradient operator, $\mathbf{D} = \varepsilon\mathbf{E}$, the “displacement vector”, and ρ the charge density. For a charge-neutral, one-dimensional plasma, Poisson’s equation becomes

$$\left[\varepsilon(\omega, x)E\right]' = 0$$

where E is the electric field in the x-direction, the prime means differentiation with respect to x, and $\varepsilon(\omega,x)$ is the plasma dielectric constant given by

$$\varepsilon(\omega, x) = 1 - \omega_p^2(x)/\left[\omega^2 - \Omega^2\right]$$
$$= 1 - \left(\omega_p/W\right)^2$$

$W = [\omega^2 - \Omega^2]^{1/2}$ and $\Omega = (eB/m_e c)$, the electron “cyclotron frequency”. We allow an infinitesimally small resistive component in order to take into account weak dissipative processes, and write the dielectric constant as[11]

$$\varepsilon(\omega, x) = 1 - \frac{\omega_p^2}{W^2 + i\omega\Gamma} \tag{2.31}$$

$\Gamma << \omega_p$. Since the plasma slab is capacitive, the spectral complex impedance of the slab can be written as

$$Z(\omega) = i(\omega C)^{-1}$$

The capacitance C in a differential length dx is, of course, $\varepsilon A/dx$, where A is the surface area of the slab. Taking A as unity, the complex impedance of the entire slab of length L is

$$Z(\omega) = \left(\frac{8\pi i}{\omega}\right) \int_0^L \frac{dx}{\varepsilon(\omega, x)} \tag{2.32}$$

where MKS units have been used and the multiplication of 4π by a factor of 2 is due to the assumption (i). $x = 0$ is the center of the slab and $x = L$, at which the dielectric constant is unity, is its edge.

Let us now consider the real part of the impedance Z, that is, the resistance of the slab. A simple calculation shows that the real part can be written as

$$\mathrm{Re}\frac{i}{\varepsilon(\omega, x)} = \frac{\omega W^2 \Gamma}{\left(W^2 - \omega_p^2\right)^2 + W^2\Gamma^2} \tag{2.33}$$

Equation 2.33 shows that as $\Gamma \to 0$, the real part remains small until $W \to \omega_p$, that is, until a resonance takes place, in which case the denominator becomes singular. This means that the real part above is negligible except at a resonance when the dielectric constant vanishes. This allows us to write

$$\mathrm{Re}\, Z(\omega) = \frac{8\pi^2}{\omega} \int_0^L dx\, \delta\left[\varepsilon(\omega, x)\right] \tag{2.34}$$

where δ is Dirac's delta function. This results from the definition of the delta sequence,[12] namely:

$$\left(\frac{1}{\pi}\right) \cdot \underset{y \to 0}{\mathrm{limit}} \left[y/\left(y^2 + x^2\right)\right] \to \delta(x)$$

From the theory of distributions,[12] we also know that

$$\delta\left[f(x)\right] = \left|f'(x)\right|^{-1} \delta(x - X), \quad f'(X) \neq 0$$

where X is the coordinate such that $f(X) = 0$. Equation 2.34 then becomes

$$\mathrm{Re}\, Z(\omega) = \frac{8\pi^2}{\omega} \left|\varepsilon'\right|^{-1}\Big|_{x(\omega)} \tag{2.35}$$

$x(\omega)$ is that coordinate x at which resonance occurs when the frequency is ω. Therefore,

$$\omega^2(x) = \Omega^2 + \omega_p^2(x) \tag{2.36}$$

$x(\omega)$ is thus determined by the zero of the plasma dielectric constant. From Equations 2.31 and 2.36, we have:

$$[\varepsilon(\omega, x)]' = -W^{-2}[\omega_p^2(x)]' = -W^{-2}[\omega^2(x)]', \quad \Gamma \to 0 \tag{2.37}$$

Substituting Equation 2.37 into Equation 2.35 and integrating, we obtain the following expression for x as a function of the frequency ω

$$x(\omega) = \frac{1}{4\pi^2} \int_{\omega}^{\omega(0)} \frac{\omega'^2}{\omega'^2 - \Omega^2} \operatorname{Re} Z(\omega')\, d\omega' \tag{2.38}$$

The upper limit ω(0) is the frequency at which resonance occurs at the center of the slab. Equation 2.38 shows that by measuring ReZ(ω′) as a function of frequency, one can determine x(ω) and hence, the electron number density at a depth of x from Equation 2.36. This, however, requires that ω(0) be known. Now, from Equations (2.34) and (2.37), ε′ is zero at the center of the slab, resulting in a singularity in ReZ(ω) at the slab center. This can be determined experimentally and, hence, also n(0).

The above discussion holds only if the approximation of a cold plasma is made. Otherwise, kinetic terms will appear in the denominator of the dielectric constant, preventing ReZ(ω) from being singular. Baldwin and Kaufmann show that even if the plasma is warm, the same conclusion still holds, provided that the plasma waves carrying the energy absorbed at the resonance point are damped either by collisions or by Landau damping. In case the plasma waves are reflected from a region such as a plasma sheath, standing wave patterns will form. In that case, the expression for the complex impedance, Equation 2.32, must be modified. The energy absorbed at resonance can also be carried away by the electron cyclotron waves propagating perpendicular to the magnetic field. These are, however, generally absorbed, resulting in the cold plasma result for the impedance.

D. EXAMPLE 4: VECTORCARDIOGRAPHY

This example is taken from the work of Bellman et al.[14] and illustrates an inverse problem in medicine. It highlights the fact that inverse problems occur not only in quantum physics, wave propagation, and so forth, but also in the more mundane matters of the human body. The previous three problems had analytical expressions for their solutions. The present example, on the other hand, will emphasize the almost indispensable necessity of numerical procedures in solving inverse problems.

It is well known that our heart is electrically active when stimulated.[15] The entire myocardium (the heart tissue) is a large electromotive surface and forms a massive electric dipole. The center of the electrical activity of the heart is a region in the left atrium called the "sino-atrial node". The electrical impulse generated at this node propagates to the left atrium and through the conducting cables known as the "Purkinje fibers" to the ventricles. As the electrical "front" sweeps over the myocardium, it depolarizes the swept region, which is subsequently repolarized with time.[16] The sweeping of the cardiac electric pulse is shown schematically in Figure 1. If the myocardial surface is divided into segments, then each of these segments behaves as an electrical dipole with the dipole moment located at its center, the direction of the moment being perpendicular to the electromotive surface of the segment. During the propagation of the "depolarization wave" described above, each of these dipoles undergoes a temporal change, the dipole moment rising from zero to a maximum and subsequently decaying to zero again. A typical-dipole moment function is shown in Figure 2. The body is usually considered as a homogeneous volume conductor, and these

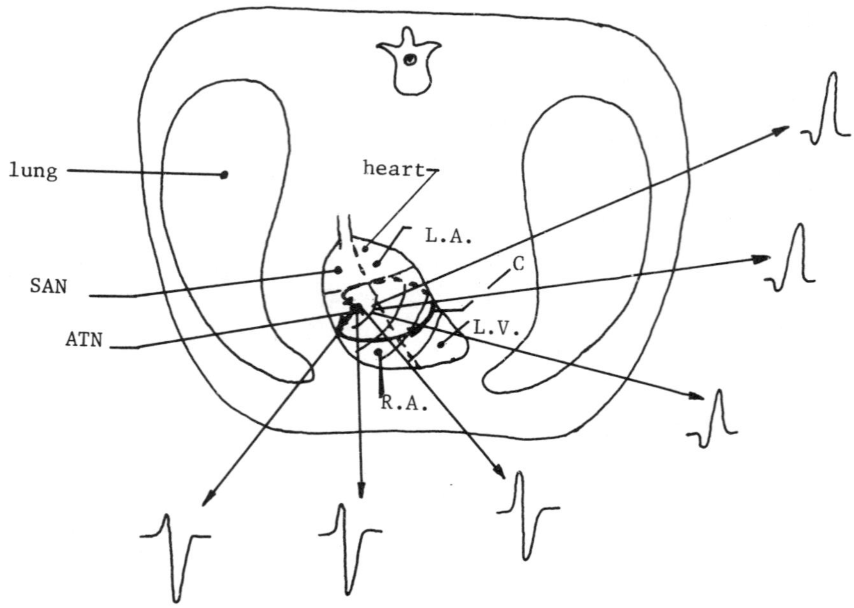

FIGURE 1. A cross-section through the chest showing lung and heart. L.A. = left atrium, R.A. = right atrium, L.V. = left ventricle, R.V. = right ventricle, SAN = sinoatrial node, ATN = atrioventricular node. The trajectory C shows the sweep of the depolarization wave starting from the ATN, veering to the left (in the figure), and circling the myocardium to the base of the heart. The pulses of voltage shown outside the thoracic area indicate the changes in the dipole moments of the ventricular segments as the depolarization wave sweeps. The concentric circles drawn in the heart schematically represent the wave-front propagation.

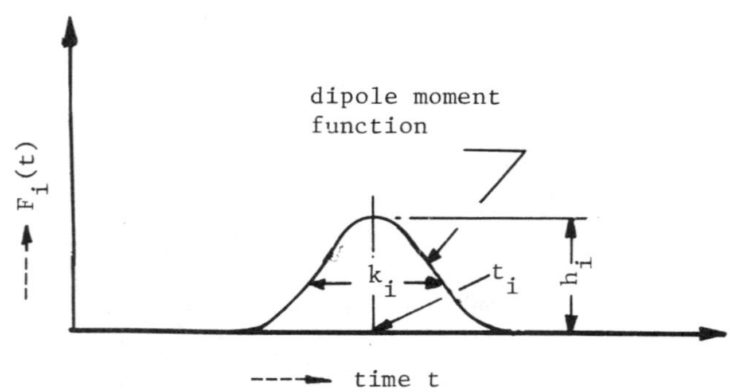

FIGURE 2. The dipole moment function $F_i(t)$ for the i-th segment of the ventricle versus time. k_i, h_i and t_i are the width, maximum height, and time for the dipole moment to reach its maximum, respectively.

time-varying dipole moments generate surface potentials which can be observed on the skin of the patient.[16] Three parameters characterize cardiac electrical activity: (1) the width k_i, (2) the maximum moment h_i, and (3) the time t_i when this maximum is reached. These are shown in Figure 2. The subscript i designates the i-th segment. The direct problem is to determine the observed skin potentials, given the time histories of the dipole moments. In other words, with k_i, h_i and t_i known for each i, determine the potentials at the points of observations on the skin. This is what is known as the "electrocardiogram". The inverse problem is the opposite and can be stated thus. Given the time histories of the observed skin potentials, determine the cardiac parameters k_i, h_i and t_i for each i. Or in other words, given the time histories of the skin potentials, determine the time histories of the individual dipole moments.

Let the i-th dipole be located at the coordinates $\mathbf{x}_i$, $\mathbf{x} = (x,y,z)$, with $\mathbf{l}_i$, $\mathbf{l} = (l,m,n)$, as the direction cosines of its axis. Let $w(\mathbf{a}_j)$, $\mathbf{a} = (a,b,c)$, be the potential due to this dipole at the observation point a_j. Assuming the body as a homogeneous conductor of infinite extent, this potential may be written as[15]

$$w_j = \left[KF_i(t)\ \cos\theta_{ij}\right] r_{ij}^{-2} \tag{2.39}$$

where,

$$w_j(t) = w\left(\mathbf{a}_j, t\right)$$

$$K = \text{a constant}$$

$$F_i(t) = \text{moment of the dipole in the i-th segment at time } t$$

$$\cos\theta_{ij} = \left[\mathbf{l}_i \cdot \left(\mathbf{a}_j - \mathbf{x}_i\right)\right] r_{ij}^{-1}$$

$$r_{ij}^{-2} = \left(\mathbf{a}_j - \mathbf{x}_i\right) \cdot \left(\mathbf{a}_j - \mathbf{x}_i\right)$$

Let M be the number of segments into which the myocardium is divided and N the number of the observation points on the skin. Let $\mathbf{X}$, $\mathbf{X} = (X_1,X_2,X_3)$, be a reference point with respect to which the surface potentials are measured. The duration of the cardiac cycle under consideration is the so-called QRS part of the ECG and is usually 80 ms long. Let $v_j(t)$ be the relative potential, that is, with respect to the reference point $\mathbf{X}$ produced at the j-th observation point. Then

$$v_j(t) = K \sum_{i=1}^{M} \left[F_i(t)\cos\theta_{ij}\right] r_{ij}^{-2} - K \sum_{i=1}^{M} \left[F_i(t)\cos\theta_i\right] r_i^{-2}$$

$$j = 1, 2 \ldots N \quad \text{and} \quad 0 \le t \le 80 \text{ ms}$$

$$\cos\theta_i = \left[l_i \cdot (\mathbf{X} - \mathbf{x}_i)\right] r_i^{-1}$$

and

$$r_i^2 = \left(\mathbf{X} - x_i\right) \cdot \left(\mathbf{X} - x_i\right),\ i = 1,\ 2\ \ldots\ M$$

The reference potential is zero if the reference point is removed to infinity.

Next, one requires the specification of the dipole moment functions $F_i(t)$, i = 1, 2, ... M. These functions are considered by Bellman et al.[14] to be the solutions of the set of the ordinary differential equations

$$\dot{F}(t) = k_i\left(t_i - t\right) F_i(t) \qquad 0 \le t \le 80 \text{ ms}$$

$$F_i(0) = h_i, \qquad i = 1,\ 2,\ \ldots M \tag{2.40}$$

where, as usual, $\cdot \equiv d/dt$. The excitation process is assumed to be completed in 80 ms and the skin potentials are measured every millisecond. It is seen that k_i, h_i, and t_i are the broadness, maximum dipole strength, and the time at which this maximum of the moment occurs, respectively, for the i-th segment.

The inverse problem then is as follows. Suppose that the heart has been observed for $0 \le t \le 80$ ms and that the surface potential time series has been recorded. That is, if $b_j(t)$ denotes the measured surface potential at the observation point j at time t, then the data obtained are $b_j(t)$, j = 1, 2, ... N, at t = 0, 1, 2, ... 80 ms. Given these data, the problem is to determine the 3M set of parameters k_i, h_i, and t_i in the differential equations (2.40). Bellman et al.[14] solve this problem by minimizing the sum S of the squares of the differences between the observed data $b_j(t)$ (the observed relative potential) and the theoretical estimates $v_j(t)$, that is, by minimizing the quantity

$$S = \sum_{p=0}^{80} \sum_{j=1}^{n} \left[v_j(t_p) - b_j(t_p) \right]^2$$

Substituting the expressions for $b_j(t)$ and $v_j(t)$, we have

$$S = \sum_{p=0}^{80} \sum_{j=1}^{N} \left[\sum_{i=1}^{M} K\left(r_{ij}^{-2} \cos\theta_{ij} - r_i^{-2} \cos\theta_i\right) F_i(t_p) - b_j(t_p) \right]^2$$

This is equivalent to minimizing

$$S = \sum_q \left\langle \mathbf{AF}(t_q) - \mathbf{b}(t_q),\ \mathbf{AF}(t_q) - \mathbf{b}(t_q) \right\rangle$$

where, **A** is an (N × M) matrix, **F** is a M-dimensional, and **b** a N-dimensional vector. Minimizing with respect to $F_i(t_q)$, that is, differentiating the quadratic form with respect to $F_i(t_q)$, and setting the result to zero, we obtain:

$$\mathbf{A}^T \mathbf{AF}(t_p) = \mathbf{A}^T \mathbf{b}(t_p), \quad p = 0,\ 1,\ 2,\ \ldots\ 80$$

This yields a formal solution:

$$\mathbf{F}(t_p) = \left[\mathbf{A}^T \mathbf{A}\right]^{-1} \cdot \mathbf{A}^T \mathbf{b}(t_p)$$

If **A** is not a square matrix, then the matrix product $\mathbf{A}^T\mathbf{A}$ may be rank deficient and its inversion may be problematic. However, by simply increasing the number of the observation points, the product can be made to be a square matrix. It will be assumed that this has been done and that N = M. Then, by inverting the matrix equation, $\mathbf{F}(t_p)$ can be obtained at each instant of time t_p.

Having known the functions $F_i(t_p)$ at all time points t_p, an appropriate interpolation scheme can be used to approximate the function. Bellman et al. use the "cubic spline"[17] interpolation. A spline is a polynomial which represents the function over a specific interval. If the degree of these polynomials is three, then the spline is called a "cubic spline". A cubic spline produces a function which, along with its first and second derivatives, remains continuous at the boundaries of the partitions of the function domain. In other words, such a function would appear to be smooth without any breaks or kinks at the end points of the partitions. Let $C_l(t_k)$ be such spline functions with unit values at the mesh points, that is,

$$C_l(t_k) = \delta_{lk}$$

In terms of these spline functions, the dipole vector $F(t_p)$ can be approximated by

$$f_i(t) = \sum_p C_p(t) F_i(t_p)$$

in each segment. Substituting this expansion and its first derivative in the set of the ordinary differential equations given by Equation 2.40 for the dipole functions, we have:

$$\sum_p \left[k_p(t_i - t_p) C_i(t_p) - C_i'(t_p)\right] F_i(t_p) = 0, \quad i = 1,\ 2,\ \ldots M$$
$$p = 1,\ 2,\ \ldots\ 80$$

Choosing three time points yields the necessary three equations to solve for the desired parameters k_i, h_i, and t_i. The process can be repeated for all the segments. The work of Bellman et al. has demonstrated excellent agreement between the solution of the inverse problem and the observed vectorcardiographic data.

E. EXAMPLE 5: WAVE PROPAGATION IN AN ELASTIC MEDIUM

This example is from Blagoveshchenskii[18] and considers wave propagation in a semiinfinite elastic medium in which the speed of propagation varies with the depth inside the medium. The inverse problem is to determine the depth-dependent speed of propagation from the Cauchy data at the boundary. The example considered is typical of inverse problems in wave propagation in elastic media and has applications in elasticity, acoustics, seismology, and vibrational studies in continuous media. The problem is often considered in frequency domain where powerful results of spectral theory of ordinary, second-order differential equations are applied. The problem is transferred to the spectral domain by taking the Fourier transform of the propagation equation with respect to time. The result is the time-independent Schrödinger or a Sturm-Liouville system. This is then solved inversely via spectral theory of the Schrödinger operator for recovering the depth-dependent inhomogeneity or for the potential, depending upon the problem at hand. The famous equations of Gel'fand and Levitan on the semiaxis or that of Marcenko on the entire real line are associated with the inverse solutions of this problem. These will be discussed in Chapter 5 and Chapter 10. The problem can also be solved directly in the time domain without appealing to the spectral theory, by converting it to a Goursat problem for a second-order hyperbolic partial differential equation. This approach will also be discussed in detail in Chapter 6. In this example, we solve the problem neither by spectral theory nor by Goursat's solution, but, rather, purely from "local" considerations, that is, by considering only the kinematics of the problem. This approach which is due to Blagoveshchenskii[18] is discussed below.

Consider wave propagation in an elastic medium, a string, for example, described by the equation:

$$\ddot{u} = c^2(x) u'', \qquad x \subset R^+ \tag{2.41}$$

The initial and boundary conditions are

$$u = 0, \qquad t < 0$$
$$u' = \delta(t), \quad x = 0$$
$$u = f(t), \quad x = 0 \tag{2.42}$$

u = u(x,t) is the amplitude of the disturbance and c(x) > 0 is the speed of propagation, which is a function of x, that is, c is depth dependent. We assume that both c(x) and f(t) $\subset C_\infty$, the class of continuously differentiable functions. The string is free to move at the boundary and is further assumed to be imbedded in an elastic medium with a stiffness of unity. In dealing with a problem of this sort, the standard technique is to define a space variable y (the so-called Liouville's transformation) such that

$$y = \int_0^x \frac{1}{c(x')} dx' \tag{2.43}$$

In terms of the new variable y, Equation 2.41, along with the initial and boundary conditions (2.42), are transformed to the form

$$\ddot{u} - u'' + K(y)u' = 0$$

$$u = 0, \quad t < 0$$

$$u' = c(0)\delta(t), \quad y = 0 \tag{2.44}$$

and

$$u = f(t), \quad y = 0$$

Here,

$$K(y) = [\ln c(y)]' \tag{2.45}$$

In these equations, ′ denotes differentiation with respect to the variable y and δ(t) is interpreted as unity when t = 0 and zero otherwise. Equation 2.44 is the Klein-Gordon equation in an elastic medium which is described by the function K(y).[26]

u′(0,t) is the force on the medium by the support at y = 0 and must be proportional to the negative of the displacement at that point which, according to the condition of the problem, is –f(t). Therefore,

$$u'(0,t) = \delta(t)c(0) = -f(t)$$

resulting in the fact that:

$$c(0) = -f(0)$$

$$f(0) = \lim_{t \to 0+} f(t)$$

Hence, the condition on u′(0,t) can be rewritten as:

$$u' = -f(0)\delta(t), \quad y = 0$$

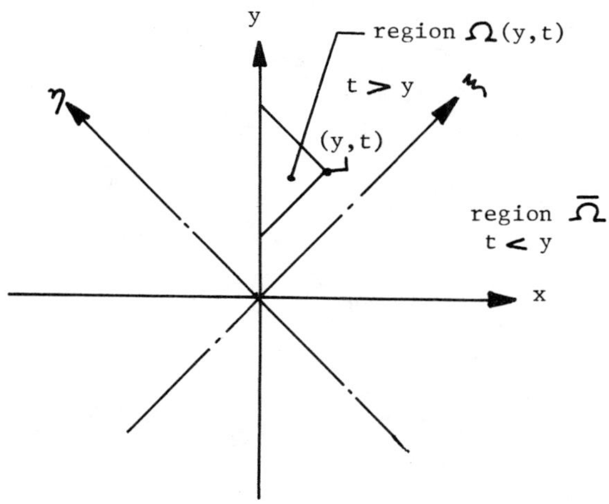

FIGURE 3. Region of integration of Equation 2.44. Also shown are the characteristics ξ and η.

Define:

$$K(y)u'(y,t) = G(y,t)$$

and consider G(y,t) as an inhomogeneous term of Equation 2.44. The homogeneous part of Equation 2.44, namely,

$$\ddot{u} - u'' = 0 \tag{2.46}$$

can be expressed as:

$$u_{\xi\eta} = 0$$

where

$$t = (1/2)(\zeta + \eta)$$

and

$$y = (1/2)(\xi - \eta)$$

The straight lines passing through a point (y,t) and parallel to ξ and η are called the "characteristics" of the hyperbolic partial differential equation (2.46). The Cauchy data for Equation 2.44 are specified at y = 0. Figure 3 shows a pair of the characteristics passing through a point (y,t) and intersecting the axis y = 0 on which the Cauchy data are specified. The solution to the boundary value problem (2.44) with the Cauchy data prescribed along the y = 0 axis is given by the well-known D'Alembert's formula[19] which can be written as:

$$u(y,t) = \frac{1}{2} \iint_{D(y,t)} G(y',t')\,dt'\,dy' + \left(\frac{1}{2}\right) f(t+y) + \left(\frac{1}{2}\right) f(t-y) \tag{2.47}$$

The region of integration D = D(y,t) is shown in Figure 3. D(y,t) means the triangle with the vertex at the point (y,t) and the base on the axis on which the Cauchy data are prescribed. The solution given by Equation 2.47 is the celebrated Riemann's method of integration along the characteristics.[19] The details of Riemann's integration are discussed in Chapters 4 and 6. Differentiating Equation 2.47 with respect to y, dividing the region of t′-integration between (t – y) to t and from t to (t + y), and changing the variable of integration t′ to t′ = t + (y – t″), we obtain:

$$u'(y,t) = \left(\frac{1}{2}\right)\int_0^y G\{t',\ t+(y-t')\}\,dt' +$$

$$+\left(\frac{1}{2}\right)\int_0^y G\{t',\ t-(y-t')\}\,dt' + \left(\frac{1}{2}\right) f'(t+y) -$$

$$-\left(\frac{1}{2}\right) f'(t-y), \quad t > y \tag{2.48}$$

For the sake of notational convenience, t′ has been written for t″.

By definition, the characteristics are the space-time (y – t in this case) trajectories along which disturbances are propagated and across which the solutions are discontinuous. Then by the initial condition that u = 0 for all y for t < 0, we conclude that u must always remain zero in the region $\overline{D}$ shown in Figure 3, that is, whenever t < y. This can also be inferred from the definition (2.43). The variable y signifies the time required by the disturbance to traverse a distance x. Thus, if time t < y, then the disturbance has not arrived at that point and, hence, u will be zero. Therefore, u(y,t) = 0, t < y, the region $\overline{D}$ of Figure 3. In the region D, however, the solution is nonzero unless the problem has only a trivial solution. Therefore u(y,t) must, undergo a jump discontinuity across the characteristic. In other words, y = t is the location of the "wavefront". The theory of propagation of singularities along and across the characteristics will be discussed in Chapter 6.

Let

$$u(y,t) = U(y,t)H(t-y)$$

be the solution across the characteristic y = t. U(y,t) is the continuous solution of the wave equation in the region D. Let us substitute this solution in the wave equation (2.44) and equate the coefficients of the singularities. The coefficient of the Heaviside function H vanishes because of Equation 2.44. Equating the coefficient of the next higher singularity δ(t – y) yields:

$$u(y,\ y+0) - u(y,\ y-0) = u(y,\ y+0) = f(0)\,\exp\{g(y)\} \tag{2.49}$$

where,

$$g(y) = \left(\frac{1}{2}\right)\int_0^y K(x')\,dx' \tag{2.50}$$

We now substitute Equation 2.49 into Equation 2.47 and obtain:

$$f(0)\exp\{g(y)\} = (1/2) \iint_{D(y,y)} G(y',t',)\,dt'\,dy' + (1/2)\left[f(2y)+f(0)\right] \tag{2.51}$$

Notice that the domain of integration has now been written as D(y,y). Let us rewrite Equation 2.51 compactly as:

$$f(0)\exp\{g(y)\} = \mathbf{T}G - P(y) \tag{2.52}$$

where **T** is the integral operator of Equation 2.51 and – P(y) denotes the quantity [(1/2){f(2y)} + f(0)]. Next, we differentiate Equation 2.51 with respect to y and use Equation 2.52 to obtain:

$$\left(\frac{1}{2}\right)K(y) = \frac{\mathbf{U}G + Q(y)}{\mathbf{T}G - P(y)} \tag{2.53}$$

where,

$$\mathbf{U}G = \int_0^y G(t', 2y - t')dt'$$

and

$$Q(y) = f'(2y)$$

Multiplying Equation 2.53 by Equation 2.48 and using the definition (preceding Equation 2.46)

$$K(y)u'(y,t) \equiv G(y,t)$$

yields the following relation for G(y,t):

$$G = -2\frac{(\mathbf{V}G + R)(\mathbf{U}G + Q)}{P - \mathbf{T}G} \tag{2.54}$$

where,

$$\mathbf{V}G = (1/2)\int_0^y G\{t',\ t + (y - t')\}\,dt' + (1/2)\int_0^y G\{t',\ t - (y - t')\}\,dt'$$

and

$$R(y) = (1/2)[f'(t + y) - f'(t - y)]$$

Let us assume for the moment that Equation 2.54 has been solved for G(y,t). Then [P(y) – **T** G] is known. From Equations 2.52 and 2.45 and using the fact that c(0) = –f(0) (following Equation 2.45), we obtain:

$$P(y) - \mathbf{T}G = -f(0)\exp\{g(y)\}$$

$$= c(0)\ \exp[(1/2)\ln\{c(y)/c(0)\}]$$

$$= [c(0)c(y)]^{1/2}$$

yielding:

$$c(y) = [P(y) - \mathbf{T}G]^2 [-f(0)]^{-1}$$

An expression for c as a function of the variable y is thus obtained in terms of quantities which are known. It only remains to express y in terms of x so that c(x) can be obtained. This is easily done using Equation 2.43 from which we can write:

$$dy/dx = 1/c(x) = 1/c(y)$$

yielding

$$x(y) = \int_0^y c(y)\, dy$$

since y = y(x) and x = x(y) describe the same curve. Inverting, we have y = y(x) and, hence, c(y) = c[y(x)] = c(x).

The inverse solution is, therefore, obtained, provided, of course, that the solution of Equation 2.54 is known. It is thus necessary to show that Equation 2.54 can be solved. The integral operators T, U, and V have the limits of y integration from 0 to y. We can then recast Equation 2.54 in a general form as:

$$G(y,t) = p(y) + \int_0^y h[y,t,s,q(G)]\, ds \tag{2.55}$$

where q(G) is a function of G. Equation 2.55, with the limit of integration as one of the variables, is known as the "nonlinear Volterra equation of the second kind".[20] The inhomogeneous term p(y) in Equation 2.55 involves the functions f and f′ and, hence, is a continuous function, $0 \le s \le y$, by the hypotheses of the problem. The function G contains K(y) and u′(y,t) and is continuous, again by the hypotheses of the problem. Hence the kernel h is continuous in all its arguments. Then from the theory of the nonlinear Volterra equations of the second kind,[20] Equation 2.55 possesses a local solution which is unique, that is, there exists some $Y > 0$, $0 \le y \le Y$, such that G(y,t), $t \ge y$ is a unique continuous solution of Equation 2.55.

The standard procedure to solve Equation 2.54 is to introduce a parameter λ and rewrite the equation as:

$$G = -2\frac{(\lambda \mathbf{V}\, G + R)(\lambda \mathbf{U}\, G + Q)}{P - \lambda^2 \mathbf{T}\, G} \tag{2.56}$$

The unknown function G is next expanded in powers of λ:

$$G = \sum_{n=0}^{\infty} \lambda^n G^{(n)} \tag{2.56a}$$

The parameter Y above is the radius of convergence of this series. Replacing this expansion in Equation 2.56, expanding the RHS in powers of λ, and then equating the coefficients of equal powers of λ, the n-th term can be obtained. Blagoveshschenskii shows that the general term of the series (2.56a) is bounded by:

$$G_n \leq \beta^n y^n$$

β being a complicated numerical coefficient and that the series has a finite radius of convergence.

F. EXAMPLE 6: INTEGRAL GEOMETRY

This example highlights an inverse problem in integral geometry, namely, that of determining a function from its known functionals. Kirillov[21] formulated the problem in an n-dimensional space (complex or real) by considering in this space a curve (not necessarily algebraic) which is intersected by straight lines. The functionals from which the function is to be reconstructed are integrals of the function along these intersecting lines. Kirillov's general solution to this problem has far-reaching significance in a wide variety of applications in reconstruction problems of diverse sorts. A more complete discussion of this problem will be presented in Chapter 12 in the context of Radon's transform. Since the main purpose of the examples is to illustrate various types of inverse problems and not to become too involved with rigorous details, we present a simplified analysis of Kirillov's problem in a real Euclidean space of three dimensions and with an algebraic curve.

Let us, therefore, consider an algebraic curve K in a real three-dimensional Euclidean space R^3. Let M be a three-dimensional manifold which consists of all straight lines intersecting the curve K. Given a real, integrable function $f(\mathbf{r})$ on R^3, let there be a mapping T: $f(\mathbf{r}) \rightarrow f(m)$, where, $m \subset M$. This means that to every function $f(\mathbf{r})$ on R^3, we can associate a function f(m) on the manifold M. The mapping T will be chosen to be the integral of $f(\mathbf{r})$ along the line $m \subset M$. The solution of the inverse problem will involve the Fourier transform of the function $f(\mathbf{r})$. In order to avoid undue complications about the legitimacy of performing the operation of Fourier transforms, it will be assumed that the function $f(\mathbf{r})$ is a "tempered distribution", that is, a "test function" of "rapid decay". This means that the function is infinitely differentiable and vanishes, along with all its derivatives, faster than the reciprocal of any polynomial. As will pointed out there, the Fourier transform pair for a tempered distribution can be treated on an equal footing. That is, the function and its transform are both tempered distributions and are elements in the Schwartz space. For details regarding tempered distributions and Schwartz space, see Chapter 8.

The inverse problem of Kirillov in a real space R^3 with an algebraic curve in the space is formulated as follows. Let M be a real manifold consisting of all straight lines intersecting an algebraic curve K in a real three-dimensional Euclidean space R^3. Let f(m), $m \subset M$, be the functional of a function $f(\mathbf{r})$ defined on R^3, obtained by integrating $f(\mathbf{r})$ along the line m. Given the functionals f(m) on the manifold M, the problem is to recover the function $f(\mathbf{r})$.

Let the curve K be described parametrically as $\mathbf{r} = \mathbf{c}(p)$, the parameter $p \subset D$, the domain $D \subset R$. Let $m = (\mathbf{a}, p)$ be a line in the direction of the vector $\mathbf{a}$ and passing through the point $\mathbf{c}(p)$. $\mathbf{a}$ can be considered to be either a unit vector in R^3 or a point on the unit sphere in R^3 with its center at the origin. If $g(\mathbf{a},p)$ is the corresponding f(m), then:

$$\begin{aligned} g(\mathbf{a},p) &= \int_R f\left[\mathbf{c}(p) + \mathbf{a}s\right] ds, \qquad s \subset R, \quad \mathbf{a} \subset R^3 \\ &= \int_{R^+} f\left[\mathbf{c}(p) + as\right] ds \end{aligned}$$

It is assumed that $f(\mathbf{r})$ is on a bounded support in R^3 which is within the curve K. Fourier transforming with respect to **a** and defining the Fourier transform of g(**a**,p) by $\hat{g}(\mathbf{k},p)$, $k \subset R^3$ we obtain:

$$\hat{g}(\mathbf{k},p) = \int_{R^3} g(\mathbf{a},p)\exp[-i\mathbf{k}\cdot\mathbf{a}]\, d\mathbf{a} = \int_{R^3}\int_{R^+} f[\mathbf{c}(p)+\mathbf{a}s]\exp[-i\mathbf{k}\cdot\mathbf{a}]\, ds\, d\mathbf{a} \tag{2.57}$$

We define the Fourier transform pair as:

$$\hat{f}(\mathbf{k}) = \int_{R^3} f(\mathbf{r})\exp[-i\mathbf{k}\cdot\mathbf{r}]\, d\mathbf{r} \tag{2.58}$$

and

$$f(\mathbf{r}) = \int_{R^3} \hat{f}(\mathbf{k})\exp[i\mathbf{k}\cdot\mathbf{r}]\, d\mathbf{k}$$

Define:

$$\mathbf{b} = \mathbf{c}(p) + \mathbf{a}s$$

Then:

$$\hat{g}(\mathbf{k},p) = \int_{R^3}\int_{R^+} f(\mathbf{b})\exp\left[-i\{\mathbf{b}-\mathbf{c}(p)\}\cdot\mathbf{k}/s\right] ds\, d\mathbf{a}$$

$$= \int_{R^3}\int_{R^+} f(\mathbf{b})\, t\, \exp\left[-i\{\mathbf{b}-\mathbf{c}(p)\}\cdot\mathbf{k}t\right] dt\, d\mathbf{b}$$

$$= \int_{R^+} t\left[\int_{R^3} f(\mathbf{b})\exp\{-i(\mathbf{b}\cdot\mathbf{k})t\}d\mathbf{b}\right]\exp\left[-i\{\mathbf{c}(p)\cdot\mathbf{k}\}t\right] dt \tag{2.59}$$

$$= \int_{R^+} t\,\hat{f}(\mathbf{k}t)\, \exp(iwt)\, dt \tag{2.59a}$$

$$= \int_{R^+} \hat{h}(\mathbf{k}t)\exp(iwt)\, dt \tag{2.60}$$

$$= \hat{F}(\mathbf{k},w) \tag{2.60a}$$

where,

$$\hat{h}(\mathbf{k}t) = t\,\hat{f}(\mathbf{k}t)$$

Therefore:

$$\hat{g}(\mathbf{k},p) = \hat{F}(\mathbf{k},w)$$

and w = $\mathbf{c}(p)\cdot k$ is the equation of a hyperplane. It may be recalled that a hyperplane P of a linear space X is a maximal proper subspace of X.[21]

The outline of the inverse solution is as follows. $\hat{F}(\mathbf{k},w)$ contains the Fourier transform of the function f($\mathbf{r}$). In order to recover f($\mathbf{r}$), it is necessary to perform the Fourier transform on $\hat{F}(\mathbf{k},w)$ with respect to the hyperplanes w. This will yield $\hat{h}(\mathbf{k}t)$, inverting which the function f($\mathbf{r}$) will be obtained from Equation (2.60a). Hence, if the function f($\mathbf{r}$) is to be recovered, $\hat{F}(\mathbf{k},w)$ must be known for all hyperplanes except the set with measure zero. Now $\hat{F}(\mathbf{k},w)$ is known from the knowledge of the Fourier transforms $\hat{g}(\mathbf{k},p)$ of the functionals g($\mathbf{a}$,p). This implies that the curve K must intersect almost all the hyperplanes.

The last two steps of Equation 2.60 show that $\hat{F}(\mathbf{k},w)$ is the Fourier transform of the function $\hat{h}(\mathbf{k}t)$ with respect to the variable t. Inverting the transform yields:

$$\hat{h}(\mathbf{k},t) = \int_{R^+} \hat{F}(\mathbf{k},w)\exp(-iwt)\,dw$$

or

$$\hat{f}(\mathbf{k},t) = \int_{R}\left[\hat{F}(\mathbf{k},w)/t\right]\exp(-iwt)\,dw \tag{2.61}$$

From the definition of the Fourier transform given by Equation 2.58, we can write:

$$\begin{aligned} f(\mathbf{r}) &= \int_{R^3} \hat{f}(\mathbf{k},t)\exp\left[i(\mathbf{k}t\cdot\mathbf{r})\right]\,d(\mathbf{k}t) \\ &= \int_{R^3} \hat{f}(\mathbf{k})\exp(i\mathbf{k}\cdot\mathbf{r})\,d\mathbf{k} \\ &= \int_{R^3}\int_{R^+} \hat{F}(\mathbf{k},w)\exp\left[i(\mathbf{k}\cdot\mathbf{r}-w)\right]\,dw\,d\mathbf{k} \\ &= \int_{R^3} d\mathbf{k}\int_{R^+} d\left[\mathbf{c}(p)\cdot\mathbf{k}\right]\hat{g}(\mathbf{k},p)\exp\left[i\{\mathbf{r}-\mathbf{c}(p)\}\cdot\mathbf{k}\right] \\ &= \int_{R^+} dp\int_{R^3} d\mathbf{k}\left[\mathbf{c}'(p)\cdot\mathbf{k}\right]\hat{g}(\mathbf{k},p)\exp\left\{+i\left(\mathbf{r}-\mathbf{c}(p)\right)\cdot\mathbf{k}\right\} \end{aligned} \tag{2.62}$$

Equations 2.60 and 2.61 have been used in obtaining the above equation. Equation 2.62 then tells us how the function f($\mathbf{r}$) can be recovered once the f(m), that is, the functionals g($\mathbf{a}$,p), are known. The above expression can be reduced to a more manageable form. However, the main features of the inverse solution are contained in the equation given above. The integration over p implies that the recovery of a function on R^3 at an arbitrary point $\mathbf{r}$ requires that the line integrals of the function be known along all m $\subset$ M passing through the point in question. This is typical of such inversions in a three-dimensional space.

Another form of f($\mathbf{r}$) may be obtained as follows. By definition:

$$f(\mathbf{r}) = \int_{R^3} d\hat{\mathbf{k}}\int_{R^+} d|\mathbf{k}|\;|\mathbf{k}|^2\;\hat{f}\left(\hat{\mathbf{k}}|\mathbf{k}|\right)\exp\left[i\left(\hat{\mathbf{k}}\cdot\mathbf{r}\right)|\mathbf{k}|\right] \tag{2.63}$$

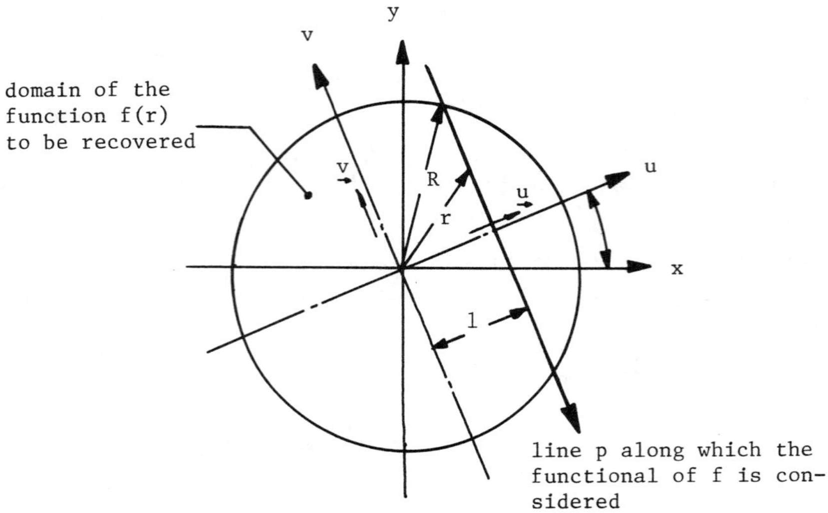

FIGURE 4. Integral geometry of the two-dimensional problem of Kirillov.

That is, we have written the wave vector $\mathbf{k}$ as $\mathbf{k} = \hat{\mathbf{k}}|\mathbf{k}|$, $\hat{\mathbf{k}}$ being a unit vector. Using Equation 2.59a one obtains:

$$\hat{g}'(\mathbf{k},p) = i[\mathbf{c}'(p)\cdot\mathbf{k}] \int_{R^+} dt\, t^2\, \hat{f}(\mathbf{k}t)\exp[i\{\mathbf{c}(p)\cdot\mathbf{k}\}t]$$

$$= i[\mathbf{c}'(p)\cdot\mathbf{k}] \int_{R^+} dt\, t^2\, \hat{f}(\mathbf{k}t)\exp[i\{\mathbf{r}\cdot\mathbf{k}\}\, t] \tag{2.64}$$

where $'$ denotes differentiation with respect to p. The last step follows from the fact that $\mathbf{c}(p)\cdot\mathbf{k} = \mathbf{r}\cdot\mathbf{k}$ for a hyperplane joining $\mathbf{c}(p)$ and $\mathbf{r}$. Comparing Equations 2.63 and 2.64 yields the reconstruction:

$$f(\mathbf{r}) = \int_{R^3} d\hat{\mathbf{k}}\left[i\mathbf{c}'(p)\cdot\hat{\mathbf{k}}\right]^{-1}\hat{g}'(\hat{\mathbf{k}},p)$$

In many of the above derivations, Fubini's theorem[21] has been tacitly used whenever the orders of integrations were interchanged such as in deriving Equation 2.59. In order to be able to perform such mathematical operations along with the operation of Fourier transforming, the functions to be recovered were postulated to be tempered distributions.

The two-dimensional version of Kirillov's solution is very useful in many practical applications and is, therefore, discussed below. A three-dimensional reconstruction can be obtained by stacking up the two-dimensional cross-sections. In two dimensions, the hyperplanes reduce to straight lines and the functionals f(m) are now specified by the line integrals along these lines. We again assume an algebraic curve of a constant order of intersection and the function $f(\mathbf{r})$ to be recovered belonging to the class of tempered distributions. The geometry is shown in Figure 4.

Let $p \subset m \subset M$ be the line identified by the direction $(l,\theta) \equiv (\mathbf{R},\mathbf{v})$, where $\mathbf{v}$ is the unit vector shown in Figure 4. The corresponding line integral is given by:

$$g(p) = \int_{R^2} f(x,y)\, d\mathbf{v}$$

$$= \int_{R^2} f(\mathbf{r})\, \delta(\mathbf{r}\cdot\mathbf{u} - l)\, d\mathbf{r}$$

where **u** is the unit vector along the coordinate u. The Fourier transform of f(**r**) is given by:

$$\hat{f}(\mathbf{k}) = \int_{R^2} f(\mathbf{r}) \exp(-i\mathbf{k}\cdot\mathbf{r}) d\mathbf{r}$$

$$= \iint_{R^2} dx\, dy\, f(x,y) \exp\left[-i\left(k_x x + k_y y\right)\right]$$

The reference frame (u,v) of Figure 4 is rotated through an angle of θ with respect to the Cartesian frame (x,y). Then:

$$\mathbf{r} = \mathbf{A}\mathbf{s}$$

where,

$$\mathbf{r} = \begin{pmatrix} x \\ y \end{pmatrix}; \quad \mathbf{s} = \begin{pmatrix} u \\ v \end{pmatrix}; \quad \mathbf{A} = \begin{pmatrix} \cos\theta & +\sin\theta \\ -\sin\theta & \cos\theta \end{pmatrix}$$

A being the matrix of rotation. The Fourier transform $\hat{f}(\mathbf{k})$ in the rotated coordinate system is written as:

$$\hat{f}(\mathbf{k}) = \int_{R^2} f(\mathbf{s}) \exp(-i\mathbf{k}\cdot\mathbf{s})\, d\mathbf{s}$$

For **k** = (|**k** |, θ):

$$\hat{f}(\mathbf{k}) = \int_{R^2} f(\mathbf{s}) \exp(-iku)\, d\mathbf{s}$$

$$= \iint_{R^2} f(u,v) \exp(-iku)\, du\, dv$$

$$= \int_{R} du\, g(u) \exp(-iku)$$

$$= \hat{g}(\mathbf{k}) = \hat{g}(|\mathbf{k}|, \theta)$$

The argument **k** of $\hat{f}$ or $\hat{g}$ is a vector in R^2, whereas the integral expression contains only the modulus k. This implies that the vector **k** in the Fourier plane lies in the direction of **u**. This is a manifestation of what is known as the "central slice theorem". By definition:

$$f(\mathbf{r}) = \int_{-\infty}^{\infty} \int_{0}^{\pi} d|\mathbf{k}|\, d\theta\, |\mathbf{k}|\, \hat{f}(\mathbf{k}) \exp(i\mathbf{k}\cdot\mathbf{r})$$

$$= \int_{-\infty}^{\infty} \int_{0}^{\pi} d|\mathbf{k}|\, d\theta\, |\mathbf{k}|\, \hat{g}(\mathbf{k}) \exp(i\mathbf{k}\cdot\mathbf{r})$$

$$= \int_{0}^{\pi} d\theta\, G\left\{\mathbf{r}\cdot\hat{\mathbf{k}}, \theta\right\}$$

$\mathbf{r} = (r,\phi)$ and $\hat{\mathbf{k}} = (\cos\theta, \sin\theta)$ and the function $G(x,\theta)$ is given by:

$$G(x,\theta) = \int_{-\infty}^{\infty} d|\mathbf{k}|\ |\mathbf{k}|\ \hat{g}(|\mathbf{k}|,\theta)\exp(ikx)$$

Invoking the inverse convolution theorem in Fourier transform theory yields:

$$G(x,\theta) = \left[|\mathbf{k}|^{\vee} * (g)^{\vee}\right](z)$$

where $z = (x,\theta)$, $*$ denotes the operation of convolution, and the superscript $^{\vee}$ denotes the inverse transform into the variable z. More explicitly,

$$f(\mathbf{r}) = \int_{-\infty}^{\infty}\int_{0}^{\pi}\left[(|\mathbf{k}|)^{\vee}(z - z')\right]\ g(z',\theta)\ dz'\ d\theta \tag{2.65}$$

Now, the generalized function $|\mathbf{k}|$ can be written as:

$$|\mathbf{k}| = k\ \mathrm{sgn}(k)$$

where

$$\mathrm{sgn}(k) = 1, \qquad k \subset R^{+}$$

$$= -1, \qquad k \subset R^{-}$$

R^{+} resp. R^{-} denoting the positive resp. negative half lines. Hence, the inverse Fourier transform of $|\mathbf{k}|$ in the variable, for example x, is i^{-1} times the x derivative of the inverse Fourier transform of sgn(k) in the variable x. From the theory of generalized functions,[22] we know that:

$$d[\mathrm{sgn}(x)]/dx = 2\delta(x)$$

Then:

$$-iy[\mathrm{sgn}(x)]^{\vee}(y) = 2$$

yielding:

$$[\mathrm{sgn}(x)]^{\vee}(y) = -\frac{2}{iy}$$

Hence,

$$|\mathbf{k}|^{\vee}(y) = \frac{1}{i}\left(-\frac{2}{i}\right)\left(-\frac{1}{y^2}\right) = -\frac{2}{y^2} \tag{2.66}$$

Substituting Equation 2.66 in Equation 2.65 results in the final solution:

$$f(\mathbf{r}) = -2\int_{-\infty}^{\infty}\int_{0}^{\pi} dz' d\theta [z - z']^{-2} g(z', \theta) \tag{2.67}$$

Integration by parts reduces Equation 2.67 to:

$$f(r) = 2\int_{-\infty}^{\infty}\int_{0}^{\pi} dz' d\theta [z - z']^{-1} g'(z', \theta) \tag{2.68}$$

The prime $'$ on g indicates differentiation with respect to z'. The factor (1/2) should be replaced by $(1/2\pi^2)$ if the Fourier transform is defined with the factor of $1/(2\pi)^{1/2}$ either in the exponent or as a multiplicative normalizing constant in front of the integral. Remembering that z is given by:

$$z = (\mathbf{k} \cdot \mathbf{r}, \theta)$$

f as a function of $\mathbf{r}$ = (x,y) is determined by either Equations 2.67 or 2.68. The inverse solution is thus obtained.

G. EXAMPLE 7: INVERSE SCATTERING TRANSFORM

This final example deals with the methodology of the "inverse scattering transform" (IST) which is probably the single-most-important advance in the research on inverse problems. The example will also highlight the conditions that must be imposed on the scattering data for a unique determination of the potential. The principal theme of the inverse scattering transform is to reconstruct a potential from the associated "scattering data". The IST is, so to speak, analogous to the Fourier transform, but in solving a certain class of special nonlinear partial differential equations known as the "evolution equations". For good reasons, the IST has been characterized as "nonlinear" Fourier transform. (See M. J. Ablowitz, D. J. Karp, A. C. Newell, and H. Segur, *Stud. Appl. Math.*, 53, 1974.) The importance of solving these equations is enormous since they govern a large number of physical phenomena ranging from shallow water waves through plasma physics to relativistic field theory. The scattering data, also referred to as the "spectral transform", are usually in the frequency domain and are analogous to the Fourier transform. Moreover, just as the dynamics in the Fourier or the k-space is simpler than that in the configuration or the x-space, so is the evolution of the system, as described by the evolution equation, simpler to analyze in the spectral space than in the x-space. The surprising fact about the inverse scattering transform is that it solves the highly nonlinear partial differential evolution equations by relating them to an equation of linear scattering, namely, the time-independent Schrödinger equation. The IST will be discussed in Chapters 9 and 10, where its similarities to and differences from the Fourier transform will be explained more rigorously. In this example, only the basic fundamentals of this remarkable discovery are illustrated. Our presentation follows that of Dodd et al.[23] in the main.

Consider the equation:

$$\dot{q} + cq' = 0, \quad x \subset R$$

$$Q(x) = q(x, t = 0) \tag{2.69}$$

The solution of Equation 2.69 subject to the prescribed initial condition, is

$$q(x,t) = Q(x - ct)$$

Let us illustrate the technique of the IST by applying the transform to this simple problem. Equation 2.69 is an evolution equation describing how q(x,t) evolves in space and time. The equation is, however, linear and may appear to be contrary to the spirit of this example. This is only for the sake of illustrating an otherwise complex methodology in as simple a way as possible. In practice, one would, of course, not use the complex procedures of the inverse scattering transform in solving such a simple equation. The use of this equation is purely pedagogical.

The first task is to relate the evolution equation to a linear scattering equation. Assume, therefore, that q(x,t) occurs in a linear eigenvalue problem such as:

$$y'' + \left[q(x,t) + k^2\right]y = 0 \tag{2.70}$$

where y = y(x,k,t) and t is a parameter in the equation. It is expected that the spectrum and the eigenfunctions y(x,k,t) will vary if the parameter t is changed. Let us further assume that the eigenfunction y evolves according to the equation:

$$\dot{y} + cy' = a(k)y \tag{2.71}$$

Taking the x-derivative of Equation 2.71 and using Equation 2.70, we obtain:

$$\left(y_t\right)_x = a(k)y' + c\left(k^2 + q\right)y \tag{2.72}$$

Similarly, differentiating Equation 2.70 with respect to t and using Equation 2.71 yields:

$$\left[y_{xx}\right]_t + y\dot{q} + \dot{k}^2 y - ay'' = c\left(k^2 + q\right)y' = 0 \tag{2.73}$$

In the above, · resp. ′ denotes, as usual, differentiation in t resp. x. However, the cross-differentiations have been written explicitly in order to emphasize which equation has been differentiated with respect to what variable. If we assume that:

$$\begin{aligned} \dot{k}^2 &= 0 \\ \dot{q} &= -cq' \end{aligned} \tag{2.74}$$

then Equation 2.73 becomes:

$$\left[\left(y_x\right)_t - c(yq) - ck^2 y - ay'\right]' = 0$$

From this, we have:

$$\left[y_x\right]_t = c\left(q + k^2\right)y + ay'$$

This is same as Equation 2.72. This means that $y_{xt} = y_{tx}$, that is, Equation 2.70 and Equation 2.71 are consistent in the sense that $y_{xt} = y_{tx}$, provided that the conditions given by

Equation 2.74 are satisfied. The first condition implies that the spectrum of Equation 2.70 is independent of the variation in t. The Schrödinger equation (2.70) is then called "isospectral". The second condition implies that q must satisfy an evolution equation, which is the wave equation given by Equation 2.69 with which we began.

Let q be a function such that $q \to 0$ as $|x| \to \infty$. Equation 2.70 then reduces to:

$$y'' + k^2 y = 0$$

asymptotically. The solution of this equation is given by:

$$y(k, x, t) = A\ \exp(-ikx) + B\ \exp(+ikx)$$

where A and B may be functions of t and k. Let us consider a solution Y of Equation 2.70 such that the following asymptotic boundary conditions are satisfied:

$$Y(x, k, t) = \begin{cases} \exp(-ikx) & \text{as } x \to -\infty \\ a\ \exp(-ikx) + b\ \exp(+ikx) & \text{as } x \to +\infty \end{cases} \tag{2.75}$$

This is a scattering solution where a plane wave of unit amplitude is incident on the potential from the right. A part of the wave, a exp(–ikx), is transmitted and the part b exp(+ikx) reflected. Using the boundary condition at $x \to -\infty$ in Equation 2.71 yields:

$$a(k) = -ikc$$

Using this result and the boundary condition at $x \to \infty$ in Equation 2.71 results in the following evolution equations for a and b:

$$\dot{b}(t, k) = -2ikc\ b(t, k)$$

$$\dot{a}(t, k) = 0 \tag{2.76}$$

Solving these:

$$b(t, k) = b_0 \exp(-2ikct)$$

$$a(t, k) = a_0 \tag{2.76a}$$

This, therefore, yields a solution Y(x,k,t), the asymptotic behavior of which is:

$$Y(x, k, t) = \begin{cases} \exp(ikx) & \text{as } x \to -\infty \\ b_0 \exp\{ik(x - 2ct)\} + a_0 \exp(-ikx) & \text{as } x \to \infty \end{cases} \tag{2.77}$$

Setting t = 0 yields:

$$Y(x, k, 0) = \begin{cases} \exp(-ikx) & \text{as } x \to -\infty \\ b_0 \exp(ikx) + a_0 \exp(-ikx) & \text{as } x \to \infty \end{cases} \tag{2.78}$$

At t = 0, q(t,x) = q(0,x) = Q(x). Hence, Equation 2.78 is the asymptotic behavior of the solution of Equation 2.70 corresponding to the potential Q(x). Replacing x by (x – ct) formally in Equation 2.77 yields:

$$Y(x,k,t)=\begin{cases}\exp(ikct)\exp(-ikx) & \text{as } x\to-\infty\\ \exp(ikct)\left[b_0\exp(ikx-2ct)+a_0\exp(-ikx)\right] & \text{as } x\to\infty\end{cases} \tag{2.79}$$

Equation 2.79 is just Equation 2.78 with an asymptotic coefficient exp(ikct). Thus, weighting the incident wave exp(–ikx) by the coefficient exp(ikct) results in a transmitted and a reflected wave which are weighted by exactly the same amount. This is a consequence of the linearity of the scattering equation (2.70). Therefore, if a wave of unit amplitude is incident on the medium, then the asymptotic solution q(x) = Q(x – ct) is given by Equation 2.77. Now, Equation 2.77 was obtained for q = q(x,t), whereas Equation 2.79 was obtained by replacing q(x,t) with Q(x – ct). If there is a one-to-one mapping T between q(x,t) and the coefficients (a(t,k),b(t,k)), then for an incident wave of unit amplitude, Equations 2.77 and 2.79 are equivalent. Hence, we conclude that:

$$q(x,t)=Q(x-ct)$$

The solution of Equation 2.69 obtained in the beginning of this section is thus recovered.

Let us pause for a moment and ponder on what is involved in the methodology described so far. Let the pair (a,b) be called the "scattering data". Then (a_0,b_0) are the scattering data for the potential Q(x), whereas (a,b) are the scattering data corresponding to the potential q(t,x). Although the pair (a,b) have been termed the scattering data, it will be seen later that more information may be necessary in specifying the exact scattering data so that a solution q(x,t) can be determined uniquely. We started out with the function Q(x), which was assumed to be given and for which the scattering data (a_0,b_0) are given by Equation 2.76a and are known (presumably by solving the direct problem with q = Q). This is the "direct" or the "forward" problem. These data (a_0,b_0) then evolved in t according to Equation 2.76, resulting in the scattering data (a,b) given by Equation 2.76a corresponding to the function q(x,t) to be determined. In order to obtain the final solution, it is necessary to go backward from (a,b) to q(x,t). In the simple case considered in this example, this inverse or backward operation was performed simply by replacing x by x – ct.

The steps involved in applying the inverse scattering transform in solving an evolution equation can be described as follows:

1. Replace the dependent variable of the evolution equation q(x,t) as the potential in the linear scattering equation (2.70).
2. Solve the direct or forward problem for Q(x) which is assumed to be specified and obtain the corresponding scattering data (a_0,b_0). This means solving the linear scattering equation for a given Q(x), along with the conditions at infinity.
3. With (a_0,b_0) as the initial conditions, solve for the evolution of the scattering data in t. In our example, this was obtained from Equation 2.76.
4. The final step is to design an inverse transform in order to go back to the function q(x,t) from the data (a,b).

Step 2 determines a mapping **T**: Q(x) → (a_0,b_0). Step 4 is the inverse map $\mathbf{T}^{-1}$: (a,b) → q(x,t). The entire procedure is schematized in Figure 5 below. The mappings **T**, $\mathbf{T}^{-1}$ are, therefore, transforms between q ↔ S[q], where S[q] implies transforming q to the scattering data (a,b). This is similar to the Fourier transform, which is a mapping between a configuration and a frequency or spectral domain.

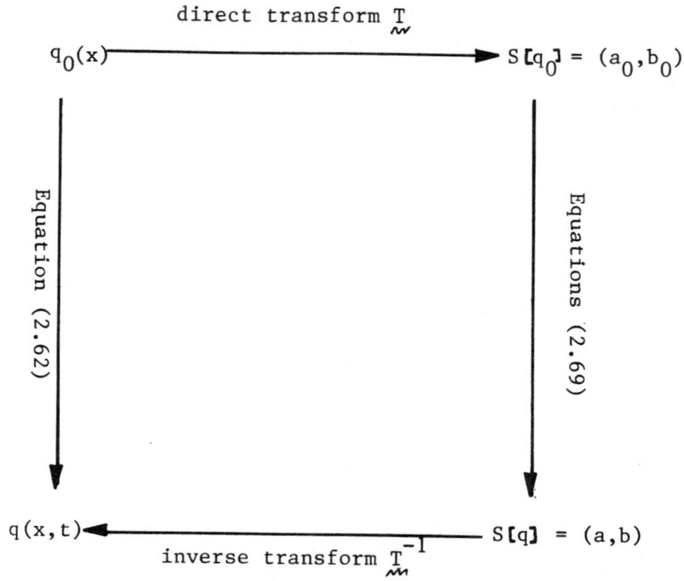

FIGURE 5. A schematic illustration of the inverse scattering transform.

Let us solve the problem by Fourier transform. The Fourier transform (in the variable x) of Equation 2.69 yields:

$$\dot{\hat{q}} + ikc\hat{q} = 0$$

$$\hat{q}(0,k) = \hat{Q}(k) \tag{2.80}$$

where $\hat{q} = \hat{q}(t,k)$. The solution of Equation 2.80 is, of course:

$$\hat{q}(t,k) = \hat{Q}(k)\ \exp(-ikct)$$

Its inverse Fourier transform is

$$q(x,t) = \int_{-\infty}^{\infty} Q(k) \exp[ik(x - ct)]\ dk$$

$$= Q(x - ct)$$

We have, of course, assumed that the function q(x,t) is such that its Fourier transform exists. The solution scheme via the Fourier transform is shown in Figure 6 below. More about Fourier transforms, scattering transforms, and their similarities and differences will be said in Chapter 9.

It was mentioned earlier that the specification of the scattering data may need more information than just the pair (a,b) if the potential is to be recovered uniquely. This crucial point is illustrated by considering a potential given by:

$$q(x) = 2\text{sech}^2 x$$

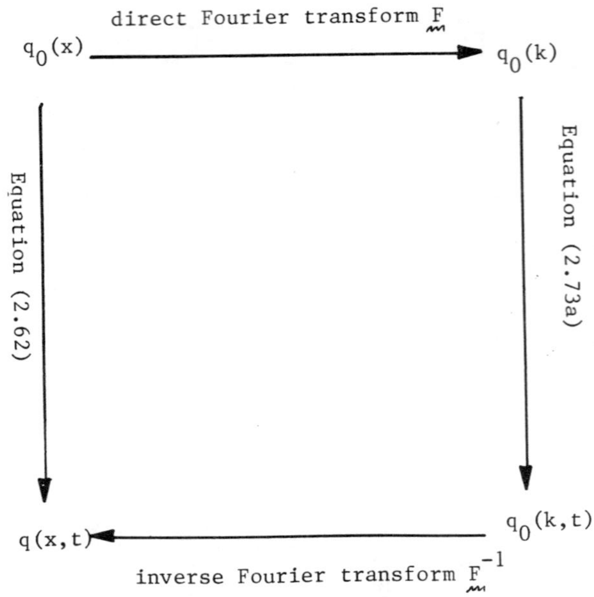

FIGURE 6. A schematic interpretation of the Fourier transform technique.

The corresponding scattering equation is

$$y'' + \left[2\operatorname{sech}^2 x + k^2\right]y = 0 \tag{2.81}$$

This equation can be rewritten as:

$$\mathbf{L}_-\mathbf{L}_+ y(x,k) = \left(k^2 + 1\right)y(x,k)$$

where

$$\mathbf{L}_\pm = (\tanh x \pm \mathbf{D}),\ \ \mathbf{D} \equiv d/dx$$

Let Y(x,k) be the solution to the auxiliary problem,

$$\mathbf{L}_+ y(x,k) = Y(x,k)\left(k^2 + 1\right)$$

such that Y(x,k) is given by the solution to the scattering free equation

$$Y'' + k^2 Y = 0$$

that is,

$$Y(x,k) = A\exp(-ikx) + B\exp(+ikx) \tag{2.82}$$

Operating the solution given by Equation 2.82 by $\mathbf{L}_-$, the general solution to Equation 2.81 is obtained:

$$y(x,k) = A(\tanh x + ik)\exp(-ikx) + B(\tanh x - ik)\exp(+ikx) \tag{2.83}$$

The solution of Equation 2.81 which is appropriate for the scattering problem must satisfy the asymptotic boundary conditions:

$$y(x,k)=\begin{cases}\exp(-ikx) & x\to-\infty\\ a(k)\exp(-ikx)+b(k)\exp(ikx), & x\to\infty\end{cases} \tag{2.84}$$

Since tanhx $\to \pm 1$ as $x \to \pm \infty$, a comparison of Equations 2.83 and 2.84 shows that the desired solution must be of the form:

$$y(x,k)=\left[(\tanh x+ik)/(ik-1)\right]\exp(-ikx) \tag{2.85}$$

From this it follows that the transmission coefficient a(k) is given by:

$$a(k)=(ik+1)/(ik-1) \tag{2.86}$$

whereas the reflection coefficient b(k) is zero. The potential 2sech^2x is thus "reflectionless". This is well known from quantum mechanics.[24,25]

It is also known[26] that every reflectionless potential must have at least one bound state. The bound state for the potential under consideration can be obtained by replacing k by iβ, β real and > 0. The bound state solution is then given by:

$$y(x,i\beta)=\left[(\beta-\tanh x)/(1+\beta)\right]\exp(\beta x)$$

and has the following asymptotic behavior:

$$y(x,i\beta)=\begin{cases}-\exp(\beta x) & x\to-\infty\\ \left[(\beta-1)/(\beta+1)\right]\exp(\beta x) & x\to\infty\end{cases}$$

By definition, a bound state must be square integrable. Since $\beta > 0$, the square integrability is possible if and only if $\beta = 1$. The bound state wavefunction is then:

$$y(x,i\beta)=(1/2)(1-\tanh x)\exp(x)$$

$$=(1/2)\text{sech}\, x$$

Computation of the L_2 norm of the bound-state wavefunction y(x,iβ) yields:

$$\int_R |y(x,i\beta)|^2\,dx=(1/4)\int_{-\infty}^{\infty}\text{sech}^2x\,dx$$

$$=(1/2)\,B(1,1)=1/2$$

B is the beta function.[27] The normalization constant is, therefore, given by:

$$C=\|y\|_2^{-2}=2 \tag{2.87}$$

Now consider a related potential given by:

$$Q(x) = 2\,K^2 \operatorname{sech}^2(Kx + c), \quad c = -\coth^{-1} K$$

where $K^2 > 1$. We will not calculate, but accept the fact that the eigenfunction for this potential which behaves as exp(–ikx) at $x \to -\infty$ is given by*

$$y^{\sim}(x,k) = -\left(1/(K + ik)\right)\left[\mathbf{D} - K - u(x)\right]y(x,k) \tag{2.88}$$

where y(x,k) is given by Equation 2.85 and the function u(x) is

$$u(x) = \operatorname{sech}^2 x / [\tanh x - K]$$

It is easy to establish that if the solution $y^{\sim}(x,k)$ is to satisfy the scattering condition at $x \to \pm\infty$ as given by Equation 2.84, we must also have a zero reflection coefficient and a transmission coefficient identical to that given by Equation 2.85. This clearly shows that specifying only the pair (a,b) is not sufficient for identifying the potential uniquely. Something more than (a,b) is required.

Let us consider the bound state for the solution $y^{\sim}(x,k)$ given by Equation 2.88. The bound-state wavefunction is determined to be

$$y(x, i\beta) = [1/(1 - K)][\mathbf{D} - u_1(x)]y(x, i\beta), \qquad u_1 = u + \mathbf{k}$$

The bound state occurs at the zeros of the coefficient a(k), which is at ik = –1 in this case, yielding $\beta = 1$. Then if C is the bound state normalization constant, we have:

$$C^{-1} = \|y^{\sim}\|_2^{-2} = \int_{-\infty}^{\infty} [(\mathbf{D} - u_1)y(x,i)][(\mathbf{D} - u_1)y(x,i)][1 - K]^{-2}\,dx$$

Integrating the right-hand side by parts yields:

$$C^{-1} = \int_{-\infty}^{\infty} y(x,i)\left[-\mathbf{D}^2 + u_1' + u_1^2\right]y(x,i)[1 - K]^{-2}\,dx$$

$$= \int_{-\infty}^{\infty} y(x,i)\left[-\mathbf{D}^2 - 2\operatorname{sech}^2 x + K^2\right]y(x,i)[1 - K]^{-2}\,dx$$

$$= \int_{-\infty}^{\infty} [(K + 1)/(K - 1)]\|y\|_2^{-2}$$

$$= (1/2)[(K + 1)/(K - 1)], \quad \text{using Equation 2.87}$$

The normalization constant is then C = 2(K – 1)/(K + 1). The bound-state normalization

* The determination of these wavefunctions is usually obtained through "Bäcklund's" transformation.

constant for the eigenfunction y~(x,k) corresponding to the potential Q(x) is, therefore, different from that for the potential $2\text{sech}^2(x)$. The specification of the scattering data must, therefore, include a(k), b(k), and C, and hence also the location of the poles, that it, the zeros of a(k).

REFERENCES

1. **J. R. Cannon:** Determination of certain parameters in heat conduction problems, *J. Math. Anal. Appl.*, 8, 1964.
2. **E. D. Rainville:** *Special Functions*, Macmillan, New York, 1967.
3. **P. Hartman and A. Wintner:** On the solutions of the equations of heat conduction, *Am. J. Math.*, 72, 1950.
4. **A. E. Taylor:** *Generalized Theory of Functions and Integration*, Dover, New York, 1985.
5. **I. Stakgold:** *Green's Functions and Boundary Value Problems*, John Wiley & Sons, New York, 1979.
6. **A. E. Taylor and D. C. Lay:** *Introduction to Functional Analysis*, John Wiley & Sons, New York, 1980.
7. **V. G. Romanov:** *Inverse Problems of Mathematical Physics*, VNU Science Press, Utrecht, The Netherlands, 1986.
8. **B. Friedman:** *Principles and Techniques of Applied Mathematics*, John Wiley & Sons, New York, 1956.
9. **N. Wiener and R. Paley:** *Fourier Transform in A Complex Domain*, American Mathematical Society, Colloquium Publications, 1939.
10. **D. E. Baldwin and A. N. Kaufman:** Determination of the density profile in a plasma slab, *Phys. Fluids*, 12, 1969.
11. **J. D. Jackson:** *Classical Electrodynamics*, John Wiley & Sons, New York, 1967.
12. **F. G. Friedlander:** *Introduction to the Theory of Distributions*, Cambridge University Press, Cambridge, 1987.
13. **R. Bellman, B. Kashef and R. Vasudevan:** The inverse problems of estimating heart parameters from cardiograms, *Math. Biosci.*, 19, 1974.
14. **A. M. Scher and A. C. Young:** The patheways of ventricular depolarization in dogs, *Circ. Res.*, 4, 1956.
15. **D. D. Durrer, R. T. Van Dam, F. L. Meijler, R. C. Arzbacher, E. J. Mueller and G. E. Freud:** *Circulation*, 34, 1966.
16. **G. H. Bell, J. N. Davidson and H. Scarborough:** *Textbook of Physiology*, Williams & Wilkins, Baltimore, 1965.
17. **C. F. Gerald and P. O. Wheatley:** *Applied Numerical Analysis*, Addison-Wesley, Reading, MA, 1984.
18. **A. S. Blagoveshschenskii:** *Tr.(Mat. Inst. Imeni V. A. Steklova)*, 65, 1971.
19. **P. R. Garabadien:** *Partial Differential Equations*, John Wiley & Sons, New York, 1964.
20. **R. K. Miller:** *Nonlinear Volterra Equations*, W. A. Benjamin, 1971.
21. **A. A. Kirillov:** On a problem of I. M. Gel'fand, *Sov. Math.*, 2, 1961.
22. **M. J. Lighthill:** *Introduction to Fourier Analysis and Generalized Functions*, Cambridge University Press, Cambridge, 1958.
23. **R. K. Dodd:** *Solitons and Nonlinear Wave Equations*, Academic Press, London, 1984.
24. **L. I. Schiff:** *Quantum Mechanics*, McGraw-Hill, New York, 1968.
25. **P. Roman:** *Advanced Quantum Theory*, Addison-Wesley, Reading, MA, 1965.
26. **P. Morse and H. Feshbach:** *Methods of Theoretical Physics*, McGraw-Hill, New York, 1953.
27. **I. S. Gradshtien and I. M. Ryzhik:** *Tables of Integrals, Series, and Products*, Academic Press, New York, 1980.

Chapter 3

THE SPECTRAL FUNCTION

I. INTRODUCTION

Some mathematical preliminaries essential to the development of inverse scattering solutions are tersely presented in this chapter. The material primarily contains the basic elements of spectral theory of a second-order ordinary differential equation. The presentation is based heavily on Levitan and Saragsjan,[1] Hille,[2] Coddington and Levinson,[3] and Titchmarsh.[4] In this chapter, we consider a Sturm-Liouville problem on the half-line R^+, i.e., on the interval $[0,\infty)$. The notion of the "spectral function" $\sigma(\lambda)$ and the motivation behind it are introduced first. Parseval's relation and an expansion theorem for a square integrable function on R^+ are discussed next. The spectrum of the Sturm-Liouville problem on the half-line is then investigated and, finally, a completeness relation is derived.

II. SPECTRAL FUNCTION

Consider the scattering equation:

$$y''(x,\lambda)+\{\lambda-q(x)\}y(x,\lambda)=0, \qquad x\subset[0,\infty) \tag{3.1}$$

and the boundary conditions:

$$y(0,\lambda)=\sin\alpha, \quad y'(0,\lambda)=-\cos\alpha$$

or

$$y(0,\lambda)\cos\alpha+y'(0,\lambda)\sin\alpha=0$$

where α is an arbitrary real parameter $0\le\alpha<\pi$. The potential q(x) is a continuous real-valued function which is summable over every finite interval. That is $\int F(x)dx<\infty$ for every compact subset of the real axis.

Equation 3.1 is a "singular" Sturm-Liouville problem, one end point being at infinity. The corresponding "regular" problem involves a finite interval [0,b] and is given by:

$$y''(x,\lambda)+\{\lambda-q(x)\}y(x,\lambda)=0, \quad x\subset[0,b] \tag{3.2}$$

with the boundary conditions:

$$y(0,\lambda)\cos\alpha+y'(0,\lambda)\sin\alpha=0 \tag{3.3}$$

and

$$y(b,\lambda)\cos\beta+y'(b,\lambda)\sin\beta=0 \tag{3.4}$$

β being another arbitrary real parameter $0\le\beta<2\pi$. The singular problem can be analyzed[1-3] as the $b\to\infty$ limit of the regular problem. It may be noticed that an ordinary boundary condition such as in the above problems can be assigned at the nonsingular left endpoint in both regular and singular problems. The point at infinity, however, poses a serious problem. In the singular case, one could consider a boundary condition such as

$$y(0,\lambda) = \lim_{x\to\infty} y(x,\lambda) = 0$$

However, such a problem is, in general, not well-posed, although the problem is solvable if special assumptions are made regarding the potential q(x).[2]

Equations 3.2 to 3.4 constitute a discrete problem. It has an infinite number of real, discrete eigenvalues $\{\lambda_{nb}\}$, n = 1, 2, ... and a complete set of orthogonal eigenfunctions $\{y_{nb}\}$. The subscript nb refers to the n-th eigenvalue or eigenfunction for the finite interval [0,b]. Passing to the limit $b \to \infty$ is facilitated by converting the discrete problem into a continuous one. This is achieved through the introduction of a "spectral function". The spectral function has, however, deeper significance. One consequence of the singular case is that the series expansions in terms of the characteristic or the eigenfunctions familiar in the regular case are to be replaced by integral expansions in the singular case. More important is the following. We are here mostly interested in L_2 solutions. In the regular case, the solutions involve unitary mappings of functions in $L_2[0,b]$ onto the Fourier coefficients which are in l_2. These two spaces are of rather different structures, although both are generalizations of a Hilbert space. It is pleasing to remove this disparity and consider the mapping to be not from L_2 to the Fourier coefficients in l_2, but, rather, onto the "generalized" Fourier transforms in another L_2 space. The study then involves transformations between two L_2 spaces. All this is accomplished through formulations in terms of the spectral function.

Let $f(x) \subset L_2[0,b]$, i.e., f(x) is a square integrable, complex-valued function over the interval [0,b]. $L_2[0,b]$ is a Lebesgue space of equivalent classes of functions. Two functions belonging to the same equivalent class are identical to each other except, perhaps, over a set of measure zero. The following inner product is defined in $L_2[0,b]$:

$$(u,v) = \int_0^b u(x)v^*(x)dx$$

"*" denotes the complex conjugate. Then,

$$\|f(x)\|_b^2 = \int_0^b |f(x)|^2 dx$$

$$= \sum_{n=0}^{\infty} r_{nb}^2 (f, y_{nb})^2 = \sum_{n=0}^{\infty} [f_{nb}]^2 \tag{3.5}$$

Here,

$$r_{nb}^2 = \|y_{nb}\|^{-2}$$

is a normalizing constant. $\|\cdot\|_b$, $|\cdot|$ denote the L_2 norm over the interval [0,b] and the absolute value, respectively. Equation 3.5 is "Parseval's relation" for the finite interval [0,b].

In the regular case, the identity (3.5) is a mapping defined by

$$T: f(x) \subset L_2[0,b] \to \{f_{nb}\} \subset l_2$$

More generally, inner products go into inner products, that is,

$$\int_0^b f(x)g^*(x)\,dx = \sum_{n=0}^{\infty} f_{nb}g_{nb}^*$$

The mapping is linear, one to one, onto and norm preserving. In other words, T is a "unitary" transformation. Thus, if $\{f_{nb}\} \subset l_2$, then there exists an $f(x) \subset L_2[0,b]$ such that $f_{nb} = r_{nb}^{-1}(f,y_{nb})$, and this f is uniquely determined except over a set of Lebesgue measure zero. The converse is also true. As mentioned earlier, these two spaces are of rather different structure, although both are realizations of Hilbert space. Let us now consider how to remove this disparity by designing the mapping to be between two L_2 spaces through the introduction of suitable measure functions which will be the spectral functions.

Toward this end, let us consider a nondecreasing, right continuous function $\sigma(t)$ on the entire real axis R. A function f(x) is "monotonically nondecreasing" if $f(x) \geq f(y)$ for every $x > y$. If only the inequality sign holds, then the function is "strictly increasing". A continuity condition imposed on $\sigma(t)$ such as

$$\sigma(t+0) = \sigma(t)$$

is called a right-hand condition. Left-hand continuity $\sigma(t-0) = \sigma(t)$ would have served just as well. The measure functions are also "normalized", that is:

$$\sigma(0+0) = \sigma(0) = 0$$

Let g(t) be a function defined on R such that it is σ-measurable and such that its integral with respect to the σ measure exists. That is,

$$\|g(t)\|_{\sigma(t)}^2 = \int_R |g(t)|^2 d\sigma(t)$$

exists. More generally, the inner product is,

$$\int_R f(t)g^*(t)d\sigma(t) = (f,g)_\sigma = (g,f)_\sigma^* \quad : \text{Hermiticity}$$

$$|(f,g)_\sigma| \leq \|f\|_\sigma \|g\|_\sigma \quad : \text{Schwartz inequality}$$

We define the space $L_2(\sigma)$ to consist of classes of σ-equivalent functions g(t), that is, of g(t)s which in a class of σ-measurable functions differ from each other at most in a set of σ-measure zero. Under the above defined norm, it can be shown[2] that the space $L_2(\sigma)$ is complete. The spaces $L_2[0,b]$ and l_2 are special instances of $L_2(\sigma)$ with the measure function $\sigma(t)$ as:

$$\sigma(t) = \begin{cases} 0 & \text{if } t<0 \\ t, & \text{if } 0 \leq t \leq b \quad \text{for } L_2|0,b| \\ b, & \text{if } t>b \end{cases}$$

and for l_2

$$\sigma(t) = \begin{cases} -1, & \text{if } t < 0 \\ n, & n \le t < n+1, \; n = 0,1,\ldots \end{cases}$$

We now return to our original question: how to replace the mapping:

$$T: f(x) \subset L_2[0,b] \to \{f_{nb}\} \subset l_2$$

by

$$T: f(x) \subset L_2[0,b] \to F(\lambda) \subset L_{2,\sigma(\lambda)}$$

where, $L_{2,\sigma(\lambda)}$ is the space of square integrable functions with respect to the measure function $\sigma(\lambda)$. For the semi-axis this is a long, drawn out process which will finally culminate into Parseval's relation on the half-line. More generally, T: $f(x) \subset L_{2,\,m(t)}$ [0,b] $\to F(\lambda) \subset L_{2,\sigma(\lambda)}$, where the measure functions m(t) and $\sigma(\lambda)$ may or may not coincide. However, here we consider only the $L_2[0,b]$ measure functions for m(t).

It was seen that l_2 was a special instance of $L_2(\sigma)$. The set $\{f_{nb}\}$ can be placed into an $L_2(\sigma)$ space that has a direct bearing on the boundary value problem. To this end, let us consider the measure function $\sigma(\lambda)$ defined by:

$$\sigma_b(\lambda) = \begin{cases} -\sum_d r_{nb}^2 & \lambda < 0 \\ 0 & \lambda = 0 \\ \sum_d r_{nb}^2 & \lambda > 0 \end{cases} \tag{3.6}$$

with right-hand continuity. Summation over d means summation over the spectral interval $-\infty < \lambda < \lambda_{nb} \le 0$ for $\lambda \le 0$ and $0 < \lambda_{nb} \le \lambda < \infty$ for $\lambda > 0$. Left-hand continuity would have done just as well. The normalized right continuous measure function $\sigma_b(\lambda)$ defined by Equation 3.6 is called the "spectral function" for the boundary value problem (3.2 to 3.4). It undergoes a jump discontinuity of r_{nb}^2 whenever $\lambda = \lambda_{nb}$, while remaining constant in between (Figure 1). The determination of the spectral function involves, therefore, two factors: (1) the distribution of the eigenvalues and (2) the normalization constants for the eigenfunctions. These two factors are reflected in the specification of the scattering data, as was seen in Example 7 of Chapter 2.

In terms of the spectral function, Parseval's relation becomes:

$$\begin{aligned} \|f(x)\|_b^2 &= \int_{-\infty}^{\infty} |F(\lambda)|^2 \, d\sigma_b(\lambda) \\ &= \|F(\lambda)\|_{\sigma_b(\lambda)}^2 \end{aligned} \tag{3.7}$$

where,

$$F(\lambda) = \big(f(x), y(x,\lambda)\big) \tag{3.8}$$

The integral on the right-hand side of Equation 3.7 is to be understood as a Stieltz integral, i.e.,

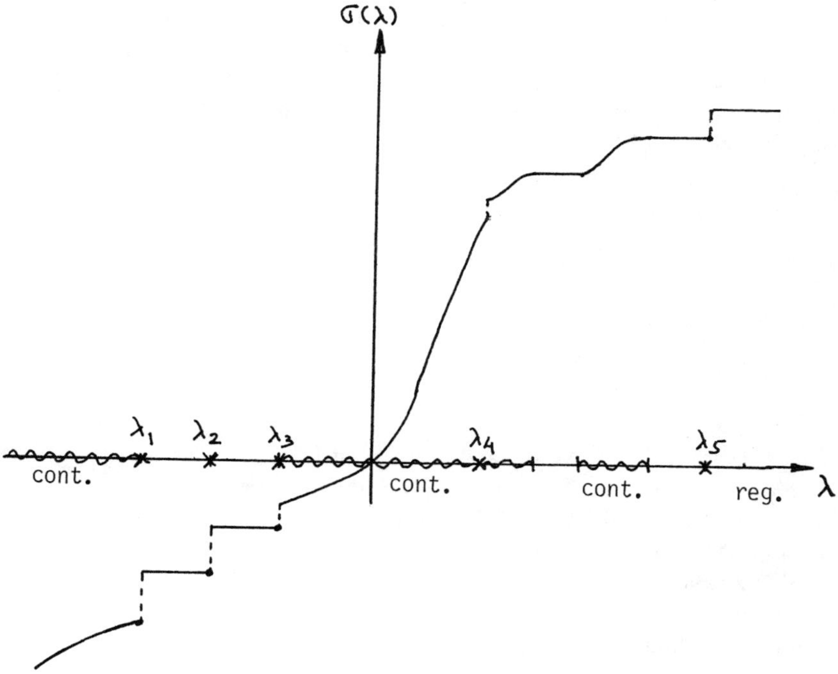

FIGURE 1. A schematic illustrating the general relation between the spectrum and the spectral function. The point spectrum denoted by λ represent the jumps in $\sigma(\lambda)$ while the continuous spectrum corresponds to the increasing part of the spectral function. The points of constancy represent regular points. A large variety of spectral behavior is possible. An example is the jump point λ_4 which indicates that an eigenvalue is embedded in the continuous spectrum. Physically, this means that a particle cannot overcome the potential barrier even though it has energy to do so. This is a longstanding problem in the theory of Schrödinger equation.

$$\|F(\lambda)\|^2_{\sigma_b(\lambda)} = \lim_{|\delta|\to 0} \sum_k |F(z_k)|^2 \left[\sigma_b(\lambda_{k+1} - \sigma_b(\lambda_k))\right]$$

where $\delta = (-\infty < \lambda_0 < \lambda_1 < \ldots < \infty)$ is an arbitrary partition of the interval and z_k is any value $\subseteq \lambda_k, \lambda_{k+1}{}^1$. More precisely, the integral of Equation 3.7 means that for every $\delta > 0$, there exists some $\varepsilon > 0$ such that:

$$\left|\|F(\lambda)\|^2_{\sigma_b(\lambda)} - \sum_k |F(z_k)|^2 \left[\sigma_b(\lambda_{k+1} - \sigma_b(\lambda_k))\right]\right| < \varepsilon$$

provided that for any arbitrary partitioning of the interval, $\max(\lambda_{k+1} - \lambda_k) < \delta$ for all admissible choices of the coordinates.* Equation 3.7 is exactly the same as Equation 3.5, but the discrete function $y(x,\lambda_{nb})$ has been replaced by a continuous function $y(x,\lambda)$. This is possible by virtue of the definition of $\sigma_b(\lambda)$. Expressing Parseval's relation as a Stieltz integral is useful in establishing appropriate limiting procedures as $b \to \infty$ and, hence, in establishing Parseval's relation on the semiaxis.

This will be established as follows. We first consider a finite interval [0,b] and a sequence of functions $\{f_n(x)\}$ defined to be nonzero only over the subsets $[0,n] \subset [0,b]$ and Parseval's equality (3.7) is considered for the sequence in terms of the spectral measure $d\sigma_b(\lambda)$. Parseval's equality on the half-line is then obtained by first passing to the limit $b \to \infty$ and,

* For details regarding a Stieltz integral, see, for example, L. Linsternik and V. Sobolev, *Elements of Functional Analysis,* Unger, 1961.

subsequently, to $n \to \infty$. The $b \to \infty$ limit is performed in two steps: first taking the limit on $\sigma_b(\lambda)$ and then passing to the limit under the integral sign (if permissible) in Equation 3.7. These limiting procedures are made possible by two theorems due to Helly.[3,5] Before these theorems can be applied, the following lemma is required. Lest the point be lost in the mathematical details, we wish to remind the reader that our main objective is to establish Parseval's relation on R^+ which will realize the mapping sought for, namely, T: $L_2(R^+) \to L_{2,\sigma(\lambda)}(R)$.

Lemma 1

For any positive integer N, there exists a positive constant C such that:

$$-\sum_{(-N,N)} r_{nb}^2 = \sigma_b(N) - \sigma_b(-N) < C(N)$$

The constant C is a function of N only and is independent of b.

A simple proof is given in Reference 1. From this lemma it follows that the set of the monotonically nondecreasing functions $\sigma_b(\lambda)$ is uniformly bounded in b for every compact spectral subinterval (–N,N]. This means that there exists a nonnegative function $H(\lambda)$ such that $\sigma_b(\lambda) \leq H(\lambda)$ independent of b for any compact spectral interval. This condition of uniform boundedness permits the $b \to \infty$ limiting operation on $\sigma_b(\lambda)$ via a theorem due to Helly called the "selection" theorem, also referred to as Helly's "first" theorem.

Selection Theorem

Let $\{\sigma_{bk}(\lambda)\}$, k = 1, 2, ... be a sequence of real, nondecreasing functions on $-\infty < \lambda < \infty$, such that $|\sigma_{bk}(\lambda)| \leq H(\lambda)$ for all k, $-\infty < \lambda < \infty$, $H(\lambda)$ continuous and nonnegative. Then there exists a nondecreasing function $\sigma(\lambda)$ and a subsequence $\{\sigma_{bk_n}(\lambda)\}$ such that:

$$\sigma(\lambda) \leq H(\lambda), \qquad -\infty < \lambda < \infty$$

and

$$\lim_{n\to\infty} \left\{\sigma_{bk_n}(\lambda)\right\} = \sigma(\lambda)$$

By the selection theorem,

$$\lim_{n\to\infty} \sigma_{b_n}(\lambda) \to \sigma(\lambda)$$

in the sense that one can select a subsequence $\{b_n\}$ for which the functions $\sigma_{b_n}\lambda)$ converge to a limiting monotonic function $\sigma(\lambda)$.

It should be pointed out that the limiting function also satisfies Lemma 1, that is, also remains bounded over every compact interval $\Delta(m) = (-m,m]$. Specifically,

$$\sigma(m) \leq R\left(1+m^2\right)$$

for every interval $\Delta(m)$, R being a constant.[3] Further discussion of the spectral function is postponed until after Parseval's relation and the expansion theorem have been established on the semiaxis.

Helly's selection theorem thus establishes the $b \to \infty$ limit $\sigma(\lambda)$ of the function $\sigma_b(\lambda)$ defined for the finite interval [0,b]. The main objective at this point is to establish Parseval's relation on the semiaxis for which it is thus necessary to pass to the limit $b \rightarrow \infty$ under the Stieltz integral of Equation 3.7. This can be achieved by applying Helly's "integral" or the "second" theorem along the subsequence $b_n(\lambda)$ of the first theorem.

Integration Theorem

Let $\{h_n(\lambda)\}$ be a real, uniformly bounded sequence of nondecreasing functions on a finite interval $a \le \lambda \le b$, and assume:

$$\lim_{n\to\infty} h_n(\lambda) \rightarrow h(\lambda), \quad a \le \lambda \le b$$

If f is any continuous function on the interval [a,b], then:

$$\lim_{n\to\infty} \int_a^b f(\lambda)\, dh_n(\lambda) = \int_a^b f(\lambda)\, dh(\lambda)$$

Now, let $\{f_n(x)\}$, $0 \le x \le n$, $n < b$, be a sequence of functions which vanish outside some interval [0,n] and possess continuous second derivatives. Then from Parseval's relation (3.5):

$$\|f_n(x)\|_b^2 = \|F_n(\lambda)\|_{\sigma_b(\lambda)}^2 \tag{3.9}$$

where $F_n(\lambda) = (f_n(x), y(x,\lambda))$ over the interval [0,n]. By virtue of Lemma 1, Helly's theorems can be applied to Equation 3.9. Passing to the limit $b \rightarrow \infty$,

$$\|f_n(x)\|_b^2 = \|F_n(\lambda)\|_{\sigma(\lambda)}^2 \tag{3.10}$$

This completes the $b \rightarrow \infty$ limit for the sequence $\{f_n(x)\}$. Taking limit $n \rightarrow \infty$ then establishes Parseval's equality on the half-line.

III. PARSEVAL'S EQUALITY ON THE HALF-LINE R^+

The sequence $\{f_n(x)\}$ described above is dense in $L_2[0,\infty)$. By definition, this is a Cauchy sequence, i.e.,

$$\lim_{n,m\to\infty} \|f_n(x) - f_m(x)\|_2^2 \rightarrow 0 \tag{3.11}$$

The norm in Equation 3.11 is that of the space $L_2[0,\infty)$. Since $L_2[0,\infty)$ is a complete space,[5] $\{f_n(x)\} \rightarrow f(x) \subset L_2[0,\infty)$. Therefore,

$$\lim_{n\to\infty} \|f(x) - f_n(x)\|_2^2 \rightarrow 0 \tag{3.12}$$

Equation 3.11 implies that:

$$\lim_{n,m\to\infty} \|F_n(\lambda) - F_m(\lambda)\|_{\sigma(\lambda)}^2 \rightarrow 0 \tag{3.13}$$

$\{F_n(\lambda)\}$ is, therefore, also a Cauchy sequence. Now, the space $L_2[R]$ is complete[5] and the completeness is carried over[1] to $L_{2,\sigma(\lambda)}$, $\lambda \subset R$, of the functions square integrable with respect to the spectral function $\sigma(\lambda)$ on R. Hence, as $n \to \infty$, $\{F_n(\lambda)\} \to F(\lambda) \subset L_{2,\sigma(\lambda)}(R)$. From Equations 3.12 and 3.13 it follows that:

$$\|f(x)\|_2^2 = \|F(\lambda)\|_{2,\sigma(\lambda)} \tag{3.14}$$

by virtue of Equation 3.10.

It may be helpful at this point to summarize. The sequence $\{f_n(x)\} \to f(x) \subset L_2[0,\infty)$. The corresponding sequence $\{F_n(\lambda)\} \to F(\lambda) \subset L_{2,\sigma(\lambda)}(R)$ such that Parseval's relation (3.14) holds for $x \subset R^+$. It remains to determine in what sense $F(\lambda)$ can be considered to be the Fourier transform of f(x).

In order to establish this, let us consider a function g(x) such that:

$$\begin{aligned} g(x) &= f(x), \qquad & 0 \le x \le n \\ &= 0 & x > n \end{aligned}$$

Then,

$$\begin{aligned} \|f(x) - g(x)\|_2^2 &= \|F(\lambda) - G(\lambda)\|_{\sigma(\lambda)}^2 \\ &= \|F(\lambda) - G_n(\lambda)\|_{\sigma(\lambda)}^2 \\ &= \|F(\lambda) - F_n(\lambda)\|_{\sigma(\lambda)}^2 \\ &= \int_n^\infty |f(x)|^2 dx \to 0 \quad \text{as} \quad n \to \infty \end{aligned}$$

This shows the limit in the mean (l.i.m.) convergence of $F_n(\lambda) \to F(\lambda)$, as $n \to \infty$.

Summing up these results together, we arrive at Parseval's equality on $[0,\infty)$.

Parseval's Equality on $[0,\infty)$

Given a function $f(x) \subset L_2[0,\infty)$, there corresponds a function $F(\lambda) \subset L_{2,\sigma(\lambda)}$, the "generalized" Fourier transform of f(x), $\sigma(\lambda)$ being a monotonically nondecreasing function independent of f(x) and bounded in each finite subinterval of λ such that:

$$\|f(x)\|_2^2 = \|F(\lambda)\|_{\sigma(\lambda)}^2 \tag{3.15}$$

The generalized Fourier transform $F(\lambda)$ is given by:

$$F(\lambda) = \big(f(x), y(x,\lambda)\big), \qquad x \subset R^+ \tag{3.16}$$

and is to be understood in the sense of norm convergence, i.e.,

$$\underset{n\to\infty}{\text{limit}} \quad \|F(\lambda)-F_n(\lambda)\|^2_{\sigma(\lambda)} = 0,$$

$F_n(\lambda)$ being given by:

$$F_n(\lambda) = \int_0^{\infty} f(x)y(x,\lambda)dx$$

If f(x) vanishes outside a finite interval, then $F(\lambda)$ exists in the ordinary sense. The mapping from $L_2[0,\infty)$ to the space $L_{2,\sigma(\lambda)}$ is onto and norm preserving, that is, unitary. Parseval's equality (3.15) finally establishes the mapping between the two L_2 spaces, namely, $L_2(R^+)$ and $L_{2,\sigma(\lambda)}$.

IV. EXPANSION THEOREM

Parseval's equality can be used to write:

$$\|f(x)+g(x)\|_2^2 = \|F(\lambda)+G(\lambda)\|^2_{\sigma(\lambda)}$$

and

$$\|f(x)-g(x)\|_2^2 = \|F(\lambda)-G(\lambda)\|^2_{\sigma(\lambda)}$$

where f(x), g(x) $\subset L_2[0,\infty)$ are continuous functions. This is due to the fact that the generalized Fourier transform of the combination $[f(x) \pm g(x)]$ is, of course, $[F(\lambda) \pm G(\lambda)]$. Subtracting the two yields:

$$\int_0^{\infty} f(x)\, g(x)dx = \int_{-\infty}^{\infty} G(\lambda)F(\lambda)d\sigma\ (\lambda) \tag{3.17}$$

Equation 3.17 is known as the "generalized" Parseval equality. This is a consequence of the unitary property of the mapping from the space $L_2[0,\infty)$ to $L_{2,\sigma(\lambda)}$.

Let us again consider the function g(x):

$$\begin{aligned} g(x) &= g(x) \qquad 0 \le x \le n \\ &= 0 \quad x > n \end{aligned}$$

Then Equation 3.17 can be written as:

$$\int_0^{n} f(x)g(x)dx = \int_{-\infty}^{\infty} F(\lambda)\left\{\int_0^{n} g(x)y(x,\lambda)\,dx\right\}d\sigma(\lambda) \tag{3.18}$$

Assuming that the integral

$$\int_{-\infty}^{\infty} F(\lambda)\, y(x,\lambda)\, d\sigma(\lambda)$$

converges absolutely, we can change the order of integration in Equation 3.18 and obtain:

$$\int_0^n f(x)g(x)dx = \int_0^n g(x)\left\{\int_{-\infty}^{\infty} F(\lambda)y(x,\lambda)d\sigma(\lambda)\right\}dx \tag{3.19}$$

Let us further assume that the integral over $(-\infty,\infty)$ in Equation 3.19 converges uniformly in x and, hence, is continuous. Since g(x) is arbitrary and f(x) is continuous over $[0,\infty)$, we have the following theorem.

Theorem 3.1: Expansion Theorem

If $f(x) \subset L_2[0,\infty)$ is continuous over $[0,\infty)$ and if the integral

$$\int_{-\infty}^{\infty} F(\lambda)\, y(x,\lambda)\, d\sigma(\lambda)$$

converges absolutely and uniformly with respect to x in every finite interval, then:

$$f(x) = \int_{-\infty}^{\infty} F(\lambda)y(x,\lambda)d\sigma(\lambda) \tag{3.20}$$

V. CONNECTION BETWEEN SPECTRAL FUNCTION AND SPECTRAL FAMILY

Let $\mathbf{E}(\lambda)$ be an operator such that:

$$\mathbf{E}(\lambda)f = \int_{-\infty}^{\lambda} F(\lambda)y(x,\lambda)d\sigma(\lambda) \tag{3.21}$$

Using the expansion Theorem 3.1:

$$\lim_{\lambda\to\infty} \mathbf{E}(\lambda)f = f, \text{i.e., } \lim_{\lambda\to\infty} \mathbf{E}(\lambda) = I, \text{ the identity operator, and } \lim_{\lambda\to-\infty} \mathbf{E}(\lambda) = 0$$

Also, $\mathbf{E}(\lambda)$ satisfies the same continuity condition as does $\sigma(\lambda)$. Thus, if $\varepsilon \to 0$, $\sigma(\lambda + \varepsilon) \to \sigma(\lambda)$, then limit $\varepsilon \to 0$, $\mathbf{E}(\lambda+ \varepsilon)f \to \mathbf{E}(\lambda)f$ for all f and ε.

Let us consider the sequence of functions g(x) again. Multiplying g(x) by $\mathbf{E}(\lambda)f$ and integrating over the interval [0,n] yields:

$$\int_0^n g(x)\mathbf{E}(\lambda)fdx = \int_0^n g(x)\left\{\int_{-\infty}^{\lambda} F(\lambda)y(x,\lambda)d\sigma(\lambda)\right\}dx$$

Since the integrals are absolutely convergent, the order of integration can be changed, yielding:

$$\int_0^n g(x)\mathbf{E}(\lambda)fdx = \int_{-\infty}^{\lambda} F(\lambda)\left\{\int_0^n g(x)y(x,\lambda)dx\right\}d\sigma(\lambda)$$

$$= \int_{-\infty}^{\lambda} F(\lambda)G_n(\lambda)d\sigma(\lambda)$$

Letting $n \to \infty$, we have,

$$\underset{n\to\infty}{\text{limit}} \int_0^n g(x)\mathbf{E}(\lambda)f dx = \int_{-\infty}^{\lambda} F(\lambda)G(\lambda)d\sigma(\lambda)$$

Identifying g with f, the following result is obtained:

$$(\mathbf{E}(\lambda)f, f) = \int_{-\infty}^{\lambda} |F(\lambda)|^2 d\sigma(\lambda) \tag{3.22}$$

Since $\sigma(\lambda)$ is a monotonically nondecreasing function, Equation 3.22 implies the following condition:

$$\mathbf{E}(\lambda) \le \mathbf{E}(\mu) \text{ whenever } \lambda \le \mu$$

Summing up these results, we have the following properties of the operator $\mathbf{E}(\lambda)$:

1. $\mathbf{E}(\lambda) \le \mathbf{E}(\lambda), \quad \lambda \le \mu$: nondecreasing
2. $\underset{\lambda\to\infty}{\text{limit}}\ \mathbf{E}(\lambda) = I$, the identity operator } : boundary conditions
3. $\underset{\lambda\to-\infty}{\text{limit}}\ \mathbf{E}(\lambda) = 0$ and }
4. $\underset{\lambda\to 0}{\text{limit}}\ \mathbf{E}(\lambda+\varepsilon)f \to \mathbf{E}(\lambda)f$ for all f and ε : right continuity

The family $\mathbf{E}(\lambda)$ is thus a family of projection operators in a Hilbert space H. This may also be viewed as E being a projection valued function on R into the class of projection operator on H. The relations 1—4 are to be understood in the sense of strong (but not necessarily uniform) operator convergence.

A family of operators $\mathbf{E}(\lambda)$ satisfying these properties is called the “spectral family” or the “resolution of the identity”. If an operator **T** possesses a resolution of the identity $\mathbf{E}(\lambda)$, then von Neumann’s abstract spectral theorem[6] yields a Stieltz integral:

$$\mathbf{T} = \int \lambda \, d\mathbf{E}(\lambda)$$

This generalizes the discrete spectral representation of a finite sequence of projection operators (finite systems) to “continuous” representation on systems not on compact support. If **T** is unbounded, then the above resolution of the identity can be applied only to those functions which are in the domain of **T**. The spectral function $\sigma(\lambda)$ is thus closely connected to the resolution of the identity through Equation 3.21. For further details, the text by Akhiezer and Glazman[5] should be consulted.

VI. SPECTRAL FUNCTIONS ON THE ENTIRE λ-AXIS

The central point in the spectral theory of a differential operator is to establish the nature of the spectrum pertaining to the operator. By the spectrum of an operator is meant the λ-set which is complement to the set of points in the neighborhood of which the spectral function remains constant. Broadly speaking, the points of discontinuity in the behavior of the spectral function constitute the “discrete” or “point” spectrum. The spectral intervals over which the spectral function increases monotonically constitute the "continuous" spectrum. These are the major divisions, each of which may admit of deeper and more subtle structures, depending on the problem. The points of constancy of the spectral function constitute the "regular" points which belong to the resolvent set. Figure 1 illustrates a very general type of relation of the spectrum to the spectral function. The fine details of the spectrum can be found in Reference 5 or in the monograph by Berthier[6] for the Schrödinger equation in particular. An excellent discussion of the subject is by the physicist Paul Roman.[9] It is, therefore, reasonable to say that the behavior of the spectral function on the λ-axis determines the nature of the spectrum. The behavior of the spectral function $\sigma(\lambda)$ of a Sturm-Liouville problem as λ varies over the real axis R will be established in this section.

Let $\lambda = k^2$, $k = u + iv$, and rewrite the Sturm-Liouville equation as:

$$y'' + k^2 y = q(x)y, \quad x \subset R^+$$

$$y(0,\lambda) = \sin\alpha \ \text{ and } \ y'(0,\lambda) = -\cos\alpha \tag{3.23}$$

or

$$y'\sin\alpha + y\cos\alpha = 0, \quad x = 0$$

In Equation 3.23, q(x)y is considered as an inhomogeneous term.

Applying the method of variation of constants to Equation 3.23, the following expression for $y(x,\lambda)$ is obtained:

$$y(x,\lambda) = \cos kx \ \sin\alpha - (1/k) \ \sin kx \ \cos\alpha + (1/k)\int_0^x \sin\{k(x-x')\}q(x')y(x',\lambda)dx' \tag{3.24}$$

It is easy to verify Equation 3.24. The last term equals

$$(1/k)\int_0^x \sin\{k(x-x')\}\left[y''(x',\lambda) + k^2 y(x',\lambda)\right]dx'$$

upon using the differential equation itself. Integrating by parts then yields:

$$\int_0^x \sin\{k(x-x')\}y''(x',\lambda)dx' =$$

$$ky(x,\lambda) + \sin kx \ \cos\alpha - k \cos kx \ \sin\alpha - k^2\int_0^x \sin\{k(x-x')\}y(x',\lambda)dx'$$

Rearranging the terms in the above equation and using the differential equation (3.23) again yields the representation given in Equation 3.24. It is shown in Appendix 2 that for a potential $q(x) \subset L_1(R^+)$ and for $|k| > 0$, $v > 0$, $y_1(x,\lambda)$ remains bounded over R^+. By the construction given in the appendix, it also follows that $y(x,\lambda) \sim O[\exp(vx)]$. The order symbol $f(z) \sim O(g(z))$ means that the function f(z) behaves as g(z) as $z \to \infty$. That is, $[f(z)/g(z)] \to$ constant $< \infty$ as $z \to \infty$. On the other hand, the order symbol $f(z) \sim o(g(z))$ means that $[f(z)/g(z)] \to 0$ as $z \to \infty$. Then, $f(z) \sim o(1)$ implies that $f(z) \to 0$ as $z \to \infty$.

A. THE SPECTRUM ON THE POSITIVE HALF-AXIS: k REAL AND POSITIVE

Let us consider the positive part of the spectrum: $k \subset R^{+}$. Let us rewrite Equation 3.24 in the form:

$$y(x,\lambda) = \cos kx \sin\alpha - (1/k)\sin kx \cos\alpha + (1/k)\int_0^\infty \sin\{k(x - x')\} \cdot q(x')y(x',\lambda)dx' -$$

$$-(1/k)\int_x^\infty \sin\{k(x - x')\}q(x')y(x',\lambda)dx' \tag{3.25}$$

For real, positive k, the asymptotic, i.e., $x \to \infty$, limit of Equation 3.25 is then given by:

$$y(x,\lambda) = \mu(\lambda)\cos kx + \upsilon(\lambda)\sin kx + o(1) \tag{3.26}$$

where,

$$\mu(\lambda) = \sin\alpha - (1/k)\int_0^\infty \sin kx' q(x')y(x',\lambda)dx' \tag{3.27}$$

and

$$\upsilon(\lambda) = -(\cos\alpha/k) + (1/k)\int_0^\infty \cos kx' q(x')y(x',\lambda)dx' \tag{3.28}$$

Equation 3.26 can be further reduced to the form:

$$y(x,\lambda) = U(\lambda)\sin[kx + \delta(\lambda)] + o(1) \tag{3.29}$$

where,

$$U(\lambda) = [\mu(\lambda) + \upsilon(\lambda)]^{1/2} \tag{3.30}$$

and

$$\delta(\lambda) = \sin^{-1}[\mu(\lambda)/U(\lambda)]$$

$$= \cos^{-1}[\upsilon(\lambda)/U(\lambda)]$$

If $k > 0$, then, according to Appendix 2, the integrals of Equations 3.27 and 3.28 converge uniformly in k. $\mu(\lambda)$, $\upsilon(\lambda)$, and, hence, the $U(\lambda)$ are continuous functions of k. This is a well-known result from function theory.[7] Therefore, $\delta(\lambda)$ is a continuous function of λ.

Let us now consider Equation 3.23 with a solution $z(x,\lambda)$ satisfying the boundary condition:

$$z(0,\lambda) = \cos\alpha, \qquad z'(0,\lambda) = -\sin\alpha$$

or

$$z(x,\lambda)\sin\alpha + z'(x,\lambda)\cos\alpha = 0$$

Equations 3.26 to 3.28, corresponding to the eigenfunction z(x,λ), are

$$z(x,\lambda) = \mu^{\sim}(\lambda)\cos kx + \upsilon^{\sim}(\lambda)\sin kx + o(1) \tag{3.31}$$

where,

$$\mu^{\sim}(\lambda) = \cos\alpha - (1/k)\int_0^{\infty} \sin kx' q(x') z(x',\lambda)dx' \tag{3.32}$$

and

$$\upsilon^{\sim}(\lambda) = -(\sin\alpha/k) + (1/k)\int_0^{\infty} \cos kx' q(x')\, z(x',\lambda)dx' \tag{3.33}$$

Next, we want to determine the Wronskian of the eigenfunctions y(x,λ) and z(x,λ). The derivatives of these functions in x are, therefore, needed. Differentiating Equations 3.25 and then taking limit $x \to \infty$ yields:

$$y'(x,\lambda) = -k\mu(\lambda)\sin kx + k\upsilon(\lambda)\cos kx + o(1) \tag{3.34}$$

the prime on y denoting differentiation in x. Similarly,

$$z'(x,\lambda) = -k\,\mu^{\sim}(\lambda)\sin kx + k\,\upsilon^{\sim}(\lambda)\cos kx + o(1) \tag{3.35}$$

y(x,λ) and z(x,λ) are linearly independent and their Wronskian determinant W(y,z) must be independent of x. Furthermore, since Equation 3.1 does not contain the first derivative, then, according to a well-known formula of Liouville,[1] W must be a constant. Evaluating this constant at x = 0 by using Equations 3.26 through 3.35, we obtain:

$$W(y,z) = yz' - zy' = W(y,z)\big|_{x=0}$$

$$= k\left[\mu(\lambda)\upsilon^{\sim}(\lambda) - \mu^{\sim}(\lambda)\upsilon(\lambda)\right] + o(1) = 1$$

This yields:

$$\mu(\lambda)\upsilon\sim(\lambda) - \mu^{\sim}(\lambda)\upsilon(\lambda) \cong \lambda^{-1/2} \tag{3.36}$$

Equation 3.36 shows that μ and υ cannot vanish simultaneously for a particular value of k > 0. Therefore, $U^2(\lambda) > 0$.

Let us now differentiate Equation 3.29 with respect to k and then take the limit $x \to \infty$. Denoting this differentiation by subscript k, we have:

$$y_k(x,\lambda) = \left[U(\lambda)\sin(kx + \delta(\lambda))\right]_k + o(1) \tag{3.37}$$

Equation 3.37 imposes a more restrictive condition on the potential than just the $L_1(R^+)$ integrability assumed thus far. This is due to the differentiation of the $\sin\{k(x - x')\}$ term in k inside the integral of Equation 3.25. The condition that the potential must satisfy for Equation 3.37 to hold is that

$$(1+x)q(x) \subset L_1(R^+)$$

instead of just $q(x) \subset L_1(R^+)$. The spectral function will be derived first under this stronger condition, which will be subsequently relaxed.

Consider the boundary value problem on an interval [0,b] with b being arbitrarily large and a boundary condition $y(b,\lambda) = 0$. The eigenvalues of this problem are then determined from the condition:

$$\sin[kb + \delta(\lambda)] = o(1)$$

The two successive roots are given by:

$$k_1 b + \delta(\lambda) = m\pi + o(1)$$

and

$$k_2 b + \delta(\lambda) = (m+1)\pi + o(1)$$

Since $\delta(\lambda)$ is a continuous function, the separation of two successive eigenvalues is given by:

$$k_2 - k_1 = (\pi/b) + o(1/b) \tag{3.38}$$

Let us consider the x-interval [0,b] and a spectral interval between $\lambda < \lambda_{nb} \leq \lambda + \Delta$ and denote this interval by d. From the definition of the spectral function given by Equation 3.6, we have:

$$\sigma_b(\lambda + \Delta) - \sigma_b(\lambda) = \sum_d r_{n,b}^2$$

$$= \sum_d \left[\left(k_{n+1,b} - k_{n,b}\right)/\left(k_{n+1,b} - k_{n,b}\right)\right] r_{n,b}^2$$

$$= \sum_d \left[\left(\lambda_{n+1,b} - \lambda_{n,b}\right)/\left(k_{n+1,b} + k_{n,b}\right)\right] r_{n,b}^2 \left[(\pi/b) + o(1/b)\right]^{-1}$$

Equation 3.38 has been used in deriving the above result. From Equation 3.29:

$$(1/b)\int_0^b \left|y\left(x,\lambda_{n,b}\right)\right|^2 dx = (1/b) r_{n,b}^{-2} = (1/2)U^2\left(\lambda_{n,b}\right) + o(1)$$

Therefore,

$$\sigma_b(\lambda+\Delta)-\sigma_b(\lambda)=(2/\pi)\sum_d\left[\left\{1/\left(k_{n+1,b}+k_{n,b}\right)\right\}U^{-2}\left(\lambda_{n,b}\right)+o(1)\right]\left(\lambda_{n+1,b}-\lambda_{n,b}\right)$$

Using Equation 3.38, the factor $(k_{n+1,b} + k_{n,b})$ is reduced to $(2k_{n,b} + \pi/b)$. Replacing in the above equation yields:

$$\sigma_b(\lambda+\Delta)-\sigma_b(\lambda)=(1/\pi)\sum_d\left[\left(1/\left(k_{n,b}+\pi/2b\right)\right)U^{-2}\left(\lambda_{n,b}\right)+o(1)\right]\left(\lambda_{n+1,b}-\lambda_{n,b}\right)$$

Passing to the limit $b \to \infty$ then results in the expression for the spectral function (or, rather, the differential spectral function):

$$\sigma(\lambda+\Delta)-\sigma(\lambda)=(1/\pi)\int_{\lambda}^{\lambda+\Delta}\lambda^{-1/2}\,U^{-2}\,d\lambda \tag{3.39}$$

In Equation 3.39, the potential satisfies the strong condition:

$$(1+x)q(x)\subset L_1\left(R^+\right) \tag{3.40}$$

This condition will now be relaxed. Consider the sequence:

$$\begin{aligned} q_n(x) &= q(x), && 0\le x\le n \\ &= 0, && x>n \end{aligned}$$

The functions $q_n(x)$ certainly satisfy the condition (3.40). Then, for the potential $q_n(x)$, we have:

$$\sigma_n(\lambda+\Delta)-\sigma_n(\lambda)=(1/\pi)\int\lambda^{-1/2}\,U_n^{-2}\,d\lambda$$

Passing to the limit $n \to \infty$, Equation 3.39 is obtained. These results are summarized in the theorem below.

Theorem 3.2

Let $q(x) \subset L_1(R^+)$ and $\lambda,\Delta > 0$. Then the spectral function of the boundary value problem (3.23) admits of the representation:

$$\sigma(\lambda+\Delta)-\sigma\ (\lambda)=(1/\pi)\int_{\lambda}^{\lambda+\Delta}\lambda^{-1/2}\,U^{-2}\,d\lambda$$

The spectrum is, therefore, continuous on R^+.

It is interesting to consider the ratio:

$$m(\lambda)=-\left[N(\lambda)/M(\lambda)\right]$$

where $N(\lambda) = \mu^{\sim}(\lambda) + i\,\upsilon^{\sim}(\lambda)$ and $M(\lambda) = \mu(\lambda) + i\,\upsilon(\lambda)$. From the Wronskian relation of Equation 3.36, it is not difficult to establish that:

$$\text{Im }m(\lambda)=-1/\left(kU^2\right) \tag{3.41}$$

Taking the limit Im k = v → 0, we can rewrite Theorem 3.2 as:

$$\sigma(\lambda+\Delta)-\sigma(\lambda)=(1/\pi)\ \lim_{v\to 0}\int_{\lambda}^{\lambda+\Delta}\{-\mathrm{Im}[m(u+iv)]\}\,du \tag{3.42}$$

This is a well-known result of the rigorous application of Weyl-Titchmarsh theory.[4] In this theory, the general solution of a Sturm-Liouville problem over an interval [a,b] is expressed as:

$$Y(x,\lambda)=y(x,\lambda)+l(\lambda,b)z(x,\lambda)$$

where l(λ,b) is a complex parameter. As b → ∞, l(λ,b) → m(λ) which is either the limit point or on the limit circle. Therefore, m(λ) has been defined above in terms of μ, μ~ and υ, υ~, which corresponds to x → ∞ limit.

B. SPECTRAL FUNCTION ON THE NEGATIVE HALF-AXIS: $k^2 \subset R^-$

We now consider complex k's, i.e., k = u + iv, v > 0. Equation 3.24 now becomes:

$$y(x,\lambda)=(1/2)\exp(-ikx)\sin\alpha+(1/2ik)\exp(-ikx)\cos\alpha+$$

$$-(1/2ik)\int_0^x\exp\{-ik(x-x')\}q(x')y(x',\lambda)dx'+$$

$$+\text{ terms}\sim\ O\{\exp(-vx)\}+$$

$$+\text{ terms}\sim\ O\left[\int_0^x\exp\{-v(x-x')\}q(x')y(x',\lambda)\,dx'\right] \tag{3.43}$$

As in Section VI.A., the integration limits are again broken into two intervals, 0 → ∞ and x → ∞. Using the results of Appendix 1, namely, |y(x,λ)| ~ {exp(vx)}, the third term in the above expression can be written as (x → ∞)

$$-(1/2ik)\int_0^\infty\exp\{-ik(x-x')\}q(x')y(x',\lambda)+o\{\exp(vx)\}$$

Using the fact that y(x,λ) ~ O{exp(vx)}, the last term can be written as:

$$O\left[\int_0^x\exp\{-v(x-x')\}|q(x')y(x',\lambda)|dx'\right]$$

$$=O\left[\int_0^x\exp\{-v(x-2x')\}|q(x')|dx'\right]$$

$$=O\left[\exp\{v(x-2d)\}\int_0^{x-d}|q(x')|dx'\right]+O\left[\exp(vx)\int_{x-d}^{x}|q(x')|dx'\right]$$

The interval [0,x] has been divided into intervals [0,x – d] and [x – d,x]. Putting, say, d = x/2 and taking the limit x → ∞, the above term is ~ O(C), where C is a constant. This we can further rewrite as:

$$\exp(-ikx)\{\exp(+ikx)O(c)\} = \exp(-ikx)o(1)$$

Similarly, o{exp(vx)} = exp(–ikx)[exp(ikx)o{exp(vx)}] ~ exp(–ikx) o(1). It also is not difficult to show that exp(ikx))O{exp(–vx)} is at most ~ o(1). For k = u + iv, v > 0, Equation 3.43 then reduces to:

$$y(x,\lambda) = \exp(-ikx)\{M^{\sim}(\lambda) + o(1)\}$$

where,

$$M^{\sim}(\lambda) = (1/2)\sin\alpha + (1/2ik)\cos\alpha - (1/2ik)\int_0^\infty \exp(ikx')q(x')y(x',\lambda)dx' \qquad (3.44)$$

The corresponding expressions for z(x,λ) is

$$z(x,\lambda) = \exp(-ikx)\{N^{\sim}(\lambda) + o(1)\}$$

where,

$$N^{\sim}(\lambda) = (1/2)\cos\alpha - (1/2ik)\sin\alpha - (1/2ik)\int_0^\infty \exp(ikx')q(x')z(x',\lambda)dx' \qquad (3.45)$$

Equations 3.44 and 3.45 show the functions $M^{\sim}(\lambda)$ and $N^{\sim}(\lambda)$ to be analytic in the upper half-plane, i.e., for k = u + iv, v > 0. In the limit u → 0, the functions $M^{\sim}(\lambda)$ and $N^{\sim}(\lambda)$ reduce to real limits:

$$M^{\sim}(\lambda) = (1/2)\sin\alpha - (1/2v)\cos\alpha + (1/2ik)\int_0^\infty \exp(-vx')q(x')\, y(x',\lambda)dx' \qquad (3.46)$$

and

$$N^{\sim}(\lambda) = (1/2)\cos\alpha + (1/2v)\sin\alpha + (1/2v)\int_0^\infty \exp(-vx')q(x')\, z(x',\lambda)dx' \qquad (3.47)$$

We now form the function $m(\lambda) = -N^{\sim}(\lambda)/M^{\sim}(\lambda)$ and apply Equation 3.42. In view of the limits given by Equations 3.46 and 3.47, Im m(λ) is zero except at the zeros of $M^{\sim}$. Now $M^{\sim}(\lambda)$ is analytic in the upper half-plane v > 0. The theory of analytic functions then tells us that these zeros must be isolated.[7] Thus, σ(λ) can only have a discrete set of points of jump discontinuity, $-\infty < \lambda < 0$. Also, from Equation 3.46, $M^{\sim} \to (1/2)\sin\alpha$ and $N^{\sim} \to (1/2)\cos\alpha$ as $k \to -\infty$. The discrete spectrum is thus bounded from below. Moreover, q(x) must belong to $L_1[0,\infty)$ as this condition is assumed in establishing the bounds for the functions considered (Appendix 1). If it so happens that $(1 + x^2)q(x) \subseteq L_1[0,\infty)$, then the number of discrete eigenvalues on the negative real axis is finite. In either case, the discreteness of the spectrum corresponds to the fact that the number of zeros of the eigenfunction is finite on the semiaxis $x \subset L_1[0,\infty)$. This is in accordance with Sturm's oscillation theorem. For a rigorous analysis of these points, the monograph of Levitan and Saragsjan[11] should be consulted.

Summarizing, we have the following theorem.

Theorem 3.3

If $q(x) \subset L_1(R^+)$, then the negative part of the spectrum of Problem 3.23 is discrete and bounded from below. If, in addition, $(1 + x^2)q(x) \subset L_1[0,\infty)$, then the number of negative eigenvalues is finite.

VII. COMPLETENESS RELATION

The contents of Sections II, III, and V can be summarized into a completeness relation for the singular Sturm-Liouville problem on the semiaxis $[0,\infty)$. For the spectral function $\sigma(\lambda)$, we have:

$$d\sigma(\lambda) = (1/\pi)\lambda^{-1/2}U^{-2}d\lambda, \qquad 0 < \lambda < \infty$$

$$= \sum_j r_j^2\, \delta(\lambda - \lambda_j)\, d\lambda, \qquad -\infty < \lambda < 0 \tag{3.48}$$

$U(\lambda)$ is defined by Equation 3.30 and r_j^2 is a suitable normalization constant. From the generalized Fourier transform $F(\lambda)$ of Equation 3.16 and the expansion Theorem 3.1, the eigenfunction expansion of an arbitrary $f(x) \subset L_2[0,\infty)$ can be written as:

$$f(x) = \int_{-\infty}^{\infty} \left\{ \int_0^{\infty} f(x')y(x',\lambda)dx' \right\} y(x,\lambda)d\sigma(\lambda) \tag{3.49}$$

Using the expressions for $d\sigma(\lambda)$ given by Equation 3.48 yields:

$$f(x) = (1/\pi)\int_0^{\infty} \lambda^{-1/2}\, U^{-2}\left[\int_0^{\infty} f(x')y(x',\lambda)dx' \right] y(x,\lambda) +$$

$$+ \sum_j r_j^2 y(x,\lambda_j) \int_0^{\infty} f(x')y(x',\lambda_j)dx' \tag{3.50}$$

Since:

$$f(x) = \int_{-\infty}^{\infty} f(x')\, \delta(x - x')dx'$$

the desired completeness relation is obtained by comparing with Equation 3.50, namely:

$$\delta(x - x') = (1/\pi)\int_0^{\infty} \lambda^{-1/2}\, U^{-2} y(x',\lambda)y(x,\lambda)d\lambda +$$

$$+ \sum_j r_j^2 y(x,\lambda_j)y(x',\lambda_j) \tag{3.51}$$

This completeness relation (Equation 3.51) will play a crucial role in the development of Gel'fand-Levitan theory.

VIII. EXAMPLES

We will consider some examples illustrating the spectral function and the spectrum. Another example will be given after the Povzner-Levitan transform is derived in Chapter 4.

Example 1

The simplest illustration of the spectral function and the spectrum is by considering a scattering-free problem, that is, a Sturm-Liouville problem with the potential q(x) set to zero. Let us, therefore, determine the spectral function and the spectrum of the following boundary value problem:

$$y'' + \lambda y = 0$$

with a general boundary condition given by either

$$z(0) = \cos\alpha, \quad z'(0) = \sin\alpha \tag{3.52}$$

or

$$y(0) = \sin\alpha, \quad y'(0) = -\cos\alpha \tag{3.53}$$

where α is arbitrary. We will first determine the function $m(\lambda)$ (Section VI.A.) and obtain the spectrum. Then, using this value of $m(\lambda)$ in Equation 3.42, the spectral function will be obtained.

Equations 3.26 and 3.31 yield the eigenfunctions $y(x,\lambda)$ and $z(x,\lambda)$, respectively. These are

$$z(x,\lambda) = \cos\alpha\cos\lambda^{1/2}x + \lambda^{-1/2}\sin\alpha\sin\lambda^{1/2}x$$

$$y(x,\lambda) = \sin\alpha\cos\lambda^{1/2}x - \lambda^{-1/2}\cos\alpha\sin\lambda^{1/2}x$$

We consider a solution which is $L_2(R^+)$. This solution is of the form:

$$Y(x,\lambda) = y(x,\lambda) + m(\lambda)z(x,\lambda) \tag{3.54}$$

We consider λ to be complex such that $\lambda = s^2 = u + iv$, $v > 0$. The solution to the scattering-free equation is $\exp(+i\lambda^{1/2}x)$. However, for $v > 0$, $\exp(-i\lambda^{1/2}x)$ cannot be a solution since this is a generalized eigenfunction, and does not belong to the Hilbert space $L_2(R^+)$. Comparing Equation 3.54 and $\exp(+i\lambda^{1/2}x)$, we obtain for $m(\lambda)$ the following expression:

$$m(\lambda) = [\sin\alpha - i\lambda^{1/2}\cos\alpha]/[\cos\alpha + i\lambda^{1/2}\sin\alpha] = [1 - i\lambda^{1/2}\cot\alpha]/[\cot\alpha + i\lambda^{1/2}] \tag{3.55}$$

Writing $h = \cot\alpha$, Equation 3.55 becomes:

$$m(\lambda) = [1 - i\lambda^{1/2}h]/[h + i\lambda^{1/2}] \tag{3.56}$$

The spectrum will depend upon whether h is greater or less than zero. If $h > 0$, then the denominator in Equation 3.56 vanishes at $\lambda = -h^2$. The scattering-free problem, therefore, has one eigenvalue $\lambda = -h^2$. The corresponding normalized eigenfunction is $[\{\exp(-hx)\}/(2h)^{1/2}]$. If $h < 0$, then the denominator is regular and for $\lambda < 0$, no eigenvalue results. Calculating for Im $m(\lambda)$ from Equation 3.56, we obtain, for $h \leq 0$:

$$-\text{Im } m(\lambda) = \lambda^{1/2}/\left[\cos^2\alpha + \lambda\sin^2\alpha\right], \quad \lambda > 0$$
$$= 0, \quad \lambda < 0$$

The negative spectrum is absent and on R^+ the spectrum is continuous. From Equation 3.42, we then obtain the spectral function or, rather, the spectral measure:

$$d\sigma(\lambda) = \lambda^{1/2}\, d\lambda/\pi\left[\cos^2\alpha + \lambda\sin^2\alpha\right]$$

The continuous spectrum can also been determined using Theorem 3.2, with U calculated from Equations 3.27, 3.28, and 3.30. In this case:

$$\mu(\lambda) = \sin\alpha \quad \text{and} \quad \upsilon(\lambda) = -\cos\alpha/\lambda^{1/2}$$

and hence:

$$d\sigma(\lambda) = \lambda^{-1/2}/\pi\left[\mu^2(\lambda) + v^2(\lambda)\right]$$

yielding the same result.

In summary, the boundary value problem has a spectrum which is continuous on the positive real axis and possesses only one eigenvalue on the negative real axis at the point $-h^2$. Its spectral measure is given by:

$$d\rho(\lambda) = \lambda^{1/2}\, d\lambda/\pi\left[\cos^2\alpha + \lambda\sin^2\alpha\right]$$

A function $f(x) \subset L_2(R^+)$ can be expanded as:

$$f(x) = (1/2h)\exp(-hx)\int_0^\infty f(t)\exp(-ht)dt + (1/\pi)\int_0^\infty F(\lambda)\{\sin\alpha\cos\lambda^{1/2}x -$$

$$-\lambda^{-1/2}\cos\alpha\sin\lambda^{1/2}x\}\left[\lambda^{1/2}\left(\cos^2\alpha + \lambda\sin^2\alpha\right)^{-1}\right]d\lambda$$

Example 2: The Hydrogen Atom[1]

Consider the equation for the radial wave function:

$$R'' + (2/r)R' - \left[l(l+1)/r^2\right]R + \left[E + (2/r)\right]R = 0, \qquad r \subset R^+$$

where the primes refer to differentiation in the radial coordinate r and R is the radial wavefunction. The angular momentum $l \subset Z^+$, the space of positive integers, and E, the energy. This is the quantum mechanical equation on the basis of which the energy levels of the hydrogen atoms are calculated.[8] It is a standard procedure to reduce the radial equation by the substitution R = y/r. In terms of r, the radial equation becomes:

$$y'' - \left[\{l(l+1)/r^2\} - \{E + (2/r)\}\right]y = 0 \tag{3.57}$$

This was encountered in Chapter 1.

Let us consider the equation for the positive values of the energy and set $E = s^2$. It is well known in quantum scattering that Equation 3.57 admits of a solution which remains bounded at the origin and which, in the asymptotic limit $s \to \infty$, reads:

$$y = \left[\{\exp(-\pi/2s)\}/s\right]\cos\left[sr + (1/s)\ln 2sr - (l+1)(\pi/2) + \alpha\right]/\left|\Gamma(l+1+i/s)\right| \tag{3.58}$$

where,

$$\alpha = \arg\Gamma(l+1+i/s)$$

This was encountered in Chapter 1.

Let us follow the procedure of Section 3.5 and consider the boundary value problem for the hydrogen atom on the half-line by first considering it as a problem on a finite interval [0,b] and then letting $b \to \infty$. As b becomes large, the wave function (3.58) becomes highly oscillatory and in the limit $b \to \infty$, the distance between the successive zeros of the solution vanishes. Thus, if s_1 and s_2 are two successive zeros, then $(s_2 - s_1) \to 0$ as $b \to \infty$. The argument of the cosine function yields:

$$(s_2 - s_1)b + (1/s_2)\ln 2bs_2 - (1/s_1)\ln 2bs_1 = \pi + o(1) \tag{3.59}$$

as in Equation 3.38. This then gives:

$$s_2 - s_1 = (\pi/b) + o(\ln b/b) \tag{3.60}$$

Since ln b is a slowly varying function compared to b itself, the order estimate o(ln b/b) ~ o(1/b). This can be seen more rigorously by substituting Equation 3.60 in Equation 3.59, which yields

$$(s_2 - s_1)b + (1/s_2 s_1)(s_2 - s_1)\ln b$$

$$= (s_2 - s_1)b + (\pi/bs_2 s_1)\ln b + o(\ln 2b/b) = \pi + o(1)$$

From this, we obtain the estimate:

$$s_2 - s_1 = (\pi/b) + o(1/b) \tag{3.61}$$

In order to calculate the spectral measure, we use the basic definition of the spectral function given by Equation 3.6. We, therefore, write:

$$\Delta\rho_b(s) = \rho_b(s+\Delta) - \rho_b(s) = \sum_d 1 \Big/ \int_a^b y^2(r, s_n)\, dr$$

$$= (1/\pi)\sum_d (s_{n+1} - s_n)\, s_n^2 \exp(\pi/s_n)\left|\Gamma(1+1+i/s_n)\right|^2 \cdot$$

$$\cdot\left(\left[b \Big/ \left\{\int_0^b \cos^2\left[s_n r + (1/s_n)\ln 2s_n r - (1+1)(\pi/2) + \alpha\right]\right\} dr\right]\right) + o(1) \qquad (3.62)$$

The summation over $d \equiv s < s_n \leq s + \Delta$. In deriving the last step, we have used the asymptotic estimate given by Equation 3.61. We use the trigonometric identity:

$$\cos^2 x = (1/2)[1 + \cos 2x]$$

The integral in Equation 3.62 then is of the form:

$$\int_a^b \cos\left[2sr + (2/s)\ln 2sr + 2\alpha\right] dr$$

We then have:

$$\int_a^b \cos\left[2sr + (2/s)\ln 2sr + 2\alpha\right] dr$$

$$= \int_a^b \cos 2sr\ \cos\left[(2/s)\ln 2sr + 2\alpha\right] dr - \int_a^b \sin 2sr\ \sin\left[(2/s)\ln 2sr + 2\alpha\right] dr$$

Since both integrals on the right-hand side can be estimated in the same way, we consider only one of them, say, the cosine part. Integrating the cosine integral on the right-hand side twice by parts yields:

$$\int_a^b \cos 2sr \cos\left[(2/s)\ln 2sr + 2\alpha\right] dr = O(1) + \left(1/2s^3\right) \cdot$$

$$\cdot \int_a^b \cos 2sr \left\{-\left(1/r^2\right)\sin\left[(2/s)\ln 2sr + \alpha\right] + \left(1/2sr^2\right)\cos\left[(2/s)\ln 2sr + \alpha\right]\right\} dr = o(1)$$

The term O(1) results from the lower limit, the integrated terms vanishing at the upper limit $b \to \infty$, being highly oscillatory there. Replacing this estimate in Equation.3.62 and using the asymptotic estimate of Equation 3.61 again then yields:

$$\rho(s+\Delta)-\rho(s)=(2/\pi)\sum_{d}(s_{n+1}-s_n)\, s_n^2 \exp(\pi/s_n)\left|\Gamma(1+1+i/s_n)\right|^2\{1+o(1)\}$$

$$\to (2/\pi)\int_{s}^{s+\Delta} s^2 \exp(\pi/s)\left|\Gamma\left(1+1+i/s\right)\right|^2 \qquad (3.63)$$

The o(1) term in Equation (3.63) results due to the fact that we are considering the difference of two quantities each of which is ~O(1). The gamma function can be reduced using the familiar identities:

$$\Gamma(z+1)=z\Gamma(z), \quad \Gamma(z)\Gamma(1-z)=\pi/\sin\pi z$$

Reducing the Γ-function, we have:

$$\left|\Gamma\{1+1+(i/s)\}\right|^2=\Gamma\{1+1+(i/s)\}\Gamma\{1+1-(i/s)\}$$

$$=(1+i/s)\ldots(1+i/s)(i/s)(1-i/s)\ldots(1-i/s)(i/s)\Gamma\{1-(i/s)\}$$

$$=(\pi/s)\left[2/\{\exp(\pi/s)-\exp(-\pi/s)\}\right]\prod_{n=1}^{1}\left(n^2+s^{-2}\right)$$

Replacing these results in Equation 3.63, we finally obtain the spectral measure:

$$\Delta\rho(s)=\int_{s}^{s+\Delta}\left[1-\exp(-2\pi/s)\right]^{-1} 4s\prod_{n=1}^{1}\left(n^2+s^{-2}\right)$$

The spectrum is, therefore, continuous. This is, of course, a familiar result from the quantum mechanical treatment of the hydrogen atom.

APPENDIX 1

From Equation 3.26:

$$y(x,\lambda)=\cos kx \sin\alpha - (1/k)\sin kx\ \cos\alpha+(1/k)\int_0^x \sin\{k(x-x')\}\, q(x')y(x',\lambda)\, dx' \qquad (A2.1)$$

Consider $k = u + iv$, $v > 0$ and set:

$$y(x,\lambda)=w(x,\lambda)\exp(vx)$$

Equation A2.1 then becomes:

$$w(x,\lambda) = \left[\cos kx \sin\alpha - (1/k)\sin kx \ \cos\alpha\right]\exp(-vx) +$$

$$+(1/k)\int_0^x \exp\{-v(x-x')\}\sin\{k(x-x')\}q(x')w(x',\lambda)\ dx' \qquad \text{(A2.2)}$$

Applying De Moiver's relations for the cosine and the sine terms, we obtain:

$$|\cos kx|, \quad \sin|kx| \le \exp(vx)$$

From Equation A2.2, the following inequality is then obtained:

$$|w(x,\lambda)| \le \left[1+(1/|k|)\right] + (1/|k|)\int_0^x |q(x')|\ w(x',\lambda)\ dx' \qquad \text{(A2.3)}$$

Writing 1 + (1/|k|) = C and |q(x′)|/|k| = p(x′), the inequality (A2.3) can be rewritten as:

$$|w(x,\lambda)| \le C + \int_0^x p(x')|w(x',\lambda)|dx' \qquad \text{(A2.4)}$$

Define:

$$z(x,\lambda) = \int_0^x p(x')|w(x',\lambda)|dx'$$

Then,

$$z'(x,\lambda) = p(x)|w(x,\lambda)| \le C\ p(x) \ + p(x)z(x,\lambda) \qquad \text{(A2.5)}$$

Let:

$$s(x,\lambda) = z(x,\lambda)\exp\left(-\int_0^x p(x')\ dx'\right) \qquad \text{(A2.6)}$$

Differentiating Equation A2.6 with respect to x and using Equation A2.5 yields:

$$s'(x,\lambda) \le C\ p(x)\ \exp\left[-\int_0^x p(x')dx'\right]$$

Integrating this inequality between 0 to x gives:

$$s(x) \le C\left[1-\exp\left(-\int_0^x p(x')dx'\right)\right] \qquad \text{(A2.7)}$$

Using the inequality (A2.7) in the definition (A2.6), and upon differentiation results in the relation:

$$z'(x) \leq Cp(x)\left[s(x)/z(x)\right]^{-1} \tag{A2.8}$$

Comparing Equations A2.5 and A2.8, we obtain:

$$p(x)\left|w(x,\lambda)\right| \leq C\, p(x)\exp\left[\int_0^x p(x')dx'\right]$$

yielding:

$$\left|w(x,\lambda)\right| \leq \left[1+(1/|k|)\right]\exp\left[(1/|k|)\int_0^x \left|q(x')\right| dx'\right]$$

If the potential $q(x) \subset L_1(R^+)$, then $w(x,\lambda)$ is bounded for all $x \subset R^+$ for $|k| > 0$, $v > 0$.

REFERENCES

1. **B. M. Levitan and I. S. Saragsjan:** *Introduction To Spectral Theory*, American Mathematical Society Monograph, Vol. 39, American Mathematical Society, 1975.
2. **E. Hille:** *Lectures on Ordinary Differential Equations*, Addison-Wesley, Reading, MA, 1968.
3. **E. A. Coddington and N. Levinson:** *Theory of Ordinary Differential Equations*, McGraw-Hill, New York, 1955.
4. **E. C. Titchmarsh:** *Eigenfunction Expansions*, Vol. 1, Oxford University Press, New York, 1946.
5. **N. I. Akhiezer and I. M. Glazman:** *Linear Operators In Hilbert Space*, Ungar, New York, 1961.
6. **A. M. Berthier:** *Spectral Theory and Wave Operators for the Schrodinger Equation*, Pitman Publishing, Boston, 1982.
7. **E. C. Titchmarsh:** *Theory of Functions*, Clarendon Press, Oxford, 1939.
8. **A. Messiah:** *Quantum Mechanics*, John Wiley & Sons, New York, 1958.
9. **P. Roman:** *Some Modern Mathematics for Physicists and Other Outsiders*, Vol. 2, Pergamon Press, 1975.

Chapter 4

THE POVZNER-LEVITAN TRANSFORM

I. INTRODUCTION

The theory of Gel'fand-Levitan uses a certain integral representation of the eigenfunction. In the previous chapter, the eigenfunction was represented by a Volterra integral equation (Equation 3.26). The kernel $\{\sin k(x-x')/k\}$ in this representation was dependent not only on the space variables, but also on k. Moreover, the integrand contained the potential as well as the eigenfunction itself. What is desired is an integral representation, preferably with a k-independent kernel operating on an eigenfunction of a given auxiliary problem. Specifically, the representation being sought is of the form

$$y(x,\lambda) = Y(x,\lambda) + \int_0^x K(x,x')Y(x',\lambda)\,dx'$$

$Y(x,\lambda)$ is the eigenfunction of an auxiliary problem. In the context of scattering, this is usually the "free" problem, i.e., one without the scattering potential. The eigenfunction $Y(x,\lambda)$ then becomes the "free" eigenfunction, which is presumably known. This representation of the eigenfunction is known as the "Povzner-Levitan transform". Such a representation is not intuitively obvious and in this section we discuss how it is obtained.

The transformation was first obtained by Povzner[1] and Levitan[2] via a "generalized displacement operator" which was introduced earlier by the French mathematician Delsartes.[3] The original derivation by Povzner and Levitan is mathematically complex and far more general than is necessary for the present purpose. A simpler derivation is due to Sabatier[4] via an application of the Paley-Wiener or its variation, the Wiener-Boas, theorem. Povzner and Levitan's original derivation will be outlined first without due consideration to mathematical rigor. This will be followed by Sabatier's derivation. Finally, we will present a derivation due to Marcenko[5] which uses neither the displacement operator nor the function theoretic results, but develops the transformation operators from the fundamental Cauchy problems associated with a Sturm-Liouville equation.

Under consideration is, of course, the Sturm-Liouville boundary value problem

$$y''(x,\lambda) + \{\lambda - q(x)\}y(x,\lambda) = 0, \quad x \subset R^+ \tag{4.1}$$

with a general boundary condition of the form

$$y'(0,\lambda) - hy(0,\lambda) = 0 \tag{4.2}$$

Two specific boundary conditions will be considered:

$$h = 0: \quad y(0,\lambda) = 1 \text{ and } y'(0,\lambda) = 0 \tag{4.3}$$

and

$$h = \infty: \quad y(0,\lambda) = 0 \text{ and } y'(0,\lambda) = 1 \tag{4.4}$$

We now discuss the derivation of the Povzner-Levitan transform via the generalized displacement operator. The basic idea is as follows. Consider an operator of the form

$$\mathbf{T}_x{}^y = \sum_{n=0}^{\infty} z_n(y)\mathbf{D}^{(n)}(x)$$

where $z_n(y)$ is a sequence of functions of y and $\mathbf{D}^{(n)}(x)$ is the n-th iteration of a linear operator $\mathbf{D}(x)$ which acts on a function of x. An operator such as $\mathbf{T}_x{}^y$, when acting on a function of x(y), results in a function of two variables x and y. The operator displaces the argument x(y) through y(x). Taylor's expansion is a special case of such a displacement operator. Now if **D** is replaced by a Sturm-Liouville operator **L** and the boundary value, say, at y = 0 is given by a function f(x), then the function $u(x,y) = \mathbf{T}_x{}^y f(x)$ satisfies a Cauchy problem:

$$\mathbf{L}(x)u(x,y) = \mathbf{L}(y)u(x,y)$$

If, furthermore, f(x) is chosen to be, say, an eigenfunction $w(x,\lambda)$ of $\mathbf{L}(x)$, then it turns out that $\mathbf{T}_x{}^y w(x,\lambda) = w(x,\lambda)w(y,\lambda)$. Equating this factorized solution to the general solution of the Cauchy problem and then manipulating the conditions of the boundary value problem such as $q(y) = 0$ or $q(x) = 0$, the Povzner-Levitan transforms and their inverses are obtained for the appropriate boundary conditions of the Cauchy problem. This will become clear as we proceed.

II. THE GENERALIZED DISPLACEMENT OPERATOR

Let A be a linear space of functions of a real variable x and **D** be a linear operator defined on A having a continuous spectrum S on some region of the complex λ-plane. For each $\lambda \subset S$, let there be a function $j(x,\lambda) \subset A$ such that

$$\mathbf{D}(x)[j(x,\lambda)] = \lambda j(x,\lambda) \tag{4.5}$$

Let us impose the condition that $j(0,\lambda) = 1$. $\mathbf{D}(x)$ has been written for **D** in order to emphasize that the operator is operating on a function of x. Let $\{z_n(x)\}$ be a sequence of functions such that

$$z_0(0) = 1 \text{ and } z_n(0) = 0 \tag{4.6}$$

Let us expand $j(x,\lambda)$ as

$$j(x,\lambda) = \sum_{n=0}^{\infty} \lambda^n z_n(x)$$

If the following recurrence relations hold good, namely,

$$\mathbf{D}(x)[z_0(x)] = 0 \text{ and } \mathbf{D}(x)[z_n(x)] = z_{n-1}(x) \tag{4.6a}$$

then the eigenvalue problem (4.5) with the condition $j(0,\lambda) = 1$ is satisfied by virtue of the initial conditions on the sequence $\{z_n(x)\}$ given by Equations 4.6. An example of these procedures is provided by the functions $\exp(\lambda x)$, $x \subset R$ with $\mathbf{D} = d/dx$. z_n in this series is

$x^n/n!$. The spectrum in this case is the entire complex plane. In general, the sequence must be selected so as to satisfy the boundary value problem for the linear operator. Another example is provided by the function

$$j(x,\lambda) = \left[2^p p!/\left(ix\lambda^{1/2}\right)^p\right] J_p\left(ix\lambda^{1/2}\right)$$

with $\mathbf{D}_x = \partial^2/\partial x^2 + ((2p + 1)/x)\partial/\partial x$, p being complex. J_p is the Bessel function of the first kind. The sequence z_n in this case is given by

$$z_n(x) = \left[p!/n!(p+n)!\right](x/2)^{2n}$$

Delsartes defines an operator $\mathbf{T}^y$ such that

$$\mathbf{T}^y = \sum_{n=0}^{\infty} z_n(y)\mathbf{D}^{(n)} \tag{4.7}$$

More specifically,

$$\mathbf{T}_x^y = \sum_{n=0}^{\infty} z_n(y)\mathbf{D}^{(n)}(x) \tag{4.7a}$$

where $\mathbf{D}^{(n)}(x)$ is the operator $\mathbf{D}(x)$ applied n times in succession. $\mathbf{D}^{(0)}(x) = \mathbf{I}$ is the identity operator. In order to simplify notations, $\mathbf{T}_x{}^y$ will be replaced by $\mathbf{T}(x,y)$. The variable y, which can be considered as a parameter in the above equation, is not to be confused with the wavefunction $y(x,\lambda)$ of the boundary value problem (4.1 and 4.2).

Let f(x) be a function of the real variable x and belong to the space L_p ($p \geq 1$). Operating on f(x) by the operator $\mathbf{T}(x,y)$ then yields

$$\mathbf{T}(x,y)f(x) = \sum_{n=0}^{\infty} z_n(y)\mathbf{D}^{(n)}(x)f(x) \tag{4.8}$$

To every function $f(x) \subset L_p$ ($p \geq 1$) of a single real variable x, $\mathbf{T}(x,y)$ associates with it another function of two variables x and y. As an example, let the sequence $z_n(y)$ be defined as

$$z_n(y) = \left[y^n/n!\right]$$

and let the operator $\mathbf{D}(x)$ be the operator of differentiation d/dx. Then Equation 4.8 yields:

$$\mathbf{T}(x,y)f(x) = \sum_{n=0}^{\infty} \left[y^n/n!\right] f^{(n)}(x) = f(x+y) \tag{4.9}$$

Thus, the action of $\mathbf{T}(x,y)$ as defined by Equation 4.8 translates the function through y, that is, displaces the argument from x to x + y. The recurrence relations (4.6a) are satisfied in this case, as can be seen by inspection. Equation 4.9 is easily recognized to be Taylor's expansion of the function f around the point x. It will be seen that Equation 4.8 results in general in displacements of the arguments. The operator $\mathbf{T}(x,y)$ of Equation (4.8) is thus a generalization of Taylor's expansion to a linear operator $\mathbf{D}(x)$. Delsartes, therefore, terms the operator **T**(x,y) the "generalized displacement operator" (g.d.o.). A wider generalization of g.d.o.s is by taking into account their algebraic nature. These more general classes of operators $\mathbf{T}^s$ satisfy some axioms which, in particular, hold for operators of group translation. Thus, for example, $\mathbf{T}^s f(t) = f(t \cdot s)$, s, t ⊂ G. G is a topological group with t·s as the group product. It may be recalled that the group multiplication t·s can be the simple addition t + s. An example is the finite Abelian group Z_n whose elements are integers 0,1,2,..., n–1 for fixed n ≥ 1. For an exhaustive discussion of the g.d.o.s, see Reference 12.

The properties of the operator **T**(x,y) are discussed in rigorous detail by Delsartes[6] and Levitan.[2] Levitan shows that if f(x) ⊂ L_p, then the function **T**(x,y)f(x) is integrable to the p-th power, that is,

$$\left[\int_R \left|\mathbf{T}(x,y)f(x)\right|^p dx\right]^{1/p} \leq A(p,y)\left[\int_R \left|f(x)\right|^p dx\right]^{1/p}$$

where A(p,y) is a positive, locally integrable function on R. **T**(x,y) possesses the properties of linearity, associativity, commutativity, continuity, boundedness, and normality. We will not go into these details, which appear in the above two references, Reference 2 in particular. The property which is important for the present purpose is that of symmetry, namely,

$$\mathbf{T}(x,y)f(x) = \mathbf{T}(y,x)f(y)$$

This property is almost intuitive since **T**(x,y) is a displacement operator. The symmetry property which will be useful in establishing the boundary value problem for **T**(x,y) can be obtained as a corollary to the properties of associativity and commutativity, which are written below.

Associativity: $\mathbf{T}(y,x')\mathbf{T}(x,y)f(x) = \mathbf{T}(x,y)\mathbf{T}(x,x')f(x)$

Commutativity: $\mathbf{T}(x,y)\mathbf{T}(x,x')f(x) = \mathbf{T}(x,x')\mathbf{T}(x,y)f(x)$

Recall that **T**(u,v)f(u) essentially displaces the argument u through v. In view of the commutativity condition, the associativity becomes

$$\mathbf{T}(y,x')\mathbf{T}(x,y)f(x) = \mathbf{T}(x,x')\mathbf{T}(x,y)f(x) \tag{4.10}$$

The family of operators $\mathbf{T}^s$ contains, by definition, a unit element e such that[2]

$$\mathbf{T}(t,e)f(t) = f(t) \text{ and } \mathbf{T}(e,s)f(t) = f(s)$$

Applying this condition to the modified associativity given by the identity (4.10), the symmetry relation results, namely,

$$\mathbf{T}(x,y)f(x) = \mathbf{T}(y,x)f(y) \tag{4.10a}$$

III. THE BOUNDARY-VALUE PROBLEM FOR T(x,y)

Let us consider the action of the displacement operator **T**(x,y) on a function f(x) ⊂ L_p, as given by Equation 4.8. This simply means that **T**(x,y)f(x) is a function of two variables x and y. The action of **T**(x,y) on a function f(x) can be considered as a function of y with x in the role of a parameter. Delsartes defines the transpose of the operation in (4.7) as

$$\mathbf{T}^{\sim}(x,y) = \sum_{n=0}^{\infty} z_n(x)\mathbf{D}^{(n)}(y) \tag{4.11}$$

From Equation 4.11, and using condition (4.6a) on $z_n(x)$, one obtains:

$$\mathbf{D}(x)\mathbf{T}^{\sim}(x,y) = \sum_{n=0}^{\infty} z_n(x)\mathbf{D}^{(n+1)}(y) \tag{4.12}$$

Next, consider the action of $\mathbf{D}(y)$ on $\mathbf{T}(y,x)$. This gives

$$\mathbf{D}(y)\mathbf{T}(x,y) = \sum_{n=0}^{\infty} z_n(x)\mathbf{D}^{(n+1)}(y) \tag{4.13}$$

Comparing Equations 4.12 and 4.13 yields

$$\mathbf{D}(x)\mathbf{T}^{\sim}(x,y) = \mathbf{D}(y)\mathbf{T}(x,y) \tag{4.14}$$

For real **T**, i.e., **T** mapping real functions into real functions, Levitan[2] proves that $\mathbf{T}^{\sim} = \mathbf{T}$. More rigorously, the transpose or the adjoint $\mathbf{T}^{s\sim}$ to $\mathbf{T}^{s}$ is defined through an inner product relationship, namely,

$$\left[(\mathbf{T}(t,s)f(t)), g(t)\right]_{m(t)} = \left[f(t), \mathbf{T}^{\sim}(t,s)g(t)\right]_{m(t)}$$

m(t) being an appropriate measure and $\mathbf{T}^{s}$: R → R. The self-adjointness then is shown to be a consequence of the commutativity property. Using this fact in Equation 4.14 and from the symmetry condition established earlier Equation 4.10, one obtains

$$\mathbf{D}(x)u(x,y) = \mathbf{D}(y)u(x,y) \tag{4.15}$$

where $\mathbf{T}(x,y)f(x)$ has been written as u(x,y).

At this point, we would like to identify the operator D with the Sturm-Liouville operator which is of interest here. Thus,

$$\mathbf{D}(x) = \mathbf{L}(x) = d^2/dx^2 - q(x)$$

q(x) being a continuous real function of the real variable x. For a Sturm-Liouville problem such as

$$\mathbf{L}(x)y = \lambda y,\ y(0,\lambda) = 1,\ y'(0,\lambda) = 0$$

the steps outlined in Equations 4.5 through 4.8 results in

$$y(x,\lambda) = \sum_{n=0}^{\infty} \lambda^n y_n(x)$$

$$\mathbf{L}(x)y_n(x) = y_{n-1}(x),\ \mathbf{L}(x)y_0(x) = 0$$

and

$$\mathbf{T}(x,y)f(x) = \sum_{n=0}^{\infty} y_n(y)[\mathbf{L}(x)]^n f(x) \tag{4.15a}$$

The following equation is obtained from Equation 4.15 for the operator **T**(x,y) or, for that matter, for **T**(x,y)f(x) = u(x,y):

$$u''(x,y) - q(x)u(x,y) = \ddot{u}(x,y) - q(y)u(x,y) \tag{4.16}$$

where $'$ and $\cdot$ denote, as before, differentiations with respect to x and y, respectively.

Now **T**(x,0) = **I**, the identity operator as can be seen by using condition (4.6) in Equation (4.15a). Then for the problem for **L**(x) described above, we have the following boundary value problem:

$$h = 0: \quad u(x,0) = f(x), \quad \dot{u}(x,0) = 0 \tag{4.17}$$

Similarly, for the second boundary condition, we obtain

$$h = \infty: \quad u(x,0) = 0, \quad \dot{u}(x,0) = f(x) \tag{4.18}$$

IV. THE SOLUTION OF THE BOUNDARY-VALUE PROBLEM

We now consider the Cauchy problems (4.16 to 4.18). The boundary conditions h = 0 and h = ∞ are considered separately. Let us summarize at this point Riemann's method of integration of the Cauchy problem with the Cauchy data u(x,0) = f(x) and $\dot{u}$(x,0) = g(x). Riemann's solution consists of determining an operator which acts upon this Cauchy data and yields the solution of the Cauchy Problem. Riemann's solution of the Cauchy problem at a point (X,Y) consists in propagating the Cauchy data along the characteristics passing through the point. The characteristics follow the equations $x \mp y = X \mp Y$. The solution is obtained in the form of an integral equation with a kernel which is a function of four parameters x, y, X, and Y. This kernel, usually denoted by R(x,y;X,Y), is known as the "Riemann function". Riemann's solution of the Cauchy problem is given by (see Appendix)

$$u(X,Y) = (1/2)[f(X+Y) + f(X-Y)] + \\ +(1/2)\int_{X-Y}^{X+Y}\left[g(x)R(x,0;X,Y) - f(x)\dot{R}(x,0;X,Y)\right]dx \tag{A.10}$$

This is the expression (A.10) of the Appendix and is reproduced here for convenience. Let the characteristics of the problem be defined as:

$$\xi = x + y \text{ and } \eta = x - y$$

In terms of the characteristics, the problem reduces to

$$u^{\sim}_{\xi\eta} - (1/4)\left[q(\xi+\eta)/2 - q(\xi-\eta)/2\right]u^{\sim} = 0 \tag{4.19}$$

where "~" indicates that u is now a function of ξ and η instead of x and y. The boundary condition h = 0 is considered first. In this case, the following conditions are obtained for $u^{\sim}$:

$$u^{\sim}(\xi,\xi) = f^{\sim}(\xi), \; u^{\sim}_{\xi} = u^{\sim}_{\eta} \text{ at } \xi = \eta \tag{4.20}$$

Equation 4.19 is a hyperbolic differential equation and, together with the conditions (4.20), can be solved by Riemann's method of integration along the characteristics. According to Equation (A.10) the solution is

$$u(x,y;f) = (1/2)[f(x+y)] + f(x-y) + (1/2)\int_{x-y}^{x+y} f(t)W_0(x,y,t)\,dt \tag{4.21}$$

where $W_0(x,y,t)$ is obtained from the Riemann function after returning from variables ε,η to the original variables of the problem (x,y). The subscript 0 to W indicates that the boundary condition $h = 0$ is being considered.

The boundary condition $h = \infty$ is considered next. Let v(x,y;f) denote the solution. The boundary condition for this case is given by Equation 4.18:

$$v(x,0) = 0 \text{ and } \dot{v}(x,y) = f(x) \text{ at } y = 0$$

The corresponding solution is

$$v(x,y;f) = (1/2)\int_{x-y}^{x+y} f(t)W_\infty(x,y,t)\,dt \tag{4.22}$$

where $W_\infty(x,y,t)$ is again obtained from the corresponding Riemann function. The subscript ∞ to W indicates that the boundary condition $h = \infty$ is being considered. The exact forms of these functions are not important at the moment.

Let us next consider the general boundary condition for **T**(x,y)f(x):

$$\mathbf{T}(x,0)f(x) = f(x) \quad \text{and} \quad \partial[\mathbf{T}(x,y)f(x)]/\partial y - h\mathbf{T}(x,y)f(x)] = 0 \quad \text{at } y = 0$$

Then it is readily seen that

$$\begin{aligned}\mathbf{T}(x,y)f(x) &= u(x,y;f) + h\,v(x,y;f)\\ &= (1/2)[f(x+y) + f(x-y)] + (1/2)\int_{x-y}^{x+y} f(t)W_h(x,y,t)\,dt\end{aligned} \tag{4.23}$$

where $W_h(x,y,t) = W_0(x,y,t) + hW_\infty(x,y,t)$.

Let $w(x,\lambda)$ be an eigenfunction which satisfies a general boundary condition:

$$w'(0,\lambda) - hw(0,\lambda) = 0$$

If it is further assumed that h is finite, then the eigenfunction is a combination of $\phi(x,\lambda)$ and $\psi(x,\lambda)$, that is,

$$w(x,\lambda) = \phi(x,\lambda) + h\psi(x,\lambda)$$

where ϕ and ψ are the eigenfunctions of Equation 4.16 corresponding to the boundary conditions $h = 0$ and $h = \infty$, respectively. If $h = \infty$, then the above condition reduces to Equation 4.18. If f(x) is now identified with $w(x,\lambda)$, then from Equation 4.23

$$\mathbf{T}(x,y)w(x,\lambda) = (1/2)[w(x+y;\lambda) + w(x-y;\lambda)] + (1/2)\int_{x-y}^{x+y} w(t,\lambda)W_h(x,y,t)\,dt \tag{4.24}$$

Let us next show that

$$\mathbf{T}(x,y)w(x,\lambda) = w(x,\lambda)w(y,\lambda) \tag{4.25}$$

It is easily verified that

$$u(x,y,\phi) = \phi(x,\lambda)\phi(y,\lambda)$$
$$u(x,y,\psi) = \psi(x,\lambda)\phi(y,\lambda)$$
$$v(x,y,\phi) = \phi(x,\lambda)\psi(y,\lambda)$$

and

$$v(x,y,\psi) = \psi(x,\lambda)\psi(y,\lambda)$$

Then

$$\begin{aligned}
\mathbf{T}(x,y)w(x,\lambda) &= u(x,y;w) + hv(x,y;w) \\
&= u(x,y;\phi + h\psi) + hv(x,y;\phi + h\psi) \\
&= u(x,y;\phi) + hu(x,y;\psi) + hv(x,y;\phi) + h^2 v(x,y;\psi) \\
&= \phi(x,\lambda)\phi(y,\lambda) + h\psi(x,\lambda)\phi(y,\lambda) + \\
&\quad + h\phi(x,\lambda)\psi(y,\lambda) + h^2\psi(x,\lambda)\psi(y,\lambda) \\
&= [\phi(x,\lambda) + h\psi(x,\lambda)][\phi(y,\lambda) + h\psi(y,\lambda)] \\
&= w(x,\lambda)w(y,\lambda)
\end{aligned}$$

Using this relation in Equation 4.24 yields

$$\begin{aligned}
\mathbf{T}(x,y)w(x,\lambda) &= w(x,\lambda)w(y,\lambda) \\
&= (1/2)[w(x+y;\lambda) + w(x-y;\lambda)] + (1/2)\int_{x-y}^{x+y} w(t,\lambda)W_h(x,y,t)\,dt
\end{aligned}$$

Let us summarize what has been done so far. We are considering the Sturm-Liouville boundary value problem

$$\mathbf{L}y(x,\lambda) + \lambda y(x,\lambda) = y''(x,\lambda) + [\lambda - q(x)]y(x,\lambda) = 0, \ x \subset R^+$$

satisfying a general boundary condition

$$y(0,\lambda) = 1 \text{ and } y'(0,\lambda) - hy(0,\lambda) = 0$$

h being a real number and q(x) a real function. Two conditions are of particular interest, namely, h is finite and $h = \infty$ when the condition passes to

$$h = \infty: \quad y(0,\lambda) = 0 \text{ and } y'(0,\lambda) = 1$$

A solution to the boundary value problem for finite h is denoted by $w(x,\lambda)$.

Next we consider the equation

$$Y''(x,\lambda) + [\lambda - q(x)]Y(x,\lambda) = \ddot{Y}(y,\lambda) + [\lambda - q(y)]Y(y,\lambda)$$

and its solution, denoted by $\mathbf{T}(x,y)f(x)$, which satisfies the boundary conditions

$$\mathbf{T}(x,0)f(x) = f(x) \quad \text{and} \quad \partial\left[\mathbf{T}(x,y)f(x)\right]/\partial y - h\mathbf{T}(x,y)f(x) = 0 \quad \text{at } y = 0$$

The solution to this problem is obtained by Riemann's method of integration along the characteristics and was found to be Equation 4.23

$$\mathbf{T}(x,y)f(x) = (1/2)\left[f(x+y) + f(x-y)\right] + (1/2)\int_{x-y}^{x+y} f(t)W_h(x,y,t)\,dt \tag{4.23}$$

The particular forms of this solution for $h = 0$ and $h = \infty$ are given by Equations 4.21 and 4.22, respectively. The Riemann function $W_h(x,y,t)$ is a combination of Riemann functions $W_0(x,y,t)$ and $W_\infty(x,y,t)$ for $h = 0$ and $h = \infty$, respectively. By substituting for $f(x)$ in (4.23) the solutions $y(x,\lambda)$ for conditions $h = 0$ and $h = \infty$, the desired Povzner-Levitan transforms will be obtained. In essence, the function of a g.d.o. is to connect the eigenfunctions of a problem with an unknown potential to that of a problem with a known potential. A prime example in quantum scattering of the radial Schrödinger equation is the so-called Newton equations for the kernel relating the l-angular momentum eigenfunction of a Schrödinger radial equation with an unknown potential to that with a known one. Details are given in Reference 53 of Chapter 1.

Next we derive the Povzner-Levitan transforms for various cases.

A. POVZNER-LEVITAN TRANSFORM: ENTIRE REAL AXIS, h = 0

Let us consider the equation:

$$\ddot{u} - q(y)u = u''$$

with the initial conditions:

$$u(x,0) = f(x) \text{ and } \dot{u}(x,y) = 0 \text{ at } y = 0$$

Denoting the Riemann function for this problem as $G_0(x,y;t)$, the solution is given by Equation 4.21, with $W_0(x,y,t)$ replaced by the function $G_0(x,y,t)$.

Let $f(x)$ be $\cos kx$, $k = \lambda^{1/2}$, which is the "free" eigenfunction of Equation 4.1 for the boundary condition $h = 0$. Then using Equation 4.25 yields

$$u(x,y;f) = w(y,\lambda)\cos kx$$

Setting $x = 0$, and from Equation 4.21 with W_0 replaced by G_0, we have

$$w(y,\lambda) = \cos ky + (1/2)\int_{-y}^{y} G_0(0,y,t)\cos kt\,dt \tag{4.26}$$

Equation 4.26 is the "Povzner-Levitan Transform" on the entire real axis for the boundary condition $h = 0$.

The inverse transformation, that is, expressing $\cos ky$ in terms of $w(y,\lambda)$ can also be written. In order to obtain this, the "reverse" problem needs to be considered:

$$u'' - q(x)u = \ddot{u}$$

with the initial conditions u(x,0) = f(x) and $\dot{u}(x,y) = 0$ at y = 0. Writing the Riemann function for this problem as $H_0(x,y,t)$, the solution is obtained from Equation 4.21.

$$u(x,y,t) = (1/2)[f(x+y) + f(x-y)] + (1/2)\int_{x-y}^{x+y} f(t)H_0(x,y,t)\,dt$$

Let f(x) be identified with w(x,λ) and f(y) with cos ky, the free eigenfunction for the boundary condition h = 0. Again setting x = 0, remembering that the case h = 0 is being considered, and using the above equation, the inverse Povzner-Levitan transformation is obtained, namely

$$\cos ky = (1/2)[w(y,k) + w(-y,k)] + (1/2)\int_{-y}^{y} w(t,k)H_0(0,y,t)\,dt \tag{4.27}$$

B. POVZNER-LEVITAN TRANSFORM: ENTIRE REAL AXIS, h = ∞

The boundary condition is now

$$v(x,y;f) = 0 \text{ and } \dot{v}(x,y;f) = f(x) \text{ at } y = 0$$

and w(y,λ) is now the solution of Equation 4.1, but with the boundary condition given by Equation 4.4. Consider the equations:

$$v'' = \ddot{v} - q(y)v \tag{4.28}$$

$$v(x,0) = 0 \text{ and } \dot{v}(x,y) = f(x) \text{ at } y = 0 \tag{4.29}$$

The function v(x,y;f) is given by

$$v(x,y;f) = (1/2)\int_{x-y}^{x+y} f(t)G_\infty(x,y,t)\,dt$$

where $G_\infty(x,y,t)$ is the Riemann function. Identifying f(x) with the corresponding free eigenfunction (sin kx/k) for the boundary condition h = ∞, and using Equation 4.25, the following expression results:

$$[\sin kx/k]w(y,\lambda) = (1/2)\int_{x-y}^{x+y} G_\infty(x,y,t)\,(1/k)\sin kt\,dt \tag{4.30}$$

Differentiating Equation 4.30 with respect to x and then setting x = 0 results in the expression:

$$w(y,\lambda) = (1/2)G_\infty(0,y,y)[\sin ky/k] + (1/2)G_\infty(0,y,-y)[\sin ky/k] +$$
$$+(1/2)\int_{-y}^{y} [\sin kt/k][G_\infty(x,y,t)]'\,\big|_{x=0}\,dt \tag{4.31}$$

We recall that G_∞ is the Riemann function. An exact calculation will show that

$$G_\infty(x,y,t) = R(\xi,\eta;\ \xi_0,\eta_0) \text{ at } \xi = \eta = t$$

For x = 0, the points $\varepsilon = \eta = t$, $t = \pm y$ lie on the characteristics passing through the point $(0,\pm t)$. But by definition, the Riemann function is unity on these characteristics. Hence, we conclude that $G_\infty(0,y,\pm y)$ is unity. Equation 4.31 then reduces to

$$w(y,\lambda) = \sin ky/k + (1/2)\int_{-y}^{y}[\sin kt/k][G_\infty(x,y,t)]'\,|_{x=0}\,dt \tag{4.32}$$

The prime, as usual, denotes differentiation with respect to x. Equation 4.32 is the "Povzner-Levitan transform" on R corresponding to the boundary condition $h = \infty$.

In order to obtain the inverse transformation, we again consider the reverse of Equation 4.28 with the x and y variables interchanged, the boundary condition (4.29) remaining the same. But now let f(x) be w(x,k) and consider the free solution (sin ky/k) to be w(y,k). Then following the same procedures as above, that is, forming the integral expression, differentiating with respect to x, and setting x = 0 as before, the following inverse Povzner-Levitan transformation is obtained:

$$\sin ky/k = (1/2)[w(y,k) - w(-y,k)] + (1/2)\int_{-y}^{y} w(t,k)[H_\infty(x,y,t)]'\,|_{x=0}\,dt \tag{4.33}$$

C. POVZNER-LEVITAN TRANSFORMATION ON THE SEMIAXIS

h = 0: In Equation 4.26, we let $G_0(y,t)$ be the even part of the Riemann function $G_0(0,y,t)$. Then, the corresponding Povzner-Levitan transformation reduces to

$$w(y,\lambda) = \cos ky + \int_{0}^{y} G_0(y,t)\cos kt\,dt \tag{4.34}$$

$h = \infty$: By similar reasoning, if $G_\infty(y,t)$ denotes the odd part of the derivative $G_\infty'(x,y,t)$ at x = 0, then the corresponding transform follows from Equation 4.32.

$$w(y,\lambda) = \sin ky/k + \int_{0}^{y}[\sin kt/k][G_\infty(x,y,t)]'\,dt \tag{4.35}$$

D. INVERSE POVZNER-LEVITAN TRANSFORM: SEMIAXIS

Equations 4.27 and 4.33 are the inverse Povzner-Levitan transforms on the entire real axis. For the semiaxis, the function w(–y,k) must be properly transformed, i.e., some relationship between w(–y,k) and w(y,k) must be obtained. The following theorem about a Volterra integral equation of the second kind will be helpful.

Theorem 4.1

If K(x,t) and f(t) are continuous, then the Volterra integral equation of the second kind, namely,

$$f(x) = g(x) + \int_{0}^{x} K(t,x)f(t)\,dt$$

has the following unique solution:

$$f(x) = g(x) + \int_0^x \Gamma(t,x)g(t)\, dt$$

where Γ(t,x) is the "resolvent" for the kernel K(x,t) and is uniformly convergent on $0 < t < x$.

For the boundary condition h = 0, and for y > 0, we obtain from Equation 4.24, for x = 0, the following equation:

$$w(y,k) = (1/2)\left[w(y,k) + w(-y,k)\right] + (1/2)\int_{-y}^{y} w(t,k)W_h(0,y,t)\, dt \tag{4.36}$$

Equation 4.36 can be recast as a Volterra equation:

$$w(-y,k) = w(y,k) - \int_0^y w(t,k)W_h(0,y,t)\, dt - \int_0^y w(-t,k)W_h(0,y,-t)\, dt$$

Then by the above theorem,

$$w(-y,k) = w(y,k) + \int_0^y A(y,t)w(k,t)\, dt \tag{4.37}$$

where A(y,t) is now the resolvent.

For h = ∞, it is already known that

$$T(x,y)w(x,\lambda) = w(x,k)w(y,k) = (1/2)\int_{x-y}^{x+y} w(t,k)W_\infty(x,y,t)\, dt$$

Differentiating with respect to x, setting x = 0 in the resulting expression, and evaluating the Riemann functions $W_\infty(0,y,y)$ and $W_\infty(0,y,-y)$ to be unity as before, we obtain

$$w(y,k) = (1/2)\left[w(y,k) - w(-y,k)\right] + (1/2)\int_{-y}^{y}\left[w_\infty(x,y,t)\right]'\Big|_{x=0} w(t,k)\, dt$$

Converting this equation to a Volterra equation and following the same line of reasoning as above, we again arrive at Equation 4.37. The following lemma is, therefore, established.

Lemma 4.1

$$w(-x,k) = w(x,k) + \int_0^x \alpha(x,t)w(t,k)\, dt, \; x > 0$$

α(x,t) being a continuous function.

This lemma can now be used to derive the inverse Povzner-Levitan transformation on the semiaxis.

h = 0. Returning to Equation 4.26, denoting by H(x,y,t) the even part of h(x,y,t), and then using Lemma 4.1, the inverse Povzner-Levitan transform for the boundary condition h = 0 is obtained,

$$\cos ky = w(y,k) + \int_0^y L(x,y,t)w(t,k)\,dt \tag{4.38}$$

where L(x,y,t) is the addition of H(y,t) and $\alpha(y\cdot t)$. Reasoning similarly, the inverse Povzner-Levitan transformation for the boundary condition $h = \infty$ is derived as

$$(1/k)\sin ky = w(y,k) + \int_0^y K(t,k)(1/k)\sin kt\,dt \tag{4.39}$$

It should be mentioned that the inverse transformations can also be derived in a more direct way. Equations 4.34 and 4.35 are Volterra equations in w(x,k) in terms of cos kx and (1/k)(sin kx), respectively. These equations could also be considered as the same Volterra equations, but in the unknown functions cos kx and (sin kx)/k. The solutions would then result in Equations 4.38 and 4.39, respectively.

V. TWO THEOREMS FROM FUNCTION THEORY[7-8]

We discuss two theorems from function theory concerning certain analytic functions. The aim is to determine the conditions that must be imposed on these functions if their restrictions on the real axis are to have certain functional behavior. The theorems under discussion are the well-known theorems of Paley-Wiener and Wiener-Boas. The functions to be considered are those of exponential types.

Let z be a complex variable and f(z) an analytic function. f(z) is said to be an "entire" function if it is analytic in the entire complex plane. For various applications, it is of interest to estimate the growth of these functions as $|z| \to \infty$. This growth is usually compared to that of an exponential such as $\exp(\alpha\,|z|)$, where α is a positive constant. If there exist two constants A and t such that an entire function f(z) of a complex variable z satisfies the inequality

$$|f(z)| \le A\exp(t|z|) \text{ for all z complex}$$

then the function f(z) is said to be of "exponential type". In English literature, these functions are also referred to as "integral functions".[13] It is of exponential type τ if the above equation holds for all $t > \tau$, but for no $t < \tau$. The functions of exponential type, therefore, have no singularities in the finite complex plane. A more precise definition is given as follows.

Define:

$$M(r) = \max_{0\le\theta\le 2\pi} |f(r\exp(i\theta))|$$

Then f(z) is of exponential type τ if and only if

$$\lim_{r\to\infty}\left[(\mathrm{Log}\,r)^{-1}\,\mathrm{Log}\,\mathrm{Log}\,M(r)\right] = 1$$

and

$$\lim_{r\to\infty}\left[r^{-1}\,\mathrm{Log}\,M(r)\right] = \tau$$

A function of exponential type can grow as fast as, but no faster than, an exponential. This growth, however, needs not be the same in all directions and the function can $\rightarrow 0$ as $|z| \rightarrow \infty$ in certain directions. Also, the rate of growth need not be the same in all directions. A polynomial is of type zero, the function exp(2z) of type 2, and the function [sin pz/z] of type p. Functions such as $\exp(z^2)$ are not of exponential type.

It is of interest to determine under what conditions the restrictions of functions of exponential type on the real axis have either square or absolutely integrability on R. These are the contents of the two theorems stated below.

Theorem 4.2. Paley-Wiener Theorem: $L_2(-\infty,\infty)$ Integrability

An entire function f(z) of exponential type τ has a $L_2(-\infty,\infty)$ restriction on the real axis if and only if its Fourier transform f^(t) is square integrable on the compact interval $[-\tau, \tau]$ on R. That is, if

$$f(z) = \int_{-\tau}^{\tau} f^{\wedge}(t)\, \exp(itz)\, dt,\ f^{\wedge}(t) \subset L_2(-,\tau,\tau)$$

This means that the Fourier transform of the restriction of an entire function of exponential type τ to the real axis R is an L_2-function of support $(-\tau, \tau)$.

Theorem 4.3. Wiener-Boas Theorem: $L_1(-\infty,\infty)$ Integrability

An entire function f(z) of exponential type τ has a $L_1(-\infty,\infty)$ restriction on the real axis if and only if its Fourier transform f^(t) is absolutely integrable on the compact interval $[-\tau,\tau]$ on R. That is, if

$$f(z) = \int_{-\tau}^{\tau} f^{\wedge}(t) \exp(itz)\, dt$$

f^(t) is such that $f^{\wedge}(-\tau) = f^{\wedge}(\tau) = 0$ and the function obtained by extending f^(t) to be zero outside the interval $[-\tau,\tau]$ has an absolutely convergent Fourier series on any larger interval $[-(\tau+\varepsilon),\tau+\varepsilon]$, $\varepsilon > 0$.

In the electrical engineering literature,[9] the functions and their restrictions having the above mentioned properties are called the "time" and "band-limited" functions. A function f(t) is band limited if its Fourier transform $f^{\wedge}(\omega)$ vanishes outside a certain spectral interval. That is, f(t) is band limited if

$$f^{\wedge}(\omega) = 0 \ \text{ for } |\omega| > \sigma$$

Its energy is, of course, required to be finite, that is,

$$E = \int_{-\infty}^{\infty} |f(t)|^2\, dt = (1/2\pi) \int_{-\infty}^{\infty} |f^{\wedge}(\omega)|^2\, d\omega < \infty$$

Fourier transform is defined here asymmetrically with a normalization factor of $(1/2\pi)$ in the inverse transform equation. This is recognized as Parseval's relation. Similarly, a function is time limited if

$$f(t) = 0 \text{ for } |t| > \tau \text{ and the energy } E < \infty$$

For a band-limited function f(t)

$$\left|f(t)\right| \le (1/2\pi)\int_{-\sigma}^{\sigma}\left|f^{\wedge}(\omega)\exp(-i\omega t)\right| d\omega$$

$$\le \left[\left\{\exp(\sigma|t|)\right\}/2\pi\right]\int_{-\sigma}^{\sigma}\left|f^{\wedge}(\omega)\right| d\omega$$

$$\le \left[\sigma E/\pi\right]^{1/2}\exp(\sigma|t|)$$

The function is clearly of exponential type σ. Theorems 4.1 and 4.2 clearly hold for a band-limited function. By the symmetry of Fourier transformations, some conclusions also apply to the time-limited functions.

VI. WIENER-BOAS THEOREM AND POVZNER-LEVITAN TRANSFORM

The derivation of the Povzner-Levitan transform using the Wiener-Boas theorem is presented after Sabatier.[4] The starting point is Equation 3.24, the integral representation of the eigenfunction derived earlier by the method of variation of parameters. This is a Volterra equation and, hence, can be solved by iteration, i.e.,

$$y(r,k) = \sum_{n=0}^{\infty} y^{(n)}(r,k) \tag{4.40}$$

where the superscript (n) means the n-th iterate.

$$y^{0}(r,k) = \sin kr/k$$
$$y^{(n)}(r,k) = \int_0^r (1/k)\sin k(r-r')q(r')y^{(n-1)}(k,r')\,dr' \tag{4.41}$$

The h = ∞ boundary condition is being considered here.

For small x, |sin x| ≤ x; otherwise, |sin x| ≤ 1. In order not to be wasteful in setting the bound, these two are combined together into a single bound[10]

$$|\sin x| \le \beta x/(1+x),\ x \ge 0$$

where β is a constant whose precise value is not important. If x is replaced by a complex variable z, then |sin z| grows as exp(|Im z|), and using the bound given above, we obtain

$$|\sin z| \le C\exp(|\mathrm{Im}\ z|)|z|/(1+|z|)$$

C being a numerical constant.

If this bound is now used in Equation 4.41, the following estimates are obtained:

$$\left|y^{(1)}(r,k)\right| \le C\exp(|\mathrm{Im}\,k|r)\left[r/(1+|k|r)\right]\int_0^r\left[r'/(1+|k|r')\right]\left|q(r')\right| dr'$$

By induction, the n-th iterate is estimated to be

$$\left|y^{(n)}(r,k)\right| \le C\ \exp(|\text{Im}\ k|r)\left[r/(1+|k|r)\right](1/n!)\left\{\int_0^r\left[r'/(1+|k|r')\right]\left|q(r')\right|\,dr'\right\}^n$$

It, therefore, follows from this estimate that y(r,k) given by the series (4.40) is absolutely and uniformly convergent in any compact set of the complex k-plane provided that the integral

$$\int_0^r\left[r'/(1+|k|r')\right]\left|q(r')\right|\,dr' < \infty \tag{4.42}$$

The estimate (4.42) requires that the potential satisfy a bound such as

$$\int_0^r r'\left|q(r')\right|\,dr' < \infty,\ r < \infty$$

From the theory of the Volterra equation, the series (4.40) uniquely satisfies Equation 3.24 for the boundary condition $h = \infty$.

The following estimates can also be established:

$$\begin{aligned}|y(r,k)| &\le C\ \exp(|\text{Im}\ k|r)\left[r/1+|k|r\right]\exp\left\{\int_0^r\left[r'/1+|k|r')\right]\left|q(r')\right|\,dr'\right\}\\ &\le C_1\ \exp(|\text{Im}\ k|r)\left[r/(1/|k|r)\right]\end{aligned}$$

and

$$\left|y-(1/k)\sin kr\right| \le C_2\exp(|\text{Im}\,k|r)\left[r/(1+|k|r)\right]\int_0^r\left[r'/(1+|k|r')\right]\left|q(r')\right|\,dr' \tag{4.43}$$

C_1 and C_2 are appropriate numerical constants.

By using the estimate given by Equation 4.43, the asymptotic behavior of y(r,k) as $|k| \to \infty$ is obtained

$$y(r,k) = (1/k)\sin kr + \left[\exp(|\text{Im}\,k|r)/|k|\right]\,o(1) \tag{4.43a}$$

where o(1) is a small term which is uniform in r for all $r \ge 0$. Since $y^{(0)}$ as well as the kernel in Equation (3.24) are entire in k, then y(r,k) is also an entire function of k, that is, an analytic function of k in the open complex k plane for every fixed finite r. Moreover, from Equation 4.43a, y(r,k) is seen to be an entire function of exponential type in k of type r. These are, of course, subject to the condition that the potential is such as to satisfy Equation 4.42. By the same condition and from the bound established in Equation 4.43, it also follows that

$$\int_0^\infty\left|y-(1/k)\sin kr\right|\,dk < \infty \tag{4.44}$$

In Equation 4.44, the space is the real half-axis since the integrand is an even function of k.

Let z(r,k) denote the function $[y(r,k) - (1/k)\sin kr]$. Then z(r,k) is an entire function of exponential in k of type r whose restriction on the real k axis is L_1. These follow from the inequalities (4.43) and (4.44). The Wiener-Boas theorem (Theorem 4.3) is then immediately applicable to z and the following Fourier series representation results:

$$\begin{aligned} z(r,k) &= \int_{-r}^{r} z^\wedge(r,r')\exp(ikr')dr' \\ &= 2\int_0^r z^\wedge(r,r')\cos kr'\,dr' \end{aligned} \tag{4.45}$$

the superscript ^ denoting the Fourier transform in its asymmetric form. The transform can be inverted and gives

$$z^\wedge(r,r') = (1/\pi)\int_0^\infty z(r,k)\ \cos\ kr'\,dk$$

This is a well-defined continuous function of r and r′ since, by virtue of the bound (4.44), the integrand is integrable. Note that $z^\wedge(r,r')$ is finite everywhere in $0 \le r' \le r$. Also, according to the condition of the Wiener-Boas theorem (Theorem 4.3.),

$$z^\wedge(r,r') = 0 \text{ at } r = \pm\, r' \tag{4.45a}$$

Integrating the right-hand side of Equation 4.45 by parts and using the condition (4.45a),

$$z(r,k) = -\int_0^r \left[z^\wedge(r,r')\right]' (1/k)\sin\ kr'\,dr'$$

where ′ denotes differentiation with respect to r′ and the differentiation is permissible. This leads to the following integral representation of y(r,k):

$$y(r,k) = (1/k)\sin kr + \int_0^r K(r,r')(1/k)\sin kr'\,dr' \tag{4.46}$$

where

$$\begin{aligned} K(r,r') &= -\left[z^\wedge(r,r')\right]' \\ &= (2/\pi)\int_0^\infty z(r,k)\,k\sin kr'\,dk \end{aligned} \tag{4.47}$$

From Section IV.C (Equation 4.35), Equation 4.46 is recognized to be the transformation of Povzner and Levitan on the semiaxis for the boundary condition $h = \infty$.

Let us now discuss the validity of this representation. In particular, the convergence of the integral at both limits will be investigated as well as the condition for the existence of $[z^\wedge(r,r')]'$. The behavior at the lower limit is easily established. From Equation 4.47, as $r' \to 0$, K(r,r′) vanishes, K(r,0) = 0. The kernel is thus well behaved at the lower limit.

The behavior at the upper limit of integration is somewhat more complex. If $r' < r$, then in the definition of the kernel K(r,r′), the integrand is an oscillating function and vanishes in the limit of large k by virtue of the Riemann-Lebesgue lemma. This is possible because of the bound established in Equation 4.43, which makes z an L_1 function on the real axis. This justifies the differentiation under the integral sign in the expression for $z^\wedge(r,r')$ in obtaining K(r,r′). $(z^\wedge(r,r'))'$ is finite everywhere for $r' < r$. Now, z(r,k) contains sin kr and by dint of the bound given by Equation 4.43, z(r,k) is only (1/k)o(1). Therefore, if $r = r'$, the integrand will keep constant sign and will not converge at the upper limit. Hence, the derivative $[z^\wedge(r,r')]'$ is finite for all $r' < r$, but infinite in general if $r = r'$. It will be shown, however, that the integral remains finite even at $r' = r$ if proper restrictions are imposed on the potential.

Let us now demonstrate what condition must be imposed on the potential for the convergence of the integral even if $r' \le r$. $z^\wedge(r,r')$ can be rewritten as

$$z^\wedge(r,r') = \int_0^\infty \cos kr'\left[y(r,k) - (1/k)\sin kr\right] dk$$

The constant factor of 2 will be ignored in what follows. Substituting Equation 3.24 for $h = \infty$ in the above equation and changing the order of integration (this is permissible since the integrals are absolutely convergent by virtue of Equation 4.44), we have

$$z^\wedge(r,r') = \int_0^r q(u)\, B(r,r';u)\, du \tag{4.48}$$

where

$$B(r,r';u) = B_1(r,r';u) + B_2(r,r';u)$$

$$B_1(r,r';u) = \int_0^\infty \cos kr'[\sin ku/k][\sin k(r-u)/k]\, dk \tag{4.49}$$

$$B_2(r,r';u) = \int_0^\infty \cos kr'[y(u,k) - \sin ku/k][\sin k(r-u)/k]\, dk \tag{4.50}$$

Therefore,

$$\left[z^\wedge(r,r')\right]' = \int_0^r [B_1(r,r';u)]'\, q(u')\, du' + \int_0^r [B_2(r,r';u)]'\, q(u')\, du'$$

Let us consider the first term $[B_1(r,r';u)]'$. Differentiating Equation 4.49

$$[B_1(r,r';u)]' \le -\int_0^\infty [\sin kr' \sin kr \sin ku/k]\, dk$$

Consider the case when $r' < r$. Then $r > r' > 0$ and for $u > 0$, the above integral vanishes.[11] In the limit $r' \to r^-$, $u \to 0^+$, the integral is $(\pi/8)$, and for $0 < u < 2r$, the integral is $(\pi/4)$.[11] Therefore, $[B_1(r,r';u)]'$ remains finite everywhere. $B_1(r,r';u)$ is, therefore, absolutely convergent. This justifies taking the derivative under the integral sign.

Let us now consider the second term. Using the bound established in the inequality (4.43), the second factor can be estimated as

$$\left|[B_2(r,r';u)]'\right| \le \text{constant } u \left[\int_0^r u'|q(u')| \left\{\int_0^\infty [\sin kr' \sin k(r-u)/(1+|k|u)(1+|k|u')]dk\right\} du'\right]$$
$$\le \text{constant } u$$

if $\int_0^r u'|q(u')|\, du'$ exists. The upper limit in the u' integral has been taken to be r rather than u since $u \le r$. Then from Equation 4.48, the r' derivative of $z^\wedge(r,r')$ exists insofar as the $[B_2(r,r';u)]'$ is concerned, provided that uq(u) is integrable at the origin.

Let us now assume that the potential is integrable at the origin, i.e.,

$$\int_0^r |q(u')|\, du' < \infty$$

In Chapter 5, an integral expression will be derived for the kernel K(r,r′) on the basis of Riemann's solution of a Cauchy problem for K. The integral representation will be given by

$$K(r,r') = (1/2)\int_B^A q(u)\,du + \int_B^A du \int_0^B q(u+u')K(u+u',u'-u)\,du'$$

where $A = (r + r')/2$ and $B = (r - r')/2$. The existence of the integral certainly poses no problem if $r' < r$. More importantly, however, the integral exists even at $r = r'$ if the potential is integrable at the origin.

Collecting these results together yields the following lemma.

Lemma 4.2

If $(1 + r)q(r)$ is integrable at the origin, that is, if $(1 + r)q(r) \subset L_1[0,r]$, then the derivative of $z^\wedge(r,r')$ with respect to r' is finite for all r' such that $0 \le r' \le r$. Otherwise, if only $rq(r) \subset L_1[0,r]$, then the derivative is finite only for $0 \le r' < r$, while being infinite at $r = r'$ in general.

To summarize, the Povzner-Levitan transform K(r,r′) is well behaved at the lower limit $r' = 0$, actually vanishing there. At the upper limit $r = r'$, the transform is generally infinite unless $[1 + rq(r)]$ is integrable at the origin. At any rate, the transform exists for all $r' < r$. However, for the representation (4.46) to be valid, what is required is only the integrability of K(r,r′) in r′ at $r = r'$. Being $z^\wedge(r,r')$, this is finite everywhere for $0 \le r' \le r$. The Povzner-Levitan representation (4.46) is, therefore, well defined.

If the potential itself is integrable at the origin as assumed above, the Povzner-Levitan transform can also be derived from the Paley-Wiener theorem (Theorem 4.2). If $q(r) \subset L_1[0,r]$, then from the bound given by the inequality (4.43), it follows that

$$z(r,k) \sim 0\left(1/k^2\right)$$

This means that the function $kz(r,k) \subset L_2(R)$. By the Paley-Wiener theorem, therefore, we can write

$$\begin{aligned} kz(r,k) &= \int_{-r}^{r} z^\wedge(r,r')\ \exp(ir'k)\ dr' \\ &= 2\int_0^r z^\wedge(r,r') \sin kr'\, dr' \end{aligned}$$

since the function is odd in k. Dividing both sides by k, the Povzner-Levitan transform as given by Equation 4.46 is again recovered. Since $kz(r,k) \subset L_2(R)$, it follows that the transform $K(r,r') \subset L_2(0,r)$ in r′ for each fixed r.

VII. DERIVATION DUE TO MARCENKO[5]

In this section, we present another derivation of the Povzner-Levitan transform which is relatively simpler and more straightforward than the previous presentations. This derivation is due to Marcenko. Consider the scattering equation:

$$y'' + [\lambda - q(x)]y = 0, \quad x \subset (-a,a), \quad a < \infty$$

The scattering-free solution of the problem, that is, the solution without the potential term, is, of course, $\exp(\pm ikx)$, $k = \lambda^{1/2}$. These constitute the so-called "fundamental" system of solutions to the equation $y'' + \lambda y = 0$. These solutions, denoted by $\mathbf{e}_0(\lambda,x)$, satisfy the following boundary conditions:

$$e_0(\lambda,0) = 1 \text{ and } e_0'(\lambda,0) = \pm\, ik$$

the prime referring to differentiation with respect to x. We will write $\mathbf{e}_0(\lambda,x) = [\exp(ikx),\exp(-ikx)]^T$, T denoting transpose, in order to indicate both solutions. Otherwise, simply e_0 will be used. Let us consider the solution $\mathbf{e}_0(\lambda,x)$ and its corresponding + ik boundary condition. Let $\mathbf{e}(\lambda,x)$ be the solution of the scattering equation with the initial conditions

$$e(\lambda,0) = 1 \text{ and } e'(\lambda,0) = ik$$

Define:

$$u(x,x') = \exp(ikx)e(\lambda,x')$$

This function u is twice differentiable and satisfies the Cauchy problem

$$\begin{aligned}
&u'' = \ddot{u} - q(x')u \\
&u(x,0) = \exp(ikx) \\
&\dot{u}(x,0) = ik\exp(ikx) = [\exp(ikx)]', \quad x \subset R, \quad -a < y < a
\end{aligned}$$

where $'$ stands for differentiation in x and $\cdot$ in x'. The Cauchy problem can be solved by using Riemann's method of integration, which is detailed in the Appendix. The solution is given by Equation A.10. Replacing g(x) by f′(x) in Equation A.10 according to the condition of the Cauchy problem, the Riemann solution takes the form

$$u(x_0,x_0') - (1/2)\int_{x_0-x_0'}^{x_0+x_0'} \{R'(x,0;x_0,x_0') + \dot{R}(x,0;x_0,x_0')\}f(x)\,dx + f(x_0,x_0')$$

Using the Riemann formula, the solution of the Cauchy problem at the point (x_0,x'_0) becomes

$$\begin{aligned}
\exp(ikx_0)e(\lambda,x_0') = \exp\{ik(x_0+x_0')\} - \\
-(1/2)\int_{x_0-x_0'}^{x_0+x_0'} \{R'(x,0;x_0,x_0') + \dot{R}(x,0;x_0,x_0')\}\exp(ikx)\,dx
\end{aligned}$$

Setting $x_0 = 0$ and denoting

$$K(x_0',x) = -(1/2)\{R'(x,0;0,x_0') + \dot{R}(x,0;0,x_0')\} \tag{4.51}$$

yields

$$\begin{aligned}
e(\lambda,x_0') &= \exp(ikx_0') + \int_{-x_0'}^{x_0'} K(x_0',x)\exp(ikx)\,dx \\
&= (\mathbf{I}+\mathbf{K})\, e_0(\lambda,x_0')
\end{aligned} \tag{4.52}$$

The operator $(\mathbf{I} + \mathbf{K})$ is a transformation operator that transforms the scattering-free solution $\mathbf{e}_0(\lambda,x)$ to an eigenfunction $e(\lambda,x)$ of the scattering equation while preserving the initial data. This holds equally for the other fundamental solution exp(–ikx).

Let us now consider the Cauchy problem with the initial data at x = 0 specified in a general form, namely,

$$\omega(\lambda,0;h)=1,\ \omega'(\lambda,0;h)=h \tag{4.53}$$

Now $\mathbf{e}_0(\lambda,x)$ forms the fundamental system of solutions. It can, therefore, be assumed that the "free" eigenfunctions which are to be transformed by the transformation operators to yield the solution $\omega(\lambda,x;h)$, while preserving the initial conditions (4.53), can be represented as linear combinations of $\mathbf{e}_0(\lambda,x)$. Denoting these free eigenfunctions by $\omega_0(\lambda,x;h)$, we can write

$$\omega_0(\lambda,x;h)=(A,B)\mathbf{e}_0(\lambda,x)$$

where A and B in the vector (A,B) are constants. Evaluating the constants from the initial conditions given by Equation 4.53, we obtain

$$\omega_0(\lambda,x;h)=\cos kx+(h/k)\sin kx$$

Equation 4.52 then yields the solution $\omega(\lambda,x;h)$:

$$\omega(\lambda,x;h)=\cos kx+(h/k)\sin kx+\int_{-x}^{x}K(x,t)\{\cos kt+(h/k)\sin kt\}\,dt$$

By dividing the integration limit (–x,x) into (–x,0) and (0,x) and changing t to –t in the interval (–x,0), the above expression reduces to

$$\begin{aligned}\omega(\lambda,x;h)&=\cos kx+\int_0^x\{h+K(x,t)+K(x,-t)\}\cos kt\,dt+\\&\quad+h\int_0^x\left[\{K(x,t)-K(x,-t)\}\int_0^t\cos kt'\,dt'\right]dt\end{aligned} \tag{4.53a}$$

The above expression can be further put in the following form:

$$\omega(\lambda,x;h)=\cos kx+\int_0^x K(x,t;h)\cos kt\,dt=(\mathbf{I}+\mathbf{K}_h)\cos kx$$

where

$$K(x,t;h)=h+K(x,t)+K(x,-t)+h\int_t^x\{K(x,t')-K(x,-t')\}\,dt'$$

This result follows on interchanging the order of integration in the second integral of Equation (4.53a).

Similarly, for the other boundary condition corresponding to $h\to\infty$, the initial conditions are

$$\omega(\lambda,0;\infty)=0 \text{ and } \omega'(\lambda,0;\infty)=1$$

and the corresponding free eigenfunction is just [sin kx/k]. Applying Equation 4.52 once more yields the solution in this case as

$$\begin{aligned}\omega(\lambda,x;\infty)&=(1/k)\sin kx+\int_0^x K(x,t;\infty)(1/k)\sin\lambda t\,dt\\&=(\mathbf{I}+\mathbf{K}_\infty)(1/k)\sin kx\end{aligned}$$

where

$$K(x,t;\infty) = K(x,t) - K(x,-t)$$

The operators $\mathbf{K}$, $\mathbf{K}_h$, and $\mathbf{K}_\infty$ are Volterra integral operators. Therefore, the operators $(\mathbf{I} + \mathbf{K})$, $(\mathbf{I} + \mathbf{K}_h)$, and $(\mathbf{I} + \mathbf{K}_\infty)$ possess inverses which are, moreover, of the same form. Defining these inverse operators by $(\mathbf{I} + \mathbf{L})$, $(\mathbf{I} + \mathbf{L}_h)$, and $(\mathbf{I} + \mathbf{L}_\infty)$, we have the following inverse solutions, that is, the transformations of the solutions $e(k,x)$, $\omega(\lambda,x;h)$, and $\omega(\lambda,x;\infty)$ to the free solutions $e_0(\lambda,x)$, $\cos kx$, and $[\sin kx/k]$:

$$\begin{aligned} e_0(\lambda,x) &= \exp(ikx) = e(\lambda,x) + \int_{-x}^{x} L(x,t)\, e(\lambda,t)\, dt \\ \cos kx &= \omega(\lambda,x;h) + \int_0^x L(x,t;h)\, \omega(\lambda,t;h)\, dt \end{aligned} \tag{4.54}$$

and

$$\sin kx/k = \omega(\lambda,x;\infty) + \int_0^x L(x,t;\infty)\, \omega(\lambda,t;\infty)\, dt \tag{4.55}$$

These results can be summarized into the theorems below.

Theorem 4.4
Consider the Cauchy problem

$$y'' + [\lambda - q(x)]y = 0$$

with the initial conditions

$$y(\lambda,0) = 1 \text{ and } y'(\lambda,0) = ik$$

The solution $y(\lambda,x)$ admits of the representation

$$y(\lambda,x) = \exp(ikx) + \int_{-x}^{x} K(x,t)\, \exp(ikt)\, dt$$

where the kernel $K(x,t)$ is a continuous function which can be obtained in terms of the Riemann function as

$$K(x,t) = -(1/2)\left\{R'(x,0;0,t) + \dot{R}(x,0;0,t)\right\}$$

Theorem 4.5
Consider the Cauchy problem with the initial conditions

$$\begin{aligned} &y(k,0;h) = 1,\ y'(k,0;h) = h \\ &y(k,0;\infty) = 0,\ y'(k,0;\infty) = 1 \end{aligned}$$

where the solutions are given by $y(k,x;h)$ and $y(k,x;\infty)$, respectively. These solutions can be represented by the integral equations

$$\begin{aligned} y(k,x;h) &= \cos kx + \int_0^x K(x,s;h) \cos ks\, ds \\ &= (\mathbf{I} + \mathbf{K}_h) \cos kx \end{aligned}$$

and

$$y(k,x;\infty) = (\sin kx)/k + \int_0^x K(x,s;\infty)(\sin ks)/k\ ds$$
$$= (\mathbf{I} + \mathbf{K}_\infty)(1/k)\sin kx$$

respectively. The kernels K(x,s;h) and K(x,s;∞) are continuous functions and can be expressed in terms of the kernel K of Theorem 4.4 as

$$K(x,s;h) = h + K(x,s) + K(x,-s) + h\int_s^x \{K(x,x') - K(x-x')\}\ dx'$$

and

$$K(x,s;\infty) = K(x,s) - K(x,-s)$$

respectively.

Theorem 4.6

The operators $(\mathbf{I} + \mathbf{K})$, $(\mathbf{I} + \mathbf{K}_h)$, and $(\mathbf{I} + \mathbf{K}_\infty)$ possess inverses $(\mathbf{I} + \mathbf{L})$, $(\mathbf{I} + \mathbf{L}_h)$, and $(\mathbf{I} + \mathbf{L}_\infty)$, in terms of which the inverse solutions are obtained as

$$\exp(ikx) = e(k,x) + \int_{-x}^{x} L(x,s)e(k,s)\ ds$$
$$\cos kx = y(k,x;h) + \int_0^x L(x,s;h)y(k,s;h)\ ds$$
$$(\sin kx)/k = y(k,x;\infty) + \int_0^x L(x,s;\infty)y(k,s;\infty)\ ds$$

The kernels $\mathbf{L}$, $\mathbf{L}_h$, and $\mathbf{L}_\infty$ are continuous solutions of the corresponding Volterra integral equations.

VIII. POVZNER-LEVITAN TRANSFORM AND THE SPECTRAL FUNCTION

We conclude this chapter by showing that the Povzner-Levitan transform or, rather, the Fourier transform of the Povzner-Levitan transform is intimately related to the spectral function of the Sturm-Liouville boundary-value problem. We follow Marcenko.[5]

Consider the Cauchy problem with the boundary condition for the eigenfunction y(λ,x;h), that is,

$$y'' + [\lambda - q(x)]y = 0, \quad x \subset R^+$$

$$y'(0) - hy(0) = 0$$

Let us consider the generalized Parseval relation (Equation 3.17) and rewrite it in the form

$$(f,g) = (FG,R) \tag{4.56}$$

where the integrals have been written as inner products. The Fourier-transformed part of Parseval's relation has been written as a R-functional, R being a distribution. In Chapter 3, the spectral function was assumed and on the basis of that Parseval's relation was derived.

Here, the reverse question is posed and it is actually demonstrated that Equation 4.56 can, indeed, be written, that is, that the distribution R does exist. Furthermore, it is shown that there is a relation between R and the Povzner-Levitan transform. To determine this distribution R is, therefore, the main objective of this section.

The basic steps are as follows. We first consider functions f(x), g(x) $\subset L_2[0,d]$, that is, square integrable functions over the interval [0,d] on the half-line. Next, we assume that a sequence of integrable functions $R_n{}^d(k)$ on R^+ exists such that

$$U_n^d(x,x') = \left(y(k,x)y(k,x'),R_n^d(k)\right) = \delta_n(x-x') \tag{4.57}$$

δ_n being a delta sequence, converges to Dirac's delta distribution $\delta(x - x')$ in the domain $0 < x < d$, $0 < y < d$, as $n \to \infty$. Notice that $U_n{}^d(x,x')$ is symmetric in the variables x and x′, that is, $U_n{}^d(x,x') = U_n{}^d(x',x)$. This fact will be used later. Multiplying f(x)g(x′) by $U_n{}^d(x,x')$, f,g $\subset L_2[0,d]$, integrating over x,x′, taking the limit $n \to \infty$, and then using Equation 4.57 yields

$$(f,g) = \lim_{n\to\infty}\left(FG,R_n^d(k)\right)$$

If limit $n \to \infty$, $R_n{}^d(k)$, exists, then from Equation 4.56,

$$\left(FG,R^d\right) = (f,g)$$

where

$$\lim_{n\to\infty} R_n^d(k) = R^d(k)$$

$R_n{}^d(k)$ is, of course, a regular distribution. The limit is in the sense of convergence of distributions. This means that

$$\lim_{n\to\infty}\left(R_n^d(k),\phi(k)\right) \to \left(R^d(k),\phi(k)\right)$$

$R_n{}^d$, $R^d \subset \mathcal{D}$, the space of distributions and the limit holds for all test functions ϕ. Finally, if in the limit $d \to \infty$, $R^d(k) \to R$ also exists, then Equation 4.56 is established for all f(x),g(x) $\subset$ to $L_2[0,d]$ with compact support d.

We begin by considering what functions qualify for $R_n{}^d(k)$. In order to do this, we set x′ to be zero in Equation 4.57 and operate both sides with the inverse Povzner-Levitan transform operator (**I** + **L**). This yields for $0 < x < d$

$$U_n^d(x,0) = \left(y(k,x),R_n^d(k)\right) = \delta_n(x) \to \delta(x) \ \text{ as } n \to \infty$$

and

$$\left(\cos kx,R_n^d(k)\right) = \delta_n(x) + \int_0^x L(x,s)\delta_n(s)\,ds \tag{4.58}$$

(cos kx,$R_n{}^d(k)$) is the integral of (cos kx $R_n{}^d(k)$) with respect to k over R^+, and Equation 4.54 has been used. In other words, the LHS is a cosine Fourier transform of $R_n{}^d(k)$. The operator (**I** + **L**) transforms y(k,x) to cos kx, and the initial value constant h has been omitted. This shows that $\alpha R_n^d(k)$ must be the cosine Fourier transform of some function which coincides with the quantity on the RHS of Equation 4.58 on the interval $0 < x < d$. The constant α depends on the definition of the Fourier transform pair. Thus, if the transform pair is defined symmetrically (see Chapter 8) with the normalization constant of $(1/2\pi)^{1/2}$ in front of both the forward and the inverse transform, then $\alpha = (\pi/2)^{1/2}$. If the transform is defined asymmetrically with the normalizing factor of $(1/2\pi)$ only in front of the inverse transform, then $\alpha = \pi/2$.

We represent $\delta_n(x)$ by a delta sequence:

$$\int_0^\infty \delta_n(x)\,dx = 1$$

$$\delta_n(x) = 0 \text{ for } x = 0 \text{ and } x \geq 1/n$$

and

$$\delta_n(x) \geq 0 \text{ for } 0 < x < 1/n$$

The compactness on x is taken into account by multiplying the RHS of Equation 4.58 by a "characteristic function" $\chi_d(x)$ over the interval 2d. This yields

$$\chi_d(x) = 1,\ 0 \leq x \leq 2d,\ \chi_d(x) = 0,\ x \geq 2d$$

Marcenko[5] shows that

$$\alpha R_n^d(k) = \int_0^\infty \left[\delta_n(x) + \int_0^x L(x,s)\delta_n(s)\,ds\right]\chi_d(x)\cos kx\,dx \tag{4.59}$$

From Equation 4.59 we obtain

$$\left(\cos kx,\ R_n^d(k)\right) = \left[\delta_n(x) + \int_0^x L(x,s)\delta_n(s)ds\right]\chi_d(x) \tag{4.60}$$

The RHS has compact support and is continuously differentiable if δ_n and χ_d be chosen sufficiently smooth. The Fourier transform $R_n^d(k)$ is then bounded and integrable on the half-line. Operating on both sides of Equation 4.60 by (**I** + **K**), which transforms cos kx back into the eigenfunction y(k,x) (Theorem 4.5), then yields the relation

$$\left(y(k,x), R_n^d(k)\right) = \delta_n(x),\ 0 \leq x \leq 2d$$

This follows from the fact that (**I** + **K**) and (**I** + **L**) are inverses of each other.

Let us now take the limit $n \to \infty$ in Equation 4.59. This results in

$$\lim_{n\to\infty} R_n^d(k) = \alpha^{-1}\left[1 + C^\wedge\right] = R^d$$

where $C^\wedge$ is the Fourier cosine transform of the function $\chi_d(x)L(x,0)$.

We now consider the final step, which consists of taking the limit $d \to \infty$. In the limit $d \to \infty$, the function $\chi_d(x)L(x,0)$ tends uniformly to L(x,0) on every bounded interval. It is known that if a sequence of locally integrable functions $\{f_n(x)\}$ converges uniformly on every bounded interval on R^+ to a function f(x), then

$$\lim_{n\to\infty} \int_0^\infty f_n(x)\cos kx\,dx \to \int_0^\infty f(x)\cos kx\,dx = C^\wedge f$$

in the sense of convergence of distributions. In other words, $C^\wedge f_n$ is a regular distribution since f_n is locally integrable. The limiting procedure means that

$$\lim C^\wedge f_n \subset \mathcal{D} \rightarrow C^\wedge f \subset \mathcal{D}$$

where $\mathcal{D}$ is the space of the distributions. Hence, in the same limit $d \rightarrow \infty$, $C^\wedge \rightarrow C^\wedge(L)$ in the sense of distribution, the Fourier cosine transform of L(x,0). Therefore,

$$\lim_{d\to\infty}\left[\lim_{n\to\infty} R_n^d(k)\right] = \alpha^{-1}\left[1 + C^\wedge(L)\right] = R \subset \mathcal{D} \tag{4.61}$$

Next, we establish Parseval's relation given by Equation 4.56. We started out by postulating identity (4.57) and then showed that for Equation 4.57 to be satisfied, $R_n^d(k)$ must be the Fourier cosine transform as given by Equation 4.59. We now demonstrate that the identity (4.57) can, indeed, be derived and by applying the result of Equation 4.61 to this identity, the desired Parseval's relation (Equation 4.56) on the half-line is obtained.

Define the function:

$$v = \int_0^N \left(y(k,x)y(k,x')R_n^d(k)\right)\, dk$$

v is twice differentiable. It is also verified that v satisfies the Cauchy problem

$$v'' - q(x)v = \ddot{v} - q(y)v$$

with the initial conditions

$$v(x,0) = \int_0^N \left(y(k,x)R_n^d(k)\right)\, dk = \delta_n^N(x)$$

$$\dot{v}(x,0) = h\int_0^N \left(y(k,x)R_n^d(k)\right)\, dk = h\delta_n^N(x)$$

The solution is given by the Riemann formula (A.10). Using Equation A.10, taking the limit $N \rightarrow \infty$, and from the symmetry of the function $U_n^d(x,x')$ in the variables x and x′, we obtain the following expression,

$$U_n^d(x,x') = (1/2)\left[\delta_n(x+x') + \delta_n(|x-x'|)\right] + T_n, \quad 0 \le x, x' \le d \tag{4.62}$$

where

$$T_n = \int_{x-x'}^{x+x'} W(x,x',s)\, \delta_n(s)\, ds$$

The function W is bounded on every bounded domain. Consequently, there exists a constant A such that A is a function of d only, that is, A = A(d), and

$$|W(x,x',s)| \le A(d), \quad 0 \le x' \le x \le d \text{ and } 0 \le s \le 2d$$

Using this inequality and the definition of the delta sequence $\delta_n(s)$, we have

$$|T_n| \le A(d)\int_0^\infty \delta_n(x)\, dx = A(d), \quad 0 \le x' \le x \le d$$

Moreover, $T_n = 0$, $|x - x'| > (1/n)$ by the definition of the delta sequence, that is, $\delta_n(s) = 0$ if $s > 1/n$.

Now consider the functional $(f(x)g(x'),U_n^d(x,x'))$, where f and g $\subset L_2[0,d]$. The integral over T_n is

$$\left| \int_0^\infty \int_0^\infty T_n\, f(x)g(x')\, dx\, dx' \right| \le A(d) \iint_{D_n} \left| f(x)g(x') \right| dx\, dx'$$

where $D_n = \{(x,x') : |x - x'| \le (1/n),\ 0 \le x \le d,\ 0 \le x' \le d\}$. In the limit $n \to \infty$, this integral, therefore, vanishes as the measure in the domain $\to 0$ and the product $f(x)g(x')$ is integrable. Thus,

$$\lim_{n\to\infty} \left| \int_0^\infty \int_0^\infty T_n f(x)g(x')dxdx' = 0 \right| \tag{4.63}$$

Using the result given by Equation 4.63, Equations 4.57 and 4.61, and from the definition of the delta sequence $\delta_n(x)$, it follows that

$$\lim_{n\to\infty} \left(FG, R_n^d(k) \right) = (f,g)$$

It may be recalled that F and G are the generalized Fourier transforms of the functions f(x) and g(x), respectively. We have, therefore, shown that there is a distribution R^d = limit $n \to \infty$ $R_n^d(k)$ such that Parseval's relation holds for all elements of $L_2[0,d]$. Equation 4.61 shows that the distribution also exists in the limit $d \to \infty$. The Parseval relation holds, therefore, for all elements in the space of the square integrable functions.

For the Sturm-Liouville problem satisfying the second boundary condition $h = \infty$ with the solution vanishing at the origin, the spectral function can be obtained as follows. The sequence $U_n^d(x,x')$ now becomes

$$U_n^d(x,x') = \left(k^2\, y(k,x;\infty)y(k,x,\infty),\ R_n^d(k) \right) \tag{4.64}$$

and the functions $R_n^d(k)$ are determined from the condition

$$\left(\lambda^2 R_n^d(k), \sin kx/k \right) = -\left[\left\{ \delta_n(x) + \int_0^x ds \int_0^\infty L(x',s;\infty)\delta_n{}'(s)ds \right\} \chi_d(x) \right]'$$

Again taking the limit $d,n \to \infty$ yields

$$\lim_{d\to\infty} \left[\lim_{n\to\infty} R_n^d(k) \right] = \alpha^{-1}\left[1 + C\hat{}(M) \right] = R \subset \not{D} \tag{4.65}$$

In the above, $C\hat{}(M)$ is the Fourier cosine transform of the function

$$M = -\int_0^\infty L'(u,0;\infty)\, du$$

From the definition of the sequence (4.64), the spectral distribution is λR, R being given by Equation 4.65. The Parseval relation in this case is then

$$(f,g) = (FG, \lambda R)$$

Summarizing, the following theorems are obtained.

Theorem 4.7

The spectral distribution R for the boundary value problem:

$$y'' + [\lambda - q(x)]y = 0$$

with the boundary condition h = finite is connected with the kernel L(x,s;h) of the inverse Povzner-Levitan transform, transforming the solution y(x,k;h) to cos kx by the formula

$$R_h = \alpha^{-1}[1 + C^\wedge(L)]$$

where C^(L) is the cosine Fourier transform of L(x,0) and the constant α is either $(2/\pi)$ or $(2/\pi)^{1/2}$, depending on the asymmetric or symmetric definition of the Fourier transform pair. The Parseval relation is given by the functional

$$(f, g) = (FG, R_h)$$

for all $f,g \subset L_2(R^+)$.

Theorem 4.8

The spectral distribution R for the boundary value problem

$$y'' + [\lambda - q(x)]y = 0$$

with the boundary condition $h = \infty$ is connected with the kernel $L(x,s;\infty)$ of the inverse Povzner-Levitan transform, taking the solution $y(x,k;\infty)$ to sin kx/k by the formula

$$R_\infty = \lambda R = \alpha^{-1}[1 + C^\wedge(M)]$$

where C^(M) is the cosine Fourier transform of the function:

$$M = -\int_0^\infty L'(u, 0; \infty)\, du$$

The Parseval relation is given by the functional

$$(f, g) = (FG, \lambda R_\infty)$$

for all $f,g \subset L_2(R^+)$.

APPENDIX

RIEMANN'S SOLUTION OF CAUCHY'S PROBLEM

Consider the Cauchy problem:

$$u''(x,y) - q_1(x)u(x,y) = \ddot{u}(x,y) - q_2(x)u(x,y) \tag{A.1}$$

$$u(x,0) = f(x) \text{ and } \dot{u}(x,0) = g(x) \tag{A.2}$$

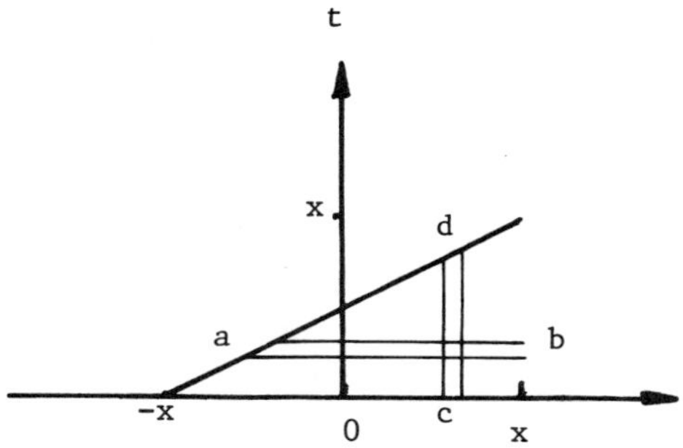

FIGURE 1. The region of integration in deriving Equation 4.59. The interchange of u and t variables is illustrated.

$x \subset R$ and $y \subset R^+$, although the latter condition is not essential. The prime $'$ denotes differentiation in x and the dot $\cdot$ that with respect to y. At the moment, the potentials $q_1(x)$ and $q_2(x)$ will be assumed to be continuously differentiable. Riemann's solution consists of determining an operator which acts upon the Cauchy data of Equation A.2 and yields the solution of the Cauchy problem. As will be seen, Riemann's solution of the Cauchy problem at a point (X,Y) consists of propagating the Cauchy data along the characteristics passing through the point. The solution is obtained in the form of an integral equation with a kernel which is a function of four parameters: x, y, X, and Y. This kernel, usually denoted by R(x,y;X,Y), is known as the "Riemann function".

Let the Riemann function be twice continuously differentiable and satisfy the equation

$$R'' - q_1(x)R = \ddot{R} - q_2(y)R \tag{A.3}$$

We are considering a solution at the point (X,Y). The lines AB and AC (Figure 1) passing through (X,Y) and following the equations

$$x \mp y = X \mp Y$$

are called the "characteristics". The integration is performed over region D of the figure. It is also assumed that the Riemann function has a value of unity on the characteristics. Multiplying Equation A.1 by R and Equation A.3 by u, subtracting, and, finally, integrating over region D yields, after some straightforward manipulation:

$$[u'R - uR']' - [\dot{u}R - u\dot{R}]y = \iint_D \left[(u'R - uR')' - (\dot{u}R - u\dot{R})_y \right] dx\, dy = 0 \tag{A.4}$$

The subscript y has been written to denote the differentiation of the bracketed quantities with respect to y. According to Green's formula, the surface integral of Equation A.4 can be converted to a line integral along the boundary Γ of region D and is written as

$$\int_\Gamma [(u'R - uR')]dy + (\dot{u}R - u\dot{R})\, dx = 0 \tag{A.5}$$

The boundary Γ consists of three segments (I, II, and III) and is to be traversed counterclockwise.

Let us evaluate the integral on segment I. On this path, y = Y – (x – X) and, therefore, dy = – dx. If F(x,y) is a differentiable function on this segment, then its derivative with respect to x is given by

$$F'(x, Y - (x - X)) = \left(F' - \dot{F}\right)\Big|_{y=Y-(x-X)} \tag{A.6}$$

Thus:

$$R'(x, Y - (x - X); X, Y) = \left(R' - \dot{R}\right)\Big|_{y=Y-(x-X)} = 0$$

since R is assumed to be unity on the characteristics. Using Equation A.6 and the fact that the Riemann function is unity on the characteristics, the integral along segment I is evaluated to be

$$\int_{I} = -u(X, Y) + f(X + Y) \tag{A.7}$$

Proceeding similarly, the following contribution is obtained for path II:

$$\int_{II} = -u(X, Y) + f(X - Y) \tag{A.8}$$

Finally, on segment III, dy = 0, u(x,y) = f(x) and $\dot{u}(x,y)$ = g(x) from the Cauchy data. Hence, the contribution from segment III is given by

$$\int_{III} = \int_{X-Y}^{X+Y} \left[g(x)R(x,0;X,Y) - f(x)\dot{R}(x,0;X,Y)\right] dx \tag{A.9}$$

Substituting Equations A.7 through A.9 in Equation A.5 yields

$$\begin{aligned} u(X, Y) = {} & (1/2)\left[f(X + Y) + f(X - Y)\right] + \\ & + 1/2 \int_{X-Y}^{X+Y} \left[g(x)R(x,0;X,Y) - f(x)\dot{R}(x,0;X,Y)\right] dx \end{aligned} \tag{A.10}$$

Equation A.10 is the Riemann solution of the Cauchy problem given by Equations A.1 and A.2. The solution exists, provided that a Riemann function R(x,y;X,Y) can be found with the properties specified above.

In order to prove the existence of the Riemann function, we change variables x and y to the "characteristics" (ξ,η) defined as

$$\xi = x + y$$
$$\eta = x - y$$

This results in

$$x = (\xi + \eta)/2$$

and

$$y = (\xi - \eta)/2$$

Define:

$$r(\xi, \eta; \xi_0 \eta_0) = R\{(\xi + \eta)/2, (\xi - \eta)/2; \xi_0, \eta_0\}$$

where $\xi_0 = X + Y$ and $\eta_0 = X - Y$. The equation satisfied by r is

$$4r_{\xi\eta} - [q_1\{(\xi + \eta)/2\} - q_2\{(\xi - \eta)/2\}]r = 0 \tag{A.11}$$

The region of integration is $\mathcal{D} = \{(\xi,\eta): \eta_0 \leq \eta \leq \xi \leq \xi_0\}$. The function r also satisfies the condition

$$r(\xi_0, \eta) = r(\xi, \eta_0) = 1$$

Notice that the Cauchy data (A.2) were not specified along the characteristics. The Cauchy data for r, on the other hand, are specified along the characteristics. In this case, the Cauchy problem is referred to as the "Goursat problem". The boundary-value problem for the function r is then a Goursat problem. Defining

$$s(\xi, \eta) = (1/4)[q_1\{(\xi + \eta)/2\} - q_2\{(\xi - \eta)/2\}]$$

and integrating Equation A.11 yields the integral equation for r:

$$r(\xi, \eta) = 1 - \int_\xi^{\xi_0} dt \int_{\eta_0}^{\eta} s(t, t')r(t, t')\, dt' \tag{A.12}$$

Equation A.12 can be solved by Picard's method of successive approximation. Expand r(t,t′) in the form of a series:

$$r(\xi, \eta) = \sum_{n=0}^{\infty} r^{(n)}(\xi, \eta) \tag{A.13}$$

Consider:

$$r^{(0)}(\xi, \eta) = 1$$
$$r^{(n)}(\xi, \eta) = -\int_\xi^{\xi_0} dt \int_{\eta_0}^{\eta} s(t, t')r^{(n-1)}(t, t')\, dt'$$

If sup |s(t,t′)| = M, then

$$\left|r^{(0)}(\xi, \eta)\right| \leq 1$$
$$\left|r^{(1)}(\xi, \eta)\right| \leq M^1(\xi_0 - \xi)(\eta_0 - \eta)$$
$$\left|r^{(2)}(\xi, \eta)\right| \leq \int_\xi^{\xi_0} dt \int_{\eta_0}^{\eta} \left|s(t, t')\right|\left|r^{(1)}(t, t')\right| dt'$$
$$\leq [M(\xi - \xi_0)(\eta - \eta_0)]^2 / (2!)^2$$

uniformly in D and by induction,

$$\left|r^{(n)}(\xi,\eta)\right| \leq \int_{\xi}^{\xi_0} dt \int_{\eta_0}^{\eta} \left|s(t,t')\right|\left|r^{(n-1)}(t,t')\right| dt'$$
$$\leq \left[M(\xi-\xi_0)(\eta-\eta_0)\right]^n/(n!)^2$$

Remembering that the region of integration is $\mathcal{D} = \{(\xi,\eta): \eta_0 \leq \eta \leq \xi \leq \xi_0\}$, the series converges uniformly and absolutely. The series representation of $r(\xi,\eta)$ given by Equation A.13 is then a continuous solution of Equation A.12.

Differentiating Equation A.12 with respect to ξ and η, we obtain

$$r_\xi(\xi,\eta) = \int_{\eta_0}^{\eta} s(\xi,t)r(\xi,t)\, dt \tag{A.14}$$

and

$$r_\eta(\xi,\eta) = -\int_{\xi}^{\xi_0} s(\eta,t)r(\eta,t)\, dt \tag{A.15}$$

respectively. If $q_1(x)$ and $q_2(x)$ are differentiable, then $s(\xi,\eta)$ is also differentiable. Consequently, $r(\xi,\eta)$ is twice continuously differentiable. Returning to the original variables (x,y), we have

$$R(x,y;X,Y) = r(x+y, x-y)$$

Since the function r has been shown to be twice differentiable, having a value of unity on the characteristics, it enjoys all the properties previously attributed to the Riemann function R. The existence of the Riemann function with the required properties, namely, twice continuous differentiability and a value of unity on the characteristics, is, therefore, established for differentiable potentials.

These results also generalize to potentials which are locally integrable and to potentials which are merely continuous. In the latter case, one considers a sequence of functions $q_1^{(n)}(x)$ and $q_2^{(n)}(y)$, which are differentiable functions and converge uniformly to $q_1(x)$ and $q_2(y)$ as $n \to \infty$, $x,y \subset R^+$. The corresponding solutions $R^{(n)}(x,y;X,Y)$ of Equation A.12 then also converge uniformly in D to the Riemann function $R(x,y;X,Y)$ for $q_1(x)$ and $q_2(y)$. The functions $R^{(n)}(x,y;X,Y)$ satisfy the same equation as Equation A.3, but with the potentials replaced by $q^{(n)}$s. The corresponding solution is given by

$$\begin{aligned} u(X,Y) &= (1/2)\left[f(X+Y) + f(X-Y)\right] \\ &+ (1/2)\int_{X-Y}^{X+Y}\left[g(x)R^{(n)}(x,0;X,Y) - f(x)\dot{R}^{(n)}(x,0;X,Y)\right] dx \\ &+ (1/2)\iint_D \left[q_1^{(n)}(x) - q_1(x) + q_2^{(n)}(y) - q_2(y)\right]R^{(n)}(x,0;X,Y)\right] dxdy \end{aligned}$$

Notice that as $n \to \infty$, Equation A.10 is recovered.

Summarizing these results, we have the following theorem.

Theorem A.1

If the potentials are continuous, then every twice continuously differentiable solution of the Cauchy problem (A.1) and Equation A.2 is given by Equation A.10, with the Riemann function R(x,y;X,Y) admitting of the integral representation of Equation A.13. Moreover, the Riemann function thus determined is also twice continuously differentiable. This theorem applies, of course, to the region of integration D.

REFERENCES

1. **A. Ya. Povzner:** On differential equations of Sturm-Liouville type on a half-axis, *Am. Math. Soc. Transl.*, No. 5, 1950.
2. **B. M. Levitan:** Application of Generalized Displacement Operators to Second Order Differential Equations, *Am. Math. Soc. Transl.*, No. 59, 1951.
3. **J. Delsartes:** Sur une extension de la formule de Taylor, *J. Math. Pures. Appl.*, 17, 1938.
4. **K. Chadan and P. C. Sabatier:** *Inverse Problems In Quantum Scattering Theory*, Springer-Verlag, New York, 1977.
5. **V. A. Marcenko:** *Sturm-Liouville Operators And Applications*, Birkhauser-Verlag, Boston, 1986.
6. **J. Delsartes:** Une extension nouvelle de la theorie des fonctions presque-peridiques de Bohr, *Acta Math.*, 69, 1938.
7. **N. Wiener And R. Paley:** *Fourier Transform In A Complex Domain*, American Mathematical Society Coll. Publ. No. 19, American Mathematical Society, 1934.
8. **R. P. Boas:** *Entire Functions*, Academic Press, New York, 1972.
9. **A. Papoulis:** *Signal Analysis*, McGraw-Hill, New York, 1977.
10. **J. R. Taylor:** *Scattering Theory*, John Wiley & Sons, New York, 1972.
11. **I. S. Gradshtein and I. M. Ryzhik:** *Table of Integrals, Series, and Products*, Academic Press, New York, 1980.
12. **B. M. Levitan:** *Generalized Translation Operators*, Israel Program for Scientific Translations, Jerusalem, 1964.
13. **E. C. Titchmarsh:** *The Theory of Functions*, Oxford University Press.

Chapter 5

THE GEL'FAND-LEVITAN EQUATION

I. DERIVATION OF THE GEL'FAND-LEVITAN EQUATION

The spectral function and its relation to the spectrum of a Sturm-Liouville boundary value problem on the half-line $x \subset R^+$ were discussed in Chapter 3. Chapter 4 introduced the direct (inverse) Povzner-Levitan transforms on the half and the entire real axis, transforming the free (scattering) eigenfunction to scattering (free) solutions of the Sturm-Liouville boundary-value problem. These results will now be used in obtaining the Gel'fand-Levitan equation,[1] the most celebrated solution of the inverse Sturm-Liouville problem on the half-line. As pointed out in Chapter 1, the method is based on the regular solution of the boundary-value problem.

Let us consider the boundary condition h = 0 first. The corresponding Povzner-Levitan transform is denoted by K(x,x′) and the solution of the Sturm-Liouville boundary value problem is given by

$$y(x,k) = \cos kx + \int_0^x K(x,x') \cos kx' \, dx', \quad x' \le x \tag{5.1}$$

A completeness relation was derived in Chapter 3 (Equation 3.51). This is reproduced below.

$$\int_{-\infty}^{\infty} y(x,k)y(x',k)d\sigma(k) = \delta(x-x')$$

For $x' < x$, this reduces to

$$\int_{-\infty}^{\infty} y(x,k)\, y(x',k)\, d\sigma(k) = \delta(x-x') = 0 \tag{5.2}$$

The eigenfunctions y(x,k) are thus orthonormal with respect to the weight given by the spectral function σ(k). From Equation 5.1, y(x′,k) are combinations of cos kx′. It thus follows from Equation 5.2 that cos kx′ is orthogonal to y(x,k) with respect to the weight function σ(k) for $x' < x$. That is,

$$\int_{-\infty}^{\infty} y(x,k) \cos kx' \, d\sigma(k) = 0 \tag{5.3}$$

Replacing y(x,k) by the representation (5.1) in Equation 5.3 yields

$$\int_{-\infty}^{\infty} \cos kx \cos kx' \, d\sigma(k) + \int_{-\infty}^{\infty} \cos kx' \left\{ \int_0^x K(x,x'') \cos kx'' \right\} d\sigma(k) = 0 \tag{5.4}$$

The spectral function σ(k) can be decomposed as follows:

$$\begin{aligned} \sigma(\lambda) &= \sigma_0(k) + \rho(\lambda), \quad k > 0, \quad \lambda = k^2 > 0 \\ &= \rho_j(\lambda), \quad k < 0, \quad \lambda = k^2 < 0, \quad k = ik_j, \quad k_j > 0 \end{aligned} \tag{5.5}$$

$\sigma_0(k)$ is the "free" spectral function, that is, the spectral function of the scattering-free problem:

$$y'' + \lambda y = 0, \quad x \subset R^+$$

$$y(0,\lambda) = 1 \quad \text{and} \quad y'(0,\lambda) = 0$$

From Chapter 3 (Equation 3.39), $d\rho(\lambda) = (2/\pi)[U^{-2}(\lambda) - 1]d\sqrt{\lambda}$. By virtue of the bounds established on the functions $\mu(\tau)$ and $\nu(\tau)$, $U^{-2}(\lambda) - 1 \to 1/|\lambda|$ as $\lambda \to \infty$. This means that the spectral function $\rho(\lambda)$ is well behaved at infinity and obeys a smoothness condition

$$\int_R |\lambda|^{1-\epsilon} d\rho(\lambda) < \infty, \epsilon > 0$$

The free spectral function for the problem above is $(2/\pi)k$ (see the first example of Chapter 3). Therefore,

$$\sigma_0(k) = \left(\frac{2}{\pi}\right)k \tag{5.6}$$

We now replace $\sigma(k)$ in Equation 5.4 by Equations 5.5 and 5.6, and after some straightforward algebra obtain

$$K(x,x') + F(x,x') + \int_0^x K(x,x'')F(x',x'')\, dx'' = 0, \quad x' < x \tag{5.7}$$

In the above derivation, the order of integration has been changed in the second integral of Equation 5.4, which is legitimate as the integrands are absolutely convergent, and the following identities have been used:

$$\left(\frac{2}{\pi}\right)\int_{-\infty}^{\infty} \cos au \cos bu\, du = \delta(a - b)$$

and

$$\int_{-\infty}^{\infty} \delta(u - v)\, f(v)\, dv = f(u)$$

The function $F(x,x')$ is given by

$$F(x,x') = \int_{-\infty}^{\infty} \cos kx \;\cos kx'\, d\rho(k) \tag{5.8}$$

where $d\rho(\lambda) = d\rho(\lambda) + d\rho_j(\lambda)$. In view of the decomposition of the spectral function as given by Equation 5.5, this can also be written as

$$F(x,x') = F_c + F_d$$

where the continuous part F_c is given by

$$F_c = \left(\frac{2}{\pi}\right)\int_0^{\infty} \cos kx \ \cos kx' \left[U^{-2}(k) - 1\right] dk \tag{5.9}$$

and the discrete part F_d is determined to be

$$F_d = \sum_j r_j^2 \cosh k_j x \cosh k_j x' \tag{5.10}$$

The function U is given by Equation 3.39 of Chapter 3 and the r_j^2 are the normalization constants. Equation 5.10 follows because of the conditions given in Equation 5.5.

Equation 5.7 is the celebrated Gel'fand-Levitan equation. The basic premise is that the spectral function $\sigma(k)$ is known. Hence $F(x,x')$ is a known function. Equation 5.7 is deceptive in appearing to be a Volterra equation since the upper limit of integration is the variable x itself. However, for each fixed x, Equation 5.7 is actually a Fredholm integral equation of the second kind in $K(x,x')$ in the variable x', with x acting as a parameter. The function $F(x,x')$ is the inhomogeneous term as well as the kernel of the integral equation.

Let us now consider the second boundary condition $h = \infty$. The Povzner-Levitan representation is now

$$z(x,k) = \left(\frac{1}{k}\right)\sin kx + \int_0^{x} L(x,x')\left(\frac{1}{k}\right) \sin kx' \, dx'$$

where $(1/k) \sin kx$ is the "free" eigenfunction for the boundary condition $h = \infty$. Again using the completeness relation, invoking the orthonormality of sin kx and $z(x',k)$ for $x' < x$, and following exactly the same line of reasoning as above in the case of $h = 0$ boundary condition, we obtain the Gel'fand-Levitan equation:

$$L(x,x') + G(x,x') + \int_0^{x} L(x,x')\, G(x',x'')\, dx'' = 0, \quad x' < x \tag{5.11}$$

In the above derivation, the inhomogeneous term $G(x,x')$, which is also the kernel of the integral equation, is given by

$$G(x,x') = \int_{-\infty}^{\infty} \left[\left(\frac{1}{k}\right)\sin kx\right]\left[\left(\frac{1}{k}\right)\sin kx'\right] d\rho(k)$$

and use has been made of the following identity:

$$\left(\frac{2}{\pi}\right)\int_0^{\infty}\left[\left(\frac{1}{k}\right)\sin kx\right]\left[\left(\frac{1}{k}\right)\sin kx'\right]k^2 dk = \delta(x - x')$$

The free spectral function $\sigma_0(k)$ in this case is given by

$$d\sigma_0(k) = \left(\frac{2}{\pi}\right) k^2 dk$$

and the continuous part of the spectral function on the positive half-axis is written as

$$d\sigma(k) = \left(\frac{2}{\pi}\right)\left[V^{-2}(k) - 1\right] k^2 dk$$

where

$$V^{-2}(k) = \left[I_1^2 + \left(1 - I_2^2\right)\right]^{-1}$$

the factors I_1 and I_2 being just the integral parts of Equations 3.27 and 3.28 for μ and υ, respectively. Using these results, we obtain for G(x,x′) the following expression,

$$G(x, x') = G_c + G_d$$

where the continuous part G_c is written as:

$$G_c = \left(\frac{2}{\pi}\right)\int_0^{\infty}\left[\left(\frac{1}{k}\right) \sin kx\right]\left[\left(\frac{1}{k}\right) \sin kx'\right]\left[V^{-2}(k) - 1\right] k^2 dk \tag{5.12}$$

and the discrete part G_d is determined to be

$$G_d = \sum_j r_j^2 k_j^{-2} \sinh k_j x \sinh k_j x' \tag{5.13}$$

The functions F and G in Equations (5.7) and (5.11) contain the functions U and V which, in turn, contain μ and υ, respectively. The boundedness of the latter was established in Chapter 3 for a potential $\subset L_1$. The continuity properties of F and G in their variables follow from the continuity and boundedness of the functions U and V. If, however, the potential is not integrable at the origin, as supposed here, complications will arise.

In the context of quantum scattering, the function U in the completeness relation given by Equation 3.51 is replaced by |F(k)| where F(k) the "Jost function" which relates the regular solution of the radial Schrödinger equation to its physical wave function. The boundary condition in quantum problem is $h = \infty$ and the Gel'fand-Levitan equation takes the form given by Equation 5.11. Using appropriate integral representations of the Jost function and the condition that $F(k) \to 1$ as $k \to \infty$, k real, and following the same line of reasoning as in Section VI of Chapter 4, it can be shown that G(x,x′) is finite for $x' < x$, whereas it is infinite in general if $x = x'$. Replacing the eigenfunction $y_\infty(k,x)$ by its asymptotic form (1/k) sin kx and again following the discussion of Section VI of Chapter 4, it can be shown that G(x,x′) remains finite even for $x' \leq x$, if the potential is integrable at the origin. A complete discussion of these points is given by Sabatier and Chadan (Reference 7).

II. UNIQUENESS OF SOLUTIONS

The Gel'fand-Levitan equation is a linear Fredholm integral equation of the second kind in K(x,x′)[L(x,x′)] considered as functions of x′ for each fixed x. In order for its solutions to be unique, Fredholm's alternative, stated below, must be satisfied.[2]

Fredholm's Alternative

A linear integral equation of the second kind

$$y = f + \mathbf{K}y$$

has a unique solution if and only if the homogeneous equation

$$y = \mathbf{K}y$$

has only a trivial solution.

Let $K_0(x,x')$ be a solution of the homogeneous Gel'fand-Levitan equation. Then:

$$K_0(x,x') + \int_0^x K_0(x,x'')\, F(x',x'')\, dx'' = 0 \tag{5.14}$$

The proof of uniqueness then lies in showing that Equation 5.14 has only a trivial solution, that is, $K_0(x,x') \equiv 0$. Toward this end, let us consider the following quadratic expression:

$$I[K_0(x)] = \int_0^x K_0^2 dx' + \int_0^x\int_0^x K_0(x,x')K_0(x,x'')F(x',x'')dx'dx'' = 0$$

Using Equation 5.8 for F in the above equation yields

$$I[K_0(x)] = \int_0^x K_0^2\, dx' + \int_{-\infty}^{\infty} \hat{K}_0^2(x,k)\, d\rho(k) = 0 \tag{5.15}$$

where

$$d\rho(k) = d\rho(k) + d\rho_j(k_j)$$

and

$$\hat{K}_0(x,k) = \int_0^x \cos kz\, K_0(x,s)\, ds$$

is the cosine Fourier transform of $K_0(x,s)$.

Using the decomposition of the spectral function $\sigma(k)$, the second integral of Equation 5.15 becomes, on the positive half-line,

$$\int_0^{\infty} \hat{K}_0^2(x,k)\, d\rho(k) = \int_0^{\infty} \hat{K}_0^2(x,k)\, d\sigma(k) - \left(\frac{2}{\pi}\right)\int_0^{\infty} \hat{K}_0^2(x,k)\, dk \tag{5.16}$$

From the definition of K_0^(x,k) given above, and using Parseval's relation for the ordinary Fourier integrals, we have

$$\left(\frac{2}{\pi}\right)\int_0^{\infty} \hat{K}_0^2(x,k)\, dk = \int_0^x K_0^2(x,s)\, ds \tag{5.17}$$

Replacing Equations 5.16 and 5.17 in Equation 5.15 yields

$$I[K_0(x)] = \int_{-\infty}^{\infty} \hat{K}_0^2(x,k)\, d\sigma(k) = 0$$

where $d\sigma(k)$ includes both the continuous and the discrete part.

Now, $K_0(x,s)$ is integrable at $x = s$ and is an $L_1(0,x)$ function. Therefore, by the Wiener-Boas theorem, $K_0\hat{}(x,k)$ is an entire function of exponential growth in k. $K_0\hat{}(x,k)$, therefore, must have only isolated zeros,[3] with the accumulation point at ∞. Since the spectrum is discrete on the negative real axis, I must be zero there, the measure of the integral being zero. On the positive axis, the spectrum is continuous and $\sigma(k)$ has infinite points of increase on any compact interval. $I(K_0) = 0$ must, therefore, imply that $K_0\hat{}(x,k) = 0$, that is, $K_0(x,s)$ must be zero. Hence, the Fredholm alternative is satisfied, establishing the uniqueness of the solutions of the Gel'fand-Levitan equation for every fixed x.

For the second boundary condition $h = \infty$, we write

$$\hat{G_0}(x,k) = \int_0^x L_0(x,s)\left(\frac{1}{k}\right)\sin ks \, ds$$

It is already known that this is also an entire function of exponential growth in k. Therefore, following exactly the same reasonings as before, the uniqueness of the corresponding solution is also established.

III. CONTINUITY PROPERTIES OF THE POVZNER-LEVITAN TRANSFORMS

It has thus been shown that for each fixed x, the Gel'fand-Levitan equation has a unique solution for the transforms $K(x,x')$ and $L(x,x')$. One can establish the continuity properties of these transforms as follows. For each fixed x, the Gel'fand-Levitan equation is a Fredholm integral equation of the second kind for $K(x,x')$ or $L(x,x')$ in the variable x'. The kernel functions F and G have been shown to be continuous and bounded. Hence, according to the Fredholm theory, the equation admits of a resolvent and can be written as

$$M(x,x') = p(x,x') + \int_0^x M^{\sim}(x,u)p(x',u) \, du$$

M has been written to designate either K or L and p indicates either F or G. $M^{\sim}(x,u)$ is the resolvent operator for M. Since $M(x,x')$ is a unique solution, then, from Fredholm's theory, the resolvent is continuous. The continuity of $M(x,x')$ in the variable x', x being a parameter, then follows from that of the functions F and G. In particular, if these functions possess m-th derivatives with respect to x', then so does $M(x,x')$.

IV. BEHAVIOR OF M(x,x′) WITH x

The behavior of $M(x,x')$ with respect to the second variable x' when the first variable x is held fixed was discussed above. The reverse question, that is, the behavior of $M(x,x')$ with respect to x can be established as follows. Since the variable x happens to be the upper limit of integration, it is convenient to change the variables such that this limit is a constant quantity. This can be achieved by substituting $x' = sx$ and $u = tx$, in terms of which the Gel'fand-Levitan equation becomes (in the notation of the paragraph above)

$$M(x,sx) + p(x,sx) + x\int_0^1 M(x,tx)\, p(sx,tx) \, dt = 0 \tag{5.18}$$

Consider the following equation:

$$g(x,a) = h(x,a) + \int_0^1 H(x,x';a)\, h(x',a)\, dx' \tag{5.19}$$

where the kernel H(x,x';a) and the inhomogeneous term g(x,a) are continuous functions of the parameter a and the independent variables. Rewrite the above integral equation formally as

$$g = (\mathbf{I} + \mathbf{H})h$$

Let the homogeneous equation have only the trivial solution at some point $a = a_0$. Then, of course, the inhomogeneous equation has a unique solution at the point a_0 of the parameter a. Then, for a in some neighborhood of a_0, we have

$$\mathbf{H}(x,x';a) = \mathbf{H}(x,x';a_0) + \mathbf{H}_1(x,x') = \mathbf{H}_0 + \mathbf{H}_1$$

where $|\mathbf{H}_1(x,x')| < \varepsilon$, $a \subset$ a sufficiently small neighborhood of a_0. Equation 5.19 can then be written as

$$g = (\mathbf{I} + \mathbf{H}_0)h + \mathbf{H}_1 h$$

yielding

$$(\mathbf{I} + \mathbf{H}_0)^{-1} g = h + (\mathbf{I} + \mathbf{H}_0)^{-1}\, \mathbf{H}_1 h$$

Since the norm of the operator $(I + \mathbf{H}_0)^{-1}\mathbf{H}_1$ can be made arbitrarily small, it follows that h(x,a) is a continuous function of x and a in some neighborhood of the point $a = a_0$. These results can be collected together in the following lemma.

Lemma 5.1

Let there be given the integral equation

$$g(x,a) = h(x,a) + \int_0^1 H(x,x';a)\, h(x',a)\, dx'$$

where the kernel and the inhomogeneous term g(x,a) are continuous functions of the parameter a and the independent variables. Let the homogeneous equation possess only a trivial solution at some point $a = a_0$. Then the solution h(x,a) is continuous in x and a in some neighborhood of the point $a = a_0$. Moreover, if the kernel H and the term g possess m continuous derivatives with respect to a, then so does the solution h(x,a).

Now consider the Gel'fand-Levitan equation as given in Equation 5.18, which is an integral equation with p(x,sx) as the free term and xp(sx,tx) as the kernel. x acts as a parameter in this equation. Since the free term is a continuous function by hypothesis, the continuity of the function M(x,x′) in x follows from Lemma 5.1. Moreover, by the same lemma, the order of continuous derivatives of M(x,x′) is the same as that of p(x,x′). Together with the results of Section 3 this implies that the function M(x,x′) is continuous in the entire set of variables.

V. RELATION BETWEEN M(x,x′) AND THE POTENTIAL q(x)

Having derived the Gel'fand-Levitan equation and investigated the properties of its solution, the next step is to solve the inverse problem. Equivalently, this means establishing the connection between M(x,x′) and the potential q(x). Recall that M(x,x′) designates either K(x,x′) or L(x,x′). Let us first consider the boundary condition h = 0, for which the solution y(x,k) of the scattering equation

$$y''(x,\lambda)+[\lambda-q(x)]y(x,\lambda)=0$$

is given by Equation 5.1 in terms of the Povzner-Levitan transform K(x,x′). We differentiate the latter in order to obtain y″. Next, we integrate the integral part of Equation 5.1 twice by parts and use the result in writing the term λy, replace Equation 5.1 for y in the term q(x)y(x,k), and finally, substitute these in the above Sturm-Liouville equation. After some straightforward algebra, the following identity results:

$$\left[K'(x,x)+(\cdot)\Big|_{x=x'}-q(x)\right]\cos kx+\dot{K}(x,x')\Big|_{x'=0}+$$

$$+\int_0^x\left[K''-\ddot{K}-q(x)K\right](x,x')\ \cos kx'\,dx'=0 \tag{5.20}$$

In Equation 5.20, (·) = K′(x,x′) + $\dot{K}$(x,x′), and ′ resp. · represents differentiation with respect to x resp. x′. From Equation 5.20, the following conditions are obtained:

$$K''(x,x')-\ddot{K}(x,x')=q(x)\,K(x,x') \tag{5.21}$$

$$\dot{K}(x,x')\Big|_{x'=0}=0 \tag{5.22}$$

and

$$2K'(x,x)=q(x) \tag{5.23}$$

Next, we derive an integral equation for K(x,x′). Let us define the characteristics

$$\xi=\frac{(x+x')}{2}\quad\text{and}\quad\eta=\frac{(x-x')}{2}$$

for the hyperbolic equation (5.21) for K. In terms of the characteristics, this equation becomes (as in the Appendix of Chapter 4)

$$4\Phi_{\xi\eta}=q(\xi+\eta)\Phi(\xi,\eta)$$

a successive integration of which yields the solution

$$\Phi(\xi,\eta)=f(\xi)+\left(\frac{1}{4}\right)\int_\eta^\xi\int_0^\eta q(\xi'+\eta')\,\Phi(\xi',\eta')\,d\xi'\,d\eta'+g(\eta) \tag{5.24}$$

In Equation 5.24, the functions f(ξ) and g(η) are to be determined, and

$$\Phi(\xi,\eta) = K[\xi+\eta,\ \xi-\eta]$$

The initial condition on x′ = x passes into the initial condition on the characteristic ξ and η = 0 for this case. Similarly, the condition on x′ = 0 passes into the condition $(\partial/\partial\eta - \partial/\partial\xi)_{\xi=\eta}$. Determining f(ξ) and g(η) from (5.24), (5.22) and (5.23) and then returning to the original variables, the integral equation for the kernel K(x,x′) is obtained as

$$K(x,x') = \left(\frac{1}{2}\right)\int_B^A q(s)\,ds + \int_B^A q(s)\,K(s+t,\ s-t)\,ds\,dt + K(0,0) \tag{5.25}$$

where A = (1/2)(x + x′), B = (1/2)(x – x′) and K(0,0) is identified with the g(η) function. From Equation 5.23:

$$K'(x,x) = \left(\frac{1}{2}\right)q(x)$$

This, then, determines the inverse problem:

$$K(x,x) = \text{constant} + \left(\frac{1}{2}\right)\int_0^x q(s)\,ds \tag{5.26}$$

Differentiating Equation 5.1 with respect to x and from the condition h = finite, namely, y′(0,k) = hy(0,k), it follows that K(0,0) = 0 if h = 0. Then

$$K(x,x) = \left(\frac{1}{2}\right)\int_0^x q(s)\,ds \tag{5.27}$$

The solution (5.25) can also be obtained alternatively by replacing the representation given by Equation 5.1 or its counterpart for the h = ∞ boundary condition on both sides of the solution of the Sturm-Liouville equation given by Equation 3.26. This yields the Fourier transform (cosine transform if the boundary condition is h = 0 and sine transform if h = ∞) of the Povzner-Levitan kernel K or L from which Equation 5.25 can be derived. This derivation is, however, lengthy and involves a substantial change of variables and orders of integration. The details can be found in Marcenko.[4] It should be pointed out that it is the representation given by Equation 5.25 that was used in Chapter 4 (Section VII) for establishing the validity of the Povzner-Levitan transform in 0 ≤ r′ ≤ r.

For the second boundary condition h = ∞, we follow the very same line of reasoning and establish the following:

$$L'' - \ddot{L} = q(x)L(x,x')$$

$$\dot{L}(x,0) = 0$$

and

$$L(x,x) = \left(\frac{1}{2}\right)\int_0^x q(s)\,ds \tag{5.28}$$

The uniqueness of the potential follows from that of solutions K(x,x′) or L(x,x′) of the Gel'fand-Levitan equation. If the potential is an ordinary continuous function, then

$$q(x) = 2K'(x, x)$$

Now, for the potential q(x) to be determined by K(x,x) or by L(x,x), it must be permissible for the second variable to be set equal to the first. We have already shown that this is valid when the potential is L_1, that is, integrable at the origin. This is certainly valid for a potential which is summable over every finite interval, and this was the basic assumption of the original work of Gel'fand and Levitan. If this condition is violated and the Povzner-Levitan kernel becomes infinite at the upper limit, then the potential cannot be determined by Equation 5.27 or 5.28.

VI. DETERMINATION OF THE BOUNDARY-VALUE PROBLEM

In the previous analysis, we started with the Sturm-Liouville equation

$$y'' + [\lambda - q(x)]y = 0$$

with the boundary condition of the form

$$y'(0) - hy(0) = 0$$

An auxiliary boundary-value problem was constructed with a known, given potential and the integral equation of Gel'fand and Levitan was derived, which was written as

$$K(x, x') + F(x, x') + \int_0^x K(x, t)\, F(x', t)\, dt = 0, \quad x' < x$$

This was shown to be a linear Fredholm integral equation of the second kind in the function K(x,x′). The inhomogeneous term F(x,x′), which is also the kernel of the integral equation, was constructed from the eigenfunctions of the auxiliary boundary-value problem on the basis of the spectral measure derived from the spectral function specified as the input data. Moreover, the Gel'fand-Levitan equation was shown to possess a unique solution. It was demonstrated that if the function K(x,x′) satisfied the Gel'fand-Levitan integral equation, then the solution of the auxiliary boundary-value problem, when transformed by this function, resulted in the solution of the original Sturm-Liouville problem. In the course of this analysis, the inverse problem was also solved, namely, that of recovering the unknown potential from the knowledge about the Povzner-Levitan transform K(x,x′). In this section, we consider, so to speak, the inverse Gel'fand-Levitan problem. That is, we start out with the Gel'fand-Levitan equation and from it determine what boundary-value problem corresponds to this integral equation. Specifically, the objective is to reconstruct the differential equation *cum* the boundary condition corresponding to the Gel'fand-Levitan equation. The analysis will also be found to yield the potential automatically. We follow Gladwell[5] here.

A few comments about the notations to be used here and after. Previously, the differentiation of a function f(x,x′) with respect to x was denoted by f′(x,x′) and that with respect to x′ by $\dot{f}(x,x')$. In what follows, we will have occasion to differentiate not only with respect to x and x′, but also in other additional variables. Differentiation of a function with respect to a variable other than x and x′ will be denoted by subscripting that variable. Furthermore, by a function such as $f_u(0,v)$ is meant $f_u(u,v)$ at $u = 0$.

Consider the following auxiliary problem with a known potential $q_0(x)$:

$$z'' + [\lambda - q_0(x)]z = 0 \tag{5.29}$$

$$z'(0) - h_0 z(0) = 0 \tag{5.30}$$

The kernel $F(x,x')$ of the Gel'fand-Levitan equation is of the form

$$F(x,x') = \int_{-\infty}^{\infty} z(\lambda,x)z(\lambda,x')d\rho(k)$$

It then follows that

$$F'(0,x') - h_0 F(0,x') = \dot{F}(x,0) - h_0 F(x,0) = 0 \tag{5.31}$$

and

$$F''(x,x') - q_0(x)F(x,x') = \ddot{F}(x,x') - q_0(x')F(x,x') \tag{5.32}$$

From the Gel'fand-Levitan integral equation,

$$F(x,x') + K(x,x') + \int_0^x K(x,t)\,F(t,x')\,dt = 0$$

we have the identities

$$K(0,x') + F(0,x') = 0 \tag{5.33}$$

and

$$\dot{F}(x,x') + \dot{K}(x,x') + \int_0^x K(x,t)\,\dot{F}(t,x')\,dt = 0 \tag{5.34}$$

From Equations 5.31, 5.33, and 5.34, we have

$$\dot{K}(x,0) - h_0 K(x,0) = 0 \tag{5.35}$$

From Equations 5.31 and 5.35,

$$K(x,t)\,F_t(t,x') - K_t(x,t)F(t,x') = 0 \quad \text{at} \quad t = 0 \tag{5.36}$$

Differentiate the Gel'fand-Levitan equation twice with respect to x and obtain

$$K''(x,x') + K'(x,x)F(x,x') + K(x,x)F'(x,x') + K'(x,t)\big|_{t=x} F(x,x') +$$

$$+ \int_0^x K''(x,t)F(t,x')\,dt + F''(x,x') = 0 \tag{5.37}$$

Similarly, differentiation of the Gel'fand-Levitan equation twice with respect to the variable x' yields

$$\ddot{K}(x,x') + \int_0^x K(x,t)\left[F_{tt}(t,x') + \{q_0(x') - q_0(t)\}F(t,x')\right] dt +$$
$$+F''(x,x') + \left[q_0(x') - q_0(x)\right]F(x,x') = 0 \tag{5.38}$$

Reducing the integral of $K(u,t)F_{tt}(t,x')$ results in

$$\int_0^x K(x,t)\, F_{tt}(t,x')\, dt = \int_0^x K_{tt}(x,t)\, F(t,x')\, dt + \left[K(x,t)F_t(t,x') - K_t(x,t)F(t,x')\right]\Big|_0^x$$

The integrated quantity at the lower limit vanishes by virtue of Equation 5.36, and we have

$$\int_0^x K(x,t)F_{tt}(t,x')\, dt = \int_0^x K_{tt}(x,t)F(t,x')\, dt + \left[K(x,t)F_t(t,x') - K_t(x,t)F(t,x')\right]\Big|_{t=x} \tag{5.39}$$

Replacing Equation 5.39 in Equation 5.38 yields

$$\ddot{K}(x,x') + \int_0^x \left[K_{tt}(x,t) + \{q_0(x') - q_0(t)\}K(x,t)F(t,x')\right] dt + F''(x,x') +$$
$$+\left[q_0(x') - q_0(x)\right] F(x,x') + \left[K(x,x)F'(x,x') - K_t(x,t)\right]\Big|_{t=x} F(x,x') = 0 \tag{5.40}$$

Subtracting Equation 5.40 from Equation 5.37 then gives

$$I(x,x') + \int_0^x I(x,t)\, F(t,x')\, dt + 2K'(x,x)F(x,x') + I^{\sim}(x,x') = 0 \tag{5.41}$$

where

$$I(x,x') = K''(x,x') - \ddot{K}(x,x') \tag{5.42}$$

$$I^{\sim}(x,x') = \int_0^x \{q_0(t) - q_0(x')\}K(x,t)F(t,x')\, dt + \left[q_0(x) - q_0(x')\right]F(x,x') \tag{5.43}$$

and

$$K'(x,t)\big|_{t=x} + K_t(x,t)\big|_{t=x} = K'(x,x)$$

is the derivative along the characteristic.

The expression for $I^{\sim}$ can be rewritten by using the Gel'fand-Levitan equation, as

$$I^{\sim}(x,x') = \int_0^x \{q_0(t) - q_0(x)\}K(x,t)F(t,x')\, dt + \left[q_0(x') - q_0(x)\right]K(x,x') \tag{5.44}$$

Replacing Equation 5.44 in Equation 5.41 yields

$$J(x,x') + \int_0^x J(x,t)\, F(t,x')\, dt + 2K'(x,x)F(x,x') = 0 \tag{5.45}$$

where the function J(x,x′) has been defined as

$$J(x,x') = I(x,x') + \{q_0(x') - q_0(x)\}K(x,x')$$

Replacing F(x,x′) in the above equation from the Gel'fand-Levitan equation,

$$F(x,x') + K(x,x') + \int_0^x K(x,t)\, F(t,x')\, dt = 0$$

J(x,x′) is obtained as

$$J(x,x') = 2K'(x,x)K(x,x')$$

Writing this in full yields

$$K''(x,x') - \ddot{K}(x,x') + \{q_0(x') - q_0(x)\}K(x,x') = 2K'(x,x)K(x,x') \tag{5.46}$$

Equation 5.46 is then the partial differential equation satisfied by the function K(x,x′). Compare with Equation 5.21. Equation 5.46 is an explicit equation for K(x,x′) in terms of the eigenfunctions of an auxiliary problem involving a known potential $q_0(x)$.

Now, the solution y of the boundary-value problem for which the Gel'fand-Levitan equation has been written is given by

$$y(x,k) = z(x,k) + \int_0^x K(x,t)\, z(x,t)\, dt \tag{5.47}$$

Differentiating in x, we obtain

$$y'(k,x) = z'(x,k) + K(x,x)z(x,k) + \int_0^x K'(x,t)\, z(k,t)\, dt$$

From this, it follows that

$$y'(0,k) = z'(0,k) + K(0,0)z(0,k) = [K(0,0) + h_0]\, y(0,k) \tag{5.48}$$

Differentiating Equation 5.47 twice with respect to x gives

$$y''(x,k) = z''(x,k) + K(x,x)\, z'(x,k) + \int_0^x K''(x,t)\, z(t,k)\, dt +$$

$$+K'(x,t)\big|_{t=x} z(x,k) + z(x,k)\, K'(x,x)$$

Using Equation 5.46 and integrating by parts twice obtains

$$y''(x,k) = z''(x,k) + K(x,x)\, z'(x,k) + \int_0^x K(x,t)\, z''(t,k)\, dt + K'(x,t)\big|_{t=x} z(x,k) +$$

$$+2K'(x,x)\int_0^x z(t,k)\, K(x,t)\, dt + \left[K_t(x,t)z(t,k) - K(x,t)z'(t,k)\right]\Big|_0^x -$$

$$-\int_0^x \{q_0(t) - q_0(x)\}K(x,t)z(t,k)\, dt + K'(x,x)z(x,k)$$

The integrated quantity is again zero at the lower limit, and we have

$$y''(x,k) = z''(x,k) + 2K'(x,x)\left[z(x,k) + \int_0^x z(t,k)K(x,t)dt\right] +$$

$$+\int_0^x K(x,t)\left[z''(t,k) - \{q_0(t) - q_0(x)\}z(t,k)\right]dt$$

From the auxiliary equation (5.29),

$$z''(x,k) = \left[q_0(x) - \lambda\right]z(x,k)$$

This finally results in the equation for y(x,k):

$$y''(x,k) = \left[\{q_0(x) + 2K'(x,x)\} - \lambda\right]\left[z(x,k) + \int_0^x z(t,k)K(x,t)dt\right]$$

Writing

$$q(x) = q_0(x) + 2K'(x,x)$$

results in the reconstructed differential equation

$$y'' + \left[\lambda - q(x)\right]y = 0$$

From Equation 5.48, the boundary condition is

$$y'(0,\lambda) - hy(0,\lambda) = 0$$

where

$$h = K(0,0) + h_0$$

The reconstructed potential is

$$q(x) = q_0(x) + 2K'(x,x)$$

The boundary-value problem is thus recovered from the Gel'fand-Levitan equation.

VII. A BRIEF SUMMARY

The important aspects of the Gel'fand-Levitan theory are briefly summarized below. Basically, an ordinary, linear, second-order differential equation in one dimension is considered on the semiaxis $[0,\infty)$, satisfying certain boundary conditions at the origin and containing a "potential" which is required to satisfy certain smoothness and integrability conditions. The equation is assumed to possess a "spectral function" $\sigma(\lambda)$, which is determined by two sequences of numbers, the eigenvalues λ_n and the normalizing constants. These sequences can be discrete, continuous, or both. The theory of Gel'fand and Levitan addresses itself to the following two equivalent problems:

1. Given $\sigma(\lambda)$, determine the differential equation *cum* boundary conditions that correspond to it and
2. Solve the "inverse" problem, i.e., determine the unknown potential. This is also equivalent to reconstructing the original boundary-value problem. The direct problem is to determine the eigenfunctions for a known potential.

Given a spectral function $\sigma(\lambda)$, the theory proceeds as follows.

1. Introduce an auxiliary problem, i.e., a problem with a known spectral function σ_0 and known eigenfunctions $\phi_0(x,\lambda)$.
2. Decompose the spectral function σ in terms of σ_0 and as given by Equation 5.5.
3. Orthogonalize $\phi_0(x,\lambda)$ with respect to $\rho(\lambda)$, that is, construct functions such as

$$F(x,x') = \int_{-\infty}^{\infty} \phi_0(x,\lambda)\phi_0(x',\lambda)\, d\rho(\lambda)$$

4. Express the eigenfunction of the original problem in terms of the following integral representation:

$$\phi(x,\lambda) = \phi_0(x,\lambda) + \int_0^x M(x,x')\phi_0(x',\lambda)dx'$$

This mapping from $M(x,x') \rightarrow \phi(x,\lambda)$ is the Povzner-Levitan transformation.

5. Derive an integral equation relating $F(x,x')$ and $M(x,x')$:

$$F(x,x') + M(x,x') + \int_0^x M(x,t)\, F(x',t)\, dt = 0, \qquad x' < x$$

This is the Gel'fand-Levitan equation.

6. Find the differential equation for $\phi(x,\lambda)$ and the boundary conditions.
7. Using the differential equation, find the conditions on the kernels $M(x,x')$.
8. From these conditions follows the solution of the inverse problem, namely:

$$\frac{dK(x,x)}{dx} = \left(\frac{1}{2}\right)q(x)$$

VIII. EXAMPLES

We will now illustrate how the Gel'fand-Levitan equation can be used in reconstructing a boundary-value problem. As mentioned in the previous section, the application of the equation requires an auxiliary boundary-value problem with a known potential. The eigenfunctions of this auxiliary problem are used in calculating the inhomogeneous term in the Gel'fand-Levitan integral equation. This potential may be taken to be zero with either cos kx or sin kx/k as eigenfunctions, depending upon whether h is zero or infinity. There is, however, no compulsory reason to choose the potential to be zero. Any other potential satisfying the required conditions of integrability, smoothness, etc., will serve just as well. In the first example, we consider the potential to be zero, whereas in the second illustration a nonzero potential is used. The first example is the one given by the authors themselves.

Example 1

The spectral function specified is of the form

$$\begin{aligned}\sigma(\lambda) &= \left(\frac{2}{\pi}\right)k + \alpha H(\lambda - \lambda_0), \quad \lambda > 0, \quad k = \lambda^{1/2} \\ &= \alpha H(\lambda - \lambda_0), \qquad \lambda < 0 \end{aligned} \tag{5.49}$$

where H is the Heaviside function, that is, $H(\lambda - \lambda_0) = 0$ if $\lambda < \lambda_0$, and $H(\lambda - \lambda_0) = 1$ if $\lambda > \lambda_0$ and α is a positive number. According to the decomposition of the spectral function given earlier by Equation 5.5,

$$\rho = \alpha H(\lambda - \lambda_0)$$

Then $d\rho(\lambda) = \alpha\delta(\lambda - \lambda_0)\, d\lambda$. From Equation 5.8 and using the measure $d\rho(\lambda)$, the inhomogeneous term $F(x,x')$ is then calculated to be

$$F(x,x') = \int_{-\infty}^{\infty} \cos kx \, \cos kx' \, d\rho(k) = \alpha \cos k_0 x \, \cos k_0 x'$$

With this expression for $F(x,x')$, the Gel'fand-Levitan equation takes the form

$$\alpha \cos k_0 x \cos k_0 x' + \alpha \int_0^x K(x,s) \cos k_0 s \, \cos k_0 x' \, ds + K(x,x') = 0 \tag{5.50}$$

The kernel $F(x,x')$ is degenerate. It is thus possible to solve the integral equation by elementary means to yield an explicit solution, which is obtained as follows.

Consider a linear integral equation of the second kind of the form

$$u(x,x') + C(\mathbf{P}\, u)(x,x') = Cf(x,x')$$

where C is a constant and **P** is an integral operator. The kernel of **P** is a function of x' and x'', and the dependent variable u inside the integral is a function of x' and x''. For a degenerate kernel $P(x',x'') = P(x')P(x'')$, with only one term in the expansion, this integral equation can be written as

$$u(x,x') + CP(x')[\mathbf{P}\, u](x) = Cf(x,x')$$

the kernel of **P** now being a function only of x''. Assuming $P(x')$ and $P(x'')$ to be of the same functional form, multiplying both sides of the above equation by P(x′) and integrating over the interval [0,x] yields

$$[\mathbf{P}\,u](x) = C[\mathbf{P}\,f](x)\left[1 + C\|\mathbf{P}\|_{2,x}^{2}\right]^{-1}$$

where the norm is L_2 over [0,x]. Comparing the integrals on both sides yields the solution

$$u(x,x') = Cf(x,x')\left[1 + C\|\mathbf{P}\|_{2,x}^{2}\right]^{-1}$$

If, now, the function f(x,x′) is also degenerate and is given by

$$f(x,x') = P(x)P(x')$$

in particular, then the solution reduces to

$$u(x,x') = CP(x)P(x')\left[1 + C\|\mathbf{P}\|_{2,x}^{2}\right]^{-1} \tag{5.51}$$

Using Equation 5.51, the solution of Equation 5.50 is given by

$$K(x,x') = -\frac{\left[\alpha \cos k_0 x \ \cos k_0 x'\right]}{\left[1 + \alpha \int_0^x \cos^2 k_0 s \, ds\right]}$$

From Equation 5.27, the potential is calculated to be

$$\frac{q(x)}{2} = K'(x,x)$$

$$= -\left\{\left[\alpha \cos^2 k_0 x\right] \Big/ \left[1 + \alpha \int_0^x \cos^2 k_0 s \, ds\right]\right\}'$$

where ′ denotes differentiation with respect to x. The corresponding eigenfunctions are given by

$$y(x,k) = \cos kx + \int_0^x K(x,t) \cos kt \, dt \tag{5.52}$$

or

$$y(x,k) = \cos kx - \left\{\frac{\left[\alpha \cos k_0 x\right]}{\left[1 + \alpha \int_0^x \cos^2 k_0 s \, ds\right]}\right\} \int_0^x \cos kt \ \cos k_0 t \, dt \tag{5.53}$$

Setting x = 0 in Equation 5.53 yields

$$y(\lambda,0) = 1$$

and

$$h = y'(\lambda,0) = -\alpha$$

Also, from the expression for K(x,x′),

$$h = K(0,0) = -\alpha$$

From the spectral function given by Equation 5.49, it is clear that $\lambda > 0$ constitutes the continuous spectrum, λ_0 being the point of the discrete spectrum. The corresponding discrete eigenfunction is then

$$y(x,\lambda_0) = \frac{\cos k_0 x}{\left[1 + \alpha \int_0^x \cos^2 k_0 s \, ds\right]}$$

Example 2

In this example, we consider an auxiliary problem with a nonzero potential. Let us consider two Sturm-Liouville problems with two different potentials:

$$y_i'' + (\lambda - q_i)y_i(x,\lambda) = 0$$

$$y_i'(0,\lambda) - h_i y_i(0,\lambda) = 0, \quad i = 1,2 \tag{5.54}$$

Let the spectral distributions of the two problems be related by

$$R_2 = R_1 + C\,\delta(\lambda - \mu)$$

Let the boundary-value problem with i = 1 be the known problem, that is, the potential $q_1(x)$. The boundary condition h_1 and the eigenfunction y_1 are known. The second problem with i = 2 is the unknown problem to be determined.

The quantity $c\delta(\lambda - \mu)$ is the spectral measure $d\rho(k)$. The inhomogeneous term F(x,x′) is then given by

$$F(x,x') = c\, y_1(x,\mu) y_1(x',\mu)$$

The corresponding Gel'fand-Levitan integral equation is then

$$c y_1(x,\mu) y_1(x',\mu) + K(x,x') + c\int_0^x K(x,t) y_1(t,\mu) y_1(x',\mu)\,dt = 0$$

Again, the kernel is degenerate and K(x,x′) can be solved explicitly, as shown in the last example. Then using Equation 5.54, we obtain for K(x,x′) the expression

$$K(x,x') = -\frac{c y_1(x,\mu) y_1(x',\mu)}{\left[1 + c\int_0^x y_1^2(t,\mu)dt\right]}$$

The eigenfunction y_2 of the unknown problem is then determined to be

$$y_2(\lambda,x) = y_1(\lambda,x) - \left\{ c y_1(x,\mu) \Big/ \left[1 + c\int_0^x y_1^2(t,\mu)dt\right]\right\} \int_0^x y_1(t,\mu) y_1(t,\lambda)dt$$

The boundary condition of the unknown problem is determined by

$$y_2(0,\lambda) = y_1(0,\lambda)$$

and

$$h_2 = y_2'(\lambda,0) = h_1 - c$$

The potential $q_2(x)$ is calculated according to Equation 5.27:

$$q_2(x) = q_1(x) - 2c\left\{ y_1^2(x,\mu) \Big/ \left[1 + c\int_0^x y_1^2(t,\mu)dt\right]\right\}'$$

It is interesting to note that for R_2 to be the spectral function of the problem, the quantity

$$1 + c\int_0^x y_1^2(t,\mu)dt$$

must not vanish on the half-line R^+.

REFERENCES

1. **I. M. Gel'fand and B. M. Levitan:** On the determination of a differential equation from its spectral function, *Am. Math. Soc. Transl.*, No. 53, 1951.
2. **J. A. Cochran:** *The Analysis of Linear Integral Equations*, McGraw-Hill, New York, 1972.
3. **E. C. Titchmarsh:** The zeros of certain integral functions, *Proc. London Math. Soc.*, 25, 1926.
4. **V. A. Marcenko:** *Sturm-Liouville Operators and Applications*, Birkhauser-Verlag, Boston, 1986.
5. **G. M. L. Gladwell:** *Inverse Problems in Vibration*, Martinus Nijhoff, 1986.
6. **E. C. Titchmarsh:** *Eigenfunction Expansions*, Vol. 1, Oxford University Press, 1962.
7. **K. Chadan and P. C. Sabatier:** *Inverse Problems in Quantum Scattering Theory*, Springer-Verlag, 1977.

Chapter 6

THE GEL'FAND-LEVITAN THEORY IN TIME DOMAIN

I. INTRODUCTION

The theory of Gel'fand and Levitan was originally formulated in the frequency domain. It is this formulation that was presented in the previous chapter. In the frequency space, the scattering equation and the boundary condition are

$$u'' + \{\lambda - q(x)\}u = 0, \quad x \subset R^{+}, \quad \lambda \subset R$$

$$-u'(0,\lambda) + hu(0,\lambda) = 0, \quad h \subset R$$

respectively. The objective is to determine this second-order differential equation, cum boundary condition, that is, to recover q(x) and h, on the real semiaxis R^+ from the knowledge of the spectral function $\sigma(\lambda)$. It is, in essence, an inverse eigenvalue problem, that of constructing a differential equation and boundary conditions, having a given spectrum. It may be recalled from Chapter 3 that the points of increase of the monotonically nondecreasing spectral function constitute the spectrum. Hence, specifying $\sigma(\lambda)$ is equivalent to specifying the spectrum. The spectral function, however, eliminates the nonuniqueness associated with solving an inverse eigenvalue problem[17] since it carries an additional information, that of the normalization constant. This was explained earlier in connection with the discussion of the spectral function and was also illustrated through the example of the "scattering transform" in Chapter 2.

It is, however, possible to formulate the theory entirely in the time domain. Indeed, the spectral equation

$$u''(x) + \{\lambda - q(x)\}u(x) = 0$$

is the Fourier transform in time of the equation of motion

$$\ddot{u} - u'' + q(x)u = 0 \tag{6.1}$$

with $\lambda = \omega^2$, ω being the variable conjugate to t in the Fourier transform, the speed of propagation being, of course, unity. Differentiations with respect to time and space are denoted by $\cdot$ and $'$, respectively. Equation 6.1 is prototypical of a wide variety of important problems of wave propagation such as in nonuniform transmission lines, nonuniform elastic media, seismology, and a host of others. In many of these problems, the data are not measured directly in the frequency domain. Also, the solution of the associated inverse problem (i.e., recovering the unknown function q(x)) in the frequency domain requires that long-term records of the data be obtained for the Fourier transformation, whereas, if formulated directly in the time domain, boundary data only on a finite segment of the time series may suffice.[1] Working directly in the time domain is, therefore, advantageous, apart from the fact that Equation 6.1 is a natural setting for propagation problems. Furthermore, the derivations of the integral equations of inverse scattering are significantly simplified in the time domain.

The occurrence of Equation 6.1 in physical problems was pointed out in Chapter 1. Here we give a physical interpretation of the boundary condition $-u'(0,t) + hu(0,t) = 0$. Equation 6.1 is equivalent to that of the one-dimensional equation of elasticity

$$\rho \ddot{U} - \left(\rho c^2 U'\right)' = 0 \tag{6.2}$$

with U(z,t) the elastic displacement and $\rho = \rho(z)$ medium density and $c = c(z)$ the wave speed. In terms of the variables

$$x = \int_0^z c^{-1}(z')\, dz' \tag{6.3}$$

$$n(x) = \left[\rho(z)c(z)\right]^{\frac{1}{2}}$$

and

$$u = nU \tag{6.4}$$

Equation 6.2 then becomes Equation (6.1) for u with with $q(x) = n''/n$, $' = \partial/\partial x$. We have

$$n^2 U' = nu' - n'u \tag{6.5}$$

The stress produced in the elastic medium is given by

$$S = -\rho(z)c^2 U' = -n^2 U'$$

Customarily, the boundary condition for Equation 6.2, $z \subset R^+$, consists of specifying the stress at $z = 0$. Using this condition in Equation 6.5 and dividing throughout by n(0), the homogeneous form of the boundary condition becomes $u'(0,t) - hu(0,t) = 0$ with $h = [n'(0)/n(0)]$.

Equation 6.1 describes an elastically braced string[4] with the units of time and length so chosen that the density and the tension are both unity. The quantity q(x) is the distributed elastic coefficient and the term $-q(x)u$ is the restoring force supplied by the elastic medium in which the string is embedded. To interpret the boundary condition, we notice that if F is the force exerted on the string by the support at $x = 0$, then

$$-u'(0,t) = F$$

If h is interpreted as the concentrated spring constant at $x = 0$, and F is separated into an elastic restoring force $-hu(0,t)$ and an externally applied force F_0 (t), then the boundary condition reduces to

$$-u'(0,t) + hu(0,t) = F_0(t)$$

The system described by Equation 6.1 and the above boundary condition closely resemble the Klein-Gordon problem.[4]

The inverse problem consists of determining a hyperbolic boundary-value problem in a half-plane from the impulse response on a boundary to a unit impulse also applied at the boundary. Specifically, the hyperbolic boundary-value problem to be determined is

$$\ddot{u} - u'' + q(x)u = 0, \quad x \subset R^+, \quad t \subset R$$

$$-u'(0,t) + hu(0,t) = 0, \quad h \subset R \tag{6.6}$$

Let $R(x_0,t)$ be the impulse response of the system for $t > t_0$ at the surface $x = x_0$ to a unit impulse applied at this surface at some initial time t_0. Then the inverse problem stated in the context of elasticity is the following. Given $R(x_0,t)$, determine the boundary value problem (6.6), that is, reconstruct the distributed elastic coefficient q(x) and the concentrated spring constant h. Stated otherwise, recover the dynamics of the system from a knowledge of its response along a surface $x = x_0$ to a unit impulse applied at the same surface at some initial time t_0, the string being quiescent $\{u(x,t) = 0\}$ for all $t < t_0$.

The corresponding formulation of the problem in the frequency domain reads:

Given the spectral function $\sigma(\lambda)$, recover the differential equation, cum boundary condition

$$u''(x) + [\lambda - q(x)]u(x) = 0, \quad x \subset R^+, \quad t \subset R$$

$$-u'(0,\lambda) + hu(0,\lambda) = 0, \quad h \subset R$$

corresponding to the spectral function specified. That is, we are to recover the potential q(x) and the parameter h from the knowledge of the spectral function. A sophisticated and illuminating discussion of this problem both on a finite and an infinite interval is presented by Chadan and Sabatier.[18] However, not being connected with the time domain analysis, we do not discuss Chadan-Sabatier's exposition here.

The problems in the time and frequency domains are equivalent. Indeed, the impulse response $R(x_0,t)$, which will be shown to be the boundary value of the Riemann function (or rather the derivative of the Riemann function; however we will continue to call it the Riemann function) for the hyperbolic system (6.6), will turn out to be none other than the Fourier transform of the spectral function $\sigma(\lambda)$ itself. Thus, the knowledge of $R(x_0,t)$ and that of $\sigma(\lambda)$ are equivalent, and so are the two problems.

It should be mentioned that what the inverse problem is capable of recovering is the quantity q(x) and hence, the characteristic impedance $\rho(z)c(z)$, and not the density and the wave speed individually. Also, q(x) must satisfy certain smoothness conditions. It will be assumed to be a locally integrable function on R^+. It may be recalled that q(x) is locally integrable whenever the integral

$$\int_\Omega |q(x)dx < \infty$$

exists for every compact subset $\Omega \subset R^+$. Local integrability is not affected by the behavior at infinity and assumes that singularities in excess of the dimension of the space are not allowed.[5] All run-of-the mill continuous and piecewise continuous functions come under this classification. It will be further assumed that $q(x) \subset H^m{}_{loc}R^+$, $m \geq 1$, which means that $q \subset C_m(R^+)$, the space of functions which are $(m - 1)$ times differentiable, whereas, the mth derivative can be identified with locally absolutely integrable functions on R^+, i.e., functions belonging to $L_1{}^{loc}[R^+]$. If q belongs to $H^m{}_{loc}$, then it will be found that $R(x_0,t)$ must belong to $H^{m+1}{}_{loc}$. The Riemann function thus possesses one more derivative than the coefficient q.

The Riemann function $R(x_0,t)$ and other closely related impulse-response functions relevant for various derivations are discussed next. It will be necessary to make use of some known results in the theory of propagation of discontinuities of a hyperbolic system. A brief outline of the theory is, therefore, presented below. The material presented in this chapter is based largely on the work of Burridge[1] and Symes.[15]

II. THEORY OF PROPAGATION OF DISCONTINUITIES

Discontinuities in the data for linear hyperbolic equations are propagated along the characteristics across which singularities or other rapid variations of the solution can occur. We are seeking solutions to the one-dimensional, hyperbolic, linear partial differential equation (6.1) in a region of interest such as at a wavefront surface by using the method of expansion behind a wave-front. Standard references are the text by Courant and Hilbert[6] and the lecture notes by Lax.[7] Detailed descriptions also appear in Friedlander,[8] Whitham,[9] Jeffrey and Taniuty,[10] Bleistein,[11] and in a recent monograph by Zauderer.[12] The discussion of this section is based on Zauderer.[12]

Let S(x,t) = 0 be a wavefront surface specifying a certain disturbance u, considering propagation in one dimension only. u = 0 for S < 0, and is finite for S > 0. This permits a representation of the form

$$u(x,t) = V(x,t)\ H[S]$$

where H is the Heaviside function that vanishes when S < 0 and is unity for S > 0. This requires that V be a solution of Equation 6.1 in the region S > 0 and that its value on S = 0 equal the jump in the solution u across the wavefront. Replacing the above representation in Equation 6.1 and equating the coefficients of singularities δ', δ, and H, $(H'[S] = \delta[S])$, to zero separately, S and V are found to satisfy the conditions:

$$\dot{S}^2 - S'^2 = 0 : \text{the “characteristic” equation} \tag{6.7}$$

and

$$2\dot{S}\dot{V} - 2S'V' + \left[\ddot{S} - S''\right]V = 0 \tag{6.8}$$

Equation 6.7 can be rewritten as

$$F(x,t,\alpha,\beta) = \alpha^2 - \beta^2 = 0$$

where $\alpha = \dot{S}$ and $\beta = S'$. The characteristic curves of Equation 6.7 are called the “bicharacteristics”, that is, the characteristics of the characteristic equation, in order to distinguish them from the characteristics of the hyperbolic equation itself. If s is a parameter such that x(s), t(s), $\alpha(s)$, $\beta(s)$, and S(s) are the parametric representations of the bicharacteristics, then the following relations hold[12] for a nonlinear, first-order hyperbolic differential equation such as Equation 6.7:

$$x_s = -2\beta, \quad t_s = 2\alpha, \quad \alpha_s = 0, \quad \text{and} \quad \beta_s = 0 \tag{6.9}$$

The subscript s denotes differentiation with respect to the parameter s. According to Equation 6.9, α (and also β) remains constant along the bicharacteristics and the parameter s can thus be replaced by t, with $dt = 2\alpha ds$. The condition for V, Equation 6.8, reduces to

$$\left(\frac{dV}{dt}\right) + \left(\frac{1}{2\alpha}\right)\left[\ddot{S} - S''\right]V = 0$$

along the bicharacteristics, where $(dV/dt) = \dot{V} - (\beta/\alpha)V'$, since $V = V[x(t),t]$, and the relations given by Equation 6.9 have been used.

The results just obtained describe the variation of the solution along the bicharacteristics. However, they can be extended in the neighborhood of the discontinuity as well by Taylor expanding V(x,t) as:

$$V(x,t) = \sum_{j=0}^{\infty} v_j(x,t)\left[\frac{S^j}{j!}\right]$$

The coefficients v_j are evaluated on the characteristic $S = 0$. Thus,

$$u(x,t) = \left\{\sum_{j=0}^{\infty} v_j\left[\frac{S^j}{j!}\right]\right\}H[S] = \sum_{j=0}^{\infty} v_j H_j[S] \tag{6.10}$$

The generalized functions $H_j[S] = (S^j/j!)H(S)$ have the property that

$$H_j'[S] = H_{j-1}[S],\ j \geq 1 \tag{6.11}$$

This can be seen as follows. Consider a "test function" $\theta(z)$. By definition, $\theta(z) \subset C^0_\infty$, the space of infinitely differentiable functions having a compact support. The inner product of $H_j(z)$ and $\theta(z)$ is defined as

$$\left(H_j(z),\ \theta(z)\right) = \int_0^{\infty}\left[\frac{1}{j!}\right] z^j\, \theta(z)\, dz$$

From the theory of generalized functions,[13] it is known that

$$\left(H_j'(z),\ \theta(z)\right) = -\left(H_j(z), \theta'(z)\right) = \left[\frac{-1}{j!}\right]\int_0^{\infty} z^j\, \theta'(z)\, dz$$

$$= -\left[\frac{1}{j!}\right] z^j\, \theta(z)\Big|_0^{\infty} + \int_0^{\infty}\left[\frac{z^{j-1}}{(j-1)!}\right]\theta(z)\, dz$$

$$= \left(H_{j-1}(z), \theta(z)\right);\ j \geq 1$$

since the test function $\theta(t)$ has a compact support.

A more general class of problems can be treated by replacing Equation 6.10 by the expansion,

$$u(x,t) = \sum_{j=0}^{\infty} v_j(x,t) P_j[S(x,t)] \tag{6.12}$$

where $P_j[S]$ is a generalized function obeying the recurrence relation of Equation 6.11, but otherwise arbitrary. For a particular problem, P_0 and P_j must be decided on the basis of the data. It is important to note that the functions P_j are increasingly less singular for increasing j.

The expansion (6.12) is a formal solution of the equation of motion (6.6) if S(x,t) satisfies certain conditions. These can be obtained by directly substituting Equation 6.12 in Equation 6.6 and then equating the coefficients of various singularities to zero, analogous to the procedure leading to Equations 6.7 and 6.8. This results in

$$\dot{S}^2 - S'^2 = 0 \quad \text{"characteristic" equation} \tag{6.13}$$

$$2\dot{S}\dot{v}_j - 2S'v_j' + \left[\ddot{S} - S''\right]v_j = -\ddot{v}_{j-1} + v''_{j-1} - q(x)v_{j-1},\ j \geq 0, \quad \text{"transport" equations} \tag{6.14}$$

It is assumed that v_{-1} is zero so that v_0 satisfies a homogeneous equation.

For one-dimensional wave propagation, it is reasonable to assume that the wavefront S(x,t) has the form

$$S(x,t) = t - f(x)$$

The characteristic equation (6.13) then leads to the "eiconal" equation, namely,

$$(f')^2 = 1: \text{"eiconal" equation} \tag{6.15}$$

There are two solutions $\tilde{s}(x,t)$ and $\hat{s}(x,t)$ corresponding to those of Equation 6.15. For each of these functions, a series (6.12) is constructed with coefficients $\tilde{v}(x)$ and $\hat{v}(x)$, respectively:

$$u(x,t) = \sum_{j=0}^{\infty} \tilde{v}_j(x)\, P_j[\tilde{S}] + \sum_{j=0}^{\infty} \hat{v}_j(x)\, P_j[\hat{S}]$$

On inserting this expansion into the prescribed data, initial conditions are obtained for the coefficients $\tilde{v}(x)$ and $\hat{v}(x)$, each of which satisfies the transport equation (6.14).

It should be mentioned that the expansion (6.12) is not restricted to describing the propagation of singularities only. For example, if

$$P_0[S] = \exp(i\omega S) \quad \text{and} \quad P_j[S] = (i\omega)^{-j} \exp(i\omega S) \tag{6.16}$$

ω real and large, then the expansion may be regarded as an asymptotic solution to the wave equation (6.6). Asymptotic solution implies large time in the unit of the inverse of the frequency considered. Thus, for example, in plasmas modeled by the Klein-Gordon equation such as 6.1, the frequency ω may be of the order of 10^{10} Hz so that large time is just the order of microseconds for which the expansion parameter $\omega t >> 1$. It would occur if the problem contained rapidly oscillating data such as

$$u(x,0) = f(x)\exp[i\omega g(x)], \quad \dot{u}(x,0) = 0$$

For a choice of $S = t - \psi(x)$ and $v_j = v_j(x)$, and with $P_j[S]$ defined by Equation 6.16, the expansion (6.12) leads to the well-known "geometrical optics expansion" for the "reduced wave equation", i.e, Equation 6.6 for U(x), where $u(x,t) = U(x)\exp(i\omega t)$. The expansion

$$U(x) = \sum_{j=0}^{\infty} v_j(x)\left(\frac{1}{i\omega}\right)^j \exp[-i\psi(x)]$$

is, indeed, a solution of the reduced wave equation.[14]

III. CAUSAL IMPULSE RESPONSE

Let the function $G_1(x,t)$ satisfy the hyperbolic equation (6.1), the boundary condition:

$$-G_1'(0,T) + hG_1(0,t) = \delta'(t) \tag{6.17a}$$

and the initial condition

$$G_1(x,t) = 0, \quad t < 0 \tag{6.17b}$$

$\delta'(t)$ is the impulsive loading at $x = 0$. Formally, $G_1(x,t)$ is represented by the expansion (6.12). A "causal" solution of the problem is desired and since the region of interest is R^+, only the positive part of the eiconal equation is considered. Therefore, $f' = +1$ and $f(x) = x +$ a constant, yielding

$$S(x,t) = t - x$$

The constant can be absorbed by shifting the time origin. With this choice of S(x,t) and considering the coefficients v_j to be functions of x only, the zero-th order ($j = 0$) transport equation (6.14) gives $v_0' = 0$. v_0 is assumed to be unity, without any loss of generality. Note that this choice of S(x,t) corresponds to the equation $u(x,t) = V(x,t)H[S]$ introduced at the beginning of the present section. The first order ($j = 1$) transport equation similarly results in

$$2v_1' = -q(x)$$

The other higher-order coefficients can be obtained in the same way.

The boundary condition (6.17a) dictates that we select P_0 in the expansion (6.12) to be a delta function. Thus,

$$P_0(z) = \delta(z) \quad \text{and} \quad P_j(z) = H_j(z), \quad j \geq 1$$

The impulse response $G_1(x,t)$ is then

$$G_1(x,t) = P_0(x,t) - K_1(x,t)$$

$$= \delta(t-x) - K_1(x,t) \tag{6.18}$$

where

$$-K_1(x,t) = v_1(x)H(t-x) + \Re(x,t) \tag{6.18a}$$

$$\Re(x,t) = \sum_{j=0}^{\infty}\left[\frac{v_j(x)}{(j-1)!}\right](t-x)^{j-1}H(t-x) \tag{6.19}$$

$\Re(x,t)$ is the remainder on truncating the sum in Equation 6.12 after two terms. It is continuous on the characteristic $t = x$ and actually vanishes there. $K_1(x,t)$ contains $v_1(x)$, which is the integral of the elastic coefficient $q(x)$. $K_1(x,t)$ thus possesses one more derivative than does q.

The form of the solution (6.18) is hardly surprising. The hyperbolic system (6.6) represents an elastically braced string on R^+ with a concentrated spring constant at the origin. It is well known[4] that such a medium is dispersive. The effect of dispersion is to "scatter" the various Fourier components constituting a pulse propagating through the medium. This results in a wake trailing behind the leading edge of the pulse. In Equation 6.18, this wake following the delta-function pulse is represented by the function $K_1(x,t)$. A more detailed discussion of δ-pulse wakes will be given in Chapter 10.

As t approaches the characteristic from above, i.e., $t \to x+0$, we have

$$K_1(x,x+0) = K_1(x,x) = -v_1(x) = \left(\frac{1}{2}\right)\int_0^x q(x)\,dx + \text{constant}$$

In view of the smoothness properties attributed to the potential q(x), namely, that its derivatives can be identified with locally integrable, absolutely continuous functions, one may also write

$$2K_1'(x,x) = -v_1'(x) = q(x)$$

It is customary to write K_1 (x,x) for K_1 (x,x+0). The solution (6.18) clearly satisfies the quiescent condition of Equation 6.17b.

It remains to be shown that the solution (6.18) can also be made to satisfy the boundary condition (6.17a). Let us rewrite Equation 6.17a as

$$G_1'(0,t) - hG_1(0,t) = -\delta'(t) = -\delta'(t) + \sum_{j=1}^{\infty} v_j'(0)\,P_j'(t) -$$

$$-\sum_{j=1}^{\infty} v_j(0)\,P_{j-1}(t) - h\sum_{j=0}^{\infty} v_j(0)\,P_j(t) \tag{6.20}$$

The expansion (6.12) along with the recurrence relation (6.11) have been used in deriving Equation 6.20. The term next in singularity to the most singular term of $\delta'(t)$, that is, the term containing $P_0(t)$, can be eliminated by choosing h to be $-v_1(0)$. This, therefore, yields

$$h = -v_1(0) = K_1(0,0)$$

The constant in the expression of $K_1(x,x)$ above is thus determined to be the concentrated spring constant h.

The elimination of the next highest singular term demands that

$$v_2(0) = -hv_1(0) + v_1'(0) = h^2 + v_1'(0) \tag{6.21}$$

From the transport equation (6.14),

$$v_2(x) = -(1/2)\int_0^x q(x)\, v_1(x)\, dx + (1/2)\, v_1'(x) + C \tag{6.22}$$

Using Equation 6.21, the constant C is evaluated to be

$$C = h^2 + \left(\frac{1}{2}\right) v_1'(0)$$

Proceeding similarly, each higher-order coefficient can be determined in terms of those of lower orders as shown by Equations 6.21 and 6.22.

Summarizing, then, the causal impulse response $G_1(x,t)$ satisfies the equation of motion (6.6), the boundary condition (6.17a), and the quiescent condition (6.17b). Its functional form is

$$G_1(x,t) = \delta(t-x) - K_1(x,t)$$

where $K_1(x,t)$ is given by Equation 6.18a. Moreover,

$$K_1(x,x) = \left(\frac{1}{2}\right)\int_0^x q(x)\, dx + h$$

$$q(x) = 2K_1'(x,x) \quad \text{and} \quad h = K_1(0,0)$$

The data $G_1(0,t)$ is supposedly known. The support of this function is shown in Figure 1. Again as earlier, once the kernel K_1 is determined, the inverse problem is essentially solved. The determination of K_1 will be discussed later.

It is interesting to consider the case when the right-hand side in the boundary condition (6.17a) is $\delta(t)$ instead of $\delta'(t)$. Denoting the corresponding impulse response by $\hat{G}_1(x,t)$, we have

$$-\tilde{G}_1'(0,t) + h\,\tilde{G}_1(0,t) = \delta(t) \quad \text{and} \quad \tilde{G}_1(x,t) = 0, \quad t < 0$$

The highest singularity in the boundary condition is now $\delta(t)$ and, accordingly, the zero-th

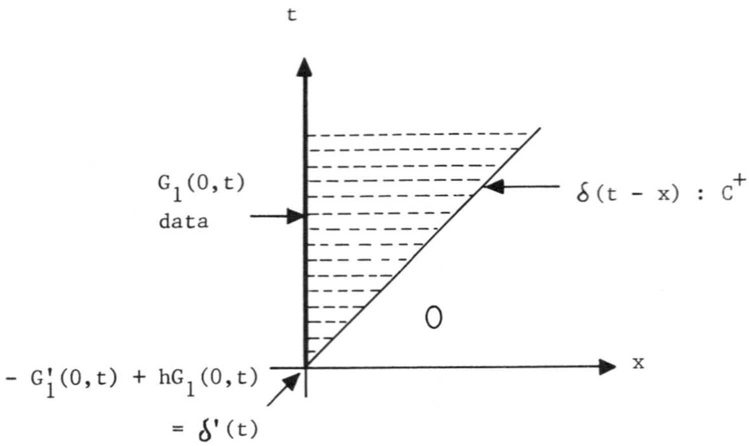

FIGURE 1. Support of the causal impulse response $G_1(x,t)$ shown by the dotted area along with the positive characteristics C^+. The measured data are on the positive t-axis indicated by the heavy line.

order term $P_0[S]$ is chosen to be the Heaviside function itself, that is,

$$P_0[t-x] = H(t-x)$$

The function $G_1(x,t)$ is found to be the time derivative of the response $\tilde{G}_1(x,t)$

$$G_1(x,t) = \dot{\tilde{G}}_1(x,t)$$

IV. THE NONCAUSAL IMPULSE FUNCTION

The noncausal impulse function G(x,t) represents the region between the characteristics $C^{\pm}$. In this domain, $|t| \leq x$. No particular physical significance should be attributed to this function. It is required in order to derive the direct Povzner-Levitan transform.

For the region under consideration, $|t| \leq x$, $x \subset R^+$, let us assume that the impulse function can be written as

$$G(x,t) = \delta(x-t) + \delta(x+t) + K(x,t) \tag{6.23}$$

We impose the following conditions:[1]

$$-G'(0,t) + h\,G(0,t) = 0 \quad \text{and} \quad G(0,t) = 2\delta(t)$$

Note that in the problems involving G and G_1, the source functions appear in part of the data instead of in the differential equation itself. This is typical of approaches involving Riemann's solution of Cauchy's problem. Indeed in Section VI, G_1 will be used in defining the Riemann functions. In the alternative approach in which the source functions appear (as distributions) in the differential equation is the technique of "Green's function". For a further discussion on this point, see Bleistein.[11] Substituting Equation 6.23 in the boundary-value problem (6.6 and 6.7), the following relation for K(x,t) is obtained:

$$\ddot{K} - K'' + q(x)K = -q(x)\,\delta(x - t) - q(x)\,\delta(x + t)$$

$$K(0,t) = 0 \tag{6.24}$$

and

$$K'(0,t) = 2h\delta(t)$$

K(x,t), therefore, possesses discontinuities across the characteristics, whereas, within the region $|t| < x$, $x \subset R^+$, K(x,t) satisfies the homogeneous wave equation.

Consider the positive characteristic C^+ first. In this region, let K be represented by

$$K(x,t) = H(x - t)\,g(x,t) \tag{6.25}$$

where H is the Heaviside function and g is a solution of the homogeneous wave equation. Equation 6.25 guarantees that K(x,t) vanishes identically for t > x. Replacing Equation 6.25 in Equation 6.24, one obtains for g(x,t)

$$\dot{g} + g' = \frac{q(x)}{2}$$

Since this is along the characteristic x = t, then

$$g'(x,t = x) = \frac{q(x)}{2}$$

yielding

$$g(x,x) = \left(\frac{1}{2}\right)\int_0^x q(x)\,dx + \text{constant} \tag{6.26}$$

Again, understanding that K(x, t = x) is the same as K(x,t → x–0), we obtain

$$K'(x,x) = g'(x,x)$$

yielding

$$K(x,x) = \left(\frac{1}{2}\right)\int_0^x q(x)\,dx + \text{constant}$$

Also, from Equation 6.25:

$$K'(0,t) = K'(0,t + 0) = \delta(t)g(0,0) = \delta(t)\text{ constant}$$

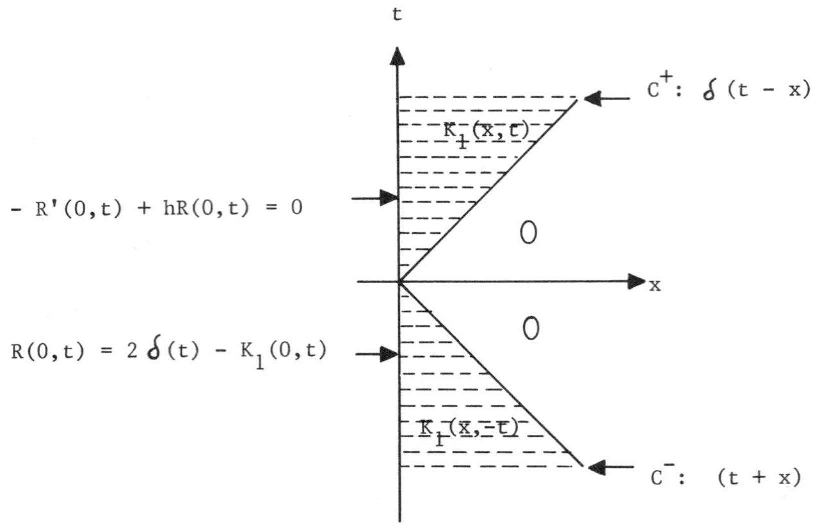

FIGURE 2. The support of the Riemann function R(x,t). Boundary values are indicated on the t-axis.

Repeating the calculations for the negative characteristic C^-, that is, $x = -t$,

$$K'(x,-x) = \left(\frac{1}{2}\right)\int_0^x q(x)\,dx + \text{constant}$$

$$K'(0, t \to 0+) = \delta(t)\ \text{constant}$$

From the boundary condition on K′ of Equation 6.24, the constant is identified with h. Replacing t by –t, the boundary value problem (6.24) remains unchanged. From this we conclude that K(x,t) is symmetric with respect to the time t and, hence, is G(x,t). Therefore,

$$G(x,t) = G(x,-t), \qquad |t| \le x, \quad x > 0$$

The support of the noncausal impulse function G(x,t) is the unshaded area of Figure 2.

V. POVZNER-LEVITAN TRANSFORM: DIRECT AND INVERSE, $[x \subset R^+]$

Let us write the solution u(x,t) of the boundary value problem (6.6) as

$$u^{\lambda}(x,t) = \phi(x,\lambda)\cos\lambda^{\frac{1}{2}}t, \quad \phi \subset C_{m+1}\left(R^+\right) \tag{6.27}$$

where $\phi(x,\lambda)$ is the eigenfunction of the boundary-value problem

$$y'' + [\lambda - q]y = 0, \quad y \subset R^+, \quad \lambda \subset R$$

$$y(0,\lambda) = 1 \ \text{ and } \ y'(0,\lambda) = -h \tag{6.27a}$$

and $\cos\lambda^{1/2}t$ is the eigenfunction of the “auxiliary” problem, that is, the problem without scattering. The significance of the auxiliary problem in formulating the Gelfand-Levitan theory was explained earlier. From previous discussions, we know that the direct and inverse Povzner-Levitan transforms are given by

$$\phi(x,\lambda) = \cos\lambda^{\frac{1}{2}} x + \int_0^x M(x,t)\cos\lambda^{\frac{1}{2}} t \, dt \tag{6.28}$$

and

$$\cos\lambda^{\frac{1}{2}} t = \phi(t,\lambda) + \int_0^x N(x,t)\,\phi(x,\lambda)\, dx \tag{6.29}$$

respectively. M(x,t) and N(x,t) are the kernels.

The transform (6.28) can also be obtained in the time domain. We notice that the left-hand side of Equation 6.28 corresponds to $t = 0$ in Equation 6.27, which is the horizontal axis in the plane of the impulse function. The kernel M(x,t), on the other hand, operates on functions on the vertical t-axis. This suggests that the direct transformation (6.28) may be obtained through appropriate transformation of the data on $x = 0$ to the space axis $t = 0$. Since $t = 0$ is in the domain of the noncausal impulse function, it is reasonable to consider G(x,t) to be the appropriate transformation. Indeed, consider the convolution of u(0,t) with [G(x,t)/2],

$$u(x,t) = \left[G(x,t) * u(o,t)\right]$$

where $*$ denotes the convolution operator. It can be verified by direct substitution that u(x,t) does solve the boundary value problem. For u even in time ($\dot{u}(x,t) = 0,\ t = 0$), using the symmetry of K(x,t) with respect to time, setting t to zero, and noting that $G(x,t) = 0$ for $|t| > x$, the convolution yields

$$u(x,0) = u(0,x) + \int_0^x K(x,t')\, u(0,t')\, dt' \tag{6.30}$$

With u(x,t) given by Equation 6.27, this results in the direct Povzner-Levitan transform of Equation 6.28.

The inverse transformation from (x,0) to (0,t) is obtained analogously using the causal form of the solution $G_1(x,t)$. The result is, on setting x to zero,

$$u(0,t) = \int_0^\infty dx_0\, G_1(x_0,t)\, u(x_0,0)$$

$$= u(t,0) - \int_0^\infty dx\, K_1(x,t)\, u(x,0) \tag{6.31}$$

Again, using Equation 6.27, the inverse Povzner-Levitan transform (6.29) is recovered.

VI. THE RIEMANN FUNCTION

The impulse responses in the regions $t \geq x$ and $|t| \leq x$, $x \subset R^+$ of the phase plane (x,t) have been considered. Let us next consider the region $|t| \geq x$. It is bounded by the characteristics C^+: $\delta(t - x)$, C^-: $\delta(t + x)$ and contains the vertical t-axis. The function R in this region can be written as

$$R(x,t) = G_1(x,t) + G_1(x,-t)$$

The progressing-wave representation of R(x,t) is given by[15]

$$R(x,t) = \delta(t - x) + \delta(t + x) - K_1(x,|t|) \tag{6.32}$$

This shows that R(x,t) is an even function of time. Moreover, the wake $K_1(x,|t|)$ vanishes identically for $|t| < x$, $x \subset R^+$. The support of the response R(x,t) is shown in Figure 2. R(x,t) satisfies Equation 6.6 and the boundary condition

$$-R'(0,t) + hR(0,t) = 0 \quad \text{for all } t$$

This is because of the fact that $G_1(x,t)$ is a solution of Equation 6.6. Since the form of the equation is invariant under time reversal, $G_1(x,-t)$ is also a solution. Being time independent, the boundary condition is satisfied. Moreover, since R(x,t) is symmetric in time, we also have $\dot{R}(x,t) = 0$ at $t = 0$ and, finally, from Equation 6.32, $R(x,t = 0) = 2\delta(x)$.

Summarizing, the function R(x,t) given by the equation

$$\delta(t - x) + \delta(t + x) - K_1(x,|t|), \; |t| \leq x \subset R^+$$

is a solution to the differential equation

$$\ddot{u} - u'' + q(x)u = 0, \quad x \subset R^+, \quad t \subset R$$

and satisfies the boundary condition

$$-u'(0,t) + h\,u(0,t) = 0 \quad \text{for all } t$$

Moreover

$$R(x,0) = 2\delta(x)$$

$$R(0,t) = 2\delta(t) - K_1(0,|t|)$$

$$\dot{R}(x,t) = 0 \quad \text{for } |t| < x$$

Now let us consider the following boundary value problem, namely

$$\ddot{u} - u'' + q(x)u = g(x,t), \quad x \subset R^+, \quad t > t_0$$

$$u'(0,t) - h\,u(0,t) = 0$$

$$u(x,t_0) = \dot{u}(x,t_0) = 0$$

Then the solution of this inhomogeneous boundary value problem is given by Duhamel's integral

$$u(x,t) = \int_{t_0}^{t} dt' \int_{t_0}^{t'} dt'' \int_{0}^{\infty} dy\, R(x,t,y,t'')\, g(y,t'')$$

In particular, if $g(x,t) = \delta(x - x_0, t - t_0)$, then

$$u(x,t) = \int_{t_0}^{t} R(x,t,x_0,t')\, dt'$$

The impulse response R(x,t) is the Riemann function for the hyperbolic system under consideration for the domain $|t| > x$. It should be noted that the function R is actually the time derivative of the Riemann function that was considered, for example, in Chapter 4. However, we will continue calling R the Riemann function in this chapter. Moreover, in the standard presentation of Riemann's method of integration (see the Appendix of Chapter 4), the boundary condition at $x = x_0$ is not considered.

From the transport equation derived earlier, the boundary value of the kernel K_1, $K_1(0,0+) = h$. The expression of the Riemann function then yields that $R(0,0+) = h$. Moreover, since $K_1(x,t)$ involves the integral of the potential, the boundary value of the Riemann function then possesses one more derivative than that of the potential. Or in the language of Symes, if $q \subset H_{loc}^{m,1(R)}$, then $R(0,t) \subset H_{loc}^{m+1,1}(R)$. These are, therefore, necessary conditions for the function R to be the Riemann function. In a later section, an inequality will be derived (Equation 6.46), imposing another condition on R. It will turn out that this inequality in time domain is directly connected with the conditions that must be imposed upon a function if that function is to be the spectral function of a Sturm-Liouville equation on the half-line. Only the necessity of these conditions will be discussed here. That these conditions are also sufficient if R(0,t) is to be the boundary value of the Riemann function for the hyperbolic system under consideration can also be established. However, for the time being, the reader is referred to Reference 15.

Summarizing the above establishes the theorem below.

Theorem 6.1

An even function R(0,t) is the boundary value of the Riemann function R(x,t), for a hyperbolic boundary value problem

$$\ddot{u} - u'' + qu = 0$$

$$-u'(0,t) + h\,u(0,t) = 0, \quad x \subset R^+, \quad t \subset R$$

with $q(x) \subset H^m{}_{loc}(R^+)$, if and only if

i) $R(0,t) \subset H^{m+1}{}_{loc}(R)$, $R(0,0) = h$, and

ii) the kernel $R(0,t,T) = (1/2)\{R(0,|t - T|) + R(0,|t + T|)$ satisfies the condition that for any $T > 0$ and for all $u \subset L_2\,[0,T]$ there exists some $\varepsilon(t) > 0$, such that the inequality of Equation 6.46 is satisfied.

The above theorem is due to Symes.[15]

We next consider the inverse problem for which the Riemann function just introduced plays a crucial role.

VII. DERIVATION OF THE LINEAR GEL'FAND-LEVITAN EQUATION

Let us consider the boundary value R(0,t) of the Riemann function R(x,t). According to Equation 6.32, R(0,t) is given by

$$R(0,t) = 2\delta(t) - K_1(0,|t|)$$

The function R(0,t) is an even function of time. The necessary and sufficient conditions for R(0,t) to be the boundary value of the Riemann function will be discussed later after the nonlinear Gel'fand-Levitan equation is introduced.

Consider the response R(x,t) shifted in time by an amount T. Superposing the ±T-shifted responses and setting x to zero, we have

$$R(0,t,T) = \left(\frac{1}{2}\right)\left[R(0,t-T) + R(0,t+T)\right]$$

$$= \delta(t-T) + \delta(t+T) + f(t,T)$$

where $f(t,T) = -(1/2)[K_1(0,|t-T|) + K_1(0,|t+T|)]$. R(0,t,T) is, therefore, symmetric in t and T. Now, from Figure 2, the support of the Riemann function is the domain $|t| \geq x$, $x \geq 0$. If t is set to zero and if $x > T > 0$, then

$$R(x,0,T) = 0$$

Application of the direct Povzner-Levitan transform of Equation 6.30 yields

$$R(x,0,T) = R(0,x,T) + \int_0^x K(x,t)\,R(0,t,T)\,dt$$

$$= f(x,T) + \int_0^x K(x,T)\,f(t,T)\,dt + K(x,T)$$

$$= 0, \quad x > T > 0 \tag{6.33}$$

That this is the linear integral equation of Gel'fand and Levitan can be recognized from earlier discussions in Chapter 5.

VIII. RELATION BETWEEN THE RIEMANN AND THE SPECTRAL FUNCTION

It was mentioned earlier that the Riemann function R(0,t) is the Fourier transform of the spectral function $\rho(\lambda)$. This can be established as follows. By definition, the Riemann function satisfies the hyperbolic differential equation (6.6), the boundary condition (6.17a), and the initial conditions

$$R(x,x_0,t=0) = \delta(x-x_0) \tag{6.34}$$

and

$$\dot{R}(x, x_0, t = 0) = 0$$

Strictly speaking, the Riemann function R(x,t) should be written as $R(x,t;x_0,0)$. The impulse is assumed to be applied at time t = 0 and at $x = x_0$.

Now, consider the solution $u^\lambda(x,t)$ of Equation 6.27. This is a solution to the boundary-value problem (6.6), along with Equation 6.27a. The initial conditions are now

$$u^\lambda(x,0) = \phi(x,\lambda)$$

and

$$\dot{u}^\lambda(x,0) = 0$$

and $u^\lambda(x,t)$ can then be formally expressed as

$$u^\lambda(x,t) = \int_0^\infty dx_0\, R(x,t;x_0,0)\, \phi(x,\lambda)$$

Multiplying both sides of this equation by $\phi(y,\lambda)$ and integrating between $-\infty$ and $+\infty$ with respect to the spectral measure, we obtain

$$\int_{-\infty}^{\infty} \phi(y,\lambda)\, u^\lambda(x,t)\, d\rho(\lambda)$$

$$= \int_{-\infty}^{\infty} \int_0^\infty dx_0\, R(x,t;x_0,0)\, \phi(x_0,\lambda)\, \phi(y,\lambda\}\, d\rho(\lambda)$$

$$= \int_0^\infty dx_0\, R(x,t;x_0,0)\, \delta(x_0 - y) = R(x,t;y,0) \tag{6.35}$$

The resolution of the identity

$$\delta(x - y) = \int_{-\infty}^{\infty} \phi(x,\lambda)\, \phi(y,\lambda)\, d\rho(\lambda)$$

has been used. Replacing $u^\lambda(x,t)$ in Equation 6.35 by Equation 6.27 and setting y = 0,

$$R(x,t;0,0) = R(x,t) = \int_{-\infty}^{\infty} \phi(x,\lambda) \cos\lambda^{\frac{1}{2}}\, t\, d\rho(\lambda) \tag{6.36}$$

If it is further assumed that x = 0, then

$$R(0,t) = \int_{-\infty}^{\infty} \cos\lambda^{\frac{1}{2}}\, t\, d\rho(\lambda) \tag{6.37}$$

Hence, R(0,t) is the Fourier transform of the spectral function as proposed. From Equation 6.35, R(x,t) is an even function of t. Also, by virtue of the resolution of the identity above, R(x,t) vanishes for t = 0 and x > 0. This is exactly the region of the Riemann function described in Figure 2.

IX. RELATION BETWEEN f(t,T) AND THE SPECTRAL FUNCTION σ(λ)

Equation 6.32 is a Fredholm integral equation of the second kind, the function f(t,T) being the inhomogeneous part. The solution of the inverse problem is crucially dependent on this function. The significance of f(t,T) derives from the fact that it represents the data on the boundary. Indeed,

$$f(t,T) = -\left(\frac{1}{2}\right)\left[K_1(0,|t-T|) + K_1(0,\ |t+T|)\right]$$

From Equation 6.32,

$$R(0,t) = 2\delta(t) + f(t)$$

where

$$f(t) = -K_1(0,|t|)$$

Following Gel'fand and Levitan again, the spectral function is decomposed as

$$\sigma(\lambda) = \rho(\lambda) + \left(\frac{2\lambda^{\frac{1}{2}}}{\pi}\right), \quad \lambda \geq 0$$

$$= \rho(\lambda), \quad \lambda < 0$$

We also note the identity

$$\delta(t) = \left(\frac{1}{\pi}\right)\int_{-\infty}^{\infty} \cos\left(\lambda^{\frac{1}{2}} t\right) d\lambda^{\frac{1}{2}}$$

On using Equation 6.37 for R(0,t) along with this decomposition and the identity, it follows that

$$f(t) = \int_{-\infty}^{\infty} \cos\lambda^{\frac{1}{2}}\ t\ d\sigma(\lambda)$$

and, hence,

$$f(t,T) = \left(\frac{1}{2}\right) \int_{-\infty}^{\infty} \left[\cos\left(\lambda^{\frac{1}{2}} |t-T|\right) + \cos\left(\lambda^{\frac{1}{2}} |t+T|\right) \right] d\sigma(\lambda)$$

$$= \int_{-\infty}^{\infty} \cos\left(\lambda^{\frac{1}{2}} t\right) \cos\left(\lambda^{\frac{1}{2}} T\right) d\sigma(\lambda)$$

This is identical to the inhomogeneous term of the Gel'fand-Levitan equation derived previously in the frequency domain.

X. UNIQUENESS OF THE SOLUTION

The uniqueness of the solution to Equation 6.33 lies again in satisfying the Fredholm alternative, that is, the triviality of the homogeneous solution. The onus of the proof is, therefore, in establishing that

$$K_1(x,T) + \int_0^x K_1(x,t)\, f(t,T)\, dt = 0 \tag{6.38}$$

has only a trivial solution. In the time-domain analysis, the proof is usually based on the nonlinear Gel'fand-Levitan equation. It may be recalled from Chapter 5 that the proof of the uniqueness of the solution in the spectral domain also involved similar nonlinearity. The nonlinear integral equation can be obtained from the inverse Povzner-Levitan transform of Equation 6.29 as follows.

Let us consider the integral

$$\int_0^{\infty} dx\, R(x,t)\, R(x,T)$$

In view of the domain of the Riemann functions (Figure 2), the integral can be written as

$$\int_0^{|\xi|} dx\, R(x,t)\, R(x,T)$$

where $|\xi| = \min(|t|,|T|)$. Let $0 < t < T$. Then,

$$\int_0^{\infty} dx\, R(x,t)\, R(x,T) = \int_0^{t} dx\, R(x,t)\, R(x,T)$$

$$= -K_1(t,T) + \int_0^{t} dx\, K_1(x,t)\, K_1(x,T)$$

In the same region, $R(t,0,T) = -K_1(t,T)$ and $R(x,0,T) = K_1(x,T)$. Then by Equation 6.29, the integral is the function R(0,t,T) which, in turn, is f(t,T), $0 < t < T$. Therefore,

$$f(t,T) = -K_1(t,T) + \int_0^t dx\, K_1(x,t)\, K_1(x,T) \tag{6.39}$$

Equation 6.39 is the nonlinear Gel'fand-Levitan equation.

It can also be obtained[15] by using the "group" property of the Riemann function. Let U(t) be an operator which maps the Cauchy data $\{u(x,t_0),\ \dot{u}(x,t_0)\}$ for a solution of the boundary-value problem under consideration at time t_0 to the Cauchy data for the same solution $\{u(x,t + t_0), \dot{u}(x, t + t_0)\}$ at time $t + t_0$. The group property is defined as:

$$U(s)U(t) = U(s+t)$$

This is a consequence of the fact that the coefficients and the boundary condition of the hyperbolic problem under consideration are independent of time. Hence, it is possible to propagate the data first from some initial time to t and then from t to t + s.

The operator U will be implemented by the Riemann function. This means the following. The Riemann function is a propagator that propagates the Cauchy data in space and time. Hence if the Cauchy data are given by

$$u(x,t) = u_0(x) \text{ and } \dot{u}(x,t) = v_0(x) \text{ at } t = t_0,\ \ u_0,\ v_0 \subset C_\infty\left(R^+\right)$$

then the solutions u(x,t) and $\dot{u}(x,t)$ can be expressed as

$$U(x,t) = \left\langle \Re(x,t;x_0,t_0)\ U_0(x)\right\rangle$$

where U is the vector $[u(x,t), \dot{u}(x,t)]^T$, and the propagator $\Re$ is a matrix-valued distribution in x_0.

$$\Re(x,t;\ x_0,t_0) = \begin{pmatrix} R(x,t;x_0,t_0) & \int_{t_0}^{t} R(x,t;x_0,t')\, dt' \\ \dot{R}(x,t;x_0,t_0) & (\partial/\partial t)\int_{t_0}^{t} R(x,t;x_0,t')\, dt' \end{pmatrix}$$

$< >$ means the evaluation of a distribution on a test function. It then follows that

$$U(x,t) = \int_0^\infty \Re(x,t;x_0,t_0)\ U_0\ dx_0$$

Written in terms of the $\Re$ distributions, the group property takes the form

$$\Re(x,t+s;x_0,t_0) = \int_0^\infty \Re(x,t+s;y,s)\,\Re(y,s;x_0,t_0) \tag{6.40}$$

Thus, for example, the solution u(x,t) is given by

$$u(x,t+s) = \int_0^\infty R(x,t+s;x',t_0)\, u_0(x')\, dx' + \int_0^\infty dx' \int_{t_0}^{t+s} R(x,t+s;x',t')\, v(x')\, dt'$$

This expression is to be contrasted with Equation A.10 of the Appendix of Chapter 4 which is written in terms of the Riemann function itself and not in terms of its derivative as in the expression above.

Moreover, the function $R(x,t;x_0,t_0)$ satisfies certain symmetry properties listed below.

$$R(x,t;x_0,t_0) = R(x_0,t_0;x,t)$$

$$= R(x,t-s;\ x_0,\ t_0 - s)$$

These reflect the fact that the propagation is linear and that the source and the field points can be interchanged. Also, $R(x,t;x_0,0) = R(x,-t;x_0,0)$.

We now use Equation 6.40 to write the functions R(0,t+s) and R(0,t–s) and use the symmetry property, especially, to note that

$$\left(\frac{\partial}{\partial s}\right) R(y,s,0,0) = -\left(\frac{\partial}{\partial s}\right) R(y,-s,0,0) \tag{6.41}$$

The addition of R(0,t+s) and R(0,t–s) using the relation (6.41) then results in the nonlinear Gel'fand-Levitan equation given by Equation 6.39.

Equation 6.39 will be used next in order to show the uniqueness of the solution of Equation 6.33. As earlier, Fredholm's alternative theorem will be used to show that the homogeneous equation

$$[(\mathbf{I}+\mathbf{K})K_0](x,t) = 0 \tag{6.42}$$

has only a trivial solution K_0. **K** is the integral operator defined of the form

$$(\mathbf{K}\, g)\ (x,t) = \int_0^T K(x,t)\, g(t,T)\ dt$$

Since x acts merely as a parameter, it will be omitted in the arguments. Taking the inner product of Equation 6.42 with $K_0(T)$ over the interval [0,x] yields

$$\left(K_0,\ (\mathbf{I}+\mathbf{K})K_0\right) = \int_0^x K_0^2(T)\, dT + \int_0^x \int_0^x K_0(t)K_0(T)f(t,T)dt\, dT \tag{6.43}$$

Equation 6.43 shows that the operator (**I** + **K**) is the operator (**I** + **F**) where, **F** is an integral operator on [0,x] with the function f(t,T) as the kernel. Let us now substitute f(t,T) in Equation 6.43 from the nonlinear Gel'fand-Levitan Equation (6.39). We note that Equation 6.39 can actually be written as

$$f(t,T) = -K_1(t,T) - K_1(T,t) + \int_0^{\min(t,T)} dx\ K_1(x,t)\, K_1(x,T) \tag{6.44}$$

This is possible since f(t,T) is symmetric in its arguments and the functions $K_1(x,y)$ vanishes for $|x| \leq y$, $y \subset R^+$. Equation 6.44 suggests that the operator $(\mathbf{I} + \mathbf{F})$ can be written as

$$(\mathbf{I}+\mathbf{F}) = \left(\mathbf{I}+\mathbf{K}^{T}\right)(\mathbf{I}+\mathbf{K}) \tag{6.45}$$

where $\mathbf{K}^T$ is the adjoint of the operator $\mathbf{K}$.

In the linear Gel'fand-Levitan equation, the interval is $0 < T < x$. We first consider the operator $(\mathbf{I} + \mathbf{F}_T)$ on the interval [0,T], where, the subscript T denotes that an interval [0,T] is being considered. The Riemann function R(y,t) defines an integral operator $(\mathbf{I} + \mathbf{K}_T)$, where,

$$(\mathbf{K}_T g)\,(y,t) = \int_0^T K(y,t)\,g(t,T)\,dt, \quad y < T$$

This is a bounded Volterra operator on [0,T] and is invertible. The nonlinear Gel'fand-Levitan equation has similar form as that of Equation 6.45, namely

$$(\mathbf{I}+\mathbf{F}_T) = \left(\mathbf{I}+\mathbf{K}_T^T\right)\,(\mathbf{I}+\mathbf{K}_T)$$

This implies that $(\mathbf{I} + \mathbf{F}_T)$ is also invertible.

Next consider a function $z \subset L_2[0,T]$. Calculating the inner product $\left[z,(\mathbf{I}+\mathbf{F}_T)z\right]\Big|_{L_2[0,T]}$ we obtain

$$\begin{aligned}\left[z,(\mathbf{I}+\mathbf{F}_T)z\right]\Big|_{L_2[0,T]} &= \int_0^T |z|^2\,dt + \int_0^T\int_0^T z^*(s)\,z(t)\,f(s,t)\,ds\,dt \\ &\leq \varepsilon \int_0^T |z|^2\,dt \end{aligned} \tag{6.46}$$

where ε is the smallest eigenvalue of $(\mathbf{I} + \mathbf{F}_T)$. Taking the limit $T \to x$ and replacing z by K_0 then yields the result that $(K_0,(\mathbf{I} + \mathbf{K})K_0)$ cannot be zero unless K_0 is zero. Fredholm's alternative is thus established proving the uniqueness of the solution.

The solution of the inverse problem is completed once the kernel K_1 is obtained. This can be obtained, for example, by solving the nonlinear Gel'fand-Levitan equation. Since the nonlinear Equation 6.39 is of Volterra type, then if it possesses a continuous solution, then that solution must be unique. An interesting iterative procedure was outlined by Symes[17] for solving the nonlinear Volterra Equation 6.39. The procedure consists of alternately solving the nonlinear equation over small intervals and the linear equations over large intervals using the preceding solution of the nonlinear equation.

Thus consider an interval t sufficiently small such that $0 \leq t' \leq t \leq t =$ a region Q(t). If $\mathbf{V}$ is the Volterra operator on $C_0(Q(t))$, then Equation 6.39 can be written as

$$\mathbf{V}K(t',t) = -f(t',t) - \int_0^{t'} dt''\,K(t'',t')\,K(t',t),\ (t',t) \subset Q(t)$$

If t is sufficiently small the standard contraction mapping argument[5] shows that the Volterra operator $\mathbf{V}$ has a fixed point in $C_0(Q(t))$. The nonlinear equation is thus solved over a small interval Q(t) in $C_0(Q(t))$.

Having determined $\phi(t',t)$ for $0 \le t' \le t \le t$, the linear equation is formed as below

$$f(t',t) = \phi(t',t) + \int_0^{t'} dt''\, \phi(t'',t')\, K(t'',t)$$

where, $\phi(t',t)$ has been written for $K(t',t)$. This is a linear Volterra equation for each $t \subset [t',T]$ and therefore, has a continuous solution since the kernel is continuous.

In the third step, the nonlinear equation is solved again on a small interval $t \le t' \le t \le t + \delta$

$$f(t',t) = K(t',t) + \int_0^{t} dt''\, \phi(t'',t')\, \phi(t'',t) + \int_t^{t'} dt''\, K(t'',t')\, K(t',t)$$

This is also solved as in the first step. The iterations are then repeated.

We now come to the final topic of this chapter, namely, the significance of the time domain inequality (6.46) and its connection with the conditions on the spectral function. We follow Symes[15] here.

XI. THE SPECTRAL FUNCTION ρ AND THE INEQUALITY (6.46)

The original paper of Gel'fand and Levitan appeared in 1951.[16] In this 1951 paper, the authors established that for a function $\rho(\lambda)$ to be the spectral function of the differential equation

$$y'' + [\lambda - q(x)]y = 0$$

certain conditions had be satisfied. These were given as

1. The integral

$$\int_{-\infty}^{0} \exp\left(|\lambda|^{1/2} x\right) d\rho(\lambda)$$

must exist for each x and

2. The function

$$\int_1^{\infty} \lambda^{-1} \cos\left(|\lambda|^{\frac{1}{2}} x\right) d\sigma(\lambda)$$

must have a continuous 4th derivative. The functions σ and ρ follow the decompositions given by Equation 5.5.

These conditions were shown to be sufficient if the potential were a continuous function of x. These were also necessary if the potential q(x) were required to have a continuous derivative.

In a later version published in 1964, Levitan and Gasymov[17] improved upon the earlier work by introducing the following necessary and sufficient conditions for the spectral function.

The non-decreasing function $\sigma(\lambda)$ is the spectral function of the boundary value problem

$$y'' + \{\lambda - q(x)\}y = 0, \quad x \subset R^+$$

$$-y'(0) + h\, y(0) = 0, \quad h \subset R$$

where q(x) is a real locally integrable function with m locally integrable derivatives if and only if:

1. $\lim_{N\to\infty} I_N(x) = \int_{-\infty}^{N} d\rho(\lambda)\cos\lambda^{1/2}x, \; N \subset Z$

$$\begin{aligned} \rho(\lambda) &= \sigma(\lambda) && \lambda < 0 \\ &= \sigma(\lambda) - \left(\frac{2}{\pi}\right)\lambda^{\frac{1}{2}} && \lambda \geq 0 \end{aligned}$$

   converges absolutely in every finite range of values of x to a function $\phi(x)$ having (m + 1) locally integrable derivatives and with $\phi(0) = h$. In the language of this section: $\lim_{N\to\infty} I_N(x) \to \phi(x) \subset W^{m+1}{}_{loc}(R^+)$ when the potential $q(x) \subset W^{m}{}_{loc}(R^+)$.
2. If $u^{\wedge}(\lambda)$ is the Fourier cosine transform of an arbitrary function $u(x) \subset L_2\,(R^+)$ having a compact support [0,T], then

$$\int_{-\infty}^{\infty} d\sigma(\lambda)\left|u^{\wedge}(\lambda)\right|^2 = 0$$

if and only if $u^{\wedge} = 0$ almost everywhere, that is, everywhere except on a set of measure zero.

The Riemann function R(0,t) satisfies condition 1 (Section VI and Equation 6.37). It is claimed that condition 2 is equivalent to the inequality given by Equation 6.46. In order to establish this, let us consider the integral

$$I = \int_0^{\tilde{T}}\int_0^{\tilde{T}} dtdT\; u(t)u(T)R(0,t,T)$$

Replacing R(0,t,T) by the expression obtained in Section IX, namely,

$$R(0,t,T) = \int_{-\infty}^{\infty} d\sigma(\lambda)\cos\lambda^{\frac{1}{2}}t\cos\lambda^{\frac{1}{2}}T$$

the integral becomes

$$I = \int_0^{\tilde{T}}\int_0^{\tilde{T}} dt\; dT\left\{\int_{-\infty}^{\infty} d\sigma(\lambda)\cos\lambda^{\frac{1}{2}}t\cos\lambda^{\frac{1}{2}}T\right\}u(t)\,u(T)$$

By the first condition, the $\sigma(\lambda)$ integral converges absolutely in the range $\{[0,\tilde{T}] \times [0,\tilde{T}]\}$. The function $u \subset L_2\,[R^+]$. Then, by Fubini's theorem, the order of integration can be interchanged, resulting in

$$I = \int_{-\infty}^{\infty} d\sigma(\lambda) \left\{ \int_0^{\tilde{T}} dt\, u(t) \cos\lambda^{1/2} t \right\} \left\{ \int_0^{\tilde{T}} dT\, u(T) \cos\lambda^{1/2} T \right\}$$

From Parseval's relation for the ordinary Fourier cosine transform, one obtains

$$I = \int_{-\infty}^{\infty} d\sigma(\lambda) \left| u^{\wedge}(\lambda) \right|^2$$

To this we add the contribution from the $(2\lambda^{1/2}/\pi)$ part of the spectral function, that is, the quantity

$$\left\| u^{\wedge} \right\|^2 = \int_0^{\infty} d\left(\frac{2\lambda^{\frac{1}{2}}}{\pi} \right) \left| u^{\wedge}(\lambda) \right|^2$$

Finally, therefore,

$$\int_{-\infty}^{\infty} d\rho(\lambda) \left| u^{\wedge}(\lambda) \right|^2 = \left\| u^{\wedge} \right\|^2 + I \geq 0$$

The inequality is strict if $\tilde{u}$ is not identically zero almost everywhere. This then establishes the equivalence between the second conditions of both the above theorems.

REFERENCES

1. **R. Burridge:** The Gel'fand-Levitan, the Marchenko, and the Gopinath-Sondhi integral equations of inverse scattering theory, regarded in the context of inverse impulse-response problems, *Wave Motion*, 2, 1980.
2. **I. S. Sokolnikoff:** *Mathematical Theory of Elasticity*, Dover, New York, 1945.
3. **P. M. Morse:** *Vibration and Sound*, Acoustical Society of America, 1976.
4. **P. M. Morse and H. Feshbach:** *Methods of Theoretical Physics*, McGraw-Hill, New York, 1953.
5. **I. Stakgold:** *Green's Functions and Boundary Value Problems*, Interscience, New York, 1979.
6. **R. Courant and D. Hilbert:** *Methods of Mathematical Physics*, Vol. 2, John Wiley & Sons, New York, 1958.
7. **P. Lax:** *Notes on Hyperbolic Differential Equations*, Stanford University, 1963.
8. **F. G. Friedlander:** *Sound Pulses*, Cambridge University Press, Cambridge, 1958.
9. **G. B. Whitham:** *Linear and Nonlinear Waves*, Interscience, New York, 1974.
10. **A. Jeffrey and T. Taniuti:** *Non-linear Wave Propagation with Applications in Physics and Magnetohydrodynamics*, Academic Press, New York, 1964.
11. **N. Bleistein:** *Mathematical Methods of Wave Phenomena*, Academic Press, New York, 1984.
12. **E. Zauderer:** *Partial Differential Equations of Applied Mathematics*, Interscience, New York, 1983.
13. **I. M. Gel'fand and G. E. Shilov:** *Generalized Functions*, Vol. 1, Academic Press, New York, 1964.
14. **J. B. Keller:** Geometrical theory of diffraction, *J. Opt. Soc. Am.*, 52, 1962.
15. **W. Symes:** *Inverse Boundary Value Problems and a Theorem of Gel'fand and Levitan*, Technical Summary Report No. 1846, Madison Mathematics Research Center, 1978.

16. **I. M. Gel'fand and B. M. Levitan:** On the determination of a differential equation from its spectral function, *Am. Math. Soc. Transl.*, 1, 1955.
17. **B. M. Levitan and M. Gasymov:** Determination of a differential equation by two of its spectra, *Russ. Math. Surv.*, 1964.
18. **K. Chadan and P. C. Sabatier:** *Inverse Problems in Quantum Scattering Theory*, 2nd ed., Springer-Verlag, 1989.

Chapter 7

JOST FUNCTIONS AND THEIR REGULARITY PROPERTIES

I. INTRODUCTION

We now consider Schrödinger's equation on the entire real axis:

$$y'' + \{\lambda - q(x)\}y = 0, \qquad \lambda = k^2 \subset C \tag{7.1}$$

the variable $x \subset R$ instead of R^+. The treatment on the entire real axis is relatively more complex than that on R^+. A Sturm-Liouville problem, Equation 7.1 can be solved within the framework of the Weyl-Titchmarsh theory with regular boundary conditions at the origin, namely,

$$y(0,\lambda) = 0, \quad y'(0,\lambda) = 1$$

and

$$y(0,\lambda) = 1, \quad y'(0,\lambda) = 0$$

The spectrum, however, can no longer be described by a single spectral function. Instead, a (2×2) "spectral matrix" (three of these elements are independent) must be considered, and the spectrum will now consist of the points of growth of all three independent spectral functions. This illustrates the complexity of the problem on the entire real axis.

In scattering theory, however, conditions are not usually imposed at the origin, but, rather, at $x \to \pm\infty$ and the potential is assumed to vanish asymptotically, i.e., $q(x) \to 0$ as $x \to \pm\infty$, at which limit the wave function $y(x,k)$ is merely a plane propagating wave described by $\exp(\pm ikx)$. We are, therefore, seeking solutions to Equation 7.1 which will propagate as plane waves as $x \to \pm\infty$. Two pairs of such solutions can be obtained:

$$\lim_{x\to+\infty} \exp(-ikx)\psi(x,k) = 1, \qquad \lim_{x\to+\infty} \exp(+ikx)\overline{\psi}(x,k) = 1$$

$$\lim_{x\to+\infty} \exp(-ikx)\psi'(x,k) = ik, \qquad \lim_{x\to+\infty} \exp(+ikx)\overline{\psi}'(x,k) = -ik$$

$$\lim_{x\to-\infty} \exp(+ikx)\phi(x,k) = 1, \qquad \lim_{x\to-\infty} \exp(-ikx)\overline{\phi}(x,k) = 1$$

$$\lim_{x\to-\infty} \exp(+ikx)\phi'(x,k) = -ik, \qquad \lim_{x\to-\infty} \exp(-ikx)\overline{\phi}'(x,k) = ik$$

The solutions ψ, ψ, ϕ, ϕ are known as the "Jost solutions". The Jost solutions with oscillations removed, that is, ψ, ψ, ϕ, ϕ with the multiplicative exponential factors as in the above expressions, are known as the "Jost functions". To discuss the Jost solutions and functions and their properties in both the physical and in the spectral domain is the objective of this chapter. In the main body of the text, these will be discussed for the entire real axis R. However, the Jost solutions and functions also play an equally important role in the inverse scattering problems on the half-line R^+. Therefore, for the sake of completeness, these functions will also be discussed on the half-line. This will be done in the Appendix at the end of the chapter.

In quantum mechanics, one works almost exclusively with the Hilbert space $L_2(R)$ on which an inner product is defined as

$$(u, v) = \int_R u(x)\, v^*(x)\, dx$$

where $*$ denotes a complex conjugate and the functions u and $v \subset L_2(R)$.

For real k, the Jost solutions do not belong to $L_2(R)$. These are the "generalized" wave functions, the primary significance of which is in providing a basis for the expansion of other eigenfunctions corresponding to the appropriate observables. However, these generalized eigenfunctions can be incorporated into the Hilbert space framework. This will be discussed later.

Subsequent chapters will deal with the problem of determining the potential from the "scattering data" on the entire line. This will involve applying complex integration and Cauchy's calculus of residues to functions containing the Jost solutions. A rigorous knowledge of the regularity properties of these functions in both x and k is, therefore, essential. These properties will be established in this section. The discussion presented here draws heavily from Deift and Trubowitz[2] Eckhaus and van Harten,[3] Fadeev[7], Calogero and Degasperis[8] and Chadan and Sabatier.[17]

II. INTEGRAL REPRESENTATIONS OF THE JOST SOLUTIONS

The Jost solutions satisfy the differential Schrödinger equation. It is, however, customary in the theory of differential equations to reformulate the problem in terms of equivalent integral equations. This has advantages from the viewpoint of establishing existence and uniqueness of a solution. It is thus convenient to express the Jost solutions in integral forms which will incorporate the boundary conditions corresponding to the differential equation. The resulting integral equations are of the Volterra type, the kernels being "triangular" for scattering in one space dimension. Since a Volterra equation can be solved by iteration, continuity and analyticity of the Jost solutions can be established via those of the iterates.

The Schrödinger equation (7.1) can be written as:

$$LY = q\sigma Y \tag{7.2}$$

where

$$L = \left(\frac{d}{dx}\right) + A$$

$$A = \begin{pmatrix} 0 & -1 \\ k^2 & 0 \end{pmatrix} \quad \text{and} \quad \sigma = \begin{pmatrix} 0 & 0 \\ 1 & 0 \end{pmatrix}$$

and $Y = (Y_1, Y_2)^T$, $Y_1 = y$, and $Y_2 = y'$, T denoting the transpose operation. This is a first-order equation in Y requiring only a one-point boundary condition. A similar procedure is also applied in solving the radial Schrödinger equation of quantum scattering, where the solution determined by the boundary conditions at two points, namely, $r = 0$ and $r = \infty$, is replaced by the "regular" solution determined by a condition only at the origin.[10] Denoting the one-point boundary condition by the subscript 0, Equation 7.2 yields

$$Y(x,k) = Y_0 \exp\{\mathbf{A}(k)\,(x_0 - x)\} - \int_{x_0}^{x} dx' \exp\{\mathbf{A}(k)(x' - x)\} q(x')\sigma Y(x',k) \qquad (7.3)$$

where $Y_0 = Y(x{=}x_0,k)$. Expanding the exponential operator, we have

$$\exp\{\mathbf{A}(k)(x' - x)\} = \mathbf{I}\ \cos k(x' - x) + \mathbf{A}(k)k^{-1}\ \sin k(x' - x)$$

where **I** is the identity operator. Replacing this in Equation 7.3, we obtain the following expression from the $Y_1 = y$ component:

$$y(x,k) = y_{01}(k)\ \cos k(x_0 - x) - k^{-1} \sin k(x_0 - x)\ y_{02}(k) -$$

$$-k^{-1} \int_{x_0}^{x} \sin k(x' - x)\, q(x')\, y(x',k)\, dx' \qquad (7.4)$$

Taking the boundary point $|x_0|$ to be at ∞ and denoting by ψ resp. ϕ the solutions that behave as $\exp(+ikx)$ resp. $\exp(-ikx)$ as $x \to +\infty$ resp. $-\infty$, Equation 7.4 yields the following integral representations of the Jost solutions:

$$\psi(x,k) = \exp(+ikx) + k^{-1} \int_{x}^{\infty} \sin k(x' - x)\, q(x')\, \psi(x',k)\, dx'$$

$$\phi(x,k) = \exp(-ikx) - k^{-1} \int_{-\infty}^{x} \sin k(x' - x)\, q(x')\, \phi(x',k)\, dx' \qquad (7.5)$$

Similar expressions can be derived for the pair $\overline{\psi}$ and $\overline{\phi}$.

Equation 7.2 can also be written as

$$\mathbf{L}Y = i\begin{pmatrix} D & -q(x) \\ 1 & -D \end{pmatrix} Y = kY$$

where

$$\mathbf{D} = \frac{d}{dx} \quad \text{and} \quad Y = (iky - y',\ \ -y)^T$$

This is in the form of a linear eigenvalue problem. If the element 1 in the above matrix is replaced by another potential function $r(x)$, then the problem is the famous "generalized problem" of Zakharov and Shabat.[5] Such an eigenvalue problem was first formulated by these authors in connection with the solution of the "nonlinear" or "cubic" Schrödinger equation in the framework of the inverse scattering transform.

It is customary to remove the oscillations from the Jost solutions and consider instead the so-called "Jost functions"

$$r(x,k) = \phi(x,k)\exp(+ikx)$$

and

$$l(x,k) = \psi(x,k)\exp(-ikx)$$

r and l being associated with waves incident from the right and from the left, respectively. The boundary value problems for r and l are

$$\begin{aligned} &r'' - 2ikr' - qr = 0 && l'' + 2ikl' - ql = 0 \\ &\lim_{x\to-\infty} r(x,k) = 1 && \lim_{x\to+\infty} l(x,k) = 1 \\ &\lim_{x\to-\infty} r'(x,k) = 0 && \lim_{x\to+\infty} l'(x,k) = 0 \end{aligned} \tag{7.6}$$

Equations 7.6 are also referred to as Schrödinger equations. Consider an arbitrary one-point boundary condition at $x_0 \subset R$, and treat qr and ql as inhomogeneous terms. r and l then have representations, the general form of which is

$$\begin{aligned} f_\pm(x,k) = f_{\pm0}(k) \mp f'_{\pm0}(k)\frac{\left[u_\pm(x,x_0,k)-1\right]}{2ik} + \\ + \int_{x_0}^{x}\left[q(x')\int_{x'}^{x} u_\pm(x'',x',k)\,dx''\right] f_\pm(x',k)\,dx' \end{aligned} \tag{7.7}$$

In Equation 7.7, $k \neq 0$ and f stands for either r or l. + resp. – corresponds to l resp. r. $f_0(k)$ means, as before, the function f(x,k) at the initial point x_0. The kernel $u_\pm(s,t,k) = \exp[\mp 2ik(s-t)]$. For k = 0, Equation 7.7 reduces to

$$f_\pm(x,0) = f_{\pm0}(0) \mp f'_{\pm0}(0)(x - x_0) + \int_{x_0}^{x} q(x')(x - x')\, f_\pm(x',0)\,dx' \tag{7.7a}$$

Taking x_0 at $\pm\infty$ and using the boundary conditions of Equation 7.6, the integral representation given by Equation 7.7 assumes the more familiar form

$$\begin{aligned} f_\pm(x,k) &= 1 + \int_\pm F_\pm(\xi_\pm,k)\, q(x')\, f_\pm(x',k)\,dx' \\ &= 1 + \int_\pm F_\pm(\xi_\pm,x',k)\, f_\pm(x',k)\,dx' \\ &= 1 + F_\pm * f_\pm = 1 + (F_\pm f_\pm)(x,k) \end{aligned} \tag{7.8}$$

where $f_+(x,k)$ resp. $f_-(x,k)$ corresponds to $l(x,k)$ resp. $r(x,k)$, ξ_+ resp. ξ_- stands for $(x' - x)$ resp. $(x - x')$, F_+ resp. $F_- = (1/2ik)[\exp(2ik\xi_+)-1]$ resp. $(1/2ik)[\exp(2ik\xi_-)-1)]$ and the kernels $F_+(\xi_+,x',k)$ resp. $F_-(\xi_-,x',k)$ indicate $[q(x')/2ik][\exp(2ik\xi_+) - 1]$ resp. $[q(x')/2ik][\exp(2ik\xi_-) - 1]$. The integrals $\int_+$ resp. $\int_-$ correspond to $\int_x^\infty$ resp. $\int_{-\infty}^x$. The convolution $F_\pm * f_\pm$ has been written as $(F_\pm f_\pm)(x,k)$ for notational convenience. It can be verified directly that Equation 7.8 solves the Schrödinger problems for r and l with the prescribed asymptotic boundary conditions, as given by Equations 7.6.

In what follows it will be assumed that the potential q(x) is a real, continuous function of x which vanishes as $|x| \to \infty$. The potential is said to satisfy a growth condition of order m if

$$\int_R \left(1 + |x|^m\right)|q(x)|\, dx < \infty$$

This will also be written symbolically as $q(x) \subset L_m^1$.

III. REGULARITY PROPERTIES OF THE JOST FUNCTIONS

Due to the factor (1/k) in the expression of the Jost functions given by Equations 7.8, it is expected that the analysis will differ depending upon whether k is zero or nonzero. The restrictions on the growth condition of the potential will also differ accordingly. For the Jost functions r and l, k will be restricted to the upper half-plane, i.e, $k \subset \overline{C}_+$, that is, Im $k \geq 0$. For $k \subset \overline{C}_+\backslash\{0\}$, that is, $\overline{C}_+$ excluding the origin, the kernels $F_\pm$ can be easily estimated as

$$|F_\pm| \leq \frac{1}{|k|} \tag{7.9}$$

This requires that the potential satisfy a growth condition of order 0. If, however, $k \subset \overline{C}_+$, then the estimate is

$$|F_\pm| \leq |\xi_\pm| \tag{7.10}$$

and the potential must be L_1^1. The bound (7.10) is also seen from the $k \to 0$ limit of Equation 7.7a. The limiting case $k \to 0$ will be seen to differ from that of $k \subset \overline{C}_+\backslash\{0\}$ in two important aspects: (1) the Jost functions become unbounded in x asymptotically and (2) analyticity at $k = 0$ requires that $q \subset L_2^1$. These two cases will be discussed separately, beginning with that of k away from zero.

A. $k \subset \overline{C}_+\backslash\{0\}$: REGULARITY IN x

We are seeking "classical" solutions to the Schrödinger equation. This means that the solutions $f_\pm$ are continuous in x (for each k) on the cylinder $(R \times \overline{C}_+\backslash\{0\})$ and remain bounded as $x \to \pm\infty$. Let $X^{\pm 0}$ denote the space of such functions +,– corresponding to $x \to +\infty$ and $-\infty$, respectively, and the superscript 0 indicating that the origin in the k space is being excluded. $X^{\pm 0}$ will be endowed with a supremum norm such that convergence in $X^{\pm 0}$

means that for a sequence $\{f_n\} \to f \subset X^{\pm 0}$ for $n \to \infty$ implies that

$$\forall\, k \subset \overline{C}_+\backslash\{0\},\ \forall\, x_0 \subset R,\ \lim_{n\to\infty}\left[\sup_{x \gtrless x_0} |f_n - f| = 0\right]$$

Let us consider the Schrödinger equation given by Equation 7.8. Since $k \subset \overline{C}_+\backslash\{0\}$, the bound given by Equation 7.9 is to be used. Then if $f_\pm \subset X^{\pm 0}$,

$$\sup_{x \gtrless x_0} |F_\pm f_\pm|\,(x,k) \le \sup_{x \gtrless x_0}\left(\frac{1}{|k|}\right)\int_\pm |q(x')|\,|f_\pm(x',k)|\,dx',\ x_0 \subset R$$

$$\le \left(\frac{1}{|k|}\right)\|f_\pm\|\int_\pm |q(x')|\,dx'$$

$$= \|f_\pm\|\left(\frac{\eta_\pm(x)}{|k|}\right) \tag{7.11}$$

where

$$\eta_\pm(x) = \int_\pm |q(x')|\,dx'$$

The equality sign in the above conditions pertains only to the limit at infinity. Otherwise, the conditions are strictly those of inequality. This will be maintained throughout the text. Assuming that the potential satisfies a growth condition of order zero, $\eta_\pm(x)$ exists. Then if $f_\pm \subset X^{\pm 0}$, then $(F_\pm f_\pm)(x,k)$ also $\subset X^{\pm 0}$. This then shows that solving the Schrödinger equation (7.6) for classical $f_\pm$, $k \subset \overline{C}_+\backslash\{0\}$, is equivalent to solving Equation 7.8 in the space $X^{\pm 0}$, provided, of course, that $\eta_\pm(x)$ exists.

Let $S^{\pm 0}$ be a linear space such that

$$S^{\pm 0} = \left\{ v(x,y,k) : v \subset C\left[R \times R \times \overline{C}_+\backslash\{0\}\right] \,\middle|\, \left|\int_\pm |v(x,y,k)|\,dy\right| \le \overline{v}(x,k) \subset X^{\pm 0} \right\}$$

Thus $v(x,y,k)$ belongs to the space of these functions which are everywhere continuous in (x,y,k) on $R \times R \times \overline{C}_+\backslash\{0\}$. It is not difficult to see that $F_\pm(x,x',k) \subset S^{\pm 0}$. The same conclusion holds for $F_\pm(x,x',k) \subset S^{\pm 0}$. From this it follows that

$$F_\pm * f_\pm \subset X^{\pm 0} \Rightarrow F_\pm(x,x',k) \subset S^{\pm 0} \quad \text{and} \quad f_\pm \subset X^{\pm 0}$$

The mapping $F_\pm{*}: X^{\pm 0} \to X^{\pm 0}$ is continuous. This justifies writing the convolution operation of Equation 7.8. The same conclusion holds for $F'_\pm(x,x',k)$, $1 = \partial/\partial x$.

In order to determine whether a solution to Equation 7.8 exists in the space $X^{\pm 0}$, Picard's method of successive approximation will be applied to obtain

$$f_\pm(x,k) = \sum_{n=0}^{\infty} f_\pm^{(n)}(x,k) \tag{7.12}$$

where $f_\pm^{(0)} = 1$, and $f_\pm^{(n+1)}(x,k) = (F_\pm f_\pm^{(n)})(x,k)$, $n \geq 0$. We first show that $f_\pm^{(n)} \subset X^{\pm 0}$ and then show that the sum given by Equation 7.12 also $\subset X^{\pm 0}$. From Equations 7.8 and 7.12

$$(F_\pm f_\pm)(x,k) = \sum_{n=1}^{\infty} f_\pm^{(n)}$$

Now, $(F_\pm f_\pm^{(0)})(x,k) \subset X^{\pm 0}$ since $F_\pm \subset S^{\pm 0}$. Note that in the notation of Equation (7.8), $F_\pm f_\pm$ is really a convolution. By repeated application, $(F_\pm f_\pm^{(n)})$ also $\subset X^{\pm 0}$. $\{f_\pm^{(n)}, n \geq 0\}$ is, therefore, a sequence $\subset X^{\pm 0}$. Moreover, if the sequence is such that

$$\forall\, k \subset \overline{C}_+ \setminus \{0\}, \quad \forall\, x_0 \subset R, \quad \sum_{0}^{\infty} \sup_{x \gtrless x_0} \left| f_\pm^{(n)}(x,k) \right| < \infty \tag{7.12a}$$

Then,

$$\sum_{n=0}^{\infty} f_\pm^{(n)}(x,k) \subset X^{\pm 0}$$

Hence, $f_\pm(x,k) \subset X^{\pm 0}$, showing that a solution to the Schrödinger equation (7.8) does exist in the space $X^{\pm 0}$.

It remains to show that

$$\forall\, k \subset \overline{C}_+ \setminus \{0\}, \quad \forall\, x_0 \subset R, \quad \sum_{n=0}^{\infty} \sup_{x \gtrless x_0} \left| f_\pm^{(n)}(x,k) \right| < \infty$$

holds. In order to demonstrate this, a bound for $f_\pm$ will be required which is derived below. Now

$$f_\pm(x,k) = 1 + \sum_{n=1}^{\infty} f_\pm^{(n)}(x,k)$$

$$\left| f_\pm^{(n)} \right|(x,k) \leq \int_{\Omega\pm} d_n \prod_{i=1}^{n} \left| F_\pm(\xi_{\pm i}, k) \right|, \quad n \geq 1$$

where

$$d_n = \prod_{i=1}^{n} dx_i, \quad \xi_i = \pm(x_i - x_{i-1}), \quad x_0 = x$$

$$\int_{\Omega+} = \int_{x \leq x_1 \leq x_2 \leq \ldots \leq x_n}$$

$$\int_{\Omega-} = \int_{x \geq x_1 \geq x_2 \geq \ldots \geq x_n}$$

Then using the estimate given by Equation 7.9

$$\left| f_\pm^{(n)} \right|(x,k) \leq \left(\frac{1}{|k|^n} \right) \int_{\Omega\pm} d_n \prod_{i=1}^{n} \left| q(x_i) \right|$$

This integral can be "majorized" as follows. The region of integration in the integrals consists of "triangular" regions. The integrands are, however, simple products of the integrals involving $|q(x_i)|$ and hence are unchanged by the permutations of the x_is. There are all together n! such permutations, summing up which would produce a "square" region with all the limits now running independently between x and $\pm\infty$. It follows that the integrals are now (1/n!) times the same integral taken over the entire larger region. Therefore,

$$\left|f_{\pm}^{(n)}\right|(x,k) \le \left(\frac{1}{n!}\right)\left[\frac{\eta_{\pm}(x)}{|k|}\right]^n$$

yielding the following bound for $f_{\pm}(x,k)$:

$$\left|f_{\pm}(x,k)\right| \le \exp\left\{\frac{\eta_{\pm}(x)}{|k|}\right\} \tag{7.13}$$

This is again subject to the condition that $\eta_{\pm}(x)$ exists. Equation 7.13 then establishes Equation 7.12a. The bound for $(F_{\pm}f_{\pm})(x,k)$ is obtained from Equation 7.8 to be

$$\left|(F_{\pm}f_{\pm})(x,k)\right| \le \left\{\frac{\eta_{\pm}(x)}{|k|}\right\} \exp\left\{\frac{\eta_{\pm}(x)}{|k|}\right\} \tag{7.14}$$

Equation 7.13 guarantees the existence of the Neumann series (7.12) for $x \subset R$ and for a fixed $k \subset \overline{C}_{+}\backslash\{0\}$. The convergence, however, is not uniform in x. For absolute and uniform convergence of the Neumann series, the potential q(x) must be absolutely integrable, that is,

$$\int_{-\infty}^{\infty}\left|q(y)\right| dy < \infty$$

The potential is thus required to satisfy a growth of order zero. Then $|f_{\pm}|(x,k) \le [\text{const.}/|k|]$ and the convergence is independent of x. This also is the sufficient condition for uniform convergence.

Similar considerations apply to $f_{\pm}'(x,k)$, $'$ denoting differentiation with respect to x. From Equation 7.8,

$$f_{\pm}'(x,k) = (F_{\pm}'f_{\pm})(x,k)$$

By the same arguments as above,

$$\left\|F_{\pm}'f_{\pm}\right\| \le \left\|f_{\pm}\right\|\eta_{\pm}(x) < \infty$$

showing that $(F_{\pm}'f_{\pm}) \subset X^{\pm 0}$ if $f_{\pm} \subset X^{\pm 0}$ and the potential is of zeroth order of growth. Iterating, we obtain

$$f_{\pm}'(x,k) = \sum_{n=0}^{\infty} f_{\pm}^{(n)\prime}(x,k) = \sum_{n=0}^{\infty}\left(F_{\pm}'f_{\pm}^{(n)}\right)(x,k)$$

By the same reasoning, $f_{\pm}'(x,k) \subset X^{\pm 0}$. From the Schrödinger equation itself it follows that

$f_\pm''(x,k)$ is also in $X^{\pm 0}$. The bound for $f_\pm^{(n)\prime}(x,k)$ is found to be

$$\left|f_\pm^{(n)'}\right|(x,k) \le \left[\frac{1}{(n+1)!}\right]\eta_\pm(x)\left[\frac{\eta_\pm(x)}{|k|}\right]^n$$

Thus the previous remarks on the uniform convergence for $x \subset R$, $k \subset \overline{C}_+\backslash\{0\}$ apply also to $f_\pm'$ and $f_\pm''$. Summarizing, the following theorem is obtained.

Theorem 7.1.

Let the potential q(x) satisfies a growth condition of order zero. Then the Jost functions r(x,k), l(x,k) and their derivatives r', r'', l', l'' are continuous in x and k on $(R \times \overline{C}_+\backslash\{0\})$ and remain bounded as $x \to \pm\infty$. Moreover, the Neumann series (7.12) is absolutely and uniformly convergent in $x \subset R$ for the above growth condition of the potential.

B. $k \subset \overline{C}_+\backslash\{0\}$: REGULARITY IN k

The inverse solution of the Schrödinger problem for the potential via Marcenko's integral equation requires complex integration of formulas involving the Jost functions. It is, therefore, necessary to establish the analyticity properties of these functions, that is, their regularity properties in the k plane. Let us first differentiate the Schrödinger equation (7.8) with respect to k and obtain

$$\dot{f}_\pm(x,k) = \left(\dot{F}_\pm f_\pm\right)(x,k) + \left(F_\pm \dot{f}_\pm\right)(x,k) \tag{7.15}$$

where (d/dk) is denoted by $\cdot$. It is not difficult to establish that the bound for $\dot{F}_\pm$ is[3]

$$\left|\dot{F}_\pm\right| \le \left(\frac{1}{2}\ |\text{Im } k|^{-2}\right)|q|$$

Then,

$$\left|\left(\dot{F}_\pm f_\pm\right)\right|(x,k) \le \left(\frac{1}{2}\ |\text{Im } k|^{-2}\right)\int_\pm |q(x')|\,|f_\pm(x',k)|\,dx' \le \left[\frac{\eta_\pm(x)}{2}\left(|\text{Im } k|^{-2}\right)\right]\exp\left\{\frac{\eta_\pm(x)}{|k|}\right\}$$

by virtue of the estimate (7.13). For each fixed x and for $k \subset \overline{C}_+\backslash\{0\}$, we then have

$$\left|\left(\dot{F}_\pm f_\pm\right)\right|(y,k) = M(x) < \infty$$

Then from Equation 7.15

$$\left|\dot{f}_\pm(x,k)\right| \le M(x) + \left|\left(F_\pm \dot{f}_\pm\right)\right|(x,k)$$

Defining $v(x,k) = |\dot{f}_\pm(x,k)|/M(x)$ and iterating on v(x,k), we obtain

$$\left|\dot{f}_\pm(x,k)\right| \le M(x)\exp\left\{\frac{\eta_\pm(x)}{|k|}\right\} < \infty$$

This bound ensures that $\dot{f}_{\pm}(x,k)$ exists for each $x \subset R$ and for all compact subsets $\Omega \subset \overline{C}_{+}\backslash\{0\}$, $k \subset \Omega$. Let $Y^{\pm}$ be a space such that

$$Y^{\pm} = \left\{ y \subset C\left(R \times C_{+}\right) \right\} \; y \text{ analytic in } k \subset C_{+} \text{ and for each } x \subset R \}$$

$$|y|\left(x_0, k\right) = \sup_{k \subset \Omega,\, x \gtrless x_0} \sup \left|y(x,k)\right| < \infty$$

Then, the Jost functions $\dot{f}_{+}$ resp. $\dot{f}_{-}$ belong to Y^{+} resp. Y^{-}.

Let us next consider the analyticity of $\dot{f}_{\pm}'(x,k)$. Now,

$$f_{\pm}'(x,k) = \left(F_{\pm}' f_{\pm}\right)(x,k)$$

Differentiation with respect to k yields

$$\dot{f}_{\pm}'(x,k) = \left(\dot{F}_{\pm}' f_{\pm}\right)(x,k) + \left(F_{\pm}' \dot{f}_{\pm}\right)(x,k)$$

It can be established that

$$\left|\dot{F}_{\pm}'\right| \le \left(\frac{1}{e \operatorname{Im} k}\right) |q|$$

From this, the following estimates are obtained:

$$\left|\left(\dot{F}_{\pm}' f_{\pm}\right)\right|(x,k) \le \left\{\frac{\eta_{\pm}(x)}{e \operatorname{Im} k}\right\} \exp\left\{\frac{\eta_{\pm}(x)}{|k|}\right\}, \quad \text{and}$$

$$\left|\left(F_{\pm}' \dot{f}_{\pm}\right)\right|(x,k) \le M(x) \exp\left\{\frac{2\eta_{\pm}(x)}{|k|}\right\}$$

Then it follows that both $(\dot{F}_{\pm}' f_{\pm})$, $(F_{\pm}' \dot{f}_{\pm})$ and, hence, $\dot{f}_{\pm}'$, belong to the space $Y^{\pm}$. The Schrödinger equation then shows that $\dot{f}_{\pm}''$ also $\subset Y^{\pm}$. Hence, the analyticity of the derivatives of the Jost functions in k is established.

The regularity of the Jost functions in k can be shown as follows. The Neumann series for $f_{\pm}$ is

$$f_{\pm} = \sum_{n=0}^{\infty} \left(F_{\pm} f_{\pm}^{(n)}\right) = 1 + \sum_{n=1}^{\infty} f_{\pm}^{(n)}$$

where we have used the relation $f_{\pm}^{(n+1)} = (F_{\pm} f_{\pm}^{(n)})$. Now, $f_{\pm}^{(1)}$ is an element of $Y^{\pm}$. If h is an element of $Y^{\pm}$, then $(f_{\pm}^{(1)},h)$ also is in the $Y^{\pm}$ space. Repeated application of this ensures that $f_{\pm}^{(n)} \subset Y^{\pm}$. Since each of the iterates is in $Y^{\pm}$, then $f_{\pm}(x,k)$ is analytic on the same domain since the series is uniformly convergent there if the potential is of zeroth order growth.[13] The same reasoning will also yield the analyticity of $f_{\pm}'$ and hence of $f_{\pm}''$ in $k \subset C_{+}$ by virtue of the Schrödinger equation itself.

These results can be summarized in the theorem below.

Theorem 7.2

Let the potential q(x) satisfy a growth condition of order zero. Then the Jost functions and their derivatives, $r, r', r'', l, l', l'', \dot{r}, \dot{r}', \dot{r}'', \dot{l}, \dot{l}', \dot{l}''$ are analytic in $k \subset C_+$ for each $x \subset R$.

C. $k \subset \overline{C}_+$: Im k ≥ 0, INCLUDING k = 0. REGULARITY IN x

In the vicinity of the origin, the bound for $F_\pm$ is given by Equation 7.10. Accordingly, in the limit $k \to 0$, Equation 7.8 for the Jost functions reduces to

$$f_\pm(x,k) = 1 + \int_\pm \xi_\pm\, q(x')\, f_\pm(x',k)\, dx', \quad k \to 0 \tag{7.16}$$

If q(x) satisfies a growth condition of order 2, an alternate representation can be obtained

$$\tilde{f}_\pm(x,k) = 1 + x^{-1} \int_\pm \xi_\pm x'\, q(x')\, \tilde{f}_\pm(x,k')\, dx' \tag{7.17}$$

That Equation 7.17 satisfies the Schrödinger problem along with the boundary condition can be verified directly.

The objective here is to establish the regularity properties in x of the functions $f_\pm(x,k)$ on $[R \times \overline{C}_+]$, i.e., in the vicinity of k = 0, including zero. This implies investigating the behavior of $f_\pm(x,k)$ in the limit $x \to \pm\infty$. A number of estimates exist in the literature for the bounds of the Jost functions as $x \to \pm\infty$, $k \subset \overline{C}_+$. These estimates were originally given by Fadeev[7] in 1964 and were later modified by Deift and Trubowitz[2] in 1979. On any half-line, i.e., for $x \subset [a,b]$, $-\infty < a < b < \infty$, the modifications are not critical except that they predict a stronger rate of increase of the bounds with x. These are, however, important if x is allowed to vary over the entire real axis. These bounds and their modifications will be discussed next.

The following bound is easily established. Iterating (7.16) and majorizing as before,

$$\left|f_\pm^{(n)}\right|(x,k) \le \int_{\Omega\pm} d_n \prod_{i=1}^{n} \xi_{\pm i} \left|q(x_i)\right| \le \frac{\{\gamma_\pm(x)\}^n}{n!}, \qquad \xi_{\pm i} = \left|x_i - x\right| \tag{7.18}$$

where

$$\gamma_\pm(x) = \int_\pm \left|\xi_\pm\right| \left|q(x')\right| dx'$$

and the estimate given by Equation 7.10 has been used. This then yields the bound

$$\left|f_\pm(x,k)\right| \le \exp\left[\gamma_\pm(x)\right] \tag{7.19}$$

Also,

$$\left|f_\pm(x,k) - 1\right| \le \gamma_\pm(x) \exp\left[\gamma_\pm(x)\right] \tag{7.19a}$$

showing that the bound increases exponentially in the nonprescribed directions, i.e., $x \to +\infty$ for f_- and $x \to -\infty$ for f_+. However, it will be shown below that the situation is actually not so severe and that the Jost functions grow only linearly with x in the nonprescribed directions. For the above bound to exist, a stronger condition must be imposed on $q(x)$, namely, that it satisfies a growth condition of order 1. This is also evident from Equation (7.16).

For the alternate representation (7.17), we again iterate and obtain

$$\left|\tilde{f}_{\pm}^{(n)}(x,k)\right| \le \left(\frac{1}{n!}\right)\{S_{\pm}(x)\}^n, \qquad x = 0$$

$$\le \left(\frac{1}{n!}\right)\{S_{\pm}(x,\delta)\}^n, \qquad x \not\subset]-\delta,\delta[$$

$$S_{\pm}(x) = \int_{\pm}\left[\left(\frac{|x'^2|}{|x|}\right) + |x'|\right]|q(x')|\, dx', \qquad x \not\subset]-\delta,\delta[$$

$$\le \int_{\pm}\left[\left(\frac{|x'^2|}{|\delta|}\right) + |x'|\right]|q(x')|\, dx'$$

Thus, any neighborhood of the origin is to be excluded.

Before discussing the modifications of Fadeev's bound by Deift and Trubowitz, let us establish an overall bound for the Jost functions. We know that

$$|f_{\pm}(x,k)| \le 1 + \int_{\pm}\xi_{\pm}\,|q(x')|\,|f_{\pm}(x',k)|\, dx'$$

$$= 1 \pm \int_{\pm} x'\,|q(x')|\,|f_{\pm}(x',k)|\, dx' \mp x\int_{\pm}|q(x')|\,|f_{\pm}(x',k)|\, dx'$$

$$\le 1 \pm \int_{\pm 0} x'\,|q(x')|\,|f_{\pm}(x',k)|\, dx' \mp x\int_{\pm}|q(x')|\,|f_{\pm}(x',k)|\, dx'$$

where

$$\int_{\pm 0} dx' = \int_{\pm} dx' \quad \text{with } x \text{ in the integration limit replaced by } 0$$

Using the bound of Equation 7.19, namely,

$$|f_{\pm}(x,k)| \le \exp\{\gamma_{\pm}(x)\}$$

the following inequality is obtained:

$$1 \pm \int_{\pm 0} x'|q(x')||f_{\pm}(x',k)|\, dx' \le 1 \pm \int_{\pm 0} x'|q(x')|\exp\{\gamma_{\pm}(0)\}\, dx'$$

$$= 1 \pm \exp[\gamma_{\pm}(0)]\int_{\pm 0} x'|q(x')|\, dx'$$

The bound can also be written as $[(1 \pm \gamma_\pm(0) \exp[\gamma_\pm(0)] \int_{\pm 0} x'|q(x')|\, dx']$ by writing $|f_\pm| = |(f_\pm - 1) + 1| \le |f_\pm - 1| + 1$ and then using the bound given by Equation 7.19a. In either case, we have

$$1 \pm \int_{\pm 0} x'|q(x')||f_\pm(x',k)|\, dx' \le K < \infty$$

These follow since q is assumed to satisfy a growth condition of order 1 and since $\gamma_+(x)$ resp. $\gamma_-(x)$ is nonincreasing resp. nondecreasing in x. Also, following Fadeev, K will be used to designate any constant quantity that depends upon the norm

$$\int_{-\infty}^{\infty} \left[1 + |x|^i\right] |q(x)|\, dx$$

Therefore,

$$|f_\pm(x,k)| \le K + \max(0, \mp x) \int_{\pm} |q(x')||f_\pm(x',k)|\, dx'$$

Defining

$$h_\pm(x,k) = \frac{f_\pm(x,k)}{K'\{1 + \max(0, \mp x)\}}$$

the above inequality can be written as

$$|h_\pm(x,k)| \le 1 + \int_{\pm} |q(x')| \{1 + \max(0, \mp x')\} |h_\pm(x',k)|\, dx'$$

Iterating,

$$|h_\pm(x,k)| \le \exp\left[\int_{\pm} |q(x')| [1 + \max(0, \mp x')]\, dx' \le K_1 < \infty\right.$$

It then follows that

$$|f_\pm(x,k)| \le K_2\{1 + \max(0, \mp x)\} \tag{7.20}$$

The constant K_2 depends on the potential only. This is one of the estimates found in the literature.[3] It is assumed that the potential is L_1^1. From Equation 7.20 it also follows that

$$|f_\pm(x,k)| \le K_2\{1 + |x|\} \tag{7.20a}$$

This is the bound most often found in the literature.[2,6,7]

Let us next consider the bound for the function $(f_\pm - 1)$ and show that it grows linearly in the nonprescribed directions, as was mentioned following the estimate (7.19).

$$\left|f_{\pm}(x,k)-1\right| \leq \pm \int_{\pm 0} x'\left|q(x')\right|\left|f_{\pm}(x',k)\right| dx' \mp x \int_{\pm}\left|q(x')\right|\left|f_{\pm}(x',k)\right| dx'$$

$$\leq \pm \int_{\pm 0} x'\left|q(x')\right|\left|f_{\pm}(x',k)\right| dx' + \max(0,\mp x)\int_{\pm}\left|q(x')\right|\left|f_{\pm}(x',k)\right| dx'$$

We replace $|f_{\pm}(x',k)|$ in the first integral on the right-hand side by the estimate given in Equation 7.19 and $|f_{\pm}(x',k)|$ in the second integral by Equation 7.20a. Then,

$$\left|f_{\pm}(x,k)-1\right| \leq \exp\{\gamma_{\pm}(0)\} \int_{\pm 0}\left|q(x')\right| x'\, dx' + K_2 \max(0,\mp x)\int_{\pm}(1+|x'|)\left|q(x')\right| dx'$$

$$\leq K_3\{1+\max(0,\mp x)\}\int_{\pm}(1+|x'|)\left|q(x')\right| dx' \tag{7.21}$$

Equation 7.21 is valid for all $x \subset R$.

The bound on $f_{\pm}'(x,k)$ can be established analogously. In the vicinity of the origin $k = 0$, $f_{\pm}'(x,k)$ is given by

$$f_{\pm}'(x,k) = \mp\int_{\pm} q(x')f_{\pm}(x',k)\, dx'$$

Then, using Equation 7.20a,

$$\left|f_{\pm}'(x,k)\right| \leq K_2 \int_{\pm}(1+|x'|)\left|q(x')\right| dx' \leq K_3 < \infty \tag{7.22}$$

This estimate is found in the literature.[3] On the positive half-line $x \subset R^+$, we can write that

$$\left|f_{\pm}'(x,k)\right| \leq K_2\, \eta_{\pm}(x)$$

using the estimate (7.20).

Inequalities 7.21 and 7.22 are also valid for k away from the origin. The kernel $F_{\pm}(\xi_{\pm},k)$ in the integral representation of the Jost functions can be written as

$$F_{\pm}(\xi_{\pm},k) = \{(1/k)\sin k\xi_{\pm}\}\exp\{ik\xi_{\pm}\} = \frac{\{\exp(2ik\xi_{\pm})-1\}}{2ik}$$

$$= \int_0^{\xi_{\pm}} \exp(2iks)\, ds$$

For k in the vicinity of zero, $|F_{\pm}(\xi_{\pm},k)| \leq \xi_{\pm}$. For k away from the origin, the integral is still bounded by $\xi_{\pm}$. Hence, the estimate also remains valid for k away from zero.

For k away from zero, Equation 7.14 gives the bound for $[f_{\pm} - 1]$. This can be combined with Equation 7.21 to yield (assuming that the factor $\{1 + \max(0,\mp x)\}$ pertains only to k in the vicinity of the origin)

$$\left|f_{\pm}(x,k)-1\right| \le K\left[\frac{\{1+\max(0,\mp x)\}}{1+|k|}\right]\cdot \int_{\pm}\left(1+|x'|\right)\left|q(x')\right| dx'$$

$$\le K_4\left[\frac{\{1+\max(0,\mp x)\}}{1+|k|}\right], \qquad k \subset \overline{C}_+ \tag{7.23}$$

For k → 0, Equation 7.23 → Equation 7.21, whereas, for |k| large, the bound falls off as 1/|k|.

Analogously from the inequality (7.22),

$$\left|f'_{\pm}(x,k)\right| \le \frac{K_5}{1+|k|}, \qquad k \subset \overline{C}_+ \tag{7.24}$$

K_5 requires that $q \subset L_1{}^1$ if $x \subset R$. Otherwise, K_5 requires that q be only $L_0{}^1$ if $x \subset R^+$, as was pointed out following Equation 7.22. The bounds originally estimated by Fadeev and also used by other authors on inverse scattering are identical to those given above in Equation 7.23 except for the factor $[1 + \max(0,\mp x)]$. This modification is due to Deift and Trubowitz.[2] Estimates 7.23 and 7.24 will be referred to as the Fadeev-Deift-Trubowitz (or FDT) bound.

For k away from zero, $k \subset C_+\backslash\{0\}$, $f_\pm(x,k)$ are bounded in x. This bound is, however, not uniform on the entire real axis unless the potential is absolutely integrable on R. On any half-line, however, the boundedness is uniform. These are the consequences of the inequality (7.13). The same is also true if the potential satisfies a growth condition of order 1, that is, if k = 0 and its vicinity are included. However, the FDT bound now indicates that the Jost functions $f_\pm$ grow linearly as $x \to \mp\infty$ for $k \to 0$, that is, in the nonprescribed directions. This is in contradistinction to the bound without the $\{1 + \max(0,\mp x)\}$ modification which predicts $f_\pm(x,k)$ to be bounded over the entire real axis even if $k \to 0$.

For $k \subset \overline{C}_+$, the iterates for the Jost functions $f_\pm{}^{(n)}$ are given by (7.18)

$$\left|f_{\pm}^{(n)}(x,k)\right| \le \left(\frac{1}{n!}\right)\{\gamma_{\pm}(x)\}^n$$

$$=\left(\frac{1}{n!}\right)\left[\eta_{\pm 1}(x)+\max(0,\mp x)\,\eta_{\pm}(x)\right]^n$$

where,

$$\eta_{\pm 1}(x)=\int_{\pm} x'\left|q(x')\right| dx'$$

In view of these bounds, and if x belongs to any half-line on the real axis, then following exactly the same procedures as before, $f_\pm(x,k) \subset X^{\pm 0}$, where $X^{\pm 0}$ is the space consisting of functions continuous on any half-line and for all $k \subset \overline{C}_+$. Similarly, $f_\pm'(x,k)$ also belongs to $X^{\pm 0}$ and so does $f_\pm''(x,k)$ by virtue of the Schrödinger equation itself.

Combining these results yields the following theorem.

Theorem 7.4.

Let the potential q(x) satisfy a growth condition of order 1. Then for all k including the origin and its vicinity, the Jost functions and their first and second derivatives are continuous functions in x on any half-line $x \subset [a,b]$, $-\infty < a < b < \infty$, for each $k \subset \overline{C}_+$. The bounds are given by

$$\left|f_{\pm}(x,k) - 1\right| \leq K \frac{\left[1 + \max(0, \mp x)\right]}{1 + |k|}$$

with the factor $\{1 + \max(0, \mp x)\}$ applying when k is in the vicinity of the origin, and

$$\left|f'_{\pm}(x,k)\right| \leq \frac{K}{1 + |k|}$$

Moreover, the bounds for $f_{\pm}$ are uniform on any half-line, but $f_{\pm}(x,0)$ increases linearly with x as $x \to \mp\infty$, that is, in the nonprescribed directions.

D. ANALYTICITY OF THE JOST FUNCTIONS FOR $k \subset \overline{C}_+$

Consider the Jost functions:

$$f_{\pm}(x,k) = 1 + \left(F_{\pm} f_{\pm}\right)(x,k)$$

Differentiation with respect to k yields

$$\dot{f}_{\pm}(x,k) = \left(\dot{F}_{\pm} f_{\pm}\right)(x,k) + \left(F_{\pm} \dot{f}_{\pm}\right)(x,k)$$

The kernel of the integral operator $\dot{F}_{\pm}$ is given by

$$\left|\dot{F}_{\pm}(\xi_{\pm}, k)\right| \leq \left| \int_0^{\xi_{\pm}} 2iz \exp(2ikz)\, dz \right| \leq (\xi_{\pm})^2 = \xi^2$$

Therefore,

$$\left|\left(\dot{F}_{\pm} f_{\pm}\right)\right|(x,k) \leq \int_{\pm} \xi^2 \left|q(x')\right| \left|f_{\pm}(x',k)\right| dx'$$

It is, therefore, clear that for the Jost functions to be analytic in the vicinity of the origin, the potential must satisfy a growth condition of order 2. We will show that

$$\left|\left(\dot{F}_{\pm} f_{\pm}\right)\right|(x,k) \leq K\{1 \mp x \max(0, \mp x)\}, \quad \text{for all } x \subset R \tag{7.25}$$

Let us prove this by considering $f_+(x,k)$. Analogous results holds for the other Jost function.

For $x \geq 0$, we have

$$\left|\left(\dot{F}_+ f_+\right)\right|(x,k) \leq \int_+ (x'-x)^2 \left|q(x')\right| \left|f_+(x',k)\right| dx'$$

$$\leq \int_+ x'^2 \left|q(x')\right| \left|f_+(x',k)\right| dx'$$

$$\leq K \int_+ x'^2 \left|q(x')\right| dx'$$

$$\leq K', \quad \text{if } q \subset L_2^1 \tag{7.26}$$

In deriving Equation 7.26, the bound given by Equation 7.20 was used.

If $x \leq 0$, then

$$\left|\left(\dot{F}_\pm f_\pm\right)\right|(x,k) \leq x^2 \int_+ \left|q(x')\right| \left|f_+(x',k)\right| dx' +$$

$$\int_+ x'^2 \left|q(x')\right| \left|f_+(x',k)\right| dx'$$

$$2x \int_+ x' \left|q(x')\right| \left|f_+(x',k)\right| dx'$$

$$\leq 2x^2 \int_+ \left|q(x')\right| \left|f_+(x',k)\right| dx' +$$

$$2 \int_+ x'^2 \left|q(x')\right| \left|f_+(x',k)\right| dx'$$

Now for $x \leq 0$

$$\int_+ x'^2 \left|q(x')\right| \left|f_+(x',k)\right| dx'$$

$$\leq \int_{+0} x'^2 \left|q(x')\right| \left|f_+(x',k)\right| dx' + x^2 \int_x^0 \left|q(x')\right| \left|f_+(x',k)\right| dx' \tag{7.27}$$

If $q \subset L_2{}^1$, then using the bound

$$\left|f_+(x,k)\right| \leq K\left[1 + \max(0,-x)\right]$$

the first integral on the right-hand side of Equation 7.27 reduces to a constant K_1. For $x \leq 0$, the quantity $[1 + \max(0,-x)]$ becomes $(1 + |x|)$, and the second integral on the right-hand side of Equation 7.27 can be written as

$$x^2 \int_x^0 \left|q(x')\right| \left|\left[1 + |x'|\right]\right| dx' \leq K_2 x^2$$

Then,

$$\int_{+} x'^2 |q(x')| |f_+(x',k)| \, dx' \le K_1 + K_2 x^2 \le K_3 \left(1 + x^2\right) \tag{7.28}$$

Also,

$$2x^2 \int_{+} |q(x')| |f_+(x',k)| \, dx' \le K_4 x^2 \tag{7.28a}$$

where the estimate of Equation 7.20 was used again. Combining the estimates of Equations 7.28 and 7.28a, we obtain

$$\left|\left(\dot{F}_\pm f_\pm\right)\right| (x,k) \le K_5 x^2 + K_3\left(1+x^2\right) \le K\left(1+x^2\right)$$

Therefore, for $x \le 0$,

$$\left|\left(\dot{F}_\pm f_\pm\right)\right| (x,k) \le K\left(1+x^2\right) = K\left[1 - x \ \max(0,-x)\right] \tag{7.29}$$

Equation 7.25 follows from Equations 7.26 and 7.29. Extending the results to $f_-(x,k)$ in a similar way, we obtain

$$\left|\dot{f}_\pm(x,k)\right| \le K(1 \mp x \ \max(0, \mp x)) + \left|\left(F_\pm \dot{f}_\pm\right)\right|(x,k) \tag{7.30}$$

Successive approximation then yields

$$\left|\dot{f}_\pm(x,k)\right| \le K\{1 \mp x \max(0, \mp x)\} \exp\{\gamma_\pm(x)\} \tag{7.31}$$

Another bound can also be obtained as follows. Inequality 7.30 can be reformulated as

$$\left|\dot{f}_\pm(x,k)\right| \le K\{1 \mp x \max(0, \mp x)\} + \int_{\pm 0} x' \left|q(x')\right| \left|\dot{f}_\pm(x',k)\right| + |x| \int_{\pm} \left|q(x')\right| \left|\dot{f}_\pm(x',k)\right| dx'$$

$$\le K\{1 \mp x \max(0, \mp x)\} + K \exp\{\gamma_\pm(0)\} \int_{\pm 0} x' \left|q(x')\right| \, dx' + |x| \int_{\pm} \left|q(x')\right| \left|\dot{f}_\pm(x',k)\right| dx'$$

Inequality 7.19 was used. We can further replace the first term by $K(1 + |x|^2)$. Then, defining a new variable

$$\left|h_\pm(x,k)\right| = \frac{\left|\dot{f}_\pm(x,k)\right|}{K\left(1+|x|^2\right)}$$

we obtain for $|h_\pm(x,k)|$ the following inequality

$$\left|h_\pm(x,k)\right| \le 1 + \int_{\pm} \left|q(x')\right| \left(1 + |x'|^2\right) \left|h_\pm(x',k)\right| dx'$$

Iterating, we find

$$\left|h_{\pm}(x,k)\right| \leq \exp\left[\int_{\pm}\left|q(x')\right|\left(1+\left|x'\right|^{2}\right)\right] dx'$$

Finally, then

$$\left|\dot{f}_{\pm}(x,k)\right| \leq K\left(1+\left|x^{2}\right|\right) \text{ for any } x \tag{7.32}$$

Since the constants Ks depend only on the integrals involving the potential on the entire line, the bounds given by Equations 7.31 and 7.32 guarantee that $\dot{f}_{\pm}(x,k)$ converges uniformly in k. $\dot{f}_{\pm}(x,k)$, therefore, exists, is analytic in $k \subset C_{+}$, and is continuous everywhere in the closed upper half-plane $k \subset \overline{C}_{+}$. It is clear that $\dot{f}_{\pm}(x,k) \subset C(R \times \overline{C}_{+})$ on any half-line in $x \subset R$. $\dot{f}_{\pm}(x,k)$ can thus be continued analytically from $C(R \times C_{+})$ to $C(R \times \overline{C}_{+})$ on any half-line. $\dot{f}_{\pm}(x,k)$ thus is a member of the space $Y^{\pm 0}$, where the subscript zero indicates the closed upper-half k-plane.

For the k-derivative of $f_{\pm}'(x,k)$, we use the relation

$$\dot{f}_{\pm}'(x,k) = \left(\dot{F}_{\pm}' f_{\pm}\right)(x,k) + \left(F_{\pm}' \dot{f}_{\pm}\right)(x,k)$$

Now,

$$\left|F_{\pm}(x,y,k)\right| \leq \left|q(y)\right| \left\{\left|y\right| + \max(0, \mp x)\right\}$$

$$\left|F_{\pm}'(x,y,k)\right| \leq \left|q(y)\right|$$

$$\left|\dot{F}_{\pm}(x,y,k)\right| \leq \left|q(y)\right| \left\{\left|y\right| + \max(0, \mp x)\right\}^{2}$$

$$\left|\dot{F}_{\pm}'(x,yk)\right| \leq 2\left|q(y)\right| \left\{\left|y\right| + \max(0, \mp x)\right\}$$

$F_{\pm}(x,k)$ and $\dot{F}_{\pm}'(x,k)$ both $\subset X^{\pm 0}$ if the growth condition of order 1 is satisfied. The same is true for $F_{\pm}'(x,k)$ for zero-order growth. On the other hand, $\dot{F}_{\pm}(x,k) \subset X^{\pm 0}$ if the potential satisfies the growth condition of order 2. By virtue of these estimates and because of the fact that $\dot{f}_{\pm}(x,k) \subset Y^{\pm 0}$, $f_{\pm}(x,k) \subset X^{\pm 0}$, and that the kernels in the above equation also $\subset X^{\pm 0}$, it is clear that $\dot{f}_{\pm}'(x,k)$ can also be extended from $(R \times C_{+})$ to $(R \times \overline{C}_{+})$. That $\dot{f}_{\pm}''(x,k) \subset (R \times \overline{C}_{+})$ follows directly from the Schrödinger equation.

Theorem 7.5.

Let the potential q(x) satisfy a growth condition of order 2. Then the Jost functions and their derivatives are continuous on $(R \times \overline{C}_{+})$, including $k = 0$. Moreover, the Jost functions are analytic in the entire upper-half plane $\text{Im } k > 0$. If $q(x) \subset L_{2}^{1}$, then the derivatives of the Jost functions with respect to k are uniformly bounded, that is,

$$\left|\dot{f}_{\pm}(x,k)\right| \leq \text{constant}\left(1+\left|x\right|^{2}\right), \quad k \subset \overline{C}_{+}$$

The importance of the above discussion is in demonstrating that the potential must satisfy a stronger, second-order growth condition if the Jost functions are to be analytic in the vicinity of $k = 0$, including zero itself. Imposition of such a stronger condition leads to the regularity at the origin of the otherwise badly singular Jost functions at $k = 0$ and its vicinity.

E. ANALYTICITY FOR Im k < 0: BARGMANN STRIP

We have seen that the Jost functions can be continued analytically off the real axis and are holomorphic in the upper half-plane $\text{Im } k > 0$. Continuation in the lower half-plane is also possible if the potential satisfies a much stronger condition, namely, if it falls off faster than any power of x. If the potential vanishes asymptotically at least exponentially,

$$\lim_{x\to\pm\infty} \left[q(x)\exp(\pm 2\alpha_{\pm}x)\right] \longrightarrow 0, \quad \alpha_{\pm} > 0$$

then from (7.8), it follows that $f_{\pm}(x,k)$ are also holomorphic in the lower half-plane $\text{Im } k < 0$, provided $\text{Im } k > -\alpha_{+}$ resp. $-\alpha_{-}$. Let α be the minimum of (α_{+},α_{-}), that is, $\alpha = \min(\alpha_{+},\alpha_{-})$. The αs in the parenthesis are the largest values compatible with the above equation for the fall-off of $q(x)$. Then the strip $-\alpha < \text{Im } k < \alpha_{+}$ is called the "Bargmann" strip. It will be shown later that the "reflection coefficient" of the scattering of a plane wave off the potential is meromorphic in this strip, while the "transmission" coefficient is meromorphic in the half-plane $\text{Im } k > -\alpha_{-}$. If the potential vanishes faster than exponentially as $x \to +\infty$, i.e., if

$$\text{limit}_{x\to\infty} \left[q(x)\exp(\mu x)\right] \longrightarrow 0$$

for any finite μ, then the Bargmann strip invades the entire upper-half plane. For these potentials, both the reflection and the transmission coefficient are meromorphic in the entire upper-half of the complex k-plane. The size of the Bargmann strip is thus determined by the asymptotic behavior of the potential. More will be said about the Bargmann strip when the scattering matrix is discussed.

F. UNIQUENESS, LINEAR INDEPENDENCE, AND REALITY PROPERTIES

Let there be more than one solution to the Schrödinger equation for the Jost functions, the difference between the two being given by $\delta f_{\pm}(x,k)$. By definition, then

$$\delta f_{\pm}(x,k) = \int_{\pm} F_{\pm}(\xi_{\pm},k)q(x')\delta f_{\pm}(x',k)dx'$$

Then from Equations 7.8 and 7.10

$$|\delta f_{\pm}(x,k)| \le \int_{\pm} \xi_{\pm}|q(x')|\,|\delta f_{\pm}(x',k)|\,dx' = S_{\pm}(x,k)$$

Differentiating $S_{\pm}(x,k)$ with respect to x and using the above inequality yields

$$\begin{aligned} S'_{\pm}(x,k) &= \mp\int_{\pm} |q(x')||\delta f_{\pm}(x',k)|\,dx' \\ &\gtrless S_{\pm}(x,k)\theta_{\pm}(x) \end{aligned} \tag{7.33}$$

where $\theta_{\pm}(x) = \mp\eta_{\pm}(x)$. The inequality $\leq$ resp. $\geq$ applies to S_- resp. S_+.

Defining

$$v_{\pm}(x,k) = S_{\pm}(x,k)\ \exp\left\{-\int_{\pm} \xi_{\pm}\left|q(x')\right|\, dx'\right\}$$

it follows from Equation 7.33 that $v_{\pm}' \gtrless 0$. The sense of $\gtrless$ is again as in Equation (7.33). Integrating v_+' between (x,∞) and v_-' between $(-\infty,x)$ yields $|\delta f_{\pm}|(x,k) \leq 0$. This implies that $\delta f_{\pm}(x,k) = 0$. Thus, the Jost functions and, hence, the Jost solutions must be unique for k in the upper half-plane, $k \subset \overline{C}_+$, if the potential satisfies a growth condition of order 1. Similar conclusions hold for the other pair of Jost solutions in the lower half-plane.

The linear independence of the Jost solutions can also be established by calculating their Wronskian. The Wronskian of two functions $y(x,k)$ and $Y(x,k)$ is defined as

$$W(y,Y) = yY' - y'Y \tag{7.34}$$

If W is nonzero, then the two functions are linearly independent of each other and vice versa. Differentiating the Wronskian expression with respect to x yields

$$W'(y,Y) = yY'' - y''Y \tag{7.35}$$

Replacing y by $\psi(x,k)$ and Y by $\phi(x,k)$ in Equation 7.35 and then using the Schrödinger equation,

$$y'' + \left(k^2 - q\right)y = 0$$

we obtain

$$W'(\psi,\phi) = 0$$

The Wronskian is thus a constant independent of x and may be a function of k establishing the linear independence of the Jost solutions $\psi(x,k)$ and $\phi(x,k)$. The x-independence of the Wronskian reflects the fact that the Schrödinger equation has no first-derivative term. The Jost solutions are, therefore, unique and linearly independent solutions of the Schrödinger equation and satisfy the asymptotic boundary conditions. These, therefore, form the basis for the solution space. Hence, an arbitrary solution of the Schrödinger equation can be expressed as a linear combination of the Jost solutions.

The reality conditions can be easily derived to be

$$f_{\pm}(x,-k) = \left[f_{\pm}^{*}\left(x,k^{*}\right)\right] = \left[f_{\pm}(x,k)\right]^{*}$$

The Jost solutions satisfy the following orthogonality conditions:

$$(2\pi)^{-1} \int_{-\infty}^{\infty} g_{\pm}(x,k)\ g_{\pm}(x,m)\ dx = G_{\pm 1}(k)\ \delta(k+m) - G_{\pm 2}(k)\ \delta(k-m)$$

and

$$(2\pi)^{-1}\int_{-\infty}^{\infty} g_{\mp}(x,k)\, g_{\pm}(x,m)\, dx = G_{\pm 3}(k)\, \delta(k-m)$$

$g_+(x,k)$ resp. $g_-(x,k)$ stand for $\psi(x,k)$ resp. $\phi(x,k)$. $G_1(k)$, $G_2(k)$, and $G_3(k)$ are functions of k.

In order to illustrate how these relations arise, let us consider the x-derivative of the Wronskian between the Jost solutions $\psi(x,k)$ and $\phi(x,m)$, that is,

$$\left[W\{\psi(x,k),\ \phi(x,m)\}\right]' = \left(k^2 - m^2\right) \psi(x,k)\, \phi(x,m)$$

Integrating between the limits x_1 and x_2, we obtain

$$\int_{x_1}^{x_2} \psi(x,k)\, \phi(x,m)\, dx = \left(k^2 - m^2\right)^{-1} \left[W\{\psi(x,k),\ \phi(x,m)\}\right]_{x_1}^{x_2}$$

Since the Wronskian is independent of x, we can evaluate it at the asymptotic limits. If we let $x_1 \to -\infty$ and $x_2 \to +\infty$, then it will be shown in Chapter 9 that the Wronskian will contain terms involving the plane waves $\exp\{\pm i(k - m)x\}$. This will result in the distributional expressions of the form

$$(k-m)^{-1} \exp\left[i(k-m)x\right] \tag{7.36}$$

According to a well-known formula in the theory of distributions,

$$\lim_{x\to\infty} \left[p^{-1} \exp(ipx)\right] = i\pi\, \delta(p)$$

in the sense of a principal value integral, Equation 7.36 will result in terms containing $\delta(k \pm m)$. The precise functional forms of G_1, G_2 and G_3 will be derived in Chapter 9.

The Schrödinger equation is invariant under translation in space. This is readily seen from the form of the equation itself. Now, suppose that the potential q(x) contains a parameter, say, a, i.e., $q = q(x,a)$. It is often important to determine the special forms of dependence of the potential on a parameter such that the quantity k remains unchanged when the parameter changes. The Schrödinger equation which satisfies such a condition is referred to as the "iso-spectral" Schrödinger equation. This fact is crucial in the inverse scattering solution of nonlinear evolution equations. The iso-spectral property is certainly satisfied if the parameter a is translational, that is, if $q(x,a) = q(x + a)$. Let $q(x + a)$ be denoted by $q^{\sim}(x)$. Then the following translational transformation properties are valid:

$$Q_{\pm}^{\sim}(x) = Q_{\pm}(x+a)$$

$$g_{\pm}^{\sim}(x,k) = \exp(\mp ika)\, g_{\pm}(x+a,\ k)$$

The symbol ~ stands for the transformed quantities and $Q_{\pm}$ is defined by

$$Q_{\pm} = \int_{\pm} q(x)\, dx$$

For infinitesimally small translations, that is, $a = \epsilon$, $\epsilon \to 0$, the variations of the various quantities are given by

$$\delta q(x) = \epsilon\, q'(x)$$

$$\delta Q_{\pm}(x) = \epsilon\, Q'_{\pm}(x) = -\epsilon\, q(x)$$

and

$$\delta g_{\pm}(x,k) = \epsilon \left[\mp ik\, g_{\pm}(x,k) + g'_{\pm}(x,k) \right]$$

The Schrödinger equation is also invariant under scale transformations. Thus, if

$$q^{\sim}(x) = \mu^2 q(\mu x)$$

then

$$Q^{\sim}_{\pm}(x) = \mu Q_{\pm}(\mu x)$$

and

$$g^{\sim}_{\pm}(x,k) = g_{\pm}\left(\mu x, \frac{k}{\mu} \right)$$

Again, for infinitesimal scale transformations, that is, $\mu = 1 + \epsilon$, $\epsilon \to 0$, the following relations hold

$$\delta q(x) = \epsilon \left[2q(x) + xq'(x) \right]$$

$$\delta Q_{\pm}(x) = \epsilon \left[Q_{\pm}(x) + xQ'_{\pm}(x) \right]$$

and

$$\delta g_{\pm}(x,k) = \epsilon \left[xg'_{\pm}(x,k) - k\dot{g}_{\pm}(x,k) \right]$$

where $\cdot$ denotes differentiation in k. These results are from Calogero and Degasperis,[8] who present an exhaustive list of such space and scale transformations.

Appendix

THE JOST SOLUTIONS AND THE JOST FUNCTIONS ON R^+

We now discuss the Jost Functions, the Jost solutions, their integral representations and their regularity properties on the semi-axis R^+. Discussions of these functions on R^+ can be found in the texts on quantum mechanics that discuss potential scattering. Monographs by Schiff,[9] Messiah,[10] Davydov,[11] Regge and D'Alfero,[12] Newton,[13] Joachin,[14] Pearson,[15] and Roman[16] contain much of the materials discussed here. From the point of view of this book, the most appropriate discussion appears in Chadan and Sabatier's text on inverse problems in quantum scattering[17] which is followed here in the main. This appendix is added for the sake of completeness of the material discussed in the main text. The presentation will, therefore, be brief and in a summary fashion.

A.1. THE JOST SOLUTIONS ON R^+

The radial Schrödinger equation is given by

$$y''(l;k,r) + \left[k^2 - l(l+1)r^{-2}\right]y(l;k,r) - q(r)y(l;k,r) = 0 \tag{A.1}$$

· denotes differentiation with respect to the radial coordinate r, l is the partial wave angular momentum, q is the potential, and $k^2 = (p^2/2m) = E$, the total energy. y(l;k,r) is the radial wavefunction for the l-th partial wave. As $r \to \infty$, Equation A.1 reduces to the "free" equation, that is, without the interaction term involving q(r). Being a differential equation of second order, the radial Schrödinger equation possesses two linearly independent solutions. If l is nonzero and/or the potential q(r) is singular, then one of the solutions is "regular", that is, vanishes at the origin, and the other solution is "irregular", that is, singular at the origin. The irregular solution for the scattering free equation is given by

$$\tilde{y}_0(l;k,r) = i\exp(i\pi l)\left(\frac{\pi kr}{2}\right)^{\frac{1}{2}} H^{(1)}_{l+\frac{1}{2}}(kr) \tag{A.2}$$

where, $H_n^{(1)}$ is the Hankel function of the first kind of order n. The subscript 0 indicates that the scattering free solution is being considered. The wavefunction $\tilde{y}_0(l;k,r)$ has the following behavior

$$\tilde{y}_0(l;k,r) \xrightarrow[r\to\infty]{} \exp\left[\left(\frac{i\pi l}{2}\right) + ikr\right] \tag{A.3a}$$

$$\xrightarrow[r\to 0]{} (-1)^l\left[(2l-1)!!\right](kr)^{-l} \tag{A.3b}$$

where $(2l-1)!! = 1.3.5....(2l-1)$. $\tilde{y}_0(l;k,r)$ is thus clearly singular at the origin if l is nonzero.

The "Jost solution" for the radial Schrödinger equation is defined to be that solution which approaches the irregular free solution $\tilde{y}_0(l;k,r)$ asymptotically as $r \to \infty$. Denoting such a solution by $f_+(l;k,r)$, we have

$$\underset{r\to\infty}{\text{limit}} \left\{f_+(l;k,r)\exp\left[-\left(\frac{i\pi l}{2}\right) - ikr\right]\right\} = 1$$

Similarly, $f_-(l;k,r) = f_+(l;-k,r)$ is also an irregular solution. We will work primarily with f_+, analogous results holding true also for the f_- Jost solution. Also, the subscript + will be dropped.

From quantum scattering theory it is known that for regular potentials, that is, potentials satisfying a growth condition of order l on R^+, the solutions f_+ and f_- exist and, moreover, satisfy the following asymptotic conditions

$$\lim_{r\to\infty}\left[f_{\pm}(l;k,r) - \exp\left(\frac{i\pi l}{2}\right)\exp(\pm ikr)\right] = 0 \tag{A.4}$$

and

$$\lim_{r\to\infty}\left[f'_{\pm}(l;k,r) \mp ik\exp\left(\frac{i\pi l}{2}\right)\exp(\pm ikr)\right] = 0 \tag{A.5}$$

Next the linear independence of f_+ and f_- is established. Since the radial Schrödinger equation does not contain the first derivative term, the Wronskian $W(f_+,f_-)$ must be independent of the radial coordinate and may be a function of k only. Evaluating W at $r \to \infty$ and using Equations A.4 and A.5 yields

$$W[f_+,f_-] = f_+f'_- - f_-f'_+ = (-1)^{l+1}2ik \tag{A.6}$$

Hence if k is nonzero, then f_+ and f_- are linearly independent solutions of the radial Schrödinger equation. This means that any third solution can be expressed as a linear combination of these two.

A.1.a. Integral Representations of the Jost Solutions

The Jost solutions on R^+ have analogous integral representations to those of their counterparts on R. Thus, for example, the S-wave Jost solution $f(l = 0;k,r) = f(k,r)$ is given by

$$f(k,r) = \exp(ikr) + \left(\frac{1}{k}\right)\int_{+r}\sin k(r'-r)\,q(r')\,f(k,r')\,dr' \tag{A.7}$$

(It was mentioned earlier in Chapter 5 that in quantum scattering theory, $h = \infty$ boundary condition is generally used). The difference is that the lower limit r, the radial coordinate, is restricted to R^+. As before, the Volterra Equation A.7 can be expressed as a series

$$f(k,r) = \sum_{n=0}^{\infty} f^{(n)}(k,r)$$

$$f^{(0)}(k,r) = \exp(ikr)$$

$$f^{(n)}(k,r) = \int_{+r} k^{-1}\sin k(r'-r)\,q(r')\,f^{(n-1)}(k,r')\,dr'$$

The series is again absolutely and uniformly convergent in any compact set in the upper half-plane $\mathrm{Im}\,k \geq 0$. By similar arguments, as in deriving the various bounds of the Jost functions on R, the following estimates can be obtained

$$\left|f^{(n)}\right| < [\text{constant}\cdot\exp(-\mathrm{Im}\,kr)]\left[\int_{r+}\left\{\frac{r'|q(r')|}{(1+|k|r')}\right\}dr'\right]^n \Big/ n!$$

yielding the estimate

$$|f(k,r)| \le \text{constant } \exp(-\text{Im } kr), \text{ Im } k \ge 0 \tag{A.8}$$

The potential is assumed to be $L^1{}_1(R^+)$. This guarantees convergence of the series for $k \subset \overline{C}_+$. Otherwise, if $k \subset \overline{C}_+\backslash\{0\}$ and $r \neq 0$, then a zeroth order growth condition will be sufficient for the convergence of the series. This is again analogous to the results obtained for the entire line.

For a fixed $r > 0$ and large $k \subset \overline{C}_+$, one obtains from Equations A.7 and A.8 the following estimates

$$|f(k,r) - \exp(ikr)| \le \left\{ \text{constant } \frac{\exp(-\text{Im } kr)}{|k|} \right\} \eta_+(r), \quad \text{Im } k \ge 0, \quad k \neq 0$$

For a fixed $k \subset \overline{C}_+$ and large r,

$$|f(k,r) - \exp(ikr)| \le \text{constant } \exp(-\text{Im } kr)\ \gamma_+(r)$$

The estimates of the first derivatives are

$$|f'(k,r) - ik \exp(ikr)| \le \text{constant } \exp(-\text{Im } kr)\ \eta_+(r), \quad k \subset \overline{C}_+, \quad r \text{ large}$$

and

$$|\dot{f}(k,r) - ik \exp(ikr)| \le \text{constant } \frac{\exp(-\text{Im } kr)}{|k|} \quad k \subset \overline{C}_+, \quad \forall r \subset R^+$$

Again, as usual, $\dot{f} = \partial f/\partial k$ and η and γ are as in the main text. The only difference is that r is in the limit $\subset R^+$. The constants involve only the integrated potential on R^+ and are independent of r and k. These bounds justify the use of Equations A.7 and A.8. Similar results hold for f_- with – replacing + in the functions γ and η.

A.1.b. Regularity Properties of the Jost Solutions

Analyticity: For $r \subset R^+$, f_+ and f_- are analytic functions of k in the open upper-half of the complex k plane, that is, for $k \subset C_+$. These are also bounded and continuous for $k \subset \overline{C}_+$, $r \subset R^+$, if the potential satisfies a growth condition of order l. Otherwise, the origin $k = 0$ and its vicinity is to be excluded.

Again, as in the case of the Jost solutions on the entire line, analytic continuation in the Bargmann strip can also be obtained for the Jost solutions on the half-line. If the potential falls off faster than $\exp(-2\mu r)$, $\mu > 0$, then the Jost solutions are also regular in the lower half-plane in the strip $\text{Im } k \ge -\mu$ (a pole of order l at $k = 0$ will be present for a partial wave of order l). If it so happens that the potential falls off faster than any exponential (perhaps a Gaussian function at large r or such that the potential has a compact support on R^+), that is, for μ arbitrarily large, then the Jost solutions become entire functions of k for all fixed r > 0.

Symmetry: $f(-k^*,r) = [f(k,r)]^*$, * denoting complex conjugate. These functions are real for k on the imaginary axis. That is, if we stay in a region in which f(k,r) is analytic, then f^*

satisfies the same equation with k^* that f satisfies with k. The asymptotic boundary conditions in the definitions of the Jost solutions are then consistent with each other for complex k, provided that the symmetry condition is applied.

Asymptotic estimates:

$$f(k,r) \xrightarrow[r\to\infty]{} \exp(ikr) + [\exp(-\mathrm{Im}\ kr)]\ o(1)$$

$$f'(k,r) \xrightarrow[r\to\infty]{} ik\ \exp(ikr) + [\exp(-\mathrm{Im}\ kr)]\ o(1)$$

and

$$f(k,r) \xrightarrow[\substack{\mathrm{Im}\,k\geq 0\\ k\to\infty}]{} \exp(ikr) + |k|^{-1}[\exp(-\mathrm{Im}\ kr)]\ o(1), \quad r>0$$

The order terms are uniform with respect to the fixed variable. If r = 0 is also considered, the above estimate is modified to be

$$f(k,r) \xrightarrow[\substack{\mathrm{Im}\,k\geq 0\\ k\to\infty}]{} \exp(ikr) + [\exp(-\mathrm{Im}\ kr)]\ o(1)$$

where again o(l) is uniform with respect to r, $r \geq 0$.

Smoothness: f is continuously differentiable with respect to k in Im $k \geq 0$, $k \neq 0$ if $q \subset L^1{}_1(R^+)$. A stronger growth condition, namely, a growth of second order, must be imposed if k = 0 is to be included.

A.2 THE JOST FUNCTIONS AND THEIR PROPERTIES

Let y(l;k,r) denote the regular solution of the radial equation (A.1) and y_0(l;k,r) the corresponding scattering free solution, again indicated by the subscript 0. The solutions y_0 have the following limiting forms, namely

$$y_0(l;kr) \xrightarrow[r\to 0]{} \frac{(kr)^{l+1}}{(2l+1)!!}$$

and

$$y_0(l;kr) \xrightarrow[r\to\infty]{} \sin\left(kr - \frac{l\pi}{2}\right)$$

For the solution when the scattering potential is present, the regular solution y(l;k,r) is defined such that

$$\underset{r\to\infty}{\mathrm{limit}}\ (2l+1)!!\ r^{-(l+1)}\ y(l;k,r) = 1$$

and

$$\underset{r\to\infty}{\mathrm{limit}}\ y(l;k,r) = |F(l;k)|\ k^{-(l+1)} \sin\left(kr - \frac{l\pi}{2} + \delta_l\right)$$

where, δ_l is the "phase shift" corresponding to the l-th partial wave which distinguishes the regular solution y(l;k,r) from the free solution y_0(l;k,r) in the asymptotic limit. The phase shifts are real quantities which are functions of energy or the momentum k, that is, δ_l = δ(l;k). In quantum scattering, a distinction is made between the solution y(l;k,r) and the "physical" solution ψ(l;k,r) of the radial Equation (A.1) in that

$$\psi(l;k,r) = \left[\frac{|F(l;k)|}{k^{l+1}}\right]^{-1} y(l;k,r)$$

The function F(l;,k) is called the "Jost function".

The solution y(l;k,r) can be expressed as a linear combination of the two linearly independent Jost solutions. It turns out that the regular solution can be expressed as

$$y(l;k,r) = \left(\frac{i}{2}\right) k^{-(l+1)} \left\{F(l;k)\, f(l;-k,r) - (-1)^l F(l;-k) f(l;k,r)\right\} \tag{A.9}$$

Since y solves the radial Schrödinger equation in which k appears only as k^2, then according to Poincare's theorem,[14] the k independence of the boundary condition implies that y(l;k,r) must be an even function in k, that is,

$$y(l;k,r) = y(l;-k,r)$$

This fact is expressed in Equation A.9 through the appearance of F in both coefficients.

Furthermore, the Jost functions can also be defined in terms of a Wronskian relation. Taking the Wronskian of, for example, f_+ and y yields

$$W[f_+(l;k,r),\ y(l;k,r)] = f_+ y' - y f_+'$$

From Equation A.9, the fact that the Jost solution behaves as exp{i(πl/2 + kr)} as r → ∞ and that the Wronskian is independent of r results in the following expression

$$(-k)^{-1} W[f_+(l;k,r), y(l;k,r)] = F(l;k) \tag{A.10}$$

For real k, it can be shown that

$$F(l;-k) = [F(l;k)]^*$$

From the theory of quantum scattering it is also known that

$$F(l;k) = |F(l;k)| \exp[-i\delta(l;k)]$$

The phase shift δ(l;k) is crucial in determining the potential and is just the negative of the phase of the Jost function, a highly remarkable connection. The scattering matrix element S(l;k) for the l-th partial wave is related to the phase shift and the Jost functions through

$$S(l;k) = \exp\{2i\delta(l;k)\} = \frac{F(l;-k)}{F(l;k)}$$

The analytic properties of the S-matrix can then be studied through those of the Jost functions. The analyticity of S is relatively more complicated than those of the Jost functions as S may be singular, not only at the poles, but also at the zeros of the Jost functions.

Let us confine our attention to S-waves only. From Equation A.10 this yields

$$F(k) = W[f(k,r),\ y(k,r)]$$

In Chapter 4, the integral representation of the regular solution was derived, namely,

$$y(k,r) = (\sin kr)\, k^{-1} + k^{-1} \int_0^r \sin k(r - r')\, q(r')\, y(kr')\, dr' \tag{A.11}$$

Replacing this expression for y(k,r) in the Wronskian above and evaluating the latter at $r \to \infty$, since the Wronskian is independent of the radial coordinate, the following integral representation is obtained for the Jost function

$$F(k) = 1 + \int_{+0} dr \exp(ikr) q(r) y(k,r) \tag{A.12}$$

In deriving Equation A.12, the discussion of asymptotic estimates in the previous section was utilized. Equation A.12 is the integral representation for the Jost functions for the continuum, that is, for real k and no bound states. When this is not the case and the bound states are to be considered, a different representation is obtained for F. This requires the knowledge of the number, multiplicity, and location of zeros of the Jost function. These are almost analogous to those for the Jost functions on the entire real line R and, hence, will be discussed summarily.

By definition, a bound state is a solution of the radial Schrödinger equation satisfying the regular boundary condition at the origin and which is square-integrable on R^+. That is, a bound state must be a function in the Hilbert space $L_2[R^+]$. From the Wronskian definition (A.10) of the Jost function and from the integral representation (A.12) for the regular solution y(k,r), it is easy to see that

$$F(k) = f(k,0)$$

Let k_0 be a value of k such that

$$F(k_0) = f(k_0,0) = 0 \tag{A.13}$$

Then it follows from the Wronskian definition (A.10) that

$$f(k_0,r) = f'(k_0,0)\, y(k_0,r) \tag{A.14}$$

where, the condition $y'(k_0,0) = 1$ was used. $f(k_0,r)$ and $y(k_0,r)$ are, therefore, not linearly independent. Since $f(k_0,r)$ vanishes asymptotically as (asymptotic estimates, Section A.1.a)

$$f(k_0,r) \xrightarrow[r \to \infty]{} \exp(ikr) + \exp(-\text{Im } kr)\, o(1)$$

then, for Im $k > 0$, Equation A.14 implies that $y(k_0,r)$ vanishes exponentially for all $r \subset R^+$.

$y(k_0,r)$ is then square-integrable and a bound state is obtained. Thus every k such that A.13 is satisfied gives rise to a bound state. That is, every zero of the Jost function, Im $k > 0$, corresponds to a bound state.

To discuss the location of these bound states: if $k \subset R$ and if, in addition, $k \neq 0$, then the Jost function cannot vanish. If it does, then from the symmetry property, both $F(k) = F(-k) = 0$, resulting in the vanishing of the regular solution $y(k,r)$ at a value of $k \neq 0$ according to (A.9). This is in contradiction to the $h = \infty$ boundary condition. Hence for any $k \neq 0$ and real, the Jost functions cannot vanish. Therefore, $k \subset C_+$ if a bound state is to be obtained. In order to locate these bound states, one combines the Schrödinger equations for $f(k_0,0)$ and $[f(k_0,0)]^*$ in order to obtain the following equation

$$f'(k_0,0)[f(k_0,0)]^* - f(k_0,0)[f'(k_0,0)]^*$$

$$= 4i\,\mathrm{Re}\,k_0\,\mathrm{Im}\,k_0 \int_{+0} |f(k_0,r)|^2\,dr \tag{A.15}$$

From Equation A.13 and the fact that $k \subset C_+$, Equation A.15 yields Re $k_0 = 0$. The zeros of the bound states must, therefore, lie on the imaginary axis. Similarly, the bound states of $y(-k,r) = [y(k,r)]^*$ correspond to $F(k) = 0$ for $k = -ik_j$, $k_j > 0$.

Next, one must address the question of the number and the order of the multiplicities of these zeros. From the integral representation of F(k) given by Equation A.12, it can be estimated that

$$|F(k) - 1| \leq \text{constant} \int_{+0} \left[\frac{r|q(r)|}{1+|k|r}\right] dr \tag{A.16}$$

Hence as $k \to \infty$, Im $k \geq 0$, $F(k) \to 1$. This means that there cannot be any zero at high values of k and no accumulation point can be at $k \to i\infty$. A similar conclusion also holds for $k = 0$. Hence the number of zeros must be finite. As to the zeros being simple or multiple depends on whether or not the first derivative of F(k) with respect to k is nonzero. If this derivative is nonzero, then the zeros are simple. The Schrödinger equation for f(k,r) is

$$f'' + (k^2 - q)f = 0$$

Differentiating this with respect to k and multiplying the resultant equation by f yields

$$0 = f\dot{f}'' + 2kf^2 + (k^2 - q)\,f\dot{f}$$

$$= f\dot{f}'' + 2kf^2 - \dot{f}f''$$

$$= [f\dot{f}' - \dot{f}f']' + 2kf^2$$

This results in

$$f\dot{f}' - \dot{f}f' = 2k \int_{+r} f^2(k,r)\,dr$$

Setting r to be zero yields

$$f(k,0)\,\dot{f}'(k,0) - \dot{f}(k,0)\,f'(k,0) = 2k \int_{+0} f^2(k,r)\,dr$$

For a bound state $k = k_0$, $f(k_0,0) = 0$ and $f(k_0,r) = f'(k_0,0)y(k_0,r)$. Then

$$\int_{+0} y^2(k_0,r)\,dr = \|y(k_0,r)\|_2^2 = -\left(\frac{1}{2k_0}\right)\frac{\dot{F}(k_0)}{f'(k_0,0)}$$

$\dot{F}(k_0)$, therefore, cannot vanish. This proves the simpleness of zeros. If $k \to 0$ along any path in the upper half-plane, F(k) does not vanish. Hence, no bound state exists at zero energy for an S-wave. For higher order partial waves, a true bound state can exist at the origin.

In the presence of the bound states, an integral representation of the Jost function can be obtained as follows. From the bound given by Equation A.16, it follows that

$$\underset{|k|\to\infty}{\text{limit}} \quad F(k) \longrightarrow 1$$

More precisely, it can be shown that

$$\underset{|k|\to\infty}{\text{limit}} \quad F(k) \longrightarrow 1 + 0(1) \tag{A.17}$$

Due to the presence of the factor of unity in this asymptotic limit, it is easy to work with [F(k) –1] instead of the function F(k) itself. This is because the dispersion relations for F(k) or, for that matter, its integral representation involves integrals over k on the entire real axis, in which case working with [F(k) –1] is easier for considerations of convergence. Since [F(k) –1] is holomorphic in Im k > 0, then by Cauchy's theorem

$$F(k) - 1 = \left(\frac{1}{2\pi i}\right)\oint \frac{[F(k') - 1]}{(k' - k)}\,dk'$$

where the contour consists of the real axis and a semicircle in the upper half-plane. By virtue of the bound A.16, the contribution from the semicircle vanishes if the contour is pushed to infinity, and one obtains

$$F(k) - 1 = \left(\frac{i}{\pi}\right) f \frac{[F(k') - 1]}{(k' - k)}\,dk', \quad k \subset R$$

f denotes the Cauchy principal value. Application of the Kramers-Krönig relations then relates the Re and the Im part of {F(k) –1] to each other through the Hilbert transform pair

$$\text{Re}\,[F(k) - 1] = -\left(\frac{1}{\pi}\right) f \left[\frac{\text{Im}\{F(k') - 1\}}{(k' - k)}\right] dk' \tag{A.18}$$

and

$$\text{Im}\left[F(k)-1\right]=\left(\frac{1}{\pi}\right) f \frac{\left[\text{Im}\{F(k')-1\}\right]}{(k'-k)} dk'$$

Now F(k) = |F(k)| exp{–iδ(k)}, and the phase shifts δ(k) (and also the bound state energies) constitute the experimental data. It is, therefore, desirable to have an integral representation of F(k) involving δ(k). An alternate method of deriving such a representation is by considering ln F(k) instead of [F(k) –1]. This does not pose any problem if only the continuous part of the spectrum is considered. However, if bound states are present, then ln F(k) has zeros and the analysis for In F(k) is not possible. In order to obviate this difficulty, the standard practice is to introduce the so-called "reduced Jost function" defined as

$$\tilde{F}(k)=\prod_{j=1}^{N}\left[\frac{(k+k_j)}{(k-k_j)}\right]F(k) \tag{A.19}$$

where, k_js are the locations of the zeros of F(k) on the positive imaginary axis and N the number of these zeros. By virtue of Equation A.19, the reduced Jost function $\tilde{F}(k)$ is nonzero for $k \subset C_+$. Moreover, $\tilde{F}(k)$ has the same analytic properties of F(k) in the absence of the bound states. Thus $\tilde{F}(k)$ is analytic for $k \subset C_+$, is continuous for $k \subset \overline{C}_+$, and has the asymptotic limit

$$\tilde{F}(k) \longrightarrow 1+o(1), \quad |k| \longrightarrow \infty$$

It follows that ln $\tilde{F}(k)$ is also holomorphic in Im k > 0 and vanishes $|\kappa| \to \infty$. One can now apply the Kramers-Krönig relations to ln $\tilde{F}(k)$. From Equation A.19

$$\text{Im}\ \ln\tilde{F}(k)=-\delta(k)+2\sum_j \arctan\left(\frac{|k_j|}{|k|}\right)$$

Then from the Hilbert transform applied to the Im part of A.19 one obtains the following integral representation

$$F(k)=\prod_{i=1}^{N}\left\{1-\left(\frac{|k_j|^2}{|k|^2}\right)\right\}\exp\left[-\left(\frac{2}{\pi}\right)\right]\int_{+0}\left[\frac{\delta(k')}{(k'^2-k^2)}\right]k'\,dk'$$

$$=\prod_{i=1}^{N}\left\{1-\left(\frac{E_j}{E}\right)\right\}\exp\left[-\left(\frac{2}{\pi}\right)\int_{+0}\left[\frac{\delta(k')}{(k'^2-k^2)}\right]k'\,dk'\right]$$

where, E_js are the bound state energies and $E_j = k_j^2 > 0$. The modulus |F(k)| will then be given by the same expression with the integral interpreted as a Cauchy principal value integral.

A.3 APPLICATION OF WIENER-LEVY THEOREM TO FUNCTIONS OF F(K)

We conclude this appendix by summarizing some important representations of various functions of the Jost functions F(k) through the application of the well-known Wiener-Levy theorem.[18] The importance of these representations derives from the fact that these are in the forms of Fourier transforms, the kernels of which are directly related to the prescribed data and which can be obtained by appropriate inverse scattering equations. Similar situations were obtained in the case of the Povzner-Levitan transform and its corresponding Gel'fand-Levitan integral equation. The importance of these representations for inverse solutions can, therefore, be easily appreciated. Only summary presentations will be given here. For extensive details reference is made to Fadeev's pioneering work.[19]

The application of the Wiener-Levy theorem depends upon two facts.

1. The Jost function F(k) must be representable in the form

$$F(k) = 1 + \int_{R^+} B(t)\exp(ikt)\,dt, \quad B(t) \subset L_1\left(R^+\right) \tag{A.20}$$

and

2. The Fourier transform of the product of two functions $f,g \subset L_1(R^+)$ is also a function $\subset L_1(R^+)$. This is a well-known theorem and is discussed by Titchmarsh.[18] The representation A.20 will be established in Chapter 8 where the Levin representations are discussed. It is worth mentioning that the representation A.20 is the point of departure for Marcenko's solution of the inverse scattering problem which will be taken up in detail in Chapter 10.

With this introduction we state the **Wiener-Levy Theorem**.

Let the curve $z = F(k)$, $k \subset R$, lie inside a domain D of the complex z-plane. Let G(z) be analytic in D and let the function F(k) have a representation of the form

$$F(k) = C + \int_{R} \Gamma(t)\exp(ikt)\,dt$$

where, $\Gamma(t) \subset L_1(R)$. Then G(F(k)) also possesses a similar representation.

Since $\Gamma(t) \subset L_1(R)$, the constant C is identical to $F(\infty)$ by virtue of the Riemann-Lebesgue lemma. We quote the results of applying the above theorem to the functions $|F(k)F(-k)|^{-1}$, the S-matrix, and the phase shifts $\delta(k)$. The function $|F(k)F(-k)|^{-1}$ has been extensively discussed by Fadeev.[19] This function and the S(k) matrix defined by

$$S(k) = \frac{F(-k)}{F(k)}$$

are directly related to the data and are, therefore, extremely important in the inverse scattering problems, not only in quantum systems, but also in the classical context. Functions such as F(k)F(–k) and S(k) occur in inverse solutions frequently (for example, see Chapter 11, Section XI). Using the two conditions above and applying the Wiener-Levy theorem then results in the following representations

$$|F(k)\,F(-k)|^{-1} = |F(k)|^{-2} = 1 + \int_R \Im(t)\exp(ikt)\,dt, \quad k \text{ real}$$

$$S(k) = 1 + \int_R \Sigma(t)\exp(ikt)\,dt$$

and

$$\delta(k) = \left(\frac{i}{2}\right)\mathrm{Log}S(k) = -\int_R \gamma(t)\,\sin kt\,dt$$

$\Im(t)$ and $\Sigma(t)$ are $L_1(R)$ and $\gamma(t) \subset L_1(R^+)$. In the above formulae, no bound states were taken into account. In the presence of the bound states, the Wiener-Levy theorem will have to be applied not to F(k) and S(k), but rather to the reduced forms

$$\tilde{F}(k) = \left[\prod_j \left[\frac{(k - ik_j)}{(k + ik_j)}\right]^{-1} F(k)\right]$$

and to

$$\tilde{S}(k) = \left[\prod_j \frac{(k - ik_j)}{(k + ik_j)}\right] S(k)$$

REFERENCES

1. **E. A. Coddington and N. Levinson:** *Theory of Ordinary Differential Equations*, McGraw-Hill, New York, 1955.
2. **P. Deift and E. Trubowitz:** Inverse scattering on the line, *Comm. Pure and Appl. Math.*, 32, 1979.
3. **W. Eckhaus and A. Van Harten:** *The Inverse Scattering Transformation And The Theory Of Solitons*, North-Holland, Amsterdam, 1981.
4. **J. R. Taylor:** *Scattering Theory*, John Wiley & Sons, New York, 1972.
5. **V. E. Zakharov and A. B. Shabat:** A scheme for integrating the nonlinear equations of mathematical physics by the method of inverse scattering problem. I, *Funct. Anal. Appl.*, 8, 1974.
6. **L. D. Fadeev:** The inverse problem in the quantum theory of scattering, *J. Math. Phys.*, 4, 1963.
7. **L. D. Fadeev:** Properties of the S-matrix of the one-dimensional Schrodinger equation, *Am. Math. Soc.*, 65, 1964.
8. **F. Calogero and A. Degasperis:** *Spectral Transform And Solitons*, North-Holland, Amsterdam, 1981.
9. **L. I. Schiff:** *Quantum Mechanics*, McGraw-Hill, New York, 1965.
10. **A. Messiah:** *Quantum Mechanics*, North-Holland, Amsterdam, 1968.
11. **A. S. Davydov:** *Quantum Theory*, Pergamon Press, Oxford, 1976.
12. **T. Regge and V. D'Alfero:** *Potential Scattering*, North-Holland, Amsterdam, 1965.
13. **R. G. Newton:** *Scattering Theory of Waves and Particles*, McGraw-Hill, New York, 1966.

14. **C. J. Joachin:** *Quantum Collision Theory*, North-Holland, Amsterdam, 1983.
15. **D. B. Pearson:** *Quantum Scattering and Spectral Theory*, Academic Press, London, 1988.
16. **P. Roman:** *Advanced Quantum Theory*, Addison Wesley, Reading, MA, 1966.
17. **K. Chadan and P. C. Sabatier:** *Inverse Problems in Quantum Scattering Theory*, Springer, New York, 1977.
18. **E. C. Titchmarsh:** *Theory of Functions*, Oxford University Press, London, 1952.

Chapter 8

FOURIER TRANSFORM OF THE JOST FUNCTIONS: THE LEVIN REPRESENTATION

I. INTRODUCTION

The Fourier transforms of the Jost functions, known as the "Levin transforms", are discussed in this chapter. On the entire real axis, the Levin transforms can be viewed as extensions of the Povzner-Levitan transforms for the radial-wave functions. The Levin transforms satisfy boundary-value problems of Goursat type for a hyperbolic partial differential equation, where the boundary conditions are specified on the characteristics of the equation. The restrictions of the solution on a characteristic is then related to the potential. Just as the Povzner-Levitan transform on the semiaxis R^+ was determined by the Gel'fand-Levitan equation, so is the Levin transform on the entire real line R given by Marcenko's integral equation, which will be discussed in some detail in Chapter 10. Levin's representation is crucial in solving inverse problems on the entire real axis. We begin by stating some well-established standard results from Fourier transform theory, which will be necessary in the development of this chapter.

II. FOURIER TRANSFORMS AND HARDY FUNCTIONS

In order that the Fourier transform of a function f(x): $\left[\text{F.T.}\{f(x)\}\right](y) = \left[f(x)\right]^{\wedge}(y)$ and the "inverse" transform $\left[\text{F.T.}^{-1}\left[f(y)\right]\right](x) = \left[f^{\wedge}(y)\right]^{\vee}(x) = f^{\vee}(x) = f(x)$ (for notational simplicity, the superscript $\vee$ on the inverse transformed quantities will be omitted) may exist, the function f(x) must satisfy certain conditions. More specifically, it is necessary to know which space f(x) belongs to. If $f(x) \subset L_1(R)$, that is, f(x) is an absolutely integrable function over the real axis, then its Fourier transform $f^{\wedge}(y)$ will exist, but will not belong to $L_1(R)$. On the other hand, for $f(x) \subset L_2(R)$, that is, for a square integrable function, the integrand of the forward transform may not exist, and certain limiting procedures may be necessary. In $L_2(R)$, the Fourier transform is defined as follows. Let $\{\phi_n\}$ be a sequence in $L_2 \cap L_1$ and let $n \to \infty$ $\{\phi_n\} \to \phi$. Since $L_2 \cap L_1$ is dense in L_2, $\phi \subset L_2$. The Fourier transform $[\{\phi_n\}]^{\wedge}$ is given by the above definition and $\phi^{\wedge}$ exists in the sense that limit $n \to \infty$ $[\{\phi_n\}]^{\wedge} \to \phi^{\wedge}$. Exactly analogous is the case for the inverse Fourier transform, which also is in L_2. However, there are two spaces on which both direct and inverse transforms can be treated on an equal footing. These are the "Schwartz space" $\mathcal{L}$ and its dual $\tilde{\mathcal{L}}$, the space of "tempered distributions". The Schwartz space is a linear vector space and consists of the "rapidly decaying test functions." That is, if f(x) is an element of $\mathcal{L}$, then (1) $f(x) \subset C_\infty(R)$, the space of infinite differentiable functions, and (2) $\|f\|_{\alpha,\beta} = \sup |x^\alpha D^\beta f| < \infty$, $D = i^{-1}\partial_j$, $j = 1,2...n$ and $i = (-1)^{1/2}$, for all pairs of multiindices α,β. This means that the function f(x) is such that for any $N \geq 0$, $N \subset Z$, $D^\beta f \to 0$ faster than $|x|^{-N}$ as $|x| \to \infty$. That is, f(x) along with its derivatives of all orders must decay at infinity faster than the reciprocal of any polynomial. The Schwartz space is complete under the seminorms $\|f\|_{\alpha,\beta}$ above. The dual space $\tilde{\mathcal{L}}$ of $\mathcal{L}$ consists of those continuous linear functionals, i.e., those distributions which act on the test functions of rapid decay belonging to the Schwartz space $\mathcal{L}$. The space $\mathcal{L}$ is larger than the space $\tilde{\mathcal{L}}$ of the ordinary distributions, whereas $\mathcal{D}$ is a subspace of $\tilde{\mathcal{D}}$, the dual of the space $\mathcal{D}$.

Some standard results from the theory of Fourier transforms will be necessary for the discussion of this chapter and are summarized below.

Theorem 8.1

The Fourier transform ^ maps the Schwartz space $\mathcal{L}(R)$ *onto* itself and its inverse map is the inverse Fourier transform. The operations ^,ˇ are linear, bicontinuous, and bijective from $\mathcal{L}(R)$ *onto* $\mathcal{L}(R)$. Moreover, Plancherel's identity holds in the L_2 norm,

$$\|f(x)\|_2 = \|f^\wedge(k)\|_2, \qquad f \subset \mathcal{L}(R)$$

Operations of the Fourier transform on $\mathcal{L}$ are continuous in the usual topology of $\mathcal{L}$.

F and F^{-1} can also be extended from $\mathcal{L}$ to $\tilde{\mathcal{L}}$ through dualization. If ϕ is a test function $\subset$ $\mathcal{L}$ and ψ is a tempered distribution $\subset \mathcal{L}$, then F and F^{-1} are defined on $\tilde{\mathcal{L}}$ through the distributions defined as

$$(F\psi, \phi) = (\psi, F\phi)$$

and

$$\left(F^{-1}\psi, \phi\right) = \left(\psi,\ F^{-1}\phi\right), \quad \phi \subset \mathcal{L}$$

Theorem 8.2

The operations of Fourier transform ^,ˇ are linear bijections from the space of the tempered distributions $\mathcal{L}(R)$ *onto* itself.

Theorem 8.3

The Fourier transform ^ and its inverse ˇ extend uniquely from $\mathcal{L}(R)$ to a unitary map of $L_2(R)$ *onto* $L_2(R)$. ^[ˇ] are isomorphisms of $L_2(R)$ *onto* itself.

Theorem 8.4

The Fourier transform extends uniquely from $\mathcal{L}(R)$ to a bounded map from $L_1(R)$ *into* $L_\infty(R)$, the space of continuous functions on R which is complete under the "supremum" norm and is bounded by the L_1 norm. Therefore,

$$\|f^\wedge\|_\infty \le \|f^\vee\|_1$$

Also,

$$\lim_{|k| \to \infty} f^\wedge(k) = 0$$

This condition is known as the "Riemann-Lebesgue Lemma".

Exhaustive accounts of the theory of Fourier transformation can be found in Reed and Simon,[1] Rudin,[2] Hörmander,[3] Bochner and Chandrasekharan,[4] Gel'fand and Shilov,[5] and Dym and McKean.[6] Another excellent and readable account of the theory of Fourier transformations is the monograph by Stein and Weiss.[7]

The Fourier transforms of the Jost functions and Jost solutions will be investigated in this chapter. We will work primarily with the functions $[f_\pm(x,k) - 1]$ and by the Fourier transforms of these functions will be meant specifically the transforms with respect to the second variable k, the first variable x acting as a parameter. The most elegant way[8] of obtaining the Fourier integral representations of these functions is perhaps by appealing to the properties of the functions in the Hardy space. These spaces, which are relatively less known, have been discussed by Beurling,[9] Hoffman,[10] Helson,[11] Duren,[12] and Dym and McKean.[6,13] The last-mentioned reference will be followed here.

The Povzner-Levitan representations of Chapter 5 were established by using the Paley-Wiener and Wiener-Boas theorems for analytic functions the Fourier transforms of the restrictions of which on the real axis were in $L_2(R)$ and $L_1(R)$, respectively, having compact supports. The Paley-Wiener theorem there dealt with the entire functions of finite type. For the Jost functions, however, analogous results are to be established in the open upper half-plane $C_+ = \{z = u + iv, v > 0\}$. This is where the Hardy spaces and the Hardy functions come into play.

A function h(k) of the complex variable k is said to belong to the Hardy class H^{2+} if

$$1)\ h(k) \text{ is analytic on } C_+, \text{ and}$$

$$2)\ \sup_{v>0} \|h(k)\|_2 = \sup_{v>0} \left[\int_R |h(u+iv)|^2 du \right] < \infty$$

Applied to the Hardy space, the Paley-Wiener theorem can be stated as follows.

Theorem 8.5: The Paley-Wiener Theorem for H^{2+}

A function $h \subset H^{2+}$ if and only if

$$h(k) = \int_{R^+} f(y) \exp(iky)\, dy, \quad k \subset C_+$$

for some f such that $f \subset L_2(R)$ and vanishes on R^-.

H^{2+} is now identified with $[L_2(0,\infty)]^\wedge$, and functions of class $[L_2(0,\infty)]^\wedge$ are called “Hardy functions of class H^{2+}”.

The theory of H^{2-} is entirely analogous to that for H^{2+}. Thus, H^{2-} is identified with $L_2(-\infty,0]^\wedge$. Moreover,

$$L_2 = H^{2-} \oplus H^{2+}$$

The operation lim $v \to 0$ h(u+iv) is a 1 : 1 length-preserving map from H^{2+} onto the class $L_2[0,\infty)^\wedge$. This is also true for H^{2-}.

Analogously, a function h(k), $k = u + iv$, $v > 0$, belongs to a Hardy class H^{1+} if

$$1.\ h(k) \text{ is analytic on } C_+, \text{ and}$$

$$2.\ \|h\|_1 = \sup_{v>0} \|h\|_1$$

$$= \sup_{v>0} \int_R |h(u+iv)|\, du \le \infty$$

The corresponding Paley-Wiener theorem is shown below.

Theorem 8.6: The Paley-Wiener Theorem for H^{1+}

A function $h \subset H^{1+}$ if

$$h(k) = \int_R f(y)\exp(iky)\,dy, \quad k \subset C_+$$

for some f such that $f \subset L_\infty$ and vanishes on R^-.

Again, the operation limit $v \to 0$ $h(u + iv)$ is unitary from the class H^{1+} onto the class $L_1[0,\infty)^\wedge$. Analogous results hold for H^{1-} Hardy functions. For H^1 classes, $H^{1-} + H^{1+}$ is direct, i.e., $H^{1-} \cap H^{1+} = 0$. However,

$$H^{1-} \oplus H^{1+} \neq L_1(R)$$

Further details can be found in Dym and McKean.[13]

III. THE LEVIN TRANSFORMS $B^\wedge_\pm(x,y)$

A. LEVIN TRANSFORM VIA PALEY-WIENER THEOREM

The functions $[f_\pm(x,k) - 1]$ are analytic on C_+, as was established in the preceding chapter. Moreover, by Theorem 7.4, these are square-integrable in C_+. Hence, $[f_\pm - 1] \subset H^{2+}$ and are, therefore, Hardy functions themselves. The Paley-Wiener Theorem 8.5 is then directly applicable to $[f_\pm(x,k) - 1]$. The Hardy functions belonging to H^{2+} can arise only in the way stated therein. Thus, as per Theorem 8.5, the Fourier representations of $[f_\pm(x,k) - 1]$ can then be written down directly as

$$f_\pm(x,k) = 1 + \int_{\pm 0} B^\wedge_\pm(x,y)\exp(\pm iky)\,dy, \quad y \subset R^\pm \tag{8.1}$$

where $B_\pm{}^\wedge(x,y)$ are known as the "Levin transforms" of $[f_\pm(x,k) - 1]$ and Equation 8.1 is the celebrated "Levin representations" of the Jost functions on the entire line and was originally given by the Russian mathematician Levin.[14] The derivation presented appears in Deift and Trubowitz.[8]

Next, we will consider the derivation of the Levin transforms from a more heuristic point of view. We will draw heavily from the treatment given by Eckhaus and van Harten.[15]

B. DERIVATION OF THE LEVIN TRANSFORMS AND THEIR STRUCTURES

Let us consider the Neumann expansion of the Jost functions

$$[f_\pm(x,k) - 1] = f_\pm^{(1)}(x,k) + \sum_{n=2}^{\infty} f_\pm^{(n)}(x,k) \tag{8.2}$$

where, as before, $f_\pm{}^{(n)} = (F_\pm f_\pm{}^{(n-1)})$ and $f_\pm{}^{(0)} = 1$. From now onward, calculations will be carried out for $f_+(x,k)$ and analogous results will be quoted for the other Jost function $f_-(x,k)$. The subscript + will be omitted, remembering that $f_+(x,k)$ is being considered. Consider the leading term $f^{(1)}$, which can be written as

$$f^{(1)}(x,k) = \int_+ F(\zeta,k)\,q(x')\,dx', \quad \zeta = x' - x$$

Then,

$$f^{(1)}(x,k) = \left(\frac{1}{2ik}\right) \int_{+} q(x') \left\{\exp(2ik\zeta) - 1\right\} dx' = \left(\frac{1}{2ik}\right) \left[-Q(x) + \int_{+} q(x') \exp(2ik\zeta)\, dx'\right]$$

where

$$Q(x) = \int_{+} q(x')\, dx' \tag{8.3}$$

This can be further reduced to

$$f^{(1)}(x,k) = \left(\frac{1}{2ik}\right) \cdot \left[\exp(2ik\zeta)\, Q(x')\right]_{+} + \left(\frac{1}{2ik}\right) \int_{+} q(x')\, \exp(2ik\zeta)\, dx'$$

$$= \int_{+} \exp(2ik\zeta)\, Q(x')\, dx' \tag{8.3a}$$

where $[\cdot]_+$ means that the bracketed quantity is to be evaluated between the lower limit of x and the upper limit of ∞. Substituting $\zeta = x' - x = y/2$, we have

$$f^{(1)}(x,k) = \left(\frac{1}{2}\right) \int_{R} H(y)\, Q\left(x + \frac{y}{2}\right) \exp(iky) dy \tag{8.4}$$

H is the Heaviside function.

It follows from the bound established earlier (Equation 7.9) that $|f^{(1)}| \sim O(|k|^{-1})$ as $|k| \to \infty$. This bound is uniform in x if the potential $q(x) \subset L^1_0$. Furthermore, from the previous chapter, we know that $f^{(1)}(x,k)$ is a continuous function of x and k on (R x R). $f^{(1)}(x,k)$ is, therefore, an element of $C[L_2(R)]$, defined as $C[L_2(R)] = \{g: g \subset L_2(R)|g \text{ continuous}\}$. Then by Theorem 8.3, its Fourier transform exists and is also $C[L_2(R)]$. We are, of course, considering the variable k and its Fourier conjugate y. The spaces $C[L_2(R)]$ refer, therefore, to the Fourier variables k and y. This practice will continue throughout this chapter unless otherwise mentioned. From Equation 8.4, the corresponding Fourier transform (symmetric form with $(1/2\pi)^{1/2}$ in both transforms) of $f^{(1)}(x,k)$ in the variable k is given by

$$f^{(1)\wedge}(x,y) = \left(\frac{\pi}{2}\right)^{\frac{1}{2}} H(y)\, Q\left(x + \frac{y}{2}\right) \tag{8.5}$$

$f^{\wedge}(x,y)$ is the direct Fourier transform of f(x,k) or f(x,k) is the inverse transform of $f^{\wedge}(x,y)$. As mentioned earlier, the transform symbol $\vee$ will be omitted for the inverse quantities.

Next consider the sum

$$S(x,k) = \sum_{n=2}^{\infty} f^{(n)}(x,k)$$

of Equation 8.2, and obtain for $|k| \to \infty$

$$|S| \le \sum_{n=2}^{\infty}\left(\frac{1}{n!}\right)\left[\frac{\eta(x)}{|k|}\right]^n \le \left(\frac{1}{2!}\right)\left(\frac{\eta}{|k|}\right)^2 \exp\left[\frac{\eta(x)}{|k|}\right]$$

Therefore, $|S| \sim O(|k|^{-2})$ as $|k| \to \infty$ again uniformly in x if $q \subset L^1{}_1$. Moreover, $S = f - 1 - f^{(1)}$ is continuous in x and k on (R x R). Together, this implies that the sum $S \subset C[L_1(R)]$. By Theorem 8.4, its Fourier transform exists and belongs to $L_\infty(R)$, the space of continuous functions vanishing at $+\infty$. That is, $L_\infty(R) = \{z \subset C(R)|\lim |k| \to \infty\ |z(k)| \to 0\}$. Furthermore, $S \subset C[L_1(R)]$ is meromorphic on C_+. It thus follows that

$$\sup_{v>0} \int_R |S(u + iv)|\, du < \infty$$

From the discussion of Section II, S is then in the Hardy space H^{1+}. By the Paley-Wiener theorem on H^{1+} (Theorem 8.6), the Fourier transform of S is in $L_\infty(R)$ and vanishes on R^-. More specifically, if we write

$$S(x,k) = \int_R s\hat{}(x,y)\exp(iky)\,dy$$

where

$$s\hat{}(x,y) = (2\pi)^{-\frac{1}{2}}\, S\hat{}(x,y) = \tilde{S}\hat{}(x,y),$$

then by virtue of the Paley-Wiener theorem,

$$s\hat{}(x,y) = H(y)\tilde{S}\hat{}(x,y) \tag{8.6}$$

Combining Equations 8.5 and 8.6 yields

$$[f(x,k) - 1]\hat{}(x,y) = I\hat{}(x,y) = \tilde{B}\hat{}(x,y), \quad y \subset R \tag{8.7}$$

where

$$\tilde{B}\hat{}(x,y) = \left(\frac{\pi}{2}\right)^{\frac{1}{2}} H(y)\, Q\left(x + \frac{y}{2}\right) + s\hat{}(x,y)$$

From Equations 8.5 and 8.6, we obtain

$$f(x,k) = 1 + \int_{+0} B\hat{}(x,y)\exp(iky)\,dy$$

$B\hat{}(x,y)$ is called the "Levin transform" of [f(x,k) – 1] and is given by

$$B\hat{}(x,y) = \left(\frac{1}{2}\right) Q\left(x + \frac{y}{2}\right) + \tilde{S}\hat{}(x,y)$$

Now, $f^{(1)\wedge} \subset C[L_2(R^+)]$, whereas $s^\wedge \subset C[L_\infty(R^+)]$. Since $L_\infty(R^+)$ is bounded by the L_1 norm, $s^\wedge$ must vanish faster than $(1/|k|)$ as $|k| \to \infty$. Hence, $\|B^\wedge\|_2$ will exist, showing that $B^\wedge(x,y) \subset C[L_2(R^+)]$. It also follows that $B^\wedge(x,y)$ belongs to the intersection $C[L_2(R^+) \cap C[L_\infty(R^+)]$. It will be found later that $\|B^\wedge\|_2$ also exist for each $x \subset R$. Moreover, $B^\wedge(x,y)$ is a continuous function on $(R \times R^+)$ which vanishes in the limit $y \to \infty$. $B^\wedge(x,y)$ is thus an element of the space $L_0(R^{+2})$, where

$$L_0\left(R^{+2}\right) = \left\{g: g \subset C\left(R \times R^+\right) \middle| \ \forall\, x \subset R: \lim_{y\to\infty} g(x,y) = 0\right\}$$

Summarizing, and quoting analogous results from the other Jost function $f_-(x,k)$, we obtain the theorem given below.

Theorem 8.7

For a potential satisfying a growth condition of order 1, the Jost functions $f_\pm(x,k)$ possess the Fourier integral representations

$$f_\pm(x,k) = 1 + \int_{\pm 0} B_\pm^\wedge(x,y)\ \exp(\pm iky)\ dy$$

$B_\pm^\wedge(x,y)$ are the "Levin transforms" of $[f_\pm - 1](x,k)$, which are given by

$$B_\pm^\wedge(x,y) = \left(\frac{1}{2}\right) Q_\pm\left(x + \frac{y}{2}\right) + s_\pm^\wedge[(x,y)]$$

$$Q_\pm(x) = \int_\pm q(x')\, dx'$$

and

$$S_\pm^\wedge(x,y) = \left[\sum_{n=2}^{\infty} f_\pm^{(n)}(x,k)\right]^\wedge (x,y)$$

The transforms $B_\pm^\wedge(x,y) \subset L_0(R^{\pm 2})]$.

IV. DIFFERENTIATIONS OF LEVIN TRANSFORMS

In this section, we consider the differentiability of Levin transforms with respect to both x and y. The existence of these derivatives is essential in formulating a boundary-value problem for $B_\pm^\wedge(x,y)$, which will be discussed in Section VII. Consider the Jost function $f_+(x,k)$ and, as before, omit the subscript +. It was established in Chapter 7 (Equation 7.20) that

$$|f(x,k)| \le \text{constant}\ (1 + |x|), \quad \text{Im}\, k \ge 0$$

From the theory of distributions,[16] it is known that if a function w(z) is such that

$$|w(z)| \le C(1 + |w|)^M, \quad z \subset R$$

where the constants C,M ≥ 0, then w(z) is a "tempered distribution". Moreover, the "structure theorem" for the space of the tempered distributions asserts that all tempered distributions can be represented in this fashion.[16] Every tempered distribution is a derivative of finite order of some continuous function of polynomial growth. Hence, considered as a function of x, the Jost function f(x,k) is a tempered distribution, i.e., $f \subset \mathcal{L}(R)$. The theory of distributions also tells us that the dual of the space $\mathcal{L}$ is stable under differentiation.[16] Hence, f′(x,k) exists and is also a tempered distribution. Its Fourier transform does, therefore, exist and, by Theorem 8.1, is also a tempered distribution. Differentiation of the Levin transforms with respect to x is, therefore, permissible.

Taking the derivative with respect to x, we obtain

$$I^{\wedge\prime}(x,y) = f^{\wedge\prime}(x,y)$$

Recall that the Fourier transform is with respect to the variable k. Now,

$$f'(x,k) = -\int_{+} q(x')\left\{\exp(2ik\zeta)\right\} f(x',k)\,dx'$$

Introducing the Neumann series for f(x,k) from Equation 7.12, we obtain

$$f'(x,k) = -\int_{+} q(x')\ \exp(2ik\zeta)\ dx' - \int_{+} q(x')\exp(2ik\zeta)\int_{+'} Q(x'')\exp(2ik\zeta')\ dx'\,dx'' +$$

$$+\sum_{n=2}^{\infty} f^{(n)\prime}(x,k) = F_1(x,k) + F_2(x,k) + F_3(x,k)$$

where $f^{(n)\prime} = (F' f^{(n-1)})$, $\xi' = x'' - x'$, and $+'$ on the inner integral of the second term on the R.H S. means that the limit goes from x' to ∞. Now, analogously to Equations (8.3a) and (8.5)

$$F_1^{\wedge}(x,y) = -\left(\frac{\pi}{2}\right)^{\frac{1}{2}} H(y)q\left(x+\frac{y}{2}\right) \subset C\left[L_2\left(R^{+}\right)\right] \tag{8.8}$$

The second term $F_2(x,k)$ can be expressed as follows.

$$F_2(x,k) = -Q(x)\int_{+} Q(x')\exp(2ik\zeta)dx' + \int_{+} Q(x')Q(x')\exp(2ik\zeta)dx'$$

$$= -\int_{+} Q(x')\left[Q(x) - Q(x')\right]\ \exp(2ik\zeta)\ dx'$$

$$= -\left(\frac{1}{2}\right)\int_{R^{+}} Q\left(x+\frac{y}{2}\right)\left[Q(x) - Q\left(x+\frac{y}{2}\right)\right]\ \exp(iky)\ dy$$

$$= -\left(\frac{1}{2}\right)\int_{R} H(y)Q\left(x+\frac{y}{2}\right)\left[Q(x) - Q\left(x+\frac{y}{2}\right)\right]\ \exp(iky)\ dy \tag{8.9}$$

Therefore,

$$F_2^{\wedge}(x,y) = -\left(\frac{\pi}{2}\right)^{\frac{1}{2}} H(y)\, Q\left(x+\frac{y}{2}\right)\left[Q(x) - Q\left(x+\frac{y}{2}\right)\right]$$

If $q \subset L_1{}^1$, then $F_2(x,k) \subset C[L_2(R)]$. Hence, by Theorem 8.3, $F_2{}^{\wedge}(x,y) \subset C[L_2(R)]$.

For the last term, we write

$$F_3(x,k) = \sum_{n=2}^{\infty} f^{(n)\prime}(x,k) = -\sum_{n=2}^{\infty} \int_{+} q(x')\exp(2ik\zeta) f^{(n)}(x',k)\, dx'$$

From before,

$$\left|\sum_{n=2}^{\infty} f^{(n)\prime}(x,k)\right| \sim O\left(|k|^{-2}\right) \text{ as } |k| \to \infty, \quad k \subset \overline{C}_+$$

F_3, therefore, $\subset L_1(R)$. Moreover, $F_3(x,k) \subset C[L_1 \cap L_2]$. Therefore, the Fourier transform of $F_3(x,k)$, namely, $F_3{}^{\wedge}(x,y)$, $\subset C[L_2 \cap L_\infty]$. Following the same procedures that led to Equation 8.8, it is seen again that

$$F_3^{\wedge}(x,y) = 0, \quad x \subset R, \quad y \subset R^-$$

Combining with this Equations 8.8 and 8.9, we have for the x-derivative of the Levin transform the following identity

$$B^{\wedge\prime}(x,y) = -\left(\frac{1}{2}\right)\left[q\left(x+\frac{y}{2}\right) + Q\left(x+\frac{y}{2}\right)\left\{Q(x) - Q\left(x+\frac{y}{2}\right)\right\}\right] + F_3^{\wedge}(x,y), \quad y \subset R^+ \tag{8.10}$$

where $F_3{}^{\wedge} \subset C[L_2 \cap L_2]$. $[B]^{\wedge\prime}$ is continuous in x and y on (R x R) and vanishes as $y \to \infty$. $[B^{\wedge\prime}(x,y)]$, therefore, $\subset L_0(R^{+2})$.

Let us next consider differentiation of the Levin transform with respect to y. Denoting $(\partial g/\partial y)$ by $\dot{g}$, we have

$$\dot{I}^{\wedge}(x,y) = \left(\frac{\partial}{\partial y}\right)\left\{(2\pi)^{-\frac{1}{2}} \int_R [f(x,k) - 1]\exp(-iky)dk\right\}$$

$$= \left(-ik[f(x,k) - 1]\right)^{\wedge}(x,y) \tag{8.11}$$

In other words, $\dot{I}^{\wedge}(x,y)$ is the Fourier transform of $(-ik[f(x,k) - 1])$. Using the Neumann expansion again,

$$\dot{I}^{\wedge}(x,y) = \left(-ikf^{(1)}\right)^{\wedge}(x,y) + \left(-ikf^{(2)}\right)^{\wedge}(x,y) + \left\{-ik\left[\sum_{n=3}^{\infty} f^{(n)}(x,k)\right]\right\}^{\wedge}(x,y)$$

Consider the first term,

$$\left(-ikf^{(1)}\right)^{\wedge}(x,y)=\left(\frac{\partial}{\partial y}\right)f^{(1)}(x,y)=\left(\frac{\pi}{2}\right)^{\frac{1}{2}}\left(\frac{\partial}{\partial y}\right)\left[H(y)Q\left(x+\frac{y}{2}\right)\right]$$

$$=\left(\frac{\pi}{2}\right)^{\frac{1}{2}}\left[\delta(y)Q\left(x+\frac{y}{2}\right)-q\left(x+\frac{y}{2}\right)H(y)\right] \tag{8.12}$$

where Equation (8.5) was used. (–ikg(x,k))^(x,y) = (–ik·{F.T.g(x,y)}(k))^(x,y) = (∂/∂y)g(x,y), by the rule of Fourier transforming a derivative. g(x,y) here is to be considered as the Fourier transform of the function g(x,k).
Similarly,

$$-\left[ikf^{(2)}\right](x,k)=-\left(\frac{1}{2}\right)\int_{+}q(x')\left[\exp(2ik\zeta)-1\right]\left(\frac{1}{2ik}\right)\left\{\int_{+'}q(x'')\left[\exp(2ik\zeta')-1\right]dx''\right\}dx'$$

$$=-\left(\frac{1}{2}\right)\int_{+}q(x')\left[\exp(2ik\zeta)-1\right]\left\{\int_{+'}Q(x'')\exp(2ik\zeta')dx''\right\}dx'$$

$$=-\left(\frac{1}{2}\right)\left(\frac{1}{2}\right)\int_{R}H(y)\,Q\left(x+\frac{y}{2}\right)\left[Q(x)-Q\left(x+\frac{y}{2}\right)\right]\cdot\exp(iky)\,dy+$$

$$+\left(\frac{1}{2}\right)\left(\frac{1}{2}\right)\int_{R}H(y)\left\{\int_{+}q(x')Q\left(x'+\frac{y}{2}\right)dx'\right\}\exp(iky)\,dy$$

where, as before, $\xi'=x''-x'$ and Equations 8.9 and 8.5 were used. From this,

$$\left(-ikf^{(2)}\right)^{\wedge}(x,y)=-\left(\frac{1}{2}\right)\left(\frac{\pi}{2}\right)^{\frac{1}{2}}H(y)\left[Q\left(x+\frac{y}{2}\right)\left\{Q(x)-Q\left(x+\frac{y}{2}\right)\right\}-\right.$$

$$\left.-\int_{+}q(x')Q\left(x'+\frac{y}{2}\right)dx'\right]\subset C\left[L_2(R)\right] \tag{8.13}$$

By similar reasonings,

$$-ik\sum_{n=3}^{\infty}f^{(n)}(x,k)\subset C\left[L_1\cap L_2\right] \tag{8.14}$$

Therefore,

$$\left[-ik\sum_{n=3}^{\infty}f^{(n)}(x,k)\right]^{\wedge}(x,k)\subset C\left[L_1\cap L_\infty\right]$$

Combining Equations 8.12 through 8.14, we obtain

$$\left[\dot{I}^{\wedge}(x,k) - \left(\frac{1}{2}\right)\delta(y)\, Q\left(x+\frac{y}{2}\right)\right] = -\left(\frac{1}{4}\right)H(y)\left[q\left(x+\frac{y}{2}\right)+Q\left(x+\frac{y}{2}\right)\cdot\right.$$

$$\left.\cdot\left\{Q(x) - Q\left(x+\frac{y}{2}\right)\right\} - \int_{+} q(x')Q\left(x'+\frac{y}{2}\right)dx'\right] + F_3^{\wedge}(x,y)$$

Again, for $y \subset R^+$, $\dot{I}^{\wedge}(x,y) = \ddot{B}^{\wedge}(x,y)$ is continuous and vanishes as $y \to \infty$. Hence, $\dot{B}^{\wedge}(x,y) \subset L_0(R^{+2})$.

V. EQUATIONS GOVERNING LEVIN TRANSFORMS

A. DIFFERENTIAL EQUATION FOR B^(x,y)

Let us reproduce the Schrödinger equation for f(x,k) (the subscript + is omitted) here for convenience:

$$f'' + 2ikf' - qf = 0$$

$$\lim_{x\to\infty} f(x,k) = 1,$$

and

$$\lim_{x\to\infty} f'(x,k) = 0$$

The equation for f(x,k) can be rewritten as

$$(f-1)'' + 2ik(f-1)' - q(f-1) - q = 0$$

Integrating both sides with respect to x from x to $+\infty$, and using the asymptotic boundary conditions above, we obtain

$$(f-1)' + 2ik(f-1) + \int_{+} q(x')[f(x',k)-1]dx' + Q(x) = 0 \tag{8.15}$$

Fourier transforming both sides of Equation 8.15 yields

$$I^{\wedge\prime}(x,y) - 2\dot{I}^{\wedge}(x,y) + \int_{+} q(x')\, I^{\wedge}(x',y)dx' + Q(x)\delta(y) = 0 \tag{8.16}$$

In deriving Equation 8.16, we have used Equations 8.7 and 8.11, and $\cdot = \partial/\partial y$.

In terms of B^(x,y), the integro-differential equation is

$$B^{\wedge\prime}(x,y) - 2\dot{B}^{\wedge}(x,y) + \int_{+} q(x')\, B^{\wedge}(x',y)dx' = 0,\ y \subset R^{+} \tag{8.17}$$

From the results of Sections III and IV, we know that the function $[(B)^{\wedge\prime} - 2\dot{B}^{\wedge}](x,y)$ is absolutely continuous in x and y. We can then differentiate this function with respect to x and obtain the wave equation

$$\left[2\dot{B}^\wedge(x,y) - B^{\wedge\prime}(x,y)\right]' + q(x)\, B^\wedge(x,y) = 0, \quad y \subset R^+ \tag{8.18}$$

Equation 8.18 is the differential equation for the Levin transform $B^\wedge(x,y)$ and can be considered to be the Fourier transform of the Schrödinger equation

$$f'' + 2ikf' - qf = 0$$

It is also customary[17,18] to consider not the Levin transform $B^\wedge(x,y)$, but, rather, its variation

$$A^\wedge(x,y) = B^\wedge(x,\, y - x)$$

This is essentially the Fourier transform of the Jost solution instead of the Jost function. In order to appreciate this, let us consider the Levin representation of the Jost solution $\psi(x,k)$ corresponding to $f(x,k)$. We then have

$$\psi(x,k) = \exp(ikx) + \int_{+0} B^\wedge(x,y)\, \exp\left[ik(y+x)\right] dy$$

Replacing $y + x$ by y' in the above equation, carrying out some straightforward algebra, and replacing y' by y yields

$$\psi(x,k) = \exp(ikx) + \int_{+} A^\wedge(x,y) \exp(iky) dy \tag{8.19}$$

where $A^\wedge(x,y) = B^\wedge(x,y - x)$. It is not difficult to show that

$$q(x) = -2\left[A^\wedge(x,x)\right]'$$

and that the kernel $A^\wedge(x,y)$ satisfies the familiar wave equation,

$$\ddot{A}^\wedge(x,y) - A^{\wedge\prime\prime}(x,y) + q(x)\, A^\wedge(x,y) = 0$$

whereas the differential equation for $B^\wedge(x,y)$, namely, Equation 8.18, is a hyperbolic partial differential equation of a more general form.

B. INTEGRAL REPRESENTATION OF $B_\pm^\wedge$ (x,y)

Equation 8.18 is the differential equation governing Levin's transforms. We now derive their integral representations, which will prove to be useful later. One way to achieve this is to appeal to the Riemann function for the hyperbolic differential equation (8.18). However, here we chose to follow an alternate, somewhat simpler method. Consider the transform $B_+^\wedge(x,y)$ first. Again, as usual, the subscript + will be dropped. Let us consider the Jost solution $\psi(x,k)$, which has the representation

$$\psi(x,k) = \exp(ikx) + \int_{+} \left[\left\{\sin k(x' - x)\right\}/k\right] q(x')\, \psi(x',k)\, dx' \tag{8.20}$$

Rearranging and using the Levin representation

$$f(x,k)=1+\int_{+0} B^{\wedge}(x,y')\ \exp(iky')\,dy'$$

in Equation 8.20 yields

$$\int_{+0} B^{\wedge}(x,y')\exp(iky')dy' = \int_{+}\left[\{\sin k(x'-x)\}/k\right]\exp\{ik(x'-x)\}q(x')dx' +$$

$$+\int_{+}\left[\{\sin k(x'-x)\}/k\right]\exp\{ik(x'-x)\}q(x')\cdot\left[\int_{+0} B^{\wedge}(x',y')\exp(iky')dy'\right]dx'$$

$$= F_1(x,k)+F_2(x,k) \tag{8.21}$$

The lower limit of the last integral in Equation 8.21 is x′. Fourier transforming both sides of Equation 8.21, that is, multiplying both sides of Equation 8.21 by exp(–iky) and then integrating with respect to k from – ∞ to + ∞, we obtain

$$B^{\wedge}(x,y)=\left[F_1(x,k)\right]^{\wedge}(x,y)+\left[F_2(x,k)\right]^{\wedge}(x,y) \tag{8.22}$$

Consider the first term on the RHS of Equation (8.22) which is

$$\left[F_1(x,k)\right]^{\wedge}(x,y)=\int_{+} q(x')\left(\int_{R}\left[\frac{\sin k(x'-x)}{k}\right]\exp\left[ik(x'-x-y)\right]dk\right)dx'$$

Using the identity

$$\int_{R} \cos kx \sin ky\ k^{-1}dk = \frac{1}{2}\left[H(x+y)-H(x-y)\right]$$

yields

$$\left[F_1(x,k)\right]^{\wedge}(x,y)=\frac{1}{2}\int_{x+\frac{y}{2}}^{\infty} q(x')\ dx' \tag{8.23}$$

Similarly,

$$\left[F_2(x,k)\right]^{\wedge}(x,y)=\int_{+0} dy'\int_{+} dx'\ q(x')\ B^{\wedge}(x',y')J(x,x',y,y')$$

where

$$J(x,x',y,y')=\int_{R}\left\{\frac{\left[\sin k(x'-x)\right]}{k}\right\}\ \exp\left\{ik(x'-x-y+y')\right\}\ dk$$

Again using the above identity, we obtain

$$\left[F_2(x,k)\right]^\wedge(x,y)=\left(\frac{1}{2}\right)\int_0^y dy' \int_{x+\frac{(y-y')}{2}}^{\infty} dx'\, q(x')\, B^\wedge(x',y') \tag{8.24}$$

From Equations 8.22 through 8.24, the final expression for B^(x,y) can be derived:

$$B^\wedge(x,y)=\left(\frac{1}{2}\right)\int_{x+\frac{y}{2}}^{\infty} q(x')dx' + \left(\frac{1}{2}\right)\int_0^y dy' \int_{x+\frac{(y-y')}{2}}^{\infty} dx'\, q(x')\, B^\wedge(x',y')$$

$$=\left(\frac{1}{2}\right)Q\left(x+\frac{y}{2}\right)+\frac{1}{2}\int_0^y dy' \int_{x+\frac{(y-y')}{2}}^{\infty} dx'\, q(x')\, B^\wedge(x',y'), \quad x\subset R,\ y\subset R^+ \tag{8.25}$$

Equation 8.25 is the desired integral equation for the Levin transform $B_+^\wedge(x,y)$. For the Jost function $f_-(x,k)$, the corresponding equation is

$$B_-^\wedge(x,y)=\left(\frac{1}{2}\right)Q_-\left(x+\frac{y}{2}\right)+\left(\frac{1}{2}\right)\int_y^0 dy' \int_{-\infty}^{x+\frac{(y-y')}{2}} dx'\, q(x')\, B_-(x',y'), \quad x\subset R,\ y\subset R^- \tag{8.25a}$$

VI. ITERATIVE SOLUTION OF THE INTEGRAL EQUATION FOR $B_\pm^\wedge(x,y)$

Equation 8.25 will be solved iteratively as follows. We write

$$B^\wedge(x,y)=\sum_{n=0}^{\infty} B^{(n)}(x,y)$$

where

$$B^{(0)}(x,y)=\left(\frac{1}{2}\right)Q\left(x+\frac{y}{2}\right)$$

and

$$B^{(n+1)}(x,y)=\left(\frac{1}{2}\right)\int_0^y dy' \int_{x+\frac{(y-y')}{2}}^{\infty} dy''q(y'')B^{(n)}(y'',y') \tag{8.26}$$

Calculations will be performed below ignoring the factor (1/2), which will be reinserted in the final results.

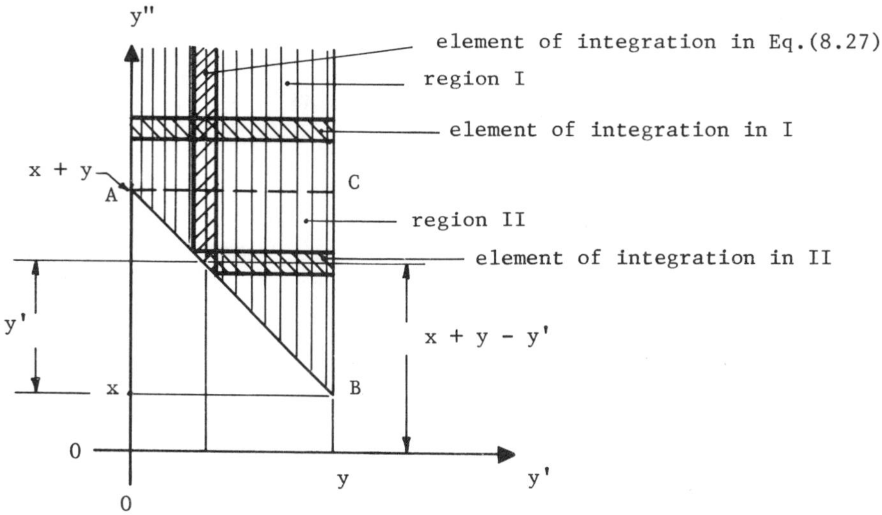

FIGURE 1. Evaluation of the integral of Equation 8.27.

Let us consider the $B^{(1)}(x,y)$ term, which is given by

$$B^{(1)}(x,y)=\int_0^y dy' \int_{x+\frac{(y-y')}{2}}^{\infty} dy''\, q(y'')\, B^{(0)}(y'',y')$$

Since

$$\left|B^{(0)}(y'',y')\right| \le \eta\left(\frac{y'}{2}+y''\right)$$

we have

$$\left|B^{(1)}(x,y)\right| \le \int_0^y dy' \int_{x+\frac{(y-y')}{2}}^{\infty} dy''\, \left|q(y'')\right| \eta\left(\frac{y'}{2}+y''\right)$$

$\eta(x)$ being nonincreasing in x, we can write

$$\left|B^{(1)}(x,y)\right| \le \eta\left(x+\frac{y}{2}\right)\int_0^y dy' \int_{x+\frac{(y-y')}{2}}^{\infty} \left|q(y'')\right| dy'' \tag{8.27}$$

The region of integration for Equation 8.27 is shown in Figure 1 and consists of the strip above the diagonal AB. The region can be decomposed into two separate regions: (1) the triangle ABC (region II) and (2) the strip above it and extending to infinity (region I). Therefore,

$$\left|B^{(1)}(x,y)\right| \le \eta\left(x+\frac{y}{2}\right)\left[\int_{I}(\cdot)+\int_{II}(\cdot)\right]$$

The integration in region I can be performed as

$$\int_{I}(\cdot)=\int_{0}^{y} dy' \int_{x+\frac{y}{2}}^{\infty} dy''\left|q(y'')\right| = \int_{x+\frac{y}{2}}^{\infty} y\left|q(y'')\right| dy''$$

Similarly, integration in region II yields

$$\int_{II}(\cdot)= \int_{x}^{x+\frac{y}{2}} dy''\left|q(y'')\right| \int_{y+2(x-y'')}^{y} dy' = 2\int_{x}^{x+\frac{y}{2}} (y''-x)\, q(y'')'' \, dy''$$

This is the integral over the rectangle ABCD. Thus,

$$\left|B^{(1)}(x,y)\right| \le \eta\left(x+\frac{y}{2}\right)\left[\int_{x+\frac{y}{2}}^{\infty} y\left|q(y'')\right| dy'' + \int_{x}^{x+\frac{y}{2}}(y''-x)\left|q(y'')\right| dy''\right]$$

$$\le \eta\left(x+\frac{y}{2}\right)\int_{+}(y''-x)\left|q(y'')\right| dy'' = \eta\left(x+\frac{y}{2}\right)\gamma(x)$$

The second term $|B^{(2)}(x,y)|$ can be calculated in a similar fashion.

$$\left|B^{(2)}(x,y)\right| \le \int_{0}^{y} dy' \int_{x+\frac{(y-y')}{2}}^{\infty} dy''\left|q(y'')\right|\gamma(y'')\eta\left(\frac{y'}{2}+y''\right)$$

$$\le \eta\left(x+\frac{y}{2}\right)\int_{0}^{y} dy' \int_{x+\frac{(y-y')}{2}}^{\infty} dy''\left|q(y'')\right|\gamma(y'')$$

Then following the exact same steps leading to the derivation of $|B^{(1)}(x,y)|$, we have

$$\left|B^{(2)}(x,y)\right| \le \left(\frac{1}{2}\right)\eta\left(x+\frac{y}{2}\right)\gamma^{2}(x)$$

Thus,

$$\left|B^{(0)}(x,y)\right| \le \eta\left(x+\frac{y}{2}\right)\gamma^{0}(x)\left(\frac{1}{0!}\right)$$

$$\left|B^{(1)}(x,y)\right| \le \eta\left(x+\frac{y}{2}\right)\gamma^{1}(x)\left(\frac{1}{1!}\right)$$

$$\left|B^{(2)}(x,y)\right| \le \eta\left(x+\frac{y}{2}\right)\gamma^{2}(x)\left(\frac{1}{2!}\right)$$

Inductively,

$$\left|B^{(n+1)}1(x,y)\right| \le \eta\left(x+\frac{y}{2}\right)\left[\frac{\gamma^{n+1}(x)}{(n+1)!}\right]$$

Reinserting the factor (1/2), the above bound is

$$\left|B^{(n)}1(x,y)\right| \le \left(\frac{1}{2}\right)\eta\left(x+\frac{y}{2}\right)\left[\frac{\gamma^{n+1}(x)}{(n+1)!}\right] \tag{8.28}$$

Alternately, let us postulate the inequality (8.28) for $|B^{(n)}(x,y)|$ to begin with and replace this in the iteration (8.26). This yields

$$\left|B^{(n+1)}(x,y)\right|$$

$$\le \int_0^y dy' \int_{x+\frac{(y-y')}{2}}^{\infty} dy'' \left|q(y'')\right| \eta\left(y''+\frac{y'}{2}\right)\gamma^n(y'')/n!$$

$$\le \eta\left(x+\frac{y}{2}\right)\int_0^y dy' \int_{x+\left(\frac{y-y'}{2}\right)}^{\infty} dy'' \left|q(y'')\right|\gamma^n(y'')/n!$$

$$= \eta\left(x+\frac{y}{2}\right)\left[\int_x^{x+\frac{y}{2}} dy''\left|q(y'')\right|\left\{\gamma^n(y'')/n!\right\}\left\{\int_{y+2(x-y'')}^{y} dy'\right\} + \int_{x+\frac{y}{2}}^{\infty} dy''\left|q(y'')\right|\left\{\gamma^n(y'')/n!\right\}\left\{\int_0^y dy'\right\}\right]$$

$$= \eta\left(x+\frac{y}{2}\right)\left[2\int_x^{x+\frac{y}{2}} dy''\left|q(y'')\right|(y''-x)\gamma^n(y'')/n! + \int_{x+\frac{y}{2}}^{\infty} dy''\left|q(y'')\right|\left\{\gamma^n(y'')/n!\right\}y\right]$$

$$\le \eta\left(x+\frac{y}{2}\right)\int_+ dy''\left|q(y'')\right|(y''-x)\gamma^n(y'')/n!$$

$$\le \eta\left(x+\frac{y}{2}\right)\frac{\gamma^{n+1}(x)}{(n+1)!}$$

Finally, taking the factor of (1/2) into account, the inequality (8.28) is again recovered. From this it also follows that

$$\left|B^{\wedge}(x,y)\right| \le \left(\frac{1}{2}\right)\eta\left(x+\frac{y}{2}\right)\exp\{\gamma(x)\}$$

Since $\eta(x + z)$ is nonincreasing in z, the L_∞ norm for B(x,y) can be given immediately as

$$\left\|B^\wedge(x,\cdot)\right\|_\infty \le \eta(x)\exp\{\gamma(x)\} \tag{8.29}$$

The L_1 norm can also be calculated, namely,

$$\left\|B^\wedge(x,\cdot)\right\|_1 \le \frac{1}{2}\exp\{\gamma(x)\}\int_{+0}\eta\left(x+\frac{y}{2}\right)dy \le \exp\{\gamma(x)\}\int_{+} dz\,\eta(z)$$

Again following the steps leading to the estimation of $|B^{(1)}(x,y)|$, we obtain

$$\left\|B^\wedge(x,\cdot)\right\|_1 \le \gamma(x)\exp\{\gamma(x)\} \tag{8.30}$$

Thus, in each $x \subset R$, the iterates are uniformly bounded in y, and the Neumann series for $B^\wedge(x,y)$ converges uniformly in y. However, the iterates vanish as $y \to \infty$. Hence, $B^\wedge(x,y)$ is an element of $L_0(R^{+2})$. In view of the existence of $\|B^\wedge(x,\cdot)\|_1$, the Levin transform $B^\wedge(x,y)$ belongs to $L_1(R)$ in y and is thus an element of $L_1(R^+) \cap L_\infty(R^+)$.

The existence of $[B^{\wedge\prime}(x,y)]$ and $\dot{B}^\wedge(x,y)$ was established earlier. In view of the inequalities established above, the bounds on these derivatives can now be estimated. From the integro-differential equation for $B^\wedge(x,y)$, Equation 8.25, we have

$$B^{\wedge\prime}(x,y) = -\left(\frac{1}{2}\right)q\left(x+\frac{y}{2}\right) - \left(\frac{1}{2}\right)\int_0^y dy'\,q\left[x+\frac{(y-y')}{2}\right]B^\wedge\left[x+\frac{(y-y')}{2},\,y'\right]$$

Therefore,

$$B^{\wedge\prime}(x,y) \le \left(\frac{1}{2}\right)\left|q\left(x+\frac{y}{2}\right)\right| + \left(\frac{1}{2}\right)\int_0^y dy'\,q\left[x+\frac{(y-y')}{2}\right]B^\wedge\left[x+\frac{(y-y')}{2},y'\right]$$

$$\le \left(\frac{1}{2}\right)\left|q\left(x+\frac{y}{2}\right)\right| + \left(\frac{1}{2}\right)\int_0^y dy' q\left[x+\frac{(y-y')}{2}\right]\cdot\eta\left\{x+\frac{(y+y')}{2}\right\}\exp\left\{\gamma\left[x+\frac{(y-y')}{2}\right]\right\}$$

$$\le \left(\frac{1}{2}\right)\left|q\left(x+\frac{y}{2}\right)\right| + \left(\frac{1}{2}\right)\eta\left(x+\frac{y}{2}\right)\exp\{\gamma(x)\}\int_0^y dy'\left|q\left[x+\frac{(y-y')}{2}\right]\right|$$

$$\le \left(\frac{1}{2}\right)\left|q\left(x+\frac{y}{2}\right)\right| + \eta\left(x+\frac{y}{2}\right)\eta(x)\exp\{\gamma(x)\}$$

again since $\eta(x)$.resp. $\gamma(x)$ is nonincreasing .resp. nondecreasing in x.

Similarly,

$$\dot{B}^\wedge(x,y) = -\frac{1}{4}q\left(x+\frac{y}{2}\right) + \frac{1}{2}\int_+ dx'\, q(x')\, B^\wedge(x',y) -$$

$$-\left(\frac{1}{4}\right)\int_0^y dy'\, q\left[x+\frac{(y-y')}{2}\right]B^\wedge\left[x+\frac{(y-y')}{2}, y'\right]$$

Therefore,

$$\left|\dot{B}^\wedge(x,y)\right| \le \left(\frac{1}{4}\right)\left|q\left(x+\frac{y}{2}\right)\right| + \left(\frac{1}{4}\right)\int_+ dx'|q(x')|\, \exp\{\gamma(x')\}\eta\left(x'+\frac{y}{2}\right) +$$

$$+\left(\frac{1}{8}\right)\eta(x)\eta\left(x+\frac{y}{2}\right)\exp\{\gamma(x)\}$$

$$\le \left(\frac{1}{4}\right) q\left(x+\frac{y}{2}\right) + \left(\frac{3}{8}\right)\eta(x)\eta\left(x+\frac{y}{2}\right)\exp\{\gamma(x)\}$$

From the various bounds, it follows that for each y, (respectively x on any half-line), the functions $B^\wedge(x,y)$ and $[2\dot{B}^\wedge(x,y) - B^{\wedge\prime}(x,y)]$ are absolutely continuous in x (respectively y). The Levin transform has an integrable first derivative with respect to both x and y, provided that the potential $q(x) \subset L^1_1$. Moreover, $[2\dot{B}^\wedge(x,y) - B^{\wedge\prime}(x,y)] \to 0$ as $x \to \infty$.

It can be verified directly that Equation 8.25 does satisfy the Schrödinger equation for the Jost function

$$f(x,k) = 1 + \int_{+0} B^\wedge(x,y)\exp(iky)\, dy$$

Differentiating once with respect to x,

$$f'(x,k) = \int_{+0} B^{\wedge\prime}(x,y)\exp(iky)dy = \int_{+0}\left[B^{\wedge\prime}(x,y) - 2\dot{B}^\wedge(x,y)\right]\exp(iky)dy +$$

$$+2\int_{+0}\dot{B}^\wedge(x,y)\exp(iky)dy = -\int_{+0}\left[\int_+ q(x')B^\wedge(x',y)dx'\right]\exp(iky)dy -$$

$$-2\dot{B}^\wedge(x,0) - 2ik\int_{+0} B^\wedge(x,y)\exp(iky)dy \tag{8.31}$$

where we have used the relation

$$\left[B^{\wedge\prime}(x,y) - 2\dot{B}^\wedge(x,y)\right]' + q(x)\, B^\wedge(x,y) = 0$$

Differentiating (8.31) once more with respect to x, we obtain

$$f''(x,y) = q(x)f(x,y) - 2ikf'(x,y)$$

leading to the Schrödinger equation

$$f''(x,k) + 2ikf'(x,k) = q(x)f(x,k)$$

The asymptotic boundary conditions, namely, $\lim_{x\to\infty} f = 1$ and $\lim_{x\to\infty} f' = 0$, are also satisfied since $B^\wedge \to 0$ in this limit.

VII. THE GOURSAT PROBLEM FOR THE LEVIN TRANSFORMS

Either from Equation 8.10 or from the integral expression (8.25) itself, it can be seen that

$$B^{\wedge\prime}(x,0+) = -\frac{q(x)}{2}$$

Similarly,

$$2B^{\wedge\prime}(x,0+) = 4\dot{B}^\wedge(x,0+) = -q(x) \tag{8.32}$$

The differential equation (8.18) for the Levin transform is in the so-called "normal form" of a hyperbolic partial differential equation. This means that x and y are themselves the characteristics. Equation 8.32 then specifies the boundary conditions along one of the characteristic, the x-axis. Equation 8.18 along with Equation 8.32 then forms a "Goursat problem" for the Levin transform. From the theory of hyperbolic partial differential equations, we know that such a problem admits of a unique solution. The Goursat problem corresponding to $f_-(x,k)$ has the boundary conditions

$$2B^{\wedge\prime}(x,0-) = 4\dot{B}^\wedge(x,0-) = q(x)$$

Let us summarize these results into the following theorem.

Theorem 8.8
The Levin transforms $B_\pm^\wedge(x,y)$ of the Jost functions $f_\pm(x,k)$ satisfy a Goursat problem

$$\left[B_\pm^{\prime\wedge} - 2\dot{B}_\pm^\wedge\right]' \pm q(x)B_\pm^\wedge = 0$$

$$2B_\pm^{\wedge\prime}(x,0\pm) = 4\dot{B}_\pm^\wedge(x,0\pm) = \mp q(x) \tag{8.33}$$

and have the following integral representations:

$$B_{\pm}^{\wedge} = \pm\left(\frac{1}{2}\right)\int_{x+\frac{y}{2}}^{\pm\infty} q(x')dx' + \left(\frac{1}{2}\right)\int_{0}^{y} dy'\left[\int_{x+\frac{(y-y')}{2}}^{\pm\infty} q(y'')B_{\pm}^{\wedge}(y'',y')dy''\right]$$

$$= \left(\frac{1}{2}\right)Q\left(x+\frac{y}{2}\right) + \left(\frac{1}{2}\right)\int_{0}^{y} dy'\left[\int_{x+\frac{(y-y')}{2}}^{\pm\infty} q(y'')\,B_{\pm}^{\wedge}(y'',y')dy''\right] \tag{8.34}$$

Moreover, $B_{\pm}^{\wedge}$ can be expanded in a Neumann series

$$B_{\pm}^{\wedge} = \sum_{n=0}^{\infty} B_{\pm}^{(n)}(x,y)$$

$$B_{\pm}^{(0)}(x,y) = \left(\frac{1}{2}\right)Q\left(x+\frac{y}{2}\right)$$

Writing η for $\eta_{\pm}$ and γ for $\gamma_{\pm}$,

$$\left|B_{\pm}^{(n)}(x,y)\right| \le \left(\frac{1}{n!}\right)\eta\left(x+\frac{y}{2}\right)\gamma^{n}(x) \tag{8.35}$$

and the series converges uniformly in y for each $x \subset R$.

Being a Goursat problem, Levin transforms are determined uniquely. Also, the Levin-transformed Jost functions, namely,

$$f_{\pm}(x,k) = 1 + \int_{\pm 0} B_{\pm}^{\wedge}(x,y)\exp(\pm iky)dy \tag{8.36}$$

satisfy the Schrödinger problem uniquely.

Finally, $B_{\pm}^{\wedge}(x,y)$ and its first derivatives obey the following bounds:

$$\left|B_{\pm}^{\wedge}\right| \le \left(\frac{1}{2}\right)\eta\left(x+\frac{y}{2}\right)\exp\{\gamma(x)\}$$

$$\left|B_{\pm}^{\wedge\prime}\right| \le \left(\frac{1}{2}\right)\left|q\left(x+\frac{y}{2}\right)\right| + \eta\left(x+\frac{y}{2}\right)\eta(x)\exp\{\gamma(x)\} \tag{8.37}$$

$$\left|\dot{B}_{\pm}^{\wedge}\right| \le \left(\frac{1}{4}\right)\left|q\left(x+\frac{y}{2}\right)\right| + \eta\left(x+\frac{y}{2}\right)\eta(x)\exp\{\gamma(x)\}$$

The following norms are established:

$$\left\|B_{\pm}^{\wedge}\right\|_{\infty} \leq \eta(x)\exp\{\gamma(x)\}$$

and

$$\left\|B_{\pm}^{\wedge}\right\|_{1} \leq \gamma(x)\exp\{\gamma(x)\}$$

Moreover, $B_{\pm}^{\wedge}(x,y)$ and $[B_{\pm}^{\wedge\prime} - 2\dot{B}_{\pm}^{\wedge}]$ are absolutely continuous in x and y.

In terms of the Jost solutions, Levin's representations take the form

$$g_{\pm}(x,k) = \exp(\pm ikx) + \int_{\pm} A_{\pm}^{\wedge}(x,y)\exp(\pm iky)\, dy \qquad (8.38)$$

where

$$A_{\pm}^{\wedge}(x,y) = B_{\pm}^{\wedge}(x,y-x)$$

In the above expressions, g_+ resp. g_- stands for the Jost solution $\psi(x,k)$ resp. $\phi(x,k)$.

The potential is found to be

$$q(x) = \mp\left[A_{\pm}^{\wedge}(x,x)\right]'$$

Moreover, representations (8.36) or (8.38) are valid for all $(x,k) \subset R \times \overline{C}_+$ provided that the potential $q \subset L_1^{\,1}$.

Many authors define the Fourier transform as

$$f^{\wedge}(y) = \left(\frac{1}{\pi}\right)\int_{R} \exp(2iky)\, f(k)dk$$

and

$$f^{\vee}(k) = \int_{R} \exp(-2iky)\, f^{\wedge}(y)dy$$

If this definition is followed, then all the expressions derived above are valid, with the following modifications: the factors of 2 and 4 disappear in Equation(s) 8.33 and the factor of (1/2) is replaced by 1 wherever it appears in Equations 8.34 through 8.37. Moreover, in the third inequality of Equation 8.37, the factor of (1/4) is to be replaced by 1 and the second term is to be multiplied by a factor of 2.

APPENDIX: THE LEVIN TRANSFORMS ON THE HALF-LINE R^+

The Levin representations were discussed in the main body of the text for the Jost functions and solutions on the entire real axis R. We now present Levin's transforms for the Jost solutions on the half-axis. This is just for completeness of our presentation. So only a very terse and summary discussion will be given here. Let us recall that the following estimate

$$|f(k,r) - \exp(ikr)| \leq \text{constant } |k|^{-1} \exp(-\operatorname{Im} kr)\, \eta(r), \quad \operatorname{Im} k \geq 0$$

was given in the Appendix of the preceding chapter for all $r > 0$. It, therefore, follows that

$$\int_R |h(r, k = u + iv)|^2 du \sim O[\exp(-2vr)]$$

where $h(k,r) = f(k,r) - \exp(ikr)$. $h(k,r)$ is then an $L_2(R)$ function of k on any line parallel to the real axis. Moreover, through the Jost solution f(k,r), the function h(k,r) is continuous for $k \subset R$. The function h is, therefore, a Hardy function. However, because of the $O[\exp(-2vr)]$ estimate, a theorem similar to that of Paley-Wiener can be applied to h(k,r). The theorem is a well known theorem of Titchmarsh[19] and can be stated as follows.

Titchmarsh Theorem

A necessary and sufficient condition for a function $g(u) \subset L_2(R)$ to be the limit $u \to 0$ of another function $g(w) = g(u + iv)$, which is analytic in the upper half-plane and such that

$$\int_R |g(u + iv)|^2 du \sim 0[\exp(-2vy)]$$

is that

$$G(t) = \int_R g(u) \exp(-itu) du = 0 \text{ for } t < y$$

Applying Titchmarsh's theorem to the h(k,r) function yields

$$A(r,t) = h^\wedge(r,t) = \int_R [f(k,r) - \exp(ikr)] \exp(-ikt)\, dk = 0, \; t < r$$

Inverting the Fourier transform then results in the Levin representation

$$f(k,r) = \exp(ikr) + \int_{+r} A(r,t) \exp(ikt) dt$$

$t \geq r$, $r > 0$, $\operatorname{Im} k \geq 0$ and the Levin transform A is an L_2 function in t. The representation given in Equation A.20 in Chapter 7 was obtained by setting r to be zero.

Some results valid for the Levin transform A(r,t) on the half-line are given below. A(r,t) satisfies the following integral equation

$$A(r,t) = \left(\frac{1}{2}\right) \int_{\frac{(r+t)}{2}}^{\infty} q(s)\, ds - \int_{\frac{(r+t)}{2}}^{\infty} \int_0^{(t-r)/2} q(s - s')\, A(s - s', \; s + s')\, ds\, ds', \; t \geq r$$

This is similar to that of the Povzner-Levitan kernel derived in Chapter 4. Analogous to the results on the entire line, A(r,t) satisfies the following bounds

$$|A(r,t)| \leq \eta\left(\frac{r+t}{2}\right) \exp(\gamma(r))$$

and

$$\left|A'(r,t) + \left(\frac{1}{4}\right) q\left(\frac{r+t}{2}\right)\right| \leq \text{constant } \eta(r)\, \eta\left(\frac{r+t}{2}\right)$$

Also, if the potential is a continuous function, then

$$q(r) = -2[A(r,r)]'$$

the prime denoting differentiation in r. Moreover, as in the case of the entire real axis, A(r,t) also satisfies a wave equation

$$A''(r,t) - \ddot{A}(r,t) - q(r)A(r,t) = 0$$

The Levin representations are more general than the Povzner-Levitan case. This is due to the fact that the bound |A(r,t)| derived above requires only that the potential be an L_1^1-function on any half-line $[b,\infty)$, $b > 0$. This means that the Levin representation of f(k,r) and the integral equation for A(r,t) are more suitable for dealing with the potentials which are singular at the origin. For more details on this point, Reference 18 should be consulted.

REFERENCES

1. **M. Reed and B. Simon:** *Methods of Mathematical Physics*, Vol. 4, Academic Press, New York, 1972.
2. **W. Rudin:** *Real and Complex Analysis*, McGraw-Hill, New York, 1966.
3. **L. Hörmander:** *An Introduction To Complex Analysis in Several Complex Variables*, Van Nostrand, Princeton, NJ, 1966.
4. **S. Bochner and K. Chandrasekharan:** *Fourier Transforms, Ann. Math. Studies*, 19, 1949.
5. **I. M. Gel'fand and G. E. Shilov:** *Generalized Functions*, Academic Press, New York, 1964.
6. **H. Dym and H. P. McKean:** *Fourier Series and Integrals*, Academic Press, New York, 1972.
7. **E. M. Stein and G. Weiss:** *Fourier Transforms in Euclidean Spaces*, Princeton University Press, Princeton, NJ, 1971.
8. **P. Deift and E. Trubowitz:** Inverse scattering on the line, *Comm. Pure and Appl. Math.*, 32, 1979.
9. **A. Beurling:** On two problems concerning linear transformations in Hilbert space, *Acta Math.*, 81, 1949.
10. **K. Hoffman:** *Banach Spaces of Analytic Functions*, Prentice-Hall, Englewood-Cliffs, NJ, 1962.
11. **H. Helson:** *Lectures On Invariant Subspaces*, Academic Press, New York, 1964.
12. **P. L. Duren:** *Theory Of H^p Spaces*, Academic Press, New York, 1970.
13. **H. Dym and H. P. McKean:** *Gaussian Processes, Function Theory, and the Inverse Spectral Problem*, Academic Press, New York, 1976
14. **B. Y. Levin:** Fourier and Laplace's types of transformations by means of solutions of differential equations of second order, *Math. Rev.*, 18, 1957.
15. **W. Eckhaus and A. van Harten:** *The Inverse Scattering Transformation andt the Theory of Solitons*, North-Holland, Amsterdam, 1981.

16. **F. G. Friedlander:** *Introduction to the Theory of Distributions*, Cambridge University Press, 1982.
17. **L. D. Fadeev:** The inverse problem in the quantum theory of scattering, *J. Math. Phys.*, 4, 1963.
18. **K. Chadan and P. C. Sabatier:** *Inverse Problems in Quantum Scattering Theory*, Springer, NY, 1977.
19. **E. C. Titchmarsh:** *The Theory of Functions*, Oxford University Press, London, 1952.

Chapter 9

SCATTERING MATRIX AND SCATTERING DATA

I. INTRODUCTION

Information about a scattering process is contained in the scattering matrix **S** and in the scattering data S. These form the core of the inverse scattering theory. For scattering in a one-space dimension, **S** is a (2×2) matrix, the elements of which consist of certain Wronskians of the Jost solutions or functions. The scattering data S, on the other hand, contain additional information about the bound states. Specifying the scattering matrix S alone is not sufficient for the unique determination of the potential. Solving inverse scattering problems using only the **S** matrix was attempted by Hyelleras[1] and Fröberg.[2] But Bargmann[3] showed by constructing counter-examples that using S alone yielded nonunique solutions. For a unique determination, the eigenvalues belonging to the point spectrum $\sigma(L)$ and the bound state normalization constants must be specified. For a detailed discussion of the inverse scattering on the line, it is, therefore, essential to study both **S** and S rigorously. This is the objective of this chapter. Much of the material presented is also scattered in the references at the end of the chapter.

II. THE FUNDAMENTAL SOLUTIONS

The Schrödinger equation does not contain a first derivative term. The Wronskian of its two linearly independent solutions must then be independent of x and may depend on k only. This simplifies the determination of the Wronskians since they can be evaluated at $|x| \to \infty$, where the Jost solutions are explicitly known to be plane waves exp $(\pm ikx)$. Consider, for example, the Jost solutions $\psi(x,k)$ and $\overline{\psi}(x,k)$. For real k, $k \neq 0$, it was shown in Chapter 7 that

$$\overline{\psi}(x,k) = \psi(x,-k)$$

Then,

$$W[\psi(x,k), \psi(x,-k)] = \lim_{x\to\infty} [\psi(x,k)\psi'(x,-k) - \psi'(x,k)\psi(x,-k)]$$

$$= \lim_{x\to\infty} [-ik \exp(ikx) \exp(-ikx) - (+ik) \exp(ikx) \exp(-ikx)] = -2ik \neq 0$$

Various Wronskian relations which are of immediate use for the present discussion are listed below.

$$W(\phi,\psi) = 2ikt_+(k) \qquad W(\overline{\phi},\overline{\psi}) = -2ik\, \overline{t}_+(k) \tag{9.1}$$

$$W(\psi,\overline{\psi}) = -2ik,\ k \neq 0 \qquad W(\phi,\overline{\phi}) = 2ik,\ k \neq 0 \tag{9.2}$$

$$W(\psi_0,\tilde{\psi}) = 1,\ k = 0 \qquad W(\phi_0,\tilde{\phi}) = 1,\ k = 0 \tag{9.3}$$

$$W(\phi,\overline{\psi}) = -2ikr_+(k) \qquad W(\overline{\phi},\psi) = 2ik\overline{r}_+(k) \tag{9.4}$$

where ψ_0, $\tilde{\psi}$, ϕ_0, and $\tilde{\phi}$ are the Jost solutions corresponding to k = 0 and were discussed in Chapter 7. The above relations define the quantities $t_+(k)$, $\overline{t}_+(k)$, $r_+(k)$, and $\overline{r}_+(k)$, respectively. We mention in passing that it is also customary in the literature to designate these parameters by the elements c_{ij}, i,j = 1,2, and also by the letters a, $\overline{a}$, b, and $\overline{b}$. The functions $t_+(k)$ resp. $\overline{t}_+(k)$ are analytic in $k \subset C_+$ resp. C_- by virtue of the analyticity of the corresponding Jost solutions involved.

From Equations 9.1 through 9.3, we conclude the following theorem.[24]

Theorem 9.1

The Jost solutions (ϕ,ψ) resp. $(\overline{\phi},\overline{\psi})$ form a basis for the solution space of the Schrödinger operator $\partial^2/\partial x^2 - qy = -k^2 y$ if Im k ≥ 0 resp. Im k ≤ 0 and k ≠ 0, unless $t_+(k) = W(\phi,\psi)/2ik$ resp. $\overline{t}_+(k) = -W(\overline{\phi},\overline{\psi})/2ik$ is zero. If Im k = 0 and k ≠ 0, then the Jost solutions ψ, $\overline{\psi}$ or ϕ, $\overline{\phi}$ are always a basis. In particular, if $\overline{r}_+(k) = W(\overline{\phi},\psi)/2ik$, $r_+(k) = -W(\phi,\overline{\psi})/2ik$, and $t_+(k)$, $\overline{t}_+(k)$ are nonzero, then any pair of the Jost solutions ψ, ϕ, $\overline{\psi}$, $\overline{\phi}$ can be used as a basis to the solution space if Im k = 0 and k ≠ 0. For k = 0, a basis is provided by the pairs of functions $(\psi_0,\tilde{\psi})$ or $(\phi_0,\tilde{\phi})$ almost everywhere.

For scattering solutions, we restrict ourselves to $k \subset R\backslash\{0\}$. Then in view of the theorem above, provided that t(k) and r(k) are nonzero, any pair of the Jost functions provide a basis for the two-dimensional solution space and form a "fundamental system" of solutions. If we consider the Schrödinger equation in the matrix form (Equation 7.2), then a fundamental system in the form of a matrix, such as

$$\phi(x,k) = \begin{pmatrix} \phi(x,k) & \overline{\phi}(x,k) \\ \phi'(x,k) & \overline{\phi}'(x,k) \end{pmatrix} \text{ or } \psi(k) = \begin{pmatrix} \psi(x,k) & \overline{\psi}(x,k) \\ \psi'(x,k) & \overline{\psi}'(x,k) \end{pmatrix}$$

is called a "fundamental matrix solution". Equation 7.2 then yields

$$\mathbf{y}'(x,k) = [q\sigma - \mathbf{A}(k)]\mathbf{y}(x,k)$$

where

$$\mathbf{y}(x,k) = \begin{pmatrix} \phi(x,k) \\ \psi(x,k) \end{pmatrix}$$

III. THE SCATTERING EQUATIONS

For $k \subset R\backslash\{0\}$, the scattering solutions are given by ψ and ϕ. In view of Theorem 9.1, these can be expressed as

$$\psi(x,k) = t_-(k)\overline{\phi}(x,k) + r_-(k)\phi(x,k) \tag{9.5}$$

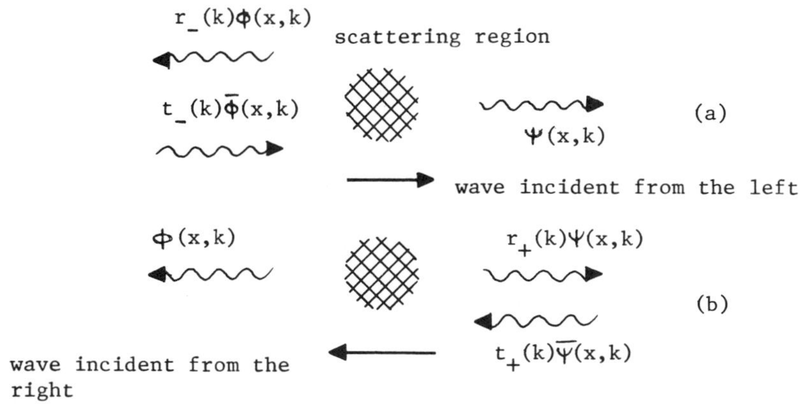

FIGURE 1. A pictorial description of scattering in the asymptotic limit $|x| \to \infty$.

and

$$\phi(x,k) = t_+(k)\overline{\psi}(x,k) + r_+(k)\psi(x,k) \tag{9.6}$$

Similarly,

$$\overline{\psi}(x,k) = \overline{t}_-(k)\phi(x,k) + \overline{r}_-(k)\overline{\phi}(x,k) \tag{9.7}$$

and

$$\overline{\phi}(x,k) = \overline{r}_+(k)\overline{\psi}(x,k) + \overline{t}_+(k)\psi(x,k) \tag{9.8}$$

Subscripts + resp. – are associated with waves incident from the right resp. from the left (see Figure 1). These equations express a very simple fact, namely, that an arbitrary solution of a second-order differential equation can be obtained as a linear combination of its two linearly independent solutions. The parameters t, $\overline{t}$, r, and $\overline{r}$ are the coefficients in these expansions. This apparently innocuous statement, however, comprises the very core of the theory of inverse scattering on the entire line. The asymptotic forms and the pictorial interpretation of the scattering processes associated with the scattering solutions Equations 9.5 and 9.6 are given below.

$$\begin{aligned} \psi(x,k) &= \exp(ikx), \ |x| \to \infty \\ &= t_-(k)\exp(ikx) + r_-(k)\exp(-ikx), \ |x| \to -\infty \end{aligned} \tag{9.9a}$$

$$\begin{aligned} \phi(x,k) &= \exp(-ikx), \ |x| \to -\infty \\ &= t_+(k)\exp(-ikx) + r_+(k)\exp(ikx), \ |x| \to \infty \end{aligned} \tag{9.9b}$$

Using Equations 9.5 and 9.6 in the Wronskian expressions in Equations 9.1 through 9.4, we obtain

$$W(\phi,\psi) = 2ik\, t_+(k) = t_-(k)W\left[\phi,\overline{\phi}\right] = 2ik\, t_-(k), \ k \subset R \setminus \{0\}$$

This yields the relation

$$t_+(k) = t_-(k) = t(k)$$

Writing these Wronskians in terms of the corresponding Jost functions $f_\pm$, we have

$$t_+(k) = t_-(k) = W[f_- \exp(-ikx),\ f_+ \exp(ikx)]/2ik$$

$$= (1/2ik)[f_+' f_- - f_-' f_+ + 2ikf_+ f_-],\ k \subset R \setminus \{0\}$$

The expression on the RHS is independent of $x \subset R$ and is well-defined for $\forall k \subset \overline{C}_+\backslash\{0\}$. Hence, the domain of definition for $t_-(k) = t_+(k) = t(k)$ can be extended from $k \subset R\backslash\{0\}$ to $k \subset \overline{C}_+\backslash\{0\}$, and one obtains the following definition.

Definition

For $k \subset \overline{C}_+\backslash\{0\}$, the coefficients $t_+(k)$ and $t_-(k)$ are defined by

$$t_+(k) = t_-(k) = t(k)$$

$$= (1/2ik)[f_+' f_- - f_-' f_+ + 2ikf_+ f_-]$$

The potential q(x) is required to satisfy a growth condition of order zero.

IV. RELATIONS BETWEEN THE COEFFICIENTS t, $\bar{t}$, r, AND $\bar{r}$

The coefficients in the expansions of the Jost solutions given in Equations 9.5 through 9.8 are related to each other in various ways. One of these relations has already been shown, namely,

$$t_+(k) = t_-(k) = t(k)$$

The Jost functions and the Jost solutions were shown in Chapter 7 to obey the following relations:

$$f_\pm^*(x, k^*) = f_\pm(x, -k^*) = \bar{f}_\pm(x, k)$$

$$\psi^*(x, k^*) = \psi(x, -k^*) = \overline{\psi}(x, k)$$

$$\phi^*(x, k^*) = \phi(x, -k^*) = \bar{\phi}(x, k)$$

$f_\pm$, ψ, and ϕ are, therefore, real functions of k on the imaginary axis, Im $k \geq 0$. Exploiting these equations and from the fact that

$$t_\pm(k) = (1/2ik)W[\phi, \psi]$$

we have for k real and $\neq 0$,

$$t_{\pm}(-k) = [t_{\pm}(k)]^* = \bar{t}_{\pm}(k) \tag{9.10}$$

and

$$r_{\pm}(-k) = [r_{\pm}(k)]^* = \bar{r}_{\pm}(k)$$

Consider Equation 9.5 in the limit $x \to \infty$ and introduce the asymptotes of ψ, $\bar{\phi}$, and ϕ given by Equations 9.9a and 9.9b in Equation 9.5. Then

$$\lim_{x\to+\infty} \psi(x,k) = t_-(k) \lim_{x\to+\infty} \bar{\phi}(x,k) + r_-(k) \lim_{x\to+\infty} \phi(x,k)$$

From this, it follows that

$$\exp(ikx) = \exp(ikx)[\bar{t}_+(k)t_-(k) + r_+(k)r_-(k)] +$$

$$+\exp(-ikx)[t_-(k)\bar{r}_+(k) + t_+(k)r_-(k)] \tag{9.11}$$

Equating the coefficients of $\exp(\pm ikx)$ on both sides of Equation 9.11 yields the identities

$$t_-(k)\bar{r}_+(k) + r_-(k)t_+(k)$$

$$= r_+^*(k)t_-(k) + r_-(k)t_+(k) = 0 \tag{9.12}$$

where Equation 9.10 has been used. This yields

$$r_-(k) = -\bar{r}_+(k) \tag{9.12a}$$

Moreover, Equation 9.11 also results in the identity

$$\bar{t}_+(k)t_-(k) + r_+(k)r_-(k)$$

$$= t_+^*(k)t(k) + r_+(k)r_-(k)$$

$$= |t(k)|^2 + r_+(k)r_-(k) = 1 \tag{9.13}$$

exploiting Equation 9.10 again and the fact that $t_+(k) = t_-(k) = t(k)$. Finally, from the compatibility between Equations 9.12, 9.12a, and 9.13, the following identities are obtained:

$$|t_+(k)|^2 = 1 - r_+(k)r_-(k) = 1 + r_+(k)r_+^*(k) = 1 + |r_+(k)|^2 \tag{9.14}$$

and

$$|t_-(k)|^2 = 1 + |r_-(k)|^2$$

V. ASYMPTOTIC BEHAVIORS

A. $|x| \to \infty$, $k \neq 0$

For $k \subset R\backslash\{0\}$, we have

$$\psi(x,k) = t_-(k)\bar{\phi}(x,k) + r_-(k)\phi(x,k)$$

and

$$\phi(x,k) = t_+(k)\bar{\psi}(x,k) + r_+(k)\psi(x,k)$$

In terms of the Jost functions, these equations can be written as

$$f_\pm(x,k)\exp(\pm ikx) = t_\mp(k)\bar{f}_\mp(x,k)\exp(\pm ikx) + r_\mp(k)f_\mp(x,k)\exp(\mp ikx)$$

Therefore,

$$f_\pm(x,k) = t_\mp(k)\bar{f}_\mp(x,k) + r_\mp(k)f_\mp(x,k)\exp(\mp 2ikx)$$

$$= t_\mp(k)f_\mp(x,-k) + r_\mp(k)f_\mp(x,k)\exp(\mp 2ikx)$$

Now the integral representations, namely,

$$f_\pm(x,k) = 1 + \int_\pm D(k,\xi_\pm)f_\pm(x',k)\, q(x')dx'$$

imply that the Jost functions satisfy the $x \to \pm\infty$ limits of the Schrödinger problem to within a term of o(1), that is,

$$\lim_{x\to\pm\infty} f_\pm(x,k) = 1 + o(1)$$

It may be recalled that $f(x) \sim o(1)$ in $\pm x$ if $f(x) \to 0$ as $x \to \pm\infty$. The potential is assumed to satisfy a growth condition of order zero. Therefore, in the nonprescribed directions

$$\lim_{x\to\mp\infty} f_\pm(x,k) = t_\mp(k) + r_\mp(k)\exp(\mp 2ikx) + o(1) \tag{9.15}$$

On the other hand, exploiting the integral representation of the Jost functions themselves, we have in the nonprescribed directions

$$\lim_{x\to\mp\infty} f_\pm(x,k) = \exp(\mp 2ikx)\left\{\left(\frac{1}{2ik}\right)\int_{-\infty}^{\infty}\exp(\pm 2ikx')q(x')f_\pm(x',k)dx'\right\} +$$

$$+\left[1-\left(\frac{1}{2ik}\right)\int_{-\infty}^{\infty}q(x')f_\pm(x',k)dx'\right] + o(1) \tag{9.16}$$

In the above equations, the o(1) term is uniform in a k-compacta $\subset R\backslash\{0\}$. Comparing Equations 9.15 and 9.16, the following integral representations of $t_\pm(x)$ and $r_\pm(k)$ are readily obtained.

$$t_{\pm}(k) = t(k) = 1 - \left(\frac{1}{2ik}\right)\int_{-\infty}^{\infty} q(x')f_{\pm}(x',k)dx' \tag{9.17}$$

and

$$r_{\mp}(k) = \left(\frac{1}{2ik}\right)\int_{-\infty}^{\infty} \exp(\pm 2ikx')q(x')f_{\pm}(x',k)dx' \tag{9.18}$$

So far, we have considered the case when $k \subset R\backslash\{0\}$. Let us extend the results to $k \subset C_+$. Consider the Jost functions $f'_{\pm}$ in the nonprescribed directions $x \rightarrow \mp\infty$. If the potential is of 0-th order growth, then from the results of Chapter 7

$$|f'_{\pm}(x,k)| = |F'_{\pm}f_{\pm}| \leq \exp[\eta(x)/|k|] \cdot \int_{\pm} |q(x')| \exp(-2\,\mathrm{Im}\,\xi_{\pm})\,dx'$$

$$\leq \exp[\eta(\infty)/|k|] \cdot \int_{\pm} |q(x')| \exp(-2\,\mathrm{Im}\,\xi_{\pm})\,dx'$$

It is not hard to show that this inequality can be rewritten as

$$|f'_{\pm}(x,k)| \leq \exp[\eta(\infty)/|k|][\int_{\pm}|q(x')|\,dx' + \exp(\pm\,\mathrm{Im}\,kx)\eta_{\pm}(x/2)$$

The limit in the first integral is between x(x/2) resp. (x/2)(x) for f'_+ resp. f'_-. Now let $x \rightarrow -\infty$ for f_+ and to $\rightarrow \infty$ for f_-. In these limits then, $|f_{\pm}'(x,k)| \rightarrow 0$ uniformly in k on any compacta $\subset C_+$. This, therefore, yields

$$\lim_{x \rightarrow \mp\infty} f'_{\pm}(x,k) = 0, \; k \subset C_+$$

the limit being uniform in $k \subset C_+$. Similar arguments will show that $f_{\pm}f_{\mp}'$ and $f_-(f_+ - 1)$ also $\rightarrow 0$ in the nonprescribed directions in k on compacta $\subset C_+$. Therefore, in the event that the k compacta $\subset C_+$, from the definition of t(k) of Section III,

$$t_{\pm}(k) = \lim_{x \rightarrow \mp\infty} f_{\pm}(x,k) \tag{9.15a}$$

If we now substitute the integral representations for the Jost functions in Equation 9.15a and apply procedures similar to those above in establishing the behavior

$$\lim_{x \rightarrow \mp\infty} f'_{\pm}(x,k) = 0$$

then it can be shown that the integral representations (9.17) also hold when $k \subset \overline{C}_+\backslash\{0\}$. In all the above equations, the potential q(x′) is required to satisfy a growth condition of at least order 0.

B. |k| → ∞

We know that

$$f_{\pm}(x,k) = 1 + \int_{\pm} D(k,\xi_{\pm})\, q(x')f_{\pm}(x',k)dx'$$

$$= 1 + (1/2ik)\int_{\pm}\{\exp(2ik\xi_{\pm}) - 1\}\, q(x')\, f_{\pm}(x',k)dx'$$

Iterating yields

$$f_{\pm}(x,k) = 1 + \int_{\pm} D(k,\xi_{\pm})q(x')dx' +$$

$$+ (1/2ik)^2 \int_{\pm} q(x')\int_{\pm x'} q(x'')dx''dx' + o\left(1/k^2\right)$$

$$= 1 + \int_{\pm} D(k,\xi_{\pm})q(x')dx' + (1/2)\{Q_{\pm}(x)/2ik\}^2 + o\left(1/k^2\right)$$

where $Q_{\pm} = \int_{\pm} q(x')dx'$, as defined earlier in Chapter 8. The potential $q(x')$ must be of first-order growth. The above expression can also be formulated as

$$\lim_{|k|\to\infty} f_{\pm}(x,k) = 1 \pm \left(\frac{1}{2ik}\right)Q_{\pm}(x) + O\left(\frac{1}{k^2}\right) \tag{9.19}$$

If the derivative $q'(x)$ exists in $L_1{}^1$, then integration by parts yields

$$f_{\pm}(x,k) = 1 - \left[q(x)/(2ik)^2\right] - (1/2ik)\, Q_{\pm}(x) + (1/2)\left[Q_{\pm}(x)/2ik\right]^2 + o\left(1/k^2\right)$$

The $|k| \to \infty$ asymptotics for $t_{\pm}$ and $r_{\pm}$ can be obtained straightforwardly. Using the integral representations of Equations 9.17 and 9.18 and the $|k| \to \infty$ asymptotics for the Jost functions given by Equation 9.19, we obtain

$$\lim_{|k|\to\infty} t_{\pm}(k) = 1 - (1/2ik)Q_{\pm}(\infty) + O\left(1/k^2\right), \quad \operatorname{Im} k \geq 0$$

and

$$\lim_{|k|\to\infty} r_{\pm}(k) = (1/2ik)\int_{-\infty}^{\infty} \exp(\pm 2ikx')q(x')dx' + O\left(1/k^2\right), \quad \operatorname{Im} k \geq 0$$

For $|k| \to \infty$, the integral containing $\exp(\pm 2ikx')$ vanishes by the Riemann-Lebesgue lemma, leading to the estimate $r_{\pm}(k) \sim o(1/k)$. Let $q(x)$ possess x-derivatives in $L_1(R)$ up to order n. Then one can integrate Equation 9.18 by parts to obtain

$$r_{\pm}(k) \sim O\left(1/k^{n+1}\right) \text{ as } |k| \to \infty$$

C. $|x| \to \infty$ AND $|k| \to \infty$

This perhaps is the most interesting case. In general, a second-order differential equation with variable coefficients cannot be solved in quadrature, that is, the solution cannot be obtained in an explicit integral form. But it is found that for $|x| \to \infty$ and large k, an asymptotic solution can be obtained for the Schrödinger equation in terms of the potential $q(x)$ and its derivatives. We assume that the potential is such that all its derivatives exist. Second, the asymptotic solution provides an infinite sequence of conserved quantities that prove to be of substantial importance in solving nonlinear evolution equations. The method is described by Verde,[4] Calogero and Degasperis,[5] and Lamb.[6] We follow Lamb here.

Let us consider the fundamental solution $\phi(x,k)$ and assume that its asymptotic expansion is of the Rytov[7] form, that is, can be written as

$$\phi(x,k) = \exp\left[-ikx + h(x,k)\right] \tag{9.20}$$

$h(x,k)$ must satisfy the condition that $h(-\infty,k) = 0$ by virtue of the integral representation of $\phi(x,k)$. Replacing Equation 9.20 in the Schrödinger equation, one obtains

$$h'' - 2ikh' + h'^2 = q \tag{9.21}$$

which is a Riccati equation in h'. Let h' be given by the series

$$h'(x,k) = \sum_{n=0}^{\infty} \frac{g_n(x)}{(2ik)^{n+1}} \tag{9.22}$$

Equation 9.22 solves Equation 9.21 in the sense that

$$\lim_{k\to\infty}\left[h'(x,k) - \sum_{n=0}^{N} g_n(x)/(2ik)^n\right]k^N = 0$$

Substituting the series (9.22) into the Riccati equation (9.21) and setting the coefficients of the powers of k to zero, we obtain

$$g_0 = -q,\ g_1 = g_0' = -q'$$

and

$$g_{n+1} = g_n' + \sum_{m=0}^{n-1} g_m g_{n-m-1},\ n \geq 1$$

The first few coefficients are

$$g_0 = -q,\ g_1 = -q',\ g_2 = -q'' + q^2,\ g3 = -q''' + 4qq',$$

$$g_4 = -q'''' + 5(q')^2 + 6qq'' - 2q^3,\ \text{etc.} \tag{9.23}$$

From Equation 9.15a,

$$\lim_{x\to\infty} f_-(x,k) \to t(k),\ \ \operatorname{Im} k > 0 \tag{9.24}$$

Combining Equation 9.24 and 9.20, we have $h(\infty,k) = \ln t(k)$. Since $h(-\infty) = 0$, the integration of Equation 9.22 with respect to x from $-\infty$ to ∞ yields

$$h(\infty,k) = \ln t(k) = \sum_{n=0}^{\infty} \frac{1}{(2ik)^{n+1}} \int_{-\infty}^{\infty} dx\, g_n(x) = \sum_{n=0}^{\infty} \frac{c_n}{k^n} \tag{9.25}$$

For large k such that $k^2 \geq q$, the wave energy is transmitted beyond the scatterer. Thus, $\lim k \to \infty$ $t(k) \to 1$, leading to the fact that $\ln t_+(k) \to 0$. The second equality in Equation 9.25 follows from the fact that $k \to \infty$, $t(k) \to 1 \Rightarrow \ln t(k) \to 0$ and the corresponding expansion for the logarithm. Comparing coefficients in Equation 9.25 gives

$$c_n = 1/(2i)^n \int_{-\infty}^{\infty} dx\, g_{n-1}(x),\ n = 1,2,3,\ldots$$

an extremely useful result in solving evolution equations.

Let us assume that the potential is "slowly varying". By this is meant that the variation of q(x) over a wavelength, that is, in a distance k^{-1} is small. It may be recalled that under the slowly varying assumption on the potential, the Schrödinger equation is solved by the WKB method,[8,9] which assumes the solution to be of the form exp(iθ), where

$$\theta = \int_{-} dx' \left(k^2 - q\right)^{\frac{1}{2}}$$

If we choose the Rytov function h to be of this form, then the WKB solution can be written as

$$\theta(x,k) = \exp\left[+i \int_{-} dx' \left(k^2 - q\right)^{1/2} - ikx\right]$$

If this WKB solution is now substituted in Equation 9.20, then the successive approximation of the series (9.22) generates the g coefficients in such a manner that $g_{2k+1} = 0$, k = 0, 1, 2 ..., whereas the even-order terms are still those given by Equation 9.23 if all the derivatives of q are neglected, that is, $g_0 = -q$, $g_2 = q^2$, $g_4 = -2q^3$, etc. Hence, the WKB solution can be recovered from the formalism presented.

On the other hand, if the potential is weak, that is, $q << k^2$, then the equation is solved in the Born approximation.[10,11] In Born approximation, the quantity q(x)ϕ(x,k) is considered to be a perturbation which is then treated as an inhomogeneous term in the Schrödinger equation. The solution is obtained by assuming ϕ(x,k) to be exp(–ikx) and then successively approximating the corresponding integral representation as given in Equation 7.5. Then repeated partial integration of [ln ϕ(x,k)]′ results in the coefficients $g_n(x)$ such that only those terms in Equation 9.23 are retained that are linear in q and its derivatives. That is,

$$g_0 = -q,\ g_1 = -q',\ g_2 = -q'',\ g_3 = -q''',\ g_4 = -q'''',\ \text{etc.}$$

D. BEHAVIOR AS k → 0

Integral Equations 9.17 and 9.18 are singular as k approaches the origin. However, the coefficients t(k) and r(k) can be made to behave regularly at the origin if a second-order growth condition is imposed upon the potential. This is analogous to the behavior of the Jost functions k → 0 established in Chapter 7.

Consider $t_\pm(k)$ first. Define a quantity $m_\pm$ as

$$m_\pm(k) = \int_{-\infty}^{\infty} q(x')f_\pm(x',k)dx'$$

Suppose that $m_\pm(0) = 0$. Then Equation 9.17 can be written as

$$t_\pm(k) = 1 - (1/2ik)\int_{-\infty}^{\infty} q(x')\left[f_\pm(x',k) - f_\pm(x',0)\right]dx'$$

Now

$$\lim_{k\to 0}(1/2ik)\left[f_{\pm}(x',k)-f_{\pm}(x',0)\right]=(1/2i)\,\dot{f}_{\pm}(x',k)\big|_{k=0}$$

From Chapter 7 (Theorem 7.5), we know that $|\dot{f}_{\pm}(x,k)| \leq$ constant $(1 + x^2)$ $\forall k$ if $q \subset L^1_2$. Hence, limit $k \to 0$ $t_{\pm}(k)$ exists and is continuous. From the consistency condition given by Equation 9.14, we have $|t_{\pm}(k)| \geq 1$, $k \subset R\backslash\{0\}$. Moreover, the Wronskian defining $t_{\pm}(k)$ is continuous on $R \times \overline{C}_+$ since the Jost solutions involved in the Wronskian, namely, ϕ and ψ, are continuous there. Thus $t_{\pm}(k)$ is continuous at the origin and is not zero. Limit $k \to 0$ $t_{\pm}(k)$, therefore, exists and is finite. In the event that $m_{\pm}(0) \neq 0$, we have

$$t_{\pm}(k)=1-\left[m_{\pm}(0)/2ik\right]-(1/2ik)\int_{-\infty}^{\infty}q(x')\left[f_{\pm}(x'k)-f_{\pm}(x',0)\right]dx'$$

$$=1-(1/2i)\int_{-\infty}^{\infty}q(x')\,\dot{f}_{\pm}(x',0)dx'-\left[m_{\pm}(0)/2ik\right],\ k\to 0$$

In this case, $t_{\pm}(k)$ is not continuous at $k = 0$.

We have considered two conditions above, namely, $m_{\pm}(0) = 0$ and $m_{\pm}(0) \neq 0$. These conditions have certain physical interpretations as discussed below. Consider the Wronskian $W[\psi(x,0),\phi(x,0)]$:

$$W\left[\tilde{\phi}(x,0),\tilde{\psi}(x,0)\right]=\tilde{\psi}'(x,0)\tilde{\phi}(x,0)-\tilde{\phi}'(x,0)\tilde{\psi}(x,0)$$

$$=\tilde{\psi}'(x)\tilde{\phi}(x,0)-\tilde{\phi}'(x)\tilde{\psi}(x,0)$$

where the alternate Jost solutions $\tilde{\psi}(x)$ and $\tilde{\phi}(x)$ were given in Chapter 7 (Equation 7.17). Evaluating the Wronskian either at $x = +\infty$ or $x = -\infty$, we obtain

$$W\left[\tilde{\phi}(x,0),\tilde{\psi}(x,0)\right]=\int_{-\infty}^{\infty}q(x')\tilde{\psi}(x',0)dx'$$

$$=\int_{-\infty}^{\infty}q(x')\tilde{\phi}(x',0)dx'=0$$

Therefore, $m_{\pm}(0) = W[\tilde{\phi}(x,0),\tilde{\psi}(x,0)$. Hence, the condition $m_{\pm}(0)$ corresponds to $\tilde{\psi}(x,0)$ and $\tilde{\phi}(x,0)$ being linearly dependent. Nonzero $m_{\pm}(0)$ then implies linear independence of either the Jost solutions or the Jost functions. The condition $m_{\pm}(0) = 0$ is highly unstable, occurs rarely, and only when another eigenvalue is to be added to the spectrum of the Schrödinger operator. This point is discussed by P. Deift and E. Trubowitz[31] whose discussion also contains the results on the asymptotics. More on this will be said later.

VI. ANALYTICITY AND THE BOUND STATES

A. THE ZEROS OF t(k)

In this section, we discuss the analyticity properties of the coefficients t, r, $\bar{t}$, and $\bar{r}$. The zeros of the t functions in the complex k-plane are most important. They determine the bound states and the point spectrum of the Schrödinger operator and enter into its closure relation. The locations, number, and order of these zeros are, therefore, important considerations.

From the consistency relation (9.14) which is valid for $k \subset R$, it follows that t(k) cannot have zeros on the real axis. The zeros, if any, must be off the real axis. Their locations can be determined as follows. Let $k_0 = u + iv$, $v > 0$, $k_0 \subset C_+$, be a zero of t(k) and $y(x,k_0)$ be a fundamental solution. Then

$$y'' + k_0^2 y = qy$$

and

$$y^{*''} + k_0^{*2} y^* = qy^*$$

where * is the complex conjugate.

$$W\left[y^*(x,k_0), y(x,k_0)\right]\Big|_{-\infty}^{\infty} = \left(k_0^{*2} - k_0^2\right)\left\|y(x,k_0)\right\|_2^2 = 0 = uv\|y\|_2^2 \tag{9.26}$$

Equation 9.26 has been obtained after multiplying the y equation by y^*, the y^* equation by y, subtracting and then integrating in x from $-\infty$ to $+\infty$, and using the fact that the Wronskian of two linearly independent solutions of the Schrödinger equation is independent of x and may depend on k only. Since $\|y\|^2_2$ is finite positive and $v > 0$, u must be zero. We, therefore, conclude that $u = \text{Re } k = 0$, and hence all zeros must lie on the positive imaginary axis. From Equation 9.10, $\bar{t}(k) = t^*(k^*)$. It thus follows that the zeros of $\bar{t}(k)$ are the complex conjugates of those of t(k), $k \subset \bar{C}_+$

Finally, from Equations 9.5 through 9.8, we note that when t(k) vanishes, the Jost solutions become linearly dependent:

$$\psi(x,k) = r_-(k)\phi(x,k) \qquad \text{Im } k \ge 0$$

and

$$\phi(x,k) = r_+(k)\psi(x,k)$$

whereas

$$\bar{\psi}(x,k) = \bar{r}_-(k)\bar{\phi}(x,k) \qquad \text{Im } k \le 0$$

and

$$\bar{\phi}(x,k) = \bar{r}_+(k)\bar{\psi}(x,k)$$

At the zeros of t(k), that is, at $k = ik_j$, $k_j > 0$, the eigenvalue $\lambda = k^2 = -k_j^2$. The eigenvalues are, therefore, strictly negative. The Schrödinger equation thus becomes an eigenvalue equation, namely,

$$\mathbf{L}y = \lambda y = -k_j^2 y, \ k_j > 0$$

At the zeros of t(k), both ψ and ϕ for Im $k \ge 0$ and $\bar{\psi}$, $\bar{\phi}$ for Im $k \le 0$ decrease exponentially on the entire real line as x tends to infinity on both sides. This is nontrivial. In the prescribed direction, the decrease is simply by the integral representations of these functions as in Chapter 7. In the nonprescribed direction, the asymptotic values of the functions are just the transmission coefficients (Section V.A) which are zero. These functions then $\subset L_2(R)$ and, by definition, are the eigenfunctions of the Schrödinger operator for the bound states. Thus, whenever the eigenvalues $\lambda \subset R^-$, the coefficients t(k) and $\bar{t}(k)$ have zeros at $\pm ik_j$, $k_j > 0$, respectively. The eigenfunctions corresponding to a single λ are nondegenerate since $y(x,+i\mu) = y(x,-i\mu) = \{y(x,i\mu^*)\}^*$.

An important question is concerning the origin, that is, is $\lambda = k^2 = 0$ an eigenvalue? If $q(x) \subset L_2^{\,1}(R)$, then the answer is negative. This can be seen as follows. From Chapter 7 we know that for $q(x) \subset L_2^{\,1}(R)$, the Jost solutions are linear in $|x|$ as $|x| \to \infty$ if $k = 0$, and hence are not in $L_2(R)$. Thus, $\lambda = k^2 = 0$ is not an eigenvalue. $k^2 = 0$, which is not an eigenvalue, gives rise to what is known as a "virtual level" of the Dirichlet operator $(-D^2 + q(x))$. More will be said on this later.

B. THE NUMBER OF ZEROS: THE UPPER AND THE LOWER BOUNDS

The locations of the zeros of t, $\bar{t}$ having been determined, the next question is: how many zeros can there be? Is their number finite or infinite? Let us note that these functions are analytic in their respective half-planes, i.e., $k \subset C_+$ for t(k) and $k \subset \overline{C}_-$ for $\bar{t}(k)$. This follows from the analyticity of the Jost solutions in these planes and their direct relations to the coefficients. The immediate consequence is that the zeros of t, $\bar{t}$ must be isolated.[12] The number of these zeros, as will be seen momentarily, are finite in number and are bounded from above as well as from below. We will follow Calogero and Degasperis[5] here and take as our starting point the well-established result[13] that the number of discrete eigenvalues of a Dirichlet operator is intimately connected to the number of zeros of the wave function in which $k = 0$. This is generalized in the lemma below. An elegant discussion of the zero of Schrödinger operator is given by M. Reed.[32]

Lemma 9.1

The number of discrete eigenvalues of the Schrödinger operator coincides with the number of (real) zeros z_j of the function y(x) defined by the equation

$$y''(x) = q(x)y(x) \tag{9.27}$$

with the boundary conditions

$$y(-\infty) = 1, \; y'(-\infty) = 0 \tag{9.28}$$

namely,

$$y(z_j) = 0, \; z_j < z_j + 1, \; j = 1,2,\ldots,N \tag{9.29}$$

Consider the function g(x) defined by

$$[\ln y(x)]' = -n^{-1} \tan g(x)$$

where n is an arbitrary positive constant. Then from Equations 9.27 through 9.29, the following boundary value problem is obtained for g(x):

$$g'(x) = -n\, q(x)\cos^2 g(x) + n^{-1} \sin^2 g(x)$$

$$g(-\infty) = 0 \tag{9.30}$$

Moreover, Equation 9.29 yields

$$g(z_j) = \left(j - \frac{1}{2}\right)\pi \tag{9.31}$$

At $z = z_j$, $y(z) = y(z_j) = 0$ and it follows from Equation 9.30 that

$$g'(x = z_j) = n^{-1} > 0$$

Combined with Equation 9.31, this implies the same ordering of the zeros as given by Equation 9.29. That is, $g(z_j)$s are in 1:1 correspondence with the zeros of y's.

As $x \to \infty$, the potential $q(x) \to 0$. Equation 9.30 implies that $g(x \to \infty) = N\pi$, N being an integer. The following lemma is thus established.

Lemma 9.2

Let g(x) be a function defined by

$$[\ln g(x)]' = -n^{-1} \tan g(x)$$

n being an arbitrary positive constant. g(x) satisfies the following boundary-value problem:

$$g'(x) = -n\, q(x) \cos^2 g(x) + n^{-1} \sin^2 g(x)$$

$$g(-\infty) = 0$$

The number of discrete eigenvalues N of the Schrödinger problem is given by

$$N = \left(\frac{1}{\pi}\right) g(+\infty) \tag{9.32}$$

We intend to show that N is bounded from above as well as from below. The lower bound will be established first. Let $q(x) = 0$ for $x > x_\infty$. Then Equation 9.31 implies that if $m\pi < g(x_\infty) \le (m + 1)\pi$, then $g(\infty) = (m + 1)\pi$. From Equation 9.32 it follows that

$$N = \text{integral part of } \left[1 + g(x_\infty)/\pi\right] \tag{9.33}$$

On the other hand, from Equation 9.30 we obtain

$$g(x_\infty) \ge -\int_{-\infty}^{x_\infty} \min\left[n^{-1}, -nq(x)\right] dx$$

since the minimum value of the function ($a \cos^2 x + b \sin^2 x$) over all real values of x is min[a,b], which is a if $a \le b$ and is b if $a \ge b$. Equivalently,

$$g(x_\infty) \ge \int_{-\infty}^{\infty} \min\left[n^{-1}, -nq(x)\right] dx$$

since min[a,0] = 0 if a is positive. The following theorem is then established regarding the lower bound of the number of discrete eigenvalues of the Schrödinger operator.

Theorem 9.2. The Lower Bound

The number N of discrete eigenvalues of the Schrödinger equation is bounded below by the formula

$$N \geq \text{integral part of} \left[1 + \pi^{-1} \int_{-\infty}^{\infty} \min\left[n^{-1}, -nq(x)\right] dx\right] \tag{9.34}$$

where n is a positive, but otherwise arbitrary, constant. The minimum of a and b is defined by

$$\min[a, b] = a \text{ if } a \leq b$$

$$= b \text{ if } a \geq b$$

Theorem 9.2 holds for potentials which do not vanish identically over the real axis. If $q(x) \equiv 0$, $x \subset R$, then, of course, N = 0. A consequence of Equation 9.34 is that if the potential $q \leq 0$, but not vanishing identically over R, then $N \geq 1$. This follows from the fact that for $q \leq 0$, $-nq(x) \geq 0$, since n is positive. Thus $\min[n^{-1}, -nq] \geq 0$ according to the definition of the minimum in the theorem. If, on the other hand, $q(x) \geq 0$, $x \subset R$, then $\min[n^{-1}, -nq] = -nq$, yielding, according to Equation 9.34, $N \leq 1$, $q \neq 0$. However, $N \equiv 0$ if q = 0. Hence, it follows that $N \equiv 0$ if $q(x) \geq 0$ and no discrete spectrum exists. This is also a consequence of the comparison theorem for a Sturm-Liouville problem, which is given in the lemma below.

Lemma 9.3

Let $N^{(1)}$ and $N^{(2)}$ be the number of eigenvalues or two potentials $q^{(1)}$ and $q^{(2)}$, respectively. Then

$$N^{(1)} \geq N^{(2)}$$

if

$$q^{(1)} \leq q^{(2)}$$

Choosing $q^{(1)} = 0$, it follows from Lemma 9.4 that N = 0 if $q \geq 0$. A generalization of this case is that if

$$q(x) \geq -Q^2, \ x \subset R, \ Q > 0$$

then choosing $n = Q^{-1}$ and from Equation 9.34:

$$N \geq \text{integral part of} \left[1 - (\pi Q)^{-1} \int_{-\infty}^{\infty} q(x) dx\right]$$

We have used the fact that

$$\min\left[n^{-1}, -nQ^2\right] = Q = -\left(-Q^2\right)/Q = -q/Q$$

The number of discrete eigenvalues is thus a measure (global) of the negativeness of the potential.

Let us now consider the upper bound. This involves moments of the potential larger than the zeroth one, that is, integrals of the form

$$I_i = \int_{-\infty}^{\infty} |x - m|^i |q(x)| dx, \ i = 0, 1, 2, \ldots \tag{9.35}$$

where m is another arbitrary (real) constant. Let us define a function $\mu(x)$ given by

$$\cot \mu(x) = [(x - m)/n] + \cot \eta(x)$$

In terms of the function $\mu(x)$, the boundary-value problem of Lemma 9.2 is rewritten as Lemma 9.4 below.

Lemma 9.4

The function $\mu(x)$ defined by

$$\cot \mu(x) = [(x - m)/n] + \cot \eta(x)$$

satisfies the following boundary value problem:

$$\mu'(x) = -yq(x)\left\{\left[\frac{(x - m)}{n}\right]\sin \mu(x) - \cos \mu(x)\right\}^2 \tag{9.36}$$

$$\mu(-\infty) = 0 \tag{9.37}$$

n (> 0) and m are arbitrary constants. The number of discrete eigenvalues is given by

$$N = \min[\text{integral part of } [1 + \mu(\infty)/\pi], 0]$$

Since the bound states are being considered, it is sufficient to work with only the nonpositive part of the potential, that is, with

$$q_-(x) = q(x) \text{ whenever } q(x) < 0$$

$$= 0 \text{ whenever } q(x) \geq 0$$

Defining

$$\cot p(x) = (m - x)/n$$

Equation 9.36 reduces to

$$\mu'(x) = |q_-(x)|\left[n + \frac{(x - m)^2}{n}\right]\sin^2[\mu(x) + p(x)] \tag{9.38}$$

From Equations 9.37 and 9.38 and using Equation 9.35, one obtains the inequality

$$\mu(\infty) \leq nI_0 + n^{-1}I_2$$

Choosing $n = (I_2/I_0)^{1/2}$, the upper bound is determined to be

$$N \leq 1 + (2/\pi)(I_2/I_0)^{\frac{1}{2}}$$

Theorem 9.3

The upper bound of the discrete eigenvalues of the Schrödinger equation is given by

$$N \leq 1 + (2/\pi)(I_2/I_0)^{\frac{1}{2}}$$

where I_m is the integral expression

$$I_i = \int_{-\infty}^{\infty} |x - m|^i |q(x)| dx, \; i = 0, \; 1, \; 2, \ldots$$

In the literature, the more familiar expression for the upper bound is given by

$$N \leq 1 + I_1$$

A heuristic derivation of such a bound can be found in Eckhaus and van Harten.[14]

The eigenvalues $\lambda = k^2$ are real, being strictly negative numbers. This reality of the eigenvalues λ implies the hermiticity of the Schrödinger operator $-d^2/dx^2 + q(x)$ [q(x) real]. The negativity of the discrete eigenvalues is necessary if the bound state wavefunction are to be square integrable on R. If k is real then the Jost solutions oscillate asymptotically and cannot, therefore, be square integrable. The finiteness of the number of discrete eigenvalues N is implied by the assumed asymptotic vanishing of q(x), namely, that q(x) must vanish faster than x^{-2}. If, on the other hand, q(x) vanishes only as $|x|^{-1}$ as $|x| \to \infty$, then there will be an infinite number of discrete eigenvalues $-k^2$, accumulating at $k = 0$. It follows from the comparison theorem (Lemma 9.3) that the number N(g) of discrete eigenvalues of a potential $Q(x) = g^2q(x)$ grows, as $g \to \infty$, proportionally to g unless Q(x) is nonnegative everywhere.

C. MULTIPLICITY OF ZEROS

It will be shown momentarily that the k-derivative of the coefficients t(k) at the discrete eigenvalues is related to the normalization constants of the bound-state wavefunctions. Hence, in order for the bound states to exist, the first-order k-derivatives must not vanish. We show in this section that this, indeed, is the case. By definition,[12] this then means that the zeros of t(k) are simple, that is, of multiplicity zero. In other words, the zeros are simple.

Let us consider the Schrödinger problem for the Jost functions $f_\pm(x,k)$, namely,

$$f''_\pm(x,k) \pm 2ik\, f'_\pm(x,k) - q(x) f_\pm(x,k) = 0$$

$$\lim_{x \to \pm\infty} f_\pm(x,k) = 1$$

and

$$\lim_{x \to \pm\infty} f'_\pm(x,k) = 0, \; k \subset \overline{C}_+$$

These limits are uniform in $k \subset \overline{C}_+$ if the potential q(x) satisfies a growth condition of order 1. On the other hand, if q(x) is of growth of zeroth order, then these limits are uniform in k $\subset \overline{C}_+\backslash\{0\}$. The k-derivatives $\dot{t}_\pm(k)$ will be considered at the zeros. Hence, all k in the derivations below will be in C_+.

It was pointed out in Section V.A that

$$t_\pm(k) = t(k) = \lim_{x\to\mp\infty} f_\pm(x,k)$$

the limits in the nonprescribed directions being uniform in k on any compacta $\subset C_+$, the potential satisfying a zero-order growth condition. Now $f_\pm(x,k)$ are analytic in $k \subset C_+$ and converge to $t_\pm(k)$ uniformly in k as $x \to \mp\infty$. This allows us to write[15]

$$\dot{t}(k) = \lim_{x\to\mp\infty} \dot{f}_\pm(x,k) \qquad \cdot = \partial/\partial k$$

Similar reasoning and the boundary conditions of the Schrödinger problems above will also permit us to write

$$\lim_{x\to\pm\infty} \dot{f}_\pm(x,k) = \lim_{x\to\pm\infty} \dot{f}'_\pm(x,k) = 0$$

In the nonprescribed direction, the results of Section V.A show that

$$\lim_{x\to\mp\infty} f'_\pm(x,k) = 0, \; k \subset C_+$$

Moreover, using the asymptotic forms of Equations 9.9a and 9.9b, it is also seen that

$$\lim_{x\to\mp\infty} \dot{f}'_\pm = 0$$

Let us now consider the function

$$w_\pm = W\left[f_\pm(x,k), \dot{f}_\pm(x,k)\right]\exp(\pm 2ikx) \tag{9.40}$$

It is not difficult to show that $w_\pm$ satisfies the following boundary-value problem:

$$w'_\pm = \mp 2if'_\pm f_\pm \exp(\pm 2ikx)$$

$$\lim_{x\to\pm\infty} w_\pm(x,k) = 0$$

Integrating w′ between x and $\pm\infty$ yields

$$w_\pm(x,k) = iy_\pm^2\Big|_\pm - 2k\textstyle\int_\pm y_\pm^2(x',k)dx' \tag{9.41}$$

where $y_+(x,k)$ resp. $y_-(x,k)$ are $\psi(x,k)$ resp. $\phi(x,k)$ and $|_\pm$ denotes limits between $(x, +\infty)$ and $(-\infty, x)$, respectively.

We are considering the bound states, that is, the eigenfunctions at $k = ik_j$, $k_j > 0$, j = 1, 2, ... N. Let us write $y_{\pm j}$ for $y_\pm(x, k = ik_j)$. Then

$$y_{\pm j} = c_{\pm j} y_{\mp j},\ c_{\pm j} = c_{\pm}(k = ikj) \tag{9.42}$$

where $c_{\pm j}$ designates the proportionality constants between the $y_{\pm j}$'s since they are linearly dependent at the zeros of t(k). Recalling that

$$y_{\pm} = f_{\pm} \exp(\pm ikx)$$

limit $x \rightarrow \pm \infty$: $f_{\pm} = 1$ and $f_{\pm}' = 0$, $k \subset C_{+}$, we have from Equation 9.42

$$\lim_{x \to \mp\infty} f_{\pm} \exp(\pm 2ikx) = c_{\pm j} \tag{9.43}$$

and

$$\lim_{x \to \mp\infty} f_{\pm}' \exp(\pm 2ikx) = \mp 2ik\ c_{\pm j} \tag{9.44}$$

In the above equations, $k = ik_j$. Using the behaviors of $f_{\pm}$ in the nonprescribed directions and using Equations 9.43 and 9.44 in Equation 9.40, we obtain

$$\lim_{x \to \mp\infty} w_{\pm}(x,k) = \pm 2ik\ \dot{t}_{\pm j} c_{\pm j}$$

$$k = ik_j \tag{9.45}$$

Equation 9.41, on the other hand, yields

$$\lim_{x \to \mp\infty} w_{\pm}(x,k) = -2k \left\| y_{\pm j} \right\|_2^2$$

$$k = ik_j \tag{9.46}$$

Comparing Equations 9.45 and 9.46, we finally obtain

$$\left\| y_{\pm j} \right\|_2^2 = \mp i c_{\pm j} \dot{t}_{\pm j}$$

or

$$\dot{t}_{\pm j} = \mp \left(i c_{\pm j} \right)^{-1} \left\| y_{\pm j} \right\|_2^2 \tag{9.47}$$

The quantities $(ic_{\pm j}\dot{t}_{\pm j})$ will also be written as $D_{\pm j}^{-1}$. $D_{\pm j}$ are then the normalization constants for the bound states of ψ and ϕ. It also follows from Equation 9.42 that $c_{-j} = c_{+j}^{-1}$. Therefore,

$$\left\langle \psi_j, \psi_j \right\rangle = \left\| \psi_j \right\|^2 = D_{+j}^{-1} = -ic_{+j}\dot{t}_{+j}$$

and

$$\langle \phi_j, \phi_j \rangle = \|\phi_j\|^2 = D_{-j}^{-1} = ic_{+j}^{-1}\dot{t}_{+j}$$

We note that $r_{\pm}(k)$ cannot vanish at $k = ik_j$ if the asymptotic boundary conditions for the Jost solutions are to be satisfied. Thus, $r_{\pm}(k) \neq 0$, $k^2 \subset R^-$. Also, for these k^2's, the $y_{\pm j}$'s are real for a real q(x) simply by virtue of their integral representations. Hence, $\|y_{\pm j}\|^2{}_2$ is real, positive, and nonzero. Hence, $D_{\pm j}$ must also be real and positive, $k^2 \subset \sigma(L)$. Therefore, as long as the potential q(x) remains real, $\dot{t}_{\pm}(k)$ cannot vanish. This directly implies that the zeros of $t_{\pm}(k)$ are simple.

VII. THE SCATTERING MATRIX S

In the scattering processes described by Equations 9.5 and 9.6, the terms $t_-(k)\overline{\phi}(x,k)$ and $t_+(k)\overline{\psi}(x,k)$ can be considered to be the incident waves giving rise to the transmitted waves $\exp(\pm ikx)$ and the reflected waves $r_-(k)\phi(x,k)$ and $r_+(k)\psi(x,k)$, respectively. Similar considerations apply to the scattering involving $\overline{\psi}$ and $\overline{\phi}$ processes. It is customary to normalize the incident waves to unity amplitudes and consider instead, the following equations

$$T_{\mp}(k)\, y_{\pm}(k) = \overline{y}_{\mp}(k) + R_{\mp}(k)\, y_{\mp}(x,k) \tag{9.48}$$

Similarly, for $\overline{y}_{\pm}$ scattering. As before, y_+ resp. y_- refer to ψ resp. ϕ. The following definitions, therefore, hold for $T_{\pm}(k)$, $R_{\pm}(k)$:

$$T_{\pm}(k) = 1/t_{\pm}(k)$$

and

$$R_{\pm}(k) = r_{\pm}(k)/t_{\pm}(k)$$

The asymptotic conditions for Equation 9.48 are

$$\begin{aligned} T_{\mp}(k)\, y_{\pm}(x,k) &\to T_{\mp}(k)\exp(\pm ikx) && x \to \pm\infty \\ &\to \exp(\pm ikx) + R_{\mp}(k)\exp(\mp ikx) && x \to \mp\infty \end{aligned} \tag{9.49}$$

These equations identify $\mp T(k)\, y_{\pm}(x,k)$ as solutions to the Schrödinger equation

$$y'' + \left[k^2 - q(x)\right]y = 0, \; x \subset R$$

to within a multiplicative constant $T_{\pm}$, which is fixed by the requirement that a wave incident from either the left or the right has an amplitude of unity.

Equation 9.49 can be expressed in terms of the Jost solutions as

$$T_{\mp}(k)\, y_{\pm}(x,k) = T_{\mp}(k)\exp(\pm ikx) + \int_{\pm} D(k,\xi_{\pm})q(x')\{T_{\mp}(k)y_{\pm}(x',k)\}dx'$$

$$= \exp(\pm ikx) + R_{\mp}(k)\exp(\mp ikx) +$$

$$+\int_{\pm} D(k,\xi_{\pm})q(x')\{T_{\mp}(k)y_{\pm}(x',k)\}dx' \tag{9.50}$$

In terms of the Jost functions $f_{\pm}(x,k)$, the scattering Equation(s) 9.48 are

$$T_{\mp}(k)f_{\pm}(x,k) = \bar{f}_{\mp}(x,k) + R_{\mp}(k)f_{\mp}(k)\exp(\mp 2ikx),\ k \neq 0$$

For obvious reasons, $T_{\mp}(k)$ and $R_{\mp}(k)$ are called the left (right) "transmission" and left (right) "reflection" coefficients for amplitudes, respectively. That is, the subscripts – and + correspond to the scattering of a plane wave of unit amplitude incident from the left and from the right, respectively.

We now define a matrix **S** with $T_{\pm}(k)$ and $R_{\pm}(k)$ as its elements such that

$$S(k) = \begin{pmatrix} T_+(k) & R_-(k) \\ R_+(k) & T_-(k) \end{pmatrix} \text{ for } k \subset R \setminus \{0\},\ q(x) \subset L_1^1$$

The matrix **S**(k) thus defined is called the "scattering matrix" and plays an all-important role in solving inverse problems.

From the relations between the coefficients $t_{\pm}$ and $r_{\pm}$, the following relation between $T_{\pm}(k)$ and $R_{\pm}(k)$ can be written down at once.

I. $\quad T_+(k) = T_-(k) = T(k)$: (symmetry)

This expresses the fact that the Schrödinger equation is invariant under time reversal.

II. $\quad |T_+(k)|^2 + |R_+(k)|^2 = |T_-(k)|^2 + |R_-(k)|^2 = 1$: (unitarity)

This expresses the conservation of probabilities or the number of particles for a real potential. This is, therefore, a statement of conservation of energy.

III. $\quad R_-(k)T_+^*(k) + R_+^*(k)T_-(k) = 0$

IV. $\quad T_{\pm}(k),\ R_{\pm}(k) < 1$

This is a consequence of unitarity. Property III is the reality condition for the coefficients, provided that the potential is real.

V. $\quad T_{\pm}^*(k) = T_{\pm}(-k) = T(-k)$: (reality)

$$R_{\pm}^*(k) = R_{\pm}(-k)$$

This implies that the moduli of $T_{\pm}$ and $R_{\pm}$ are even functions of k

$$|T_{\pm}(-k)| = |T_{\pm}(k)|$$

$$|R_{\pm}(-k)| = |R_{\pm}(k)|$$

whereas their phases $\alpha_{\pm}(k)$ and $\beta_{\pm}(k)$, defined by

$$T_{\pm}(k)=|T_{\pm}(k)|\exp[i\alpha_{\pm}(k)]$$

$$R_{\pm}(k)=|R_{\pm}(k)|\exp[i\beta_{\pm}(k)]$$

are instead odd in k

$$\alpha_{\pm}(-k)=-\alpha_{\pm}(k)$$

$$\beta_{\pm}(-k)=-\beta_{\pm}(k)$$

The phases are defined to only mod(2π).

In the discussion that follows, q(x) will be assumed to be $L^1{}_1$. The potential will be required to satisfy a stronger growth condition of $L^1{}_2$ only when it is necessary to prove the continuity of the **S** matrix at k = 0. These relations are certainly valid for $k \subset R\backslash\{0\}$ and generally have continuations off the real axis.

From Equations 9.17 and 9.18, the integral representations for $T_{\pm}(k)$ and $R_{\pm}(k)$ can also be written down directly:

$$\left(1/T_{\pm}(k)\right)=1-(1/2ik)\int_{-\infty}^{\infty}q(x')f_{\pm}(x',k)dx' \tag{9.51}$$

and

$$\left[R_{\pm}(k)/T_{\pm}(k)\right]=\left(\frac{1}{2ik}\right)\int_{-\infty}^{\infty}q(x')f_{\mp}(x',k)\exp(\mp 2ikx')dx' \tag{9.52}$$

VIII. ASYMPTOTICS

Again, as in the case of the coefficients t and r,

$$1/T_{\pm}(k)=1-(1/2ik)Q_{\pm}(\infty)+O\left(1/k^2\right) \text{ as } |k|\to\infty$$

Since $Q_{\pm}(\infty)$ is finite, then in the limit $|k| \to \infty$, we can write

$$\begin{aligned} T_{\pm}(k)&=1+\left\{\frac{Q_{\pm}(\infty)}{2ik}\right\}+O\left(\frac{1}{k^2}\right)\\ &=1+O\left(\frac{1}{k}\right),\ |k|\to\infty,\ \operatorname{Im}k\ge 0 \end{aligned} \tag{9.53}$$

For the reflection coefficients, we have analagous expressions.

$$R_{\pm}(k)=\left\{T_{\pm}(k)/2ik\right\}\int_{-\infty}^{\infty}q(x')\exp(\pm 2ikx')dx'+O\left(1/k^2\right),\ k \text{ real}$$

The order term $O(1/k^2)$ results from the fact that the Jost functions $f_{\pm} \to 1 + O(1/k)$ as $|k| \to \infty$. From Equation 9.53 for $T_{\pm}(k)$, $R_{\pm}(k) \sim O(1/k)$. However, due to the Riemann-Lebesgue theorem, $R_{\pm}(k) \sim o(1/k)$. Moreover, exactly as in the case of the coefficients t and r, if q(x) has N derivatives $\subset L_1(R)$, then $R_{\pm}(k) \sim O(1/k^{N+1})$. k is, of course, real when considering reflection coefficients. Extension of $R_{\pm}(k)$ off the real axis will be discussed later.

In addition, following procedures similar to those of Section V.B, we also obtain

$$\underset{|k|\to\infty}{T_{\pm}(k)} = 1 + (1/2ik)Q_{\pm}(\infty) + [Q_{\pm}(\infty)/(2ik)]^2/2 + o(1/k^2) \quad q \subset L_1^1$$

A. LIMIT $k \to 0$

This limit is relatively more complicated. Continuity of the **S** matrix at the origin requires a stronger growth condition on the potential, as will be shown below. We again consider two different cases, $m_{\pm}(k = 0) = 0$ and $m_{\pm}(k = 0) \neq 0$, as was done in Section V.D. For $m_{\pm} \neq 0$, we have

$$1/T_{\pm}(k) = \left\{1 - (1/2i)\int_{-\infty}^{\infty} q(x')\dot{f}_{\pm}(x',0)dx'\right\} - (m_{\pm}/2ik)$$

For $q \subset L^1{}_2$, the quantity $\{\cdot\}$ is finite, and we have

$$1/T_{\pm}(k) = \text{constant} - m_{\pm}/2ik$$

yielding

$$T_{\pm}(k) = -2ik/[m_{\pm} - 2ik \text{ constant}] \approx -2ik/m_{\pm} + o(1)$$

Hence, as $k \to 0$

$$T_{\pm}(k) = \alpha_{\pm}k + o(1) \tag{9.54}$$

where $\alpha_{\pm}(k) = -[2i/m_{\pm}(k = 0)]$ and is purely imaginary. Moreover, $\alpha_{\pm}(k) \neq 0$ since $m_{\pm}(k = 0) \neq 0$.

Similarly, from the integral representation (9.52), we have for real k,

$$\{1 + R_{\pm}(k)\}/T_{\pm}(k) = 1 + \int_{-\infty}^{\infty}\left[\{\exp(\mp 2ikx') - 1\}/2ik\right]\cdot q(x')\, f_{\mp}(x',k)dx'$$

Thus:

$$\lim_{k\to 0}\{1 + R_{\pm}(k)\}/T_{\pm}(k) = 1 + \int_{-\infty}^{\infty} x'q(x')f_{\mp}(x',0)dx' + o(1)$$

$$= \{\cdot\} + o(1)$$

Therefore,

$$\lim_{k\to 0}\{1 + R_{\pm}(k)\} = \alpha'_{\pm}k + o(k) \tag{9.55}$$

where

$$\alpha'_{\pm} = \alpha_{\pm}\{\cdot\}$$

The estimates (Equations 9.54 and 9.55) for $T_{\pm}(k)$ and $R_{\pm}(k)$ are given for a potential $q(x) \subset L^1{}_2$. If the potential is only $L^1{}_1$, then the following estimates obtain. From Equation 9.17,

$$[1/T_\pm(k)] = 1 - (1/2ik)\int_{-\infty}^{\infty} q(x')f_\pm(x',k)dx'$$

Multiplying both side by $2kT_\pm(k)$ and taking the absolute value yields

$$|2k| \le |T_\pm(k)|\left[|2k| + \int_{-\infty}^{\infty}|q(x')||f_\pm(x',k)|dx'\right]$$

$$\le |T_\pm(k)|\left[|2k| + \text{constant}\right]$$

The constant term is proportional to $\int^{\infty}_{-\infty}[1 + |x'|]|q(x')|\, dx'$, where we have used the bound derived in Chapter 7 that $|f_\pm| \le K(1 + |x|)$, $k \ge 0$. Therefore,

$$|k| \le \text{constant}\, |T_\pm(k)|,\ q \subset L_1^1 \text{ and } k \to 0$$

If $k \ne 0$ and Im $k \ge 0$, then$(1/T_\pm(k))$ is continuous. But from the property of unitarity (Property II of Section VII), $(1/T_\pm(k)) \ge 1$ for $k \subset R/\{0\}$. This then implies that $(1/T_\pm)$ is also continuous near the Re-axis if $k \ne 0$. As a matter of fact,

$$|T_\pm(k)| > 0,\ q \subset L_1^1 \text{ and } k \subset \overline{C}_+ / \{0\}$$

The case for $R_\pm(k)$ is relatively more complicated as it is not obvious from either the integral representation of Equation 9.18 or from the Wronskian expression, namely,

$$R_\pm(k) = [T_\pm(k)/2ik]W[\phi(\overline{\phi}), \psi(\overline{\psi})]$$

that $R_\pm(k)$ can be continued off the real axis. The possibility of such analytic continuation depends on how fast the potential falls off with $|x| \to \infty$. This will be considered in Section IX, where the analyticity properties of T and R will be discussed.

IX. VIRTUAL LEVELS OF THE ZERO-ENERGY BOUND STATES

The bound states correspond to the zeros of the coefficients $t_\pm(k)$ or, what is the same, to the poles of the transmission coefficient $T_\pm(k)$. No zeros (poles) of $t_\pm(k)(T_\pm(k))$ exist for $k \subset R$, except possibly at $k = 0$. This is a special case and will be discussed later. The zeros (poles) of $t_\pm(T_\pm)$ are simple and lie on the positive Im k axis. Their number coincides with the number of the zeros of the wavefunction of the Dirichlet operator $(-\partial^2/\partial x^2 + q(x))$.[13] Moreover, from Equation 9.17 we see that the Jost functions $f_\pm(x,k)$ have zeros only at the poles of $T_\pm(k)$. Hence, $f_\pm(x,k)$ have a finite number of simple zeros, all on the positive imaginary k-axis, and wherever $k = ik_j$, $(k_j > 0)$ is a zero of $f_\pm(x,k)$ in Im $k > 0$, $k^2 = -k^2_j$ is an eigenvalue (nondegenerate) of the Schrödinger operator.

It is thus established that the poles of T(k), that is, zeros of $t_\pm(k)$, cannot be on the real axis. We have also shown that $\lambda = k^2 = 0$ cannot be an eigenvalue. Yet it is possible for $f_\pm(x,0)$ to be zero, which, as was demonstrated earlier, leads to $m_\pm(0) = 0$, yielding a linear dependence between $f_+(x,0)$ and $f_-(x,0)$, thereby mimicking a bound state. Since $\lambda = k^2 = 0$ is certainly not an eigenvalue, such bound states are called either the "virtual levels" or simply the "zero-energy" bound states.[16]

From the previous section, we know that the transmission and the reflection coefficients have very simple forms as $k \to 0$, namely,

$$T_{\pm}(k) = \alpha_{\pm}k + o(1) \text{ and } 1 + R_{\pm}(k) = \alpha'_{\pm}k + o(1)$$

Hence, in the limit $k \to 0$,

$$T_{\pm}(0) \to 0$$

and

$$R_{\pm}(0) \to -1 \tag{9.56}$$

These, of course, were derived under the condition that $m_{\pm}(0) \neq 0$, i.e., the zero-energy bound state was not obtained and $f_{\pm}$ were linearly independent. On the other hand, we observed that if a virtual state is obtained at $k = 0$, i.e., $m_{\pm}(0) = 0$ or $f_{\pm}(x,0)$ were, in fact, linearly related to each other, then the above $k \to 0$ relations do not hold for $T_{\pm}(0)$ and $R_{\pm}(0)$. Indeed, it is found that $0 <$ constant $\leq |T(k)|$. In this case, then, we may write instead

$$T_{\pm}(0) = \sin\theta$$

and (9.57)

$$R_{\pm}(0) = \cos\theta$$

so that the unitarity relation is satisfied. Thus, the objective is to determine under what condition Equations 9.56 through 9.57 apply.

Let us rewrite the integral representations (Equations 9.51 and 9.52) for $T_+(k)$ and $R_+(k)$, respectively, as

$$T(k) = 1 + (1/2ik)\int_{-\infty}^{\infty} \exp(+ikx')q(x')T(k)\phi(x',k)dx'$$

and

$$R_+(k) = (1/2ik)\int_{-\infty}^{\infty} \exp(-ikx')q(x')T(k)\phi(x',k)dx'$$

Rearranging terms, we have

$$T(k) = 2ik\, G(k)$$

and (9.58)

$$R_+(k) = G_1(k)G(k)$$

where

$$G_1(k) = A(p,k)\big|_{p=k}$$

(9.59)

$$G(k) = \left[2ik - A(p,k)\right]^{-1}\Big|_{p=-k}$$

and

$$A(p,k) = \int_{-\infty}^{\infty} \exp(-ipx')q(x')\phi(x',k)dx', \ k \subset R$$

In the neighborhood p = k = 0, we expand A(p,k) as

$$A(p,k) = \int_{-\infty}^{\infty} g(x',0)dx' - ip\int_{-\infty}^{\infty} x'g(x',0)dx' + k\int_{-\infty}^{\infty} \dot{g}(x',0)dx' + O\left(k^2, p^2, pk\right) \quad (9.60)$$

where g(x′,0) = φ(x′,0) q(x′) and "·" stands for d/dk.

In order to evaluate the integrals, we pose the following problem

$$y''(x) = q(x)y(x)$$

$$y(-\infty) = 1 \text{ and } y'(-\infty) = 0 \quad (9.61)$$

Also,

$$\lim_{x \to \infty} y(x) = C + \beta x$$

when C and β are constants. y stands for φ. The BVP (9.61) is the zero energy BVP for φ. (From Chapter 7, the Jost functions grow linearly [asymptotically] in the nonprescribed directions as k → 0.)

Now the Jost solution φ(x,k) obeys the following asymptotic boundary conditions:

$$\lim_{x \to -\infty} \phi(x,k) = \exp(-ikx) + o(1)$$

$$\lim_{x \to -\infty} \dot{\phi}(x,k) = -ix \ \exp(-ikx) + o(1)$$

The alternate integral representation $\tilde{\phi}$ of the Jost equation at k = 0 given by Equation 7.17, on the other hand, has the boundary conditions

$$\lim_{x \to -\infty} \tilde{\phi}(x) = x + o(1)$$

$$\lim_{x \to -\infty} \dot{\tilde{\phi}}(x) = 1 + o(1) \quad (9.62)$$

Hence, in the limit k → 0, we have the relation

$$\dot{\phi}(x,0) = -i\tilde{\phi}(x) \quad (9.63)$$

From Equation 9.61,

$$y(x) = \phi(x,0)$$

$$\dot{y}(x) = -i\tilde{\phi}(x) \quad (9.64)$$

We now insert Equation 9.64 in Equation 9.60 and use Equation 9.61 and 9.62 to obtain

$$A(p,k) = \beta - ip[1 - C] - ik\left[\tilde{C} - 1\right] + O\left(k^2, p^2, pk\right) \tag{9.65}$$

where $\overline{C} = \tilde{\phi}'(\infty)$. It was assumed throughout that the potential q(x) is L^1_2 in order to ensure the analyticities of T and R in k near the origin. From Equations 9.58, 9.59 and 9.64, we have in the limit $k \to 0$

$$T(0) = 0 \text{ if } \beta \text{ does not vanish}$$

$$= 2/\left(C + \tilde{C}\right) \text{ if } \beta \text{ does vanish}$$

Similarly,

$$R_+(0) = -1 \text{ if } \beta \text{ does not vanish}$$

$$= \left(C - \tilde{C}\right)/\left(C + \tilde{C}\right) \text{ if } \beta \text{ does vanish}$$

Therefore, if β does not vanish, then the relations (9.56) are obtained, i.e., the Jost functions are linearly independent of each other at k = 0 and no virtual level results. On the other hand, if β does vanish, then Equation(s) 9.57 is obtained if C is chosen to be cot(α/2) and $\tilde{C}$ = tan(α/2), $|\alpha| \le \pi$. This case corresponds to the linear dependence of the Jost solutions, resulting in a zero-energy bound state. Indeed, in the event that β = 0, limit $x \to \infty$ φ(x) is constant. Then from the Wronskian relation derived earlier, namely,

$$m_\pm(0) = W\left[f_+(x,0), f_-(x,0)\right]$$

we see that $m_\pm(0)$ must be zero, which is the condition for a zero-energy bound state.

Physically then, a virtual level corresponds to the disappearance of the linearly growing term x in Equation 9.67. This can happen if a small attractive potential is sitting somewhere away from the potential q(x). Although a highly unusual phenomenon under normal circumstances, it is almost a norm for the potentials generating multisoliton solutions. The discussion of this section was based on Reference 5.

X. ANALYTICITY OF $T_\pm(k)$ AND $R_\pm(k)$

The analyticity properties of $T_\pm(k)$ are obtained directly from those of the Jost solutions in the Wronskian expression

$$\left[1/T_\pm(k)\right] = (1/2ik)\, W[\phi, \psi]$$

The analyticity properties of $R_\pm(k)$ are somewhat more complex. These depend on the ability to continue $R_\pm$ analytically off the real axis. This is due to the fact that the Wronskians involved in the expression for $R_\pm(k)$ involve the Jost functions (ψ,φ) as well as $(\overline{\psi}, \overline{\phi})$ which have different half-planes as their domains of analyticity. Hence, continuation off the real axis depends entirely on the possibility of continuing analytically from one characteristic half-plane to another. For example, $(\overline{\psi}, \overline{\phi})$, which are analytic in the lower half-plane, must be analytically continuable above the real axis if $R_\pm(k)$ is to be analytic there. Such continuations depend on how fast the potential q(x) falls off as $|x| \to \infty$, and are usually discussed in terms of "Bargmann's strip", to which we now turn.

Assume that the potential q(x) vanishes exponentially, that is,

$$\lim_{x \to \pm\infty} \left[q(x) \exp(\pm 2\mu_{\pm} x) \right] = 0, \ \mu_{\pm} > 0$$

A factor of two appears in the exponential in order to be consistent with the integral representations of the Jost functions or solutions given in Chapter 7 in which the kernel has the factors of $\exp(2ik\xi_{\pm})$. It is clear that $f_+(x,k)$ and $f_-(x,k)$ are holomorphic in the half-planes $\text{Im } k > -\mu_+$ and $\text{Im } k > -\mu_-$, respectively. Let μ be the minimum of μ_+ and μ_-. Since the transmission coefficient $T_{\pm}(k)$ is given by

$$T_{\pm}(k) = 2ik/W[\phi, \psi]$$

we conclude that $T_{\pm}(k)$ is analytic in the half-plane $\text{Im } k > -\mu$. The reflection coefficient $R_{\pm}(k)$, on the other hand, involves Wronskians which, as mentioned earlier, contain the Jost solutions that demand analyticity in both the upper and the lower half-plane. For example, $R_+(k)$ has the expression

$$R_+(k) = -W[\phi, \overline{\psi}]/W[\phi, \psi] \tag{9.66}$$

This shows that $R_+(k)$ is analytic in the strip

$$-\mu < \text{Im } k < \mu_+$$

This is called the "Bargmann strip".[16] This restriction on the domain of analyticity in the upper half-plane for the reflection coefficient is due to the presence of the Jost solution $\overline{\psi}(x,k)$ in the numerator Wronskian in Equation 9.66.

For $R_{\pm}(k)$, the matter is complicated by the fact that the expression (9.66) can be singular not only at the zeros of the Wronskian in the denominator, but also at the singularities of the Wronskian in the numerator. From Equation 9.66, it can be concluded that inside the Bargmann strip the poles of $R_+(k)$ are due to the zeros of the denominator and, hence, correspond one-to-one to the discrete eigenvalues which, as we already know, are situated on the positive imaginary axis. Thus, if there are singularities of the reflection coefficient inside the Bargmann strip, these correspond one-to-one to the discrete eigenvalues on the positive imaginary axis. Similar conclusion results for R(k) on the negative imaginary axis.

Inside the strip, any pole $k = ik_j$, $k_j > 0$, of the transmission coefficient must be a pole of $R_+(k)$ also. Assume that there is such a pole inside the strip and yet it does not correspond to the singularity of $R_+(k)$. This may mean that either the numerator has no analytic continuation in the strip or that it is zero at the pole. The former assumption is, of course, not true since, by definition, analyticity prevails inside the strip. The second assertion is also not valid since the residue of $R_+(k)$ at $k = ik_j$ inside the strip is simply $i\|\phi\|^{-1}$. Hence, all poles of $T_{\pm}(k)$ inside the strip must also be the poles of the reflection coefficient. These two assertions, however, fail outside or at the boundary of the Bargmann strip.

If the potential vanishes faster than exponentially as $x \to \infty$ for any finite μ, then the Bargmann strip invades the entire upper half-plane. Then both the transmission and the reflection coefficient are meromorphic in the upper half-plane, with singularities only at the zeros of $t_{\pm}(k)$ lying on the positive imaginary axis and again in one-to-one correspondence. The width of the Bargmann strip is thus determined by the asymptotic properties of the potential in x.

XI. DETERMINATION OF S: HILBERT RELATIONS

It was pointed out by Fadeev[17] that all three independent elements of the **S** matrix should enjoy a single relation among themselves. Hence, the entire matrix should be determined from the knowledge of only one of its elements, assuming that the information about the isolated point spectrum is given. The rationale behind this derives from the simple fact that all the elements of the **S** matrix are determined by the same potential. It turns out that given one of the reflection coefficients and the locations of the bound states, the rest of the **S** matrix can be determined. The reflection coefficient has two parts, the modulus and the phase. It will be shown in Chapter 10 that both $|R(k)|$ and $\arg R(k)$ are needed to reconstruct the potential. This is, two real functions, namely, $q(x)$, $x \subset R^-$ and $q(x)$, $x \subset R^+$ are determined from two other real functions $|R(k)|$ and $\arg R(k)$.

Assume that one of the reflection coefficients, say $R_-(k)$, is given along with the locations of the bound states. The symmetry of the **S** matrix yields $T_+(k) = T_-(k) = T(k)$. Moreover, for real k, $k \neq 0$,

$$R_-(k) = -R_+(-k)T(k)/T(-k)$$

This follows from the relations given by Equation 9.10, Equation 9.12a, and the definitions of R and T in Section VII. Hence, if T(k) can be constructed from the knowledge of $R_-(k)$, then $R_+(k)$ is determined as well and, thus, the rest of the **S**-matrix.

The construction of T(k) from R(k) is well known. A particularly simple derivation is due to Lamb[6] and is followed here. Consider a function Z(k) which is analytic in C_+ and possessing neither zeros nor poles in the upper half-plane. The logarithm of this function, log Z(k), also enjoys the same properties and can, therefore, be represented by the Cauchy formula

$$\log Z(k) = \left(\frac{1}{2\pi i}\right)\oint_{\Gamma}\left[\log Z(k')/(k'-k)\right]dk' \tag{9.67}$$

Moreover, let $Z(k) \to 1$ as $k \to \infty$. The contour Γ in Equation 9.67 can then be closed in the upper half-plane, yielding

$$\log Z(k) = \left(\frac{1}{2\pi i}\right)\int_{-\infty}^{\infty}\left[\log Z(k')/(k'-k)\right]dk' \tag{9.68}$$

If now k is replaced by its complex conjugate k^*, then the Cauchy theorem gives us

$$\log Z\left(k^*\right) = \left(\frac{1}{2\pi i}\right)\int_{-\infty}^{\infty}\left[\log Z(k')/\left(k'-k^*\right)\right]dk' = 0 \tag{9.69}$$

since the point k^* is not enclosed by the contour Γ. Taking the complex conjugate of Equation 9.69 and subtracting from Equation 9.68 yields

$$\log Z(k) = \left(\frac{1}{2\pi i}\right)\int_{-\infty}^{\infty}\left[\log\left|Z(k')\right|^2/(k'-k)\right]dk' \tag{9.70}$$

From Equation 9.70,

$$Z(k) = \exp[(1/2\pi i)]\int_{-\infty}^{\infty}\left[\log|Z(k')|^2/(k'-k)\right]dk' \tag{9.71}$$

Let us now consider the transmission coefficient T(k) which has poles at $k = ik_j$, $k_j > 0$. Moreover, $|T| \to 1$ as $k \to \infty$. Hence, if we define a function h(k) such that

$$h(k) = T(k)\prod_{1}^{N}(k - k_j)/(k + k_j) \tag{9.72}$$

then h(k) enjoys the same properties as those of the function Z(k). Identifying $h(k) \equiv Z(k)$ in Equation 9.71 and according to property II of Section VII, we obtain

$$T(k) = \left[\prod_{1}^{N}(k + k_j)/(k - k_j)\right]\cdot \exp\left[(1/2\pi i)\int_{-\infty}^{\infty}\left[\log|1 - R(k')|^2/(k'-k)\right]dk'\right] \tag{9.73}$$

The same result and more can be obtained in another, more elegant way. The treatment is due to Deift and Trubowitz.[31] As defined in Equation 9.72, h(k) is analytic in C_+ and $|h(k)| \le 1$. h(k), therefore $\subset$ the Hardy class $H^{\infty+}$. According to the properties of the function in $H^{\infty+}$, it can be written down immediately that[18]

$$\log|h(k)| \le (v/\pi)\int_{-\infty}^{\infty}\left[\log h(k')/\left\{(k'-u)^2 + v^2\right\}\right]dk'$$

$$k = u + iv, \; v > 0 \tag{9.74}$$

Deift and Trubowitz then consider the function $h_\beta(k) = h(k + i\beta)$, $\beta > 0$, Im $k \ge 0$. Following the same procedures used in establishing Equation 9.70, they obtain

$$\log|h(k)| = (v/\pi)\int_{-\infty}^{\infty}\left[\log h(k')/\left\{(k'-u)^2 + v^2\right\}\right]dk' \tag{9.75}$$

Using the results of Section V.D, namely,

$$\lim_{k\to 0} T(k) \ge k \cdot \text{constant}$$

and letting $\beta \to 0$ in Equation 9.75, h(k) is actually the "outer function", that is, the inequality sign $\le$ in Equation 9.74 is replaced by the equality sign =. In the course of the derivation, it is also obtained that

$$T(k) = \lim_{\beta\to 0} T(k + i\beta), \; \text{Im}\, k \ge 0$$

The rest of the proof follows by defining a function g(k) such that

$$g(k) = \exp[(1/i\pi)\int_{-\infty}^{\infty}\left[\log|T(k')|/(k'-k)\right]dk'$$

Then

$$\log|g(k)| = (v/\pi)\int_{-\infty}^{\infty}\left[\log|T(k')|/\left\{(k'-u)^2+v^2\right\}\right]dk'$$

$$= \log|h(k)|$$

This implies that $|h(k)| = |g(k)|$. The rational function $g(k)/h(k)$ is then an analytic function of constant modulus,[18] yielding

$$h(k) = \text{constant}\cdot g(k)$$

Since $h(k) \to 1 + O(1/k)$, $k \to \infty$, the constant is 1. Finally, replacing $\log|T(k')|$ by $(1/2) \log|T(k')|^2$, the representation (9.73) is obtained. In summary, then

$$T(k) = \left[\prod_{1}^{N}(k+k_j)/(k-k_j)\right]$$

$$\exp\left[(1/2\pi i)\int_{-\infty}^{\infty}\left[\log|1-R(k')|^2/(k'-k)\right]dk', \quad \text{Im}\, k > 0\right.$$

and

$$T(k) = \lim_{\beta\to 0} T(k+i\beta), \quad \text{Im}\, k = 0$$

Next, we show that the modulus and the phase of the transmission coefficient T(k) are related to each other. In other words, we establish a dispersion relation for the function T(k). In order to do this, the Re and the Im part of Equation 9.73 are to be evaluated. Evaluating the imaginary part yields

$$\theta(k) = 2\sum_{n=1}^{N}\tan^{-1}(k_n'/k) - (1/\pi)\int_{-\infty}^{\infty}(k'-k)^{-1}\log|T(k')|dk' \tag{9.76}$$

We are using the definition

$$T(k) = |T(k)|\exp[i\theta(k)], \quad \text{Im}\, k = 0$$

where $\theta(k)$ is only mod(2π). From the symmetry considerations (see Section VII, Property V),

$$|T(-k)| = |T(k)|$$

and

$$\theta(-k) = -\theta(k), \quad \text{Im}\, k = 0$$

The oddness of $\theta(k)$ implies that Equation 9.74 can be rewritten as

$$\theta(k) = 2\sum_{n=1}^{N}\tan^{-1}(k_n'/k) - (k/\pi)\int_{-\infty}^{\infty}\left(k'^2-k^2\right)^{-1}\log|T(k')|dk', \quad \text{Im}\, k = 0$$

The mod(2π) ambiguity can be removed by constraining $\tan^{-1}$ to its principal part Tan^{-1} and obtain

$$\theta(k) = 2\sum_{n=1}^{N} \mathrm{Tan}^{-1}(k_n'/k) - (k/\pi)\int_{-\infty}^{\infty}\left(k'^2 - k^2\right)^{-1} \log\left|T(k')\right|dk', \quad \mathrm{Im}\,k = 0 \qquad (9.77)$$

For the number of discrete eigenvalues $N > 0$, $\theta(k)$ has a discontinuity at $k = 0$, namely,

$$\theta(k = 0\,-) = -N\pi$$

and

$$\theta(k = 0\,+) = N\pi$$

giving

$$\theta(k = 0\,-) - \theta(k = 0\,+) = 2N\pi$$

This, of course, is "Levinson's theorem".[19]

Similarly, the evaluation of the real part of Equation 9.74 results in an expression for the modulus of T(k):

$$\left|T(k)\right| = \prod_{1}^{N}\left[1 + (k_n/k)\right]^{-1} \cdot \exp\left\{(1/\pi)\int_{-\infty}^{\infty} k'\left(k'^2 - k^2\right)\theta(k')dk'\right\} \qquad (9.78a)$$

Similar relations for $|T(k)|$ and $\theta(k)$ also obtain for $k \subset C_+$, that is, $|T(k)|$, $\theta(k)$, $k \subset C_+$ are still given by Equations 9.75 and 9.76, respectively, the only difference being that the principal value $⨍_{-\infty}^{\infty}$ must now be replaced by $\int_{-\infty}^{\infty}$. Equations 9.75 and 9.76 express the absolute value $|T(k)|$ of T(k) in terms of the phase $\theta(k)$ and vice versa. Such relations are known as the "Kramers-Krönig" or the "Hilbert" relations.[20] We have, therefore, obtained the "dispersion relations" for the transmission coefficient T(k). The transmission coefficient T(k) is, therefore, obtained by the reflection coefficient R(k), according to the unitarity relations $|T(k)|2 = 1 - |R(k)|^2$, and from the phase $\theta(k)$ determined by the locations of the discrete eigenvalues k_j.

It is useful to note that neither the phase of R(k) nor the specifications of the normalization constants D_j are needed in order to determine T(k). These are, however, absolutely necessary if the potential is to be reconstructed. This is where the scattering data S comes into the picture. In summary, then, we have the following theorem.

Theorem 9.4

The scattering matrix **S** given by

$$\mathbf{S} = \begin{pmatrix} T_+(k) & R_-(k) \\ R_+(k) & T_-(k) \end{pmatrix}$$

can be reconstructed from the specifications of either of the reflection coefficients $R_\pm(k)$ and the point spectrum $\sigma(L)$.

XII. FOURIER TRANSFORM OF $R_\pm(K)$ AND $T_\pm(K)$

The integral representations for $T_\pm(K)$ and $[R_\pm(K)/T_\pm(K)]$ are given by Equations 9.17 and 9.18, which are reproduced below.

$$1/T_\pm(k) = 1 - (1/2ik)\int_{-\infty}^{\infty} q(x')f_\pm(x',k)dx'$$

and

$$\left[R_\pm(k)/T_\pm(k) = (1/2ik)\int_{-\infty}^{\infty} \exp(\mp 2ikx')q(x')f_\mp(x',k)dx'\right.$$

The Levin transforms for $f_\pm(x,k)$ are given by (omitting the symbol $\wedge$ from the transforms).*

$$f_\pm(x,k) = 1 + \int_{\pm 0} \exp(\pm 2ikx')B_\pm(x',x'')dx''$$

where

$$\int_{0+} \text{ and } \int_{=0} \equiv \int_0^{\infty} \text{ and } \int_{-\infty}^{0}$$

respectively. Substituting in the expression for $T_\pm^{-1}(k)$, we obtain

$$T_\pm^{-1}(k) = 1 - (1/2ik)Q(\infty) - (1/2ik)\int_{-\infty}^{\infty} q(x')\left\{\int_{\pm 0} \exp(\pm 2ikx'')B_\pm(x',x'')dx''\right\}dx'$$

Since q is at least $L^1{}_1$ and $B_\pm(x,y) \subset L_1(y \subset R)$, the order of integration can be interchanged, yielding

$$T_\pm^{-1}(k) = 1 - \{Q(\infty)/2ik\} - (1/2ik)\int_{\pm 0} \exp(\pm 2ikx')\Pi_\pm^{(1)}(x')dx' \qquad (9.78b)$$

where

$$\Pi_\pm^{(1)}(x') = \int_{-\infty}^{\infty} q(x'')B_\pm(x',x'')dx''$$

Similarly, substituting the Levin transforms for $f_\pm(x,k)$ in the expression for $[R_\pm(k)/T_\pm(k)]$, one obtains

$$\left[R_\pm(k)/T_\pm(k)\right] = (1/2ik)\int_{-\infty}^{\infty} q(x')\left[\int_{\pm 0} \exp\{\mp 2ik(x'+x'')\}B_\mp(x',x'')dx''\}\right]dx' + Q(\infty)/2ik$$

Substituting $s = x' + x''$,

$$\left[R_\pm(k)/T_\pm(k)\right] = (1/2ik)\int_{-\infty}^{\infty} q(x')\left\{\int_{\mp x'} \exp(\pm 2iks)B_\mp(x',s-x')ds\right\}dx' +$$

$$+ Q(\infty)/2ik \qquad (9.79)$$

* Here the argument of the exponential contains $\pm\, 2ikx'$ instead of $\pm\, ikx'$ of Chapter 8. This is because of the fact that the same factor appears in the representation of R/T. However, changing $\pm\, ikx' \rightarrow \pm\, 2ikx'$ in the Levin representation is just a matter of how the Fourier transforms are defined. This was discussed at the end of Chapter 8.

We again change the order of integration in Equation 9.79, which then reduces to the form familiar in the literature,

$$\left[R_{\pm}(k)/T_{\pm}(k)\right] = (1/2ik)\int_{-\infty}^{\infty} \exp(\mp 2iks)\Pi_{\pm}^{(2)}(s)ds \tag{9.80}$$

where

$$\Pi_{\pm}^{(2)}(s) = q(s) + \int_{\pm s} q(t)B_{\mp}(t, s-t)dt$$

Equation 9.80 is in the form of a Fourier transform. However, the integral may not exist since from Chapter 8 we know that

$$\left|B_{\pm}(x,y)\right| \le \eta_{\pm}(x+y)\exp\{\gamma_{\pm}(x)\}$$

and $\gamma_{\pm}(x)$ may grow linearly as $|x| \to \infty$. In order to obtain an estimate for $\Pi_{\pm}^{(2)}(s)$ so that the validity of the Fourier transform can be ascertained, some other alternate procedure must be found. Deift and Trubowitz[31] obtain such an estimate by following a simple, elegant method proposed originally by Tanaka.[21] We outline Tanaka's computation as presented by Deift and Trubowitz. The method uses the Wronskian definition of the quantity $(R_{\pm}/T_{\pm})$.

Now consider $(R_{\pm}/T_{\pm})$ again. From the Wronskian definitions (Equations 9.1 through 9.4,

$$R_{\pm}/T_{\pm} = r_{\pm} = \mp(1/2ik)W\left[\phi(\bar{\phi}), \bar{\psi}(\psi)\right]$$

$$= \mp(1/2ik)\exp(\mp 2ikx)W\left[f_{-}(\bar{f}_{-}), \bar{f}_{+}(f_{+})\right]$$

This yields

$$\exp(\pm 2ikx)2ik\left(R_{\pm}/T_{\pm}\right) = \mp W\left[f_{-}(\bar{f}_{-}), \bar{f}_{+}(f_{+})\right] = f_{-}(\bar{f}_{-})\bar{f}'_{+}(f'_{+}) - f'_{-}(\bar{f}'_{-})\bar{f}_{+}(f_{+})$$

$$= \bar{f}'_{+}(f'_{+}) - f'_{-}(\bar{f}'_{-}) - \bar{f}'_{-}(f'_{-})\left\{\bar{f}_{+}(f_{+}) - 1\right\} + \bar{f}'_{+}(f'_{+})\left\{f_{-}(\bar{f}_{-}) - 1\right\}$$

Using the Levin transforms for $f_{\pm}$ and from the reality condition $\bar{f}_{\pm}(x,k) = f_{\pm}(x,-k)$, we obtain

$$\exp(\pm 2ikx)2ik\left(R_{\pm}/T_{\pm}\right) = \int_{+0} \exp(\mp 2ikx')B'_{+}(x,x')dx' - \int_{-0} \exp(\mp 2ikx')B'_{-}(x,x')dx' +$$

$$+\int_{+0}\int_{-0} \exp\left\{\mp 2ik(x'+x'')\right\}B'_{+}(x,x')B_{-}(x,x'')dx''\,dx' -$$

$$-\int_{-0}\int_{+0} \exp\left\{\mp 2ik(x'+x'')\right\}B'_{-}(x,x')B_{+}(x,x'')dx'\,dx''$$

Changing $(x'+x'')$ to z and remembering that $B_{+}(x,x') = 0$, $x' < 0$, where $B_{-}(x,x') = 0$ when $x' > 0$, the above equation can be rewritten as

$$\exp(\pm 2ikx)2ik(R_\pm/T_\pm) = \int_{+0}\exp(\mp 2ikx')B'_+(x,x')dx' - \int_{-0}\exp(\mp 2ikx')B'_-(x,x')dx' +$$

$$+\int_0\left\{\int_{-x'}\exp(\mp 2ikz)B'_+(x,x')B_-(x,z-x')dz\right\}dx' -$$

$$-\int_{-0}\left\{\int_{x'}\exp(\mp 2ikz)B'_-(x,x')B_+(x,z-x')dz\right\}dx'$$

Upon interchanging the orders of integration, this reduces to

$$\int_{-\infty}^{\infty}\exp(\mp 2ikz)\Pi_\pm^{(3)}(x,z)dz$$

where

$$\Pi_\pm^{(3)}(x,z) = B'_+(x,z) - B'_-(x,z) + \int_{+z}B'_+(x,x')B_-(x,z-x')dx' -$$

$$-\int_{-z}B'_-(x,x')B_+(x,z-x')dx'$$

Therefore,

$$\left[R_\pm(k)/T_\pm(k)\right] = (1/2ik)\int_{-\infty}^{\infty}\tilde{\Pi}_\pm^{(3)}(z)\exp(\mp 2ikz)dz$$

where

$$\tilde{\Pi}_\pm^{(3)}(z) = \exp(\mp 2ikx)\Pi_\pm^{(3)}(x,z) \text{ for any } x \subset R$$

In Equation 9.80, we now identify

$$\Pi_\pm^{(2)}(z) = \Pi_\pm^{(3)}(z)\big|_{x=0} = \Pi_\pm^{(3)}(0,z)$$

Then, for $z \geq 0$,

$$\left|\Pi_\pm^{(2)}(z)\right| \leq \left|B'_+(0,z)\right| + \left|\int_{+z}B'_+(0,x')B_-(0,z-x')dx'\right| + \left|\int_{+z}B'_-(0,x')B_+(0,z-x')dx'\right|$$

Using the results of Chapter 8,

$$\left|B_+(x,y)\right| \leq K\eta_+(x+y)$$

and

$$\left|B'_+(x,y) + q(y)\right| \leq K\eta_+(x+y),\ y \geq 0$$

K being a constant. Then

$$\left|\Pi_\pm^{(2)}(z)\right| \leq \left|q(z)\right| + K\eta_+(z) + \int_{+z}\left|B'_+(0,x')\right|\left|B_-(0,z-x')\right|dx' +$$

$$+\int_{-z}\left|B'_-(0,x')\right|\left|B_+(0,z-x')\right|dx'$$

Writing

$$|B'_{\pm}(0,x')| = |B'_{\pm}(0,x') + q(x') - q(x')| \le |B'_{\pm}(0,x') + q(x')| + |q(x')| \le K\eta_{\pm}(x') + |q(x')|$$

it is not difficult to show that

$$\left|\Pi_{\pm}^{(2)}(z)\right| \le |q(z)| + K_1\eta_{+}(z)$$

Similarly, if $z \le 0$, then

$$\left|\Pi_{\pm}^{(2)}(z)\right| \le |q(z)| + K_1\eta_{-}(z)$$

We, therefore, have

$$\left|\Pi_{\pm}^{(2)}(s)\right| \subset L_1(R)$$

if the potential is of zero-order growth. The estimate for $\Pi_{\pm}^{(1)}(s)$ can be calculated analogously and we obtain $|\Pi_{\pm}^{(1)}(z)| \subset L_1(R^{\pm})$.

XIII. SPECTRAL REPRESENTATION ON R

In the previous sections, we have defined the scattering matrix **S** and have studied the properties of its elements. The matrix **S** is unitary and describes the physical process of the scattering of a plane wave. In quantum scattering, the **S** matrix is defined as a "unitary normalization operator". Let **L**, $\mathbf{L_0}$ be two linear operators on the Hilbert space $L_2(R)$. Assume that these operators are "unitarily equivalent", that is, there is a unitary operator **U** called the "intertwining operator" such that

$$\mathbf{LU} = \mathbf{UL_0} \tag{9.81}$$

There may exist more than one unitary operator for which Equation 9.81 holds. If **V** is another such operator, then we also have

$$\mathbf{LV} = \mathbf{VL_0} \tag{9.82}$$

Premultiplying Equation 9.82 by $\mathbf{U^+}$, the adjoint of **U**, we have from Equations 9.81 and 9.82

$$\mathbf{U^+VL_0} = \mathbf{U^+LV} = \mathbf{U^+LUU^+V} = \mathbf{U^+UL_0U^+V} = \mathbf{L_0U^+V}$$

where the unitarity relation $\mathbf{UU^+} = \mathbf{I}$ has been used. Thus,

$$\left[\mathbf{L_0}, \mathbf{U^+V}\right] = 0$$

The consistency between Equations 9.81 and 9.82 requires that $\mathbf{L_0}$ and $\mathbf{U^+V}$ commute with each other.

The operator $\mathbf{U^+V}$ is called a "unitary normalization operator". In quantum mechanics, the scattering matrix **S** is, indeed, a unitary normalization operator which involves the well-known "Mϕller operators" Ω_+ and Ω_-[22,23,25] as the intertwining operators. The objective here is to establish this fact and, en route, "derive" the form of **S**. In order to accomplish this, the spectral representation of the Schrödinger operator will be required on the entire real axis R which, in turn, requires the use of the resolvent operator. These are discussed below.

A. THE RESOLVENT OPERATOR

One of the objectives is to establish a completeness relation for the Schrödinger operator on R. An analogous relation was obtined (Equation 3.33) in Chapter 3 for the scattering on the semiaxis R^+. The Schrödinger equation is given by

$$\mathbf{L}\psi(x,k) = \lambda\psi(x,k)$$

or

$$(\mathbf{L} - \lambda\mathbf{I})\psi(x,k) = 0, \; x \subset R, \; \lambda = k^2 \subset C$$

$$\mathbf{L} = -d^2/dx^2 + q(x)$$

If λ, ($\lambda \subset C$), is not in the point spectrum of **L**, then the inverse operator $(\mathbf{L} - \lambda\mathbf{I})^{-1}$ is called the "resolvent operator" or simply the "resolvent" of **L**, and will be denoted by $\mathbf{R}_L$. If $\lambda \subset C$ is such that $\mathbf{R}_L$ is unbounded, then $\lambda \subset$ "continuous spectrum" of **L**. Otherwise, $\lambda \subset$ the "resolvent set" for the operator **L**. The boundedness or unboundedness is, of course, with respect to the domain of definition of **L**. For real potentials, the operator **L** is "formally self-adjoint", i.e.,

$$(\mathbf{L}u, v) = [uv' - vu']_{-\infty}^{\infty} + (u, \mathbf{L}v)$$

as obtained by partial integration of the Schrödinger equation. $(\cdot,\cdot)$ denotes an inner product and u,v $\subset L_2(R)$. Let us restrict the domain to a subspace of $L_2(R)$ such that the boundary terms within the braces vanish and that if u $\subset L_2(R)$, then **L**u also $\subset L_2(R)$. The domain of definition thus restricted renders the Schrödinger operator "self-adjoint". Acting on a square integrable function, the operator then yields another function which is also square integrable. This, of course, requires that u' be locally integrable and be absolutely continuous. This, in turn, imposes a restriction as to the spatial fall off of the potential q(x). It was shown in Chapter 7 that for u' to be absolutely continuous on every finite interval, q(x) must be an L^1_1 function.

Let $G(x,x';\lambda)$ be "Green's function" for the operator $(\mathbf{L} - \lambda\mathbf{I})$. Then

$$(\mathbf{L} - \lambda\mathbf{I})G(x,x';\lambda) = \delta(x - x')$$

or

$$\left[-\partial^2/\partial x^2 + q(x) - \lambda\right]G(x,x';\lambda) = \delta(x - x')$$

Then by definition, the resolvent $\mathbf{R}_L$ is

$$(\mathbf{R}_L u)(x,\lambda) = \int_{-\infty}^{\infty} G(x,x';\lambda)u(x')dx', \; u(x) \subset D_L$$

D_L is the domain of the self-adjoint operator **L**. The boundary conditions for $G(x,x';\lambda)$ are

$$\lim_{|x|\to\infty} G(x,x';\lambda)=0$$

$$\lim_{|x'|\to\infty} G(x,x';\lambda)=0$$

We require that G(x,x′;λ) be continuous at the point of discontinuity x = x′. Moreover, the first derivative of Green's function must undergo a jump discontinuity of –1 at x = x′.* This can easily be shown by integrating the differential equation of G over a small interval (x′ – ε) and (x′ + ε) around x = x′. We finally obtain for Green's function the following expressions:

$$G\left(x,x';\lambda^{\frac{1}{2}}\right)=i\left\{2\lambda^{\frac{1}{2}}t\left(\lambda^{\frac{1}{2}}\right)\right\}^{-1}\phi\left(x',\lambda^{\frac{1}{2}}\right)\psi\left(x,\lambda^{\frac{1}{2}}\right),\quad x\ge x'$$

$$=i\left\{2\lambda^{\frac{1}{2}}t\left(\lambda^{\frac{1}{2}}\right)\right\}^{-1}\cdot\left\{\psi\left(x',\lambda^{\frac{1}{2}}\right)\phi\left(x,\lambda^{\frac{1}{2}}\right)\right\},\quad x'\ge x$$

$$\operatorname{Im}k\ge 0,\ k\ne 0,\ \text{and } t(k)\ne 0 \tag{9.83}$$

The Jost solutions appear in the expressions for G by virtue of the fact that for x ≠ x′, δ(x–x′) = 0 and G satisfies the Schrödinger equation whose solutions are ϕ and ψ. In deriving Equation 9.83, the Wronskian relation W[ϕ,ψ] = 2ikt(k) was used. It can be easily verified that Equation 9.83 fulfills all the conditions for G to be Green's function. From Theorem 7.1, it follows that Equation 9.83 is defined for a potential $q\subset L^1{}_1$.

From Equation 9.83,

$$(\mathbf{R}_L u)(x,\lambda)=i\left\{2\lambda^{\frac{1}{2}}t\left(\lambda^{\frac{1}{2}}\right)\right\}^{-1}$$

$$\left\{\psi\left(x,\lambda^{\frac{1}{2}}\right)\int_{-}\phi\left(x',\lambda^{\frac{1}{2}}\right)u(x')dx'+\phi\left(x,\lambda^{\frac{1}{2}}\right)\int_{+}\psi\left(x',\lambda^{\frac{1}{2}}\right)u(x')dx'\right\}$$

For Im k = v ≥ δ > 0, the bounds for the Jost functions are

$$\left|f_{\pm}\left(x,\lambda^{\frac{1}{2}}\right)\right|\le\exp\left[\left|\lambda^{\frac{1}{2}}\right|^{-1}\eta_{\pm}(x)\right],\ \lambda\ne 0$$

For $q(x)\subset L_1{}^1$, we have

$$\sup_{|\lambda|,|x|}\left|f_{\pm}(x,\lambda)\right|\le C_{\delta}$$

* These are the fundamental properties of a Green's function. See, for example, I. Stakegold, *Boundary Value Problems of Mathematical Physics*, Vol. 1, MacMillan, New York, 1968.

where C_δ is a real number, depending on δ. Hence,

$$|\phi(x,k)| \le C_\delta \exp(vx)$$

and

$$|\psi(x,k)| \le C_\delta \exp(-vx), \quad \left|\lambda^{\frac{1}{2}}\right| = k \ge \delta > 0$$

Then,

$$|\mathbf{R}_L u|(x,k) \le \left|\left\{2\lambda^{\frac{1}{2}} t\left(\lambda^{\frac{1}{2}}\right)\right\}^{-1}\right| C_\delta \cdot$$

$$\cdot \left\{\exp(vx)\textstyle\int_+ \exp(-vy)|u(y)|dy + \exp(-vx)\int_- \exp(vy)|u(y)|dy\right\}$$

$$= 2\left|\left\{\lambda^{\frac{1}{2}} t\left(\lambda^{\frac{1}{2}}\right)\right\}^{-1}\right| C_\delta \left\{|P_+ u|(x) + |P_- u|(x)\right\} \tag{9.84}$$

where P_+ resp. P_- represents the intergral operations $\int_+$ resp. $\int_-$.

Consider $P_- u(x)$, $u(x) \subset L_2(R)$. Then

$$|P_- u|(x) = \exp(-vx)\textstyle\int_- \exp(vy)|u(y)|dy$$

$$= \exp(-vx)\left[\textstyle\int_{-\infty}^{\frac{x}{2}} \exp(vy)|u(y)|dy + \int_{\frac{x}{2}}^{x} \exp(vy)|u(y)|dy\right]$$

Since $u(y) \subset L_2(R)$, we can use the Cauchy-Schwartz inequality to obtain

$$|P_- u|(x) \le \exp(-vx)\left[\left\{\textstyle\int_{-\infty}^{x/2}\exp(2vy)dy\int_{-\infty}^{x/2}|u(y)|^2 dy\right\}^{\frac{1}{2}} + \left\{\int_{x/2}^{x}\exp(2vy)dy\int_{x/2}^{x}|u(y)|^2 dy\right\}^{\frac{1}{2}}\right]$$

$$= \left\{\exp(-vx)/(2v)^{\frac{1}{2}}\right\} \cdot$$

$$\cdot \left\{\exp(vx/2)\left[\textstyle\int_{-\infty}^{x/2}|u(y)|^2 dy\right]^{\frac{1}{2}} + \left[\exp(2vx) - \exp(vx)\right]^{\frac{1}{2}}\left[\int_{x/2}^{x}|u(y)|^2 dy\right]^{\frac{1}{2}}\right\}$$

Since $u \subset L_2(R)$, we conclude that

$$\lim_{x\to\infty} |P_- u|(x) \to 0$$

The same conclusion also holds true for $|P_+ u|(x)$ in the limit $x \to -\infty$. Thus,

$$\lim |P_{\pm}u|(x) \to 0 \quad \text{for } x \to +\infty \quad \text{resp.} -\infty$$

Also,

$$\left[|P_{\pm}u|\right]'(x) = \pm v|P_{\pm}u|(x) \mp |u(x)|$$

Then

$$\left[|P_{\pm}u|^2\right]' = \left\{(P_{\pm}u)(P_{\pm}u)^*\right\}'$$

$$= \pm 2v|P_{\pm}u|^2(x) \mp u(x)(P_{\pm}u)(x) \mp u(x)(P_{\pm}u)^*(x)$$

Integrating the last equation over the real axis and remembering that limit $|P_{\pm}u|(x) \to 0$ as $x \to \pm\infty$, we have

$$0 = \pm 2v\|P_{\pm}u\|_2^2 \mp (P_{\pm}u, u) \mp (u, P_{\pm}u)$$

By the Cauchy-Schwartz inequality, the inner product terms are

$$(P_{\pm}u, u) + (u, P_{\pm}u) \le 2\|P_{\pm}u\|_2 \|u\|_2$$

This yields

$$\|P_{\pm}u\|_2 \le \|u\|_2 / v$$

Thus, the operator norm

$$\|P_{\pm}\|_2 = \sup \|P_{\pm}u\|_2 / \|u\|_2 < \infty, v > 0, \quad u \subset D_L$$

$P_{\pm}$ are, therefore, bounded and hence are continuous linear operators. Returning to Equation 9.84,

$$|\mathbf{R}_L\|_2(x, \lambda) \le 2\left|\left\{\lambda^{\frac{1}{2}} t\left(\lambda^{\frac{1}{2}}\right)\right\}^{-1}\right| C_\delta / v, \ \operatorname{Im}\lambda^{\frac{1}{2}} = v > 0 \tag{9.85}$$

The resolvent operator $\mathbf{R}_L$ is thus bounded for $\lambda \not\subset \sigma(\mathbf{L})$, $\sigma(\mathbf{L})$ being the point spectrum.

In the above, we have assumed that $t(\lambda^{1/2}) \ne 0$. If k is such that $t(\lambda^{1/2}) = 0$, then $\mathbf{R}_L$ does not exist. These values of $\lambda = k^2$ then $\subset \sigma(\mathbf{L})$, in conformity with the fact that

$$\sigma(\mathbf{L}) = \left\{\lambda : t\left(\lambda^{\frac{1}{2}}\right) = 0, \ \lambda^{\frac{1}{2}} \subset C\right\}$$

Equation 9.83 is valid for Im $\lambda^{1/2} \ge 0$, whereas Equation 9.85 is for Im $\lambda^{1/2} > 0$. Although the resolvent $\mathbf{R}_L$ exits for Im $\lambda^{1/2} \ge 0$ according to Equation 9.83, it may not be bounded.

That is, although the inverse $(\mathbf{L} - \lambda\mathbf{I})^{-1}$ may exist for λ such that Im $\lambda^{1/2} \geq 0$, the inverse may not be bounded. Consider, for example, a sequence of functions $g_n(x, \lambda^{1/2})$ such that

$$g_n\left(x, \lambda^{\frac{1}{2}}\right) = \phi^*\left(x, \lambda^{\frac{1}{2}}\right), \quad |x| \leq n$$

$$= 0 \qquad\qquad |x| > n$$

g_n is, therefore, ϕ^* multiplied by a characteristic function over the interval $[-n,n]$. Then,

$$(\mathbf{R}_L g_n)\left(x, \lambda^{\frac{1}{2}}\right) = i\left\{2\lambda^{\frac{1}{2}} t\left(\lambda^{\frac{1}{2}}\right)\right\}^{-1} \psi\left(x, \lambda^{\frac{1}{2}}\right)\|\phi\|_n^2 \quad |x| \geq n$$

From this,

$$\|\mathbf{R}_L g_n\|_2^2 \geq \left\{\left|2\lambda^{\frac{1}{2}} t_{\pm}\left(\lambda^{\frac{1}{2}}\right)\right|\right\}^{-2} \|\phi\|_n^4 \int_n^{\infty} \left|\psi\left(x, \lambda^{\frac{1}{2}}\right)\right|^2 dx$$

Let $n \to N$, where N is a large number. Then since $\psi(x,\lambda^{1/2}) \to \exp(i\lambda^{1/2}x)$ as $x \to \infty$,

$$\int_{n \to N}^{\infty} \left|\psi\left(x, \lambda^{\frac{1}{2}}\right)\right|^2 dx = \|\psi\|_N^2 \to \exp(-2vN)/2v, \quad \mathrm{Im}\,\lambda^{\frac{1}{2}} \geq 0$$

Or,

$$\|\psi\|_{n \to N} \to \left\{1/(2v)^{\frac{1}{2}}\right\}\exp(-vN)$$

This then yields

$$\|\mathbf{R}_L g_n\|_2 \geq \|\phi\|_n^2 \exp(-vN)\left[2\left|\lambda^{\frac{1}{2}} t\left(\lambda^{\frac{1}{2}}\right)\right|\right]^{-1} \Big/ (2v)^{\frac{1}{2}}$$

The norm of the operator $\mathbf{R}_L$ is, therefore,

$$\|\mathbf{R}_L\| \geq \left[1/2\left|\lambda^{\frac{1}{2}}\right|\left|t\left(\lambda^{\frac{1}{2}}\right)\right|(2v)^{\frac{1}{2}}\right]^{-1} \exp(-vN)$$

This shows that $\|\mathbf{R}_L\| \to \infty$ as Im $\lambda^{1/2} \to 0$, $\lambda^{1/2} \geq 0$. The R^+ axis of the λ-plane then belongs to the continuous spectrum of the Schrödinger operator **L**.

These results can be summarized in the following theorem.

Theorem 9.5

Let the potential $q(x)$ be an $L^1{}_1$ function. Then the resolvent operator $\mathbf{R}_L = (\mathbf{L} - \lambda\mathbf{I})^{-1}$ for the Schrödinger operator **L** is given by

$$\mathbf{R}_L u\left(x,\lambda^{\frac{1}{2}}\right) = \int_{-\infty}^{\infty} G(x,x';\lambda)u\left(x',\lambda^{\frac{1}{2}}\right)dx'$$

where

$$G(x,x';\lambda) = i\left[2\lambda^{\frac{1}{2}} t\left(\lambda^{\frac{1}{2}}\right)\right]^{-1} \phi\left(x',\lambda^{\frac{1}{2}}\right)\psi\left(x,\lambda^{\frac{1}{2}}\right), \; x \geq x'$$

$$= i\left[2\lambda^{\frac{1}{2}} t\left(\lambda^{\frac{1}{2}}\right)\right]^{-1} \psi\left(x',\lambda^{\frac{1}{2}}\right)\phi\left(x,\lambda^{\frac{1}{2}}\right), \; x' \geq x$$

and

$$\lambda \subset C, \operatorname{Im}\lambda^{\frac{1}{2}} \geq 0, \; t\left(\lambda^{\frac{1}{2}}\right) \neq 0$$

$$G(x,x';\lambda) = \left[G\left(x,x';\lambda^*\right)\right]^* \text{ if } \operatorname{Im}\lambda^{\frac{1}{2}} < 0$$

$\lambda \subset C$, Im $\lambda^{1/2} > 0$, $t(\lambda^{1/2}) \neq 0$ is the "resolvent set" of **L**.
$\lambda \subset C$, Im $\lambda^{1/2} > 0$, $t(\lambda^{1/2}) = 0$ constitutes the $\sigma(\mathbf{L})$.
$\lambda \geq 0$, Im $\lambda^{1/2} = 0$ constitutes the "continuous spectrum" of **L**. Moreover, $\mathbf{R}_L$ is bounded:

$$\|\mathbf{R}_L\|_2 \leq C_\delta\left[\left|\lambda^{\frac{1}{2}}\right|\left|t_\pm\left(\lambda^{\frac{1}{2}}\right)\right|\operatorname{Im}\lambda^{\frac{1}{2}}\right]^{-1}, \operatorname{Im}\lambda^{\frac{1}{2}} > 0, \; |\lambda| \geq \delta \text{ for every } \delta > 0$$

This section was based on Reference 24.

XIV. THE SPECTRAL REPRESENTATION

Consider the integral

$$I\left(x,x';\tilde{\lambda}\right) = (1/2\pi i)\int_\Gamma \left[G(x,x';\lambda)/\left(\lambda - \tilde{\lambda}\right)\right]d\lambda$$

where the contour Γ is shown in Figure 2, $\tilde{\lambda}$ being an arbitrary point in the complex λ-plane. G contains $\lambda^{1/2}$ and, hence, has a branch point at the origin. The spectral representation of **L** is usually based on the integral representation of $I(x,x';\tilde{\lambda})$ as given above. G undergoes a jump discontinuity across the Re λ axis. By Cauchy's integral theorem, the integral over Γ yields

$$I\left(x,x';\tilde{\lambda}\right) = G\left(x,x';\tilde{\lambda}\right) + \sum_{i=1}^{M} \operatorname{Res}\left[G\left(x,x';\tilde{\lambda}\right)/\left(\lambda - \tilde{\lambda}\right)\right] \tag{9.86}$$

It is, however, more convenient to consider the integral in the k-plane instead of the λ-plane, k^2 being λ. The integral I then takes the form

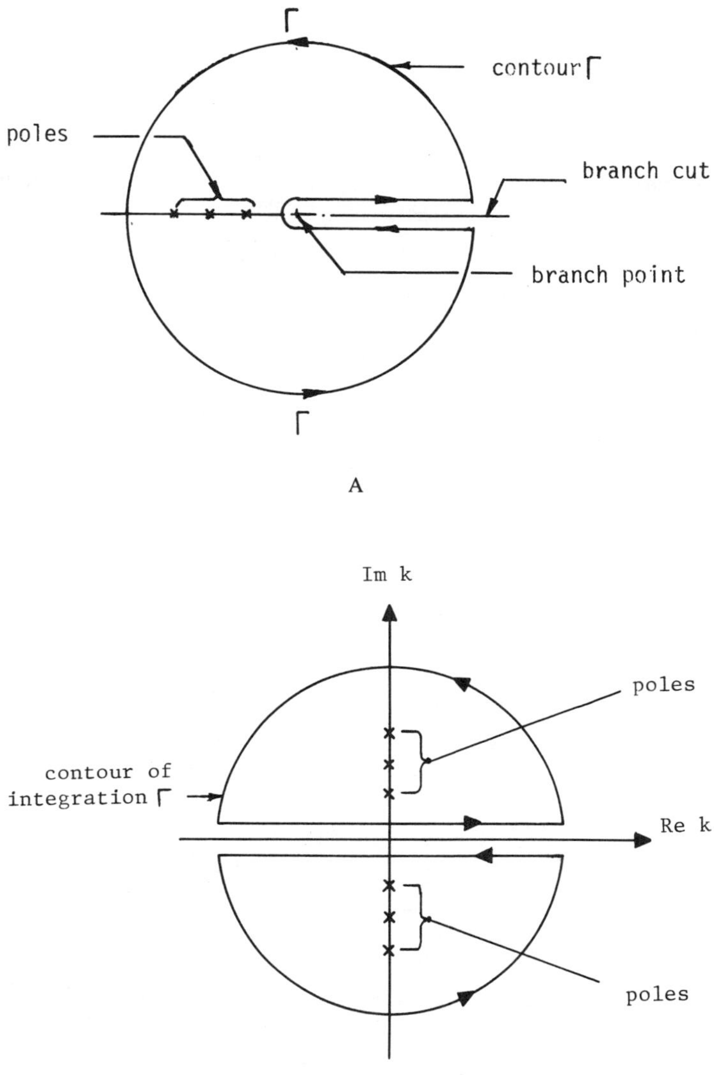

FIGURE 2. Contour for complex integration in the λ-plane.

$$I\left(x, x'; \tilde{k}\right) = (1/2\pi i)\int_{\Gamma'} \left\{\hat{G}\left(x, x'; k, \tilde{k}\right) / \left(k - \tilde{k}\right)\right\} dk$$

where Γ′ is the contour in the k-plane and

$$\hat{G}\left(x, x'; k, \tilde{k}\right) = 2G(x, x'; k)k / \left(k + \tilde{k}\right)$$

The spectrum is doubly degenerate. The modified Green function $\hat{G}$ is analytic for $|\text{Im } k| \geq |\varepsilon|$, except for a finite number of singularities which lie on the imaginary axis. For a fixed $k = u + iv$, $v > 0$,

$$\left|\hat{G}\left(x, x'; k, \tilde{k}\right) \leq |k|^{-1} \exp\left\{-v|x - x'|\right\} \to 0 \text{ as } |k| \to \infty\right.$$

Then

$$\lim_{R\to\infty} \int_\gamma \left[\hat{G}/\left(k-\tilde{k}\right)dk \to 0\right.$$

The contour γ consists of a circle in the upper half-plane. Thus, contributions from the circle in the limit $|k| \to \infty$, that is, $\lambda \to \infty$, vanish.

Let $g(x,x';\tilde{\lambda})$ be the contribution to the integral $I(x,x';\tilde{\lambda})$ from the horizontal contours $\lambda_\pm = \lambda_R \pm i0$ (Figure 2). Then $g(x,x';\tilde{\lambda})$ is the jump discontinuity in the Cauchy formula (9.86) across the real axis. Recalling that

$$G(x,x';\lambda) = \left[G\left(x,x';\lambda^*\right)\right]^* \text{ for } \operatorname{Im}\lambda < 0$$

and using the expression 9.83 for the resolvent, we obtain the jump contribution as

$$g\left(x,x';\tilde{k}\right) = (1/4\pi)(g_1 + g_2)$$

where

$$g_1 = \int_{-\infty}^{\infty}\left[1/t(k)\right]\left\{\phi(x',k)\psi(x,k)/\left(k^2-\tilde{k}^2\right)\right]dk$$

and

$$g_2 = \int_{-\infty}^{\infty}\left[1/t^*(k)\right]\left[\phi(x',k)\psi(x,k)/\left(k^2-\tilde{k}^2\right)\right]dk$$

Using Equations 9.6 and 9.8 for ϕ and $\overline{\phi}$ in the above expressions for g_1 and g_2 and finally adding them yields for $g(x,x';\tilde{k})$ the following expression,

$$g\left(x,x';\tilde{k}\right) = (1/4\pi)\int_{-\infty}^{\infty}\left[\begin{matrix}\left\{\overline{\psi}(x',k)\psi(x,k) + \left\{(r_+/t)(k)\right\}\psi(x',k)\psi(x,k) + \right.\\ \left. +\psi(x',k)\overline{\psi}(x,k) + \left\{\left(r_+^*/t^*\right)(k)\right\}\overline{\psi}(x',k)\psi(x,k)\right\}/\left(k^2-\tilde{k}^2\right)\end{matrix}\right]dk$$

$$= \frac{1}{4\pi}[\cdot]$$

where $[\cdot]$ is the term within the braces, and the relation $s_\pm(-k) = [s_\pm(k)]^*$, $s = r, t$ was used. This, therefore, yields for $g(x,x';\tilde{k})$ the following expression.

$$g\left(x,x';\tilde{k}\right) = (1/4\pi)\int_{-\infty}^{\infty}\left\{[\cdot]/\left(k^2-\tilde{k}^2\right)\right\}dk$$

Next, we calculate the contributions from the discrete spectrum:

$$\sum_{j=1}^{M} \operatorname{Res}\, G\left(x,x';\lambda_j,\tilde{\lambda}\right)/\left(\lambda_j-\tilde{\lambda}\right),\ \lambda = k^2,\ \ k = \pm ik_j\ \ k_j > 0$$

$$= \sum_{j=1}^{M} ic_-(k_j)\psi(x',k_j)\psi(x,k_j)\big/2\dot{t}_+(k_j)\left(k_j^2 - \tilde{k}^2\right) +$$

$$+ \sum_{i=1}^{M} i\bar{c}_-(k_i)\bar{\psi}(x,k_i)\bar{\psi}(x',k_i)\big/2\dot{t}_-(k_i)\left(k_i^2 - \tilde{k}^2\right) = \sum_{j=1}^{M}(\cdot)_j + \sum_{i=1}^{M}(\cdot)_i$$

the sums over j and i representing the residues from the poles in the upper and the lower half-plane, respectively, and M being the number of zeros of the transmission coefficient t(k) and $\bar{t}(k)$.

Collecting these results together in the Cauchy integral formula 9.86, applying the operator ($\mathbf{L} - k^2\mathbf{I}$) to both sides of this equation, and using the operational definition of Green's function given by Equation 9.83, we have

$$\delta(x - x') = (1/4\pi)\int_{-\infty}^{\infty}[\cdot]dk - \sum_{j=1}^{M}(\cdot)_j\left(k_j^2 - \tilde{k}^2\right) - \sum_{j=1}^{M}(\cdot)_i\left(k_i^2 - \tilde{k}^2\right)$$

In terms of the normalization constants D_{+j} and D_{+j} of Section VI.C,

$$ic_-/\dot{t} = i/c_+\dot{t} = -D_{+j}$$

and

$$i\bar{c}_-/\dot{\bar{t}} = -\bar{D}_{+j}$$

the resolution of the identity becomes

$$\delta(x - x') = (1/4\pi)\int_{-\infty}^{\infty}[\bullet]dk + \sum_{j=1}^{M} D_{+j}\psi(x,k_j)\psi(x',k_j)/2 + \sum_{i=1}^{M}\bar{D}_{+i}\bar{\psi}(x,k_i)\bar{\psi}(x',k_i)/2$$

However, from Section VI.A we know that for $\lambda = k_j^2$, $k_j = ik_j$, $k_j > 0$, the eigenfunctions are nondegenerate. Therefore, $\psi(x,k) = \bar{\psi}(x,k)$, $k = ik_j$. Hence, finally we obtain

$$\delta(x - x') = (1/4\pi)\int_{-\infty}^{\infty}[\cdot]dk + \sum_{i=1}^{M} D_{+j}\psi(x,k_j)\psi(x',k_j) \tag{9.86a}$$

The resolution of the identity (9.86a) can also be recast in a more compact form by writing the integrals g_1 and g_2 as

$$g_1 = \int_{-\infty}^{\infty}\left[T(k)\phi(x',k)\psi(x,k)\big/\left(k^2 - \tilde{k}^2\right)\right]dk$$

and

$$g_2 = \int_{-\infty}^{\infty}\left[T^*(k)\bar{\phi}(x',k)\bar{\psi}(x,k)\big/\left(k^2 - \tilde{k}^2\right)\right]dk$$

From Equation 9.48,

$$T(k)\phi(x,k) = \overline{\psi}(x,k) + R_+(k)\psi(x,k)$$

Using this in the above expressions and after some simple algebra, we obtain

$$g_1 + g_2 = \int_{-\infty}^{\infty} |T(k)|^2 \left[\psi(x',k)\overline{\psi}(x,k) + \phi(x',k)\overline{\phi}(x,k)\right] dk$$

The resolution of the identity is then

$$\delta(x - x') = (1/4\pi)\int_{-\infty}^{\infty} |T(k)|^2 \left[\psi(x',k)\overline{\psi}(x,k) + \phi(x',k)\overline{\phi}(x,k)\right] dk + \sum_{j=1}^{M} D_{+j}\psi(x,k_j)\psi(x',k_j)$$

$$= (1/4\pi)\int_{-\infty}^{\infty} \begin{bmatrix} \{\phi(x,k)/t(k)\}\{\phi^*(x',k)/t^*(k)\} + \\ \{\psi(x,k)/t(k)\}\{\psi^*(x',k)/t^*(k)\} \end{bmatrix} dk +$$

$$+ \sum_{j=1}^{M} D_j\psi(x,ik_j)\psi(x',ik_j) \tag{9.87}$$

XV. KREIN FUNCTIONALS AND MØLLER OPERATORS

It was mentioned earlier that the S-matrix or the scattering matrix arises in quantum mechanics as a unitary normalization operator which involves the so-called "Møller" operators. These operators describe the forward (backward) evolution of a freely evolving in-asymptote (out-asymptote) wave packet to the scattering-state vector at an arbitrary time, usually taken to be zero. The Møller operators and their relation to the S-matrix will be discussed in Section XVI.B as they arise in quantum scattering. These will next be generalized in Section XVII in the context of the scattering problems of this monograph. In Section XVII we will also derive the form of the scattering matrix already encountered in this chapter (Theorem 9.4). All this requires that the role of the improper eigenfunctions be examined. This is necessary since the Møller operators act only on the proper, normalized vectors in a Hilbert space. This is where the "Krein functionals", which will be introduced next, come into the picture.

A. THE KREIN FUNCTIONALS

Equation 9.86 expresses the completeness relation on the entire real axis. An arbitrary vector $u(x) \subset L_2(R)$ can be written as

$$u(x) = \int_{-\infty}^{\infty} u(x')\delta(x - x')dx' \tag{9.88}$$

with $\delta(x - x')$ given by Equation 9.87. This then represents the expansion of an arbitrary $u(x) \subset L_2(R)$ in terms of the eigenfunctions of the Schrödinger operator **L**. That is, the scattering states together with the bound states span the Hilbert space $L_2(R)$.

The Hilbert space $L_2(R)$ is a space the elements of which are square-integrable functions. Such a space must be spanned only by the "proper" eigenfunctions, which are vectors having finite lengths. If v is such a vector, i.e., if $v \subset L_2(R)$, then in Dirac's notation

$$\langle v'|v\rangle = \delta_{vv'}$$

δ being the Kronecker delta. We have, of course, considered v to be normalized. An arbitrary state vector $|\psi>$ in H can be expanded as

$$|\psi\rangle = \sum_{v} \psi_v |v\rangle$$

where $\psi_v = <v|\psi>$ is the "wave function" in the |v> representation. In the completeness relation above, the self-adjoint operator **L** has been treated as if its eigenvectors ("improper" vectors included) were a basis for H. The normalization conditions have been reproduced so as to resemble as closely as possible those involving an orthonormal basis, as given in the above equations in Dirac's notation. This is the point of view of Dirac, which is commonly held by physicists. Another point of view is that of Von Neuman's treatment in terms of the spectral family, generally favored by mathematicians. For further details on this point, Taylor[22] should be consulted.

From Equations 9.87 and 9.88,

$$u(x) = (1/4\pi)\int_{-\infty}^{\infty}\left\{(u,\hat{\phi})\hat{\phi}(x,k) + (u,\hat{\psi})\hat{\psi}(x,k)\right\}dk +$$

$$+\sum_{j=1}^{M} D_{+j}(u,\psi_j)\psi(x,ik_j),\; k_j > 0 \tag{9.89}$$

where $\hat{\phi}(\hat{\psi}) = \phi(\psi)(x,k)/t(k)$. (u,v) is the inner product appropriate to the Hilbert space $L_2(R)$, that is,

$$(u,v) = \int_{-\infty}^{\infty} u(x',k)v^*(x',k)dx',\; u,v \subset L_2(R)$$

and ψ_j in the inner product inside the sum over the discrete spectrum stands for $\psi(x',ik_j)$. We are concerned here essentially with the continuum. The sum over the point spectrum σ(L) will therefore be dropped in the rest of the discussion. Let us rewrite Equation 9.89 as

$$u(x) = \int_{-\infty}^{\infty} \hat{u}(k)\xi(x,k)d\rho(k),\; u(x) \subset L_2(R) \tag{9.90}$$

where,

$$\hat{\mathbf{u}}(k) = [\hat{u}_1(k), \hat{u}_2(k)]$$

$$\hat{u}_1(k) = (u,\hat{\phi})$$

$$\hat{u}_2(k) = (u,\hat{\psi})$$

$$\xi(x,k) = \begin{pmatrix} \hat{\phi}(x,k) \\ \hat{\psi}(x,k) \end{pmatrix} \tag{9.91}$$

and

$$d\rho(k) = dk/4\pi$$

The integral of Equation 9.90 is with respect to the ρ-measure. ρ(k) is the spectral function for the continuum and was encountered earlier in Chapters 3 and 5. Equation 9.90 is the generalization of the ordinary Fourier transform of $u(x) \subset L_2(R)$. $\hat{u}(k)$ is the "generalized" Fourier transform and ξ(x,k) the "generalized" Fourier kernel. It is known in the literature[24] that $u(k) \subset L_2(\rho,R)$. This means that

$$\|\hat{\mathbf{u}}(k)\|_\rho < \infty$$

As in Chapter 3, $L_2(\rho,R)$ is the space of square-integrable functions on R with ρ-measure. $\|\cdot\|_\rho$ is the L_2 norm in this space. Thus, $\hat{u}(k)$ is square integrable over R with respect to the measure ρ(k) of the continuum. Equation 9.90 is then a transformation **K** such that

$$\mathbf{K}\text{: } u(x) \subset L_2(R) \mapsto \hat{\mathbf{u}}(k) \subset L_{2,\rho}(R)$$

In a more convenient notation,

$$\mathbf{K}\text{: } u(x) \subset L_x \mapsto \hat{\mathbf{u}}(k) \subset L_\rho$$

$L_2(R)$ resp. $L_2(\rho,R)$ have been written $L_x(R)$ resp. $L_{2,\rho}(R)$ in order to explicitly distinguish the spaces on which u(x) resp. $\hat{\mathbf{u}}(k)$ are defined. ξ(k), which consists of the improper or generalized eigenfunctions ϕ(x,k) and ψ(x,k) normalized by t(k), is the kernel of this mapping known as[24] the "Krein direction functionals". Through ξ, the generalized eigenfunctions are incorporated into the Hilbert space framework as the kernels of the map **K**. The mapping is analogous to that derived in Chapter 3 for the case of the scattering on the semiaxis R_+. Stated otherwise, the improper eigenfunctions do not represent physically realizable states and many conclusions of scattering theory which are valid for proper vectors may not apply to the improper ones. The significance of the latter is in spanning the space $L_2(R)$ for an expansion of a square-integrable physical state u(x). This is what is meant by saying that the improper eigenfunctions are incorporated into the framework of the Hilbert space $L_2(R)$.

The mapping **K** is norm preserving, which can be shown as follows.

$$\|u\|_x^2 = \int_{-\infty}^{\infty} u(x)u^*(x)dx$$

$$= \int_{-\infty}^{\infty} dx\, u^*(x)\left[\int_{-\infty}^{\infty} \hat{\mathbf{u}}(k)\xi(x,k)d\rho(k)\right]$$

$$= \int_{-\infty}^{\infty} d\rho(k)\hat{\mathbf{u}}(k)\left[\int_{-\infty}^{\infty} \mathbf{u}^*(x)\xi(x,k)dx\right], \qquad u^* = \begin{pmatrix} u^* & 0 \\ 0 & u^* \end{pmatrix}$$

$$= \int_{-\infty}^{\infty} d\rho(k)\hat{\mathbf{u}}(k)\hat{\mathbf{u}}^*(k)$$

$$= \int_{-\infty}^{\infty} d\rho(k)|\hat{\mathbf{u}}(k)|^2$$

$$= \|\hat{\mathbf{u}}(k)\|_\rho^2$$

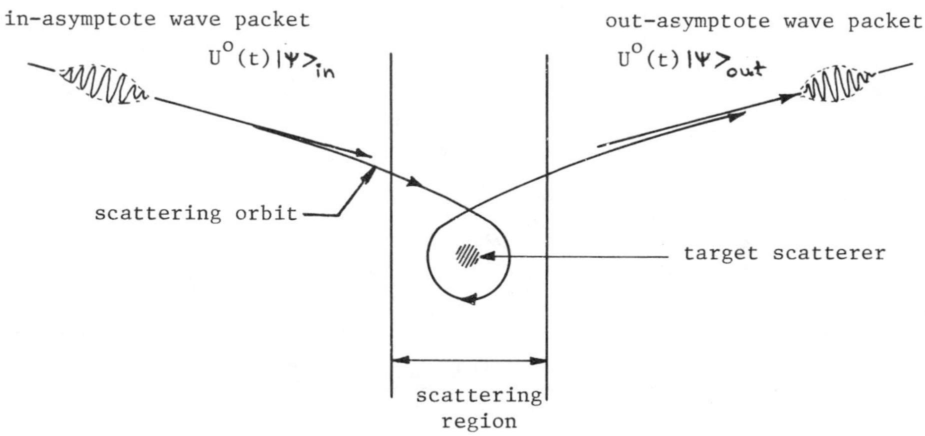

FIGURE 3. Scattering of wave packets from a potential.

Thus, **K** is a norm preserving operator with the Hilbert space L_2 (R) as its domain. Therefore,

$$D(\mathbf{K}) = H \equiv L_2(R) \text{ and } \|\mathbf{K}v\|_2 = \|v\|_2, \ \forall v \subset H$$

By definition, the mapping **K** is "isometric". Thus, if $\mathbf{K}^+$ denotes the adjoint of **K**, then $\mathbf{K}^+\mathbf{K} = \mathbf{I}$, but $\mathbf{K}\mathbf{K}^+ \neq \mathbf{I}$. Otherwise, unitarity will result. The space considered is, of course, infinite dimensional. For a finite dimensional space, the distinction between isometry and unitarity disappears. However, for an isometric operator $\mathbf{K}^+ = \mathbf{K}^{-1}$ on the range R(**K**). Hence, $\mathbf{K}^+\mathbf{K} = \mathbf{K}^{-1}\mathbf{K} = \mathbf{I}$ on L_x, whereas $\mathbf{K}\mathbf{K}^{-1} = \mathbf{I}$ acting on L_ρ. In this sense, **K** is an isometric operator which is also unitary.

B. THE MØLLER OPERATORS

Let us consider a typical scattering orbit shown in Figure 3. $|\psi>_{in}$ and $|\psi>_{out}$ are some arbitrary vectors in an appropriate Hilbert space. In quantum mechanics, the scattering is described by the time-dependent Schrödinger equation

$$i|\dot{\psi}\rangle = \mathbf{H}|\psi\rangle$$

or

$$|\psi\rangle = \exp(-i\mathbf{H}t)|\psi\rangle_\phi = \mathbf{U}(t)|\psi\rangle_\phi$$

H is the full Hamiltonian, $\mathbf{H} = \mathbf{H}_0 + V$, $\mathbf{H}_0$ being the "free" Hamiltonian and V, the potential of interaction. $|\psi>$ is the Dirac-state vector at any time t, whereas $|\psi>_0$ represents the state vector at t = 0. **U**(t), which is only formally exp(–i**H**t), is the "evolution operator". $\mathbf{U}(t)|\psi>_0$ is the scattering "orbit", although no such orbit actually exists in the Euclidean space. A wave packet which was "freely" evolving in the remote past, $t \to -\infty$, is represented by $\mathbf{U}^0(t)|\psi>_{in}$, $\mathbf{U}^0(t) = \exp(-i\mathbf{H}_0t)$. It forms the "in asymptote orbit" which evolves into the scattering "orbit" $\mathbf{U}(t)|\psi>_0$. This, in turn, evolves in the remote future, $t \to +\infty$, into another "freely" evolving wave packet $\mathbf{U}^0(t)|\psi>_{out}$, the "out asymptote". The "remote" past and future, $t \to \mp\infty$, are only mathematical, actual collision time being, of course, extremely small — of the order of less than a nanosecond. Thus,

$$\mathbf{U}(t)|\psi\rangle_0 \xrightarrow[t\to-\infty]{} \mathbf{U}^0(t)|\psi\rangle_{in};\ |\psi\rangle_{in},\ |\psi\rangle_{out},\ |\psi\rangle_0 \subset H$$

$$\mathbf{U}(t)|\psi\rangle_0 \xrightarrow[t\to+\infty]{} \mathbf{U}^0(t)|\psi\rangle_{out}$$

The convergence simply means that the difference $\to 0$ as $t \to \pm\infty$ in norm.

In quantum scattering, the so-called "asymptotic condition"[22,23] states that if the potential q is "regular" (see Chapter 7 for regularity conditions), then for every $|\psi\rangle \subset H$ there is a $|\psi\rangle_{in}$ such that

$$\mathbf{U}(t)|\psi\rangle_0 \to \mathbf{U}^0(t)|\psi\rangle_{in} \xrightarrow[t\to-\infty]{} 0$$

and likewise for every $|\psi\rangle_{out}$ in H as $t \to +\infty$. This condition guarantees that for every $|\psi\rangle_{in}, |\psi\rangle_{out} \subset H$, there exists a state $|\psi\rangle_0 \subset H$ such that the $\mathbf{U}^0(t)|\psi\rangle_{in}$ and $\mathbf{U}^0(t)|\psi\rangle_{out}$ do describe the asymptotes to an "actual" orbit $\mathbf{U}(t)|\psi\rangle_0$. On the other hand, the converse is not true in general. Now the Hilbert space H is

$$H = R \oplus B$$

R and B being spaces spanned by the continuum and the bound states, respectively, and $\oplus$, the "direct sum". Then the theorem of "asymptotic completeness"[22,23] guarantees that for every $|\psi\rangle_0 \subset R$, $\mathbf{U}(t)|\psi\rangle_0$ does describe a scattering orbit with in- and out- asymptotes. These conclusions are valid if the potential is well behaved and hold for the sort of potentials encountered so far.

The evolution operator $\mathbf{U}(t)$ is unitary and reflects the conservation of probability. Then $\mathbf{U}^+\mathbf{U} = \mathbf{U}\mathbf{U}^+ = \mathbf{I}$, and we have

$$|\psi\rangle_0 = \lim_{t\to-\infty} \mathbf{U}^+(t)\mathbf{U}^0(t)|\psi\rangle_{in}$$

that is, the wave packet $\mathbf{U}^0(t)|\psi\rangle_{in}$ evolving freely in the remote past $t \to -\infty$ develops through $\mathbf{U}^+(t)$ into the state vector $|\psi\rangle_0$ at $t = 0$. Similarly,

$$|\psi\rangle_0 = \lim_{t\to\infty} \mathbf{U}^+(t)\mathbf{U}^0(t)|\psi\rangle_{out}$$

This says that the state vector $|\psi\rangle_0$ at $t = 0$ will evolve as $t \to \infty$ into another "freely" evolving wave packet $\mathbf{U}^0(t)|\psi\rangle_{out}$. Denoting

$$\lim_{t\to\pm\infty} \mathbf{U}^+(t)\mathbf{U}^0(t) = \Omega_{\mp}(t) \tag{9.92}$$

we have

$$|\psi\rangle_0 = \mathbf{\Omega}_+(t)|\psi\rangle_{in}$$

and

$$|\psi\rangle_0 = \mathbf{\Omega}_-(t)|\psi\rangle_{out}$$

The subscript + indicates a time-forward evolution *into* the state $|\psi>_0$, whereas the subscript – is associated with the backward process, that is, time-backward evolution of the out-asymptote into the scattering state $|\psi>_0$. Thus, the operators $\Omega\mp$ act on any vector $|\psi> \subset H$ and yield a state at t = 0 that would either evolve to or from the asymptote represented by $|\psi>$.

In Equation 9.92, the operators $\Omega_\mp$ are defined as the limits involving unitary operators and, hence, are not necessarily unitary themselves. Now,

$$\mathbf{\Omega}_\pm|\psi\rangle = \lim_{t\to\mp\infty} \mathbf{U}^+(t)\mathbf{U}^0(t)|\psi\rangle$$

Then

$$\|\Omega_\pm\psi\| = \lim_{t\to\mp\infty}\left\|\mathbf{U}^+(t)\mathbf{U}^0(t)\psi\right\| = \|\psi\|$$

Thus, the operators $\Omega_\pm$ are norm preserving. As mentioned before, although $\Omega_\pm$ may act on the entire Hilbert space, their range is a subspace of H, namely, the subspace R spanned by the continuum states. Then by definition, the operators $\Omega_\pm$ are isometric, and not unitary in general. However, in the absence of the bound states, B = 0 and $\Omega_\pm$ become unitary.

In summary, for every $|\psi>$ in R the orbit $\mathbf{U}(t)|\psi>$ describes a scattering process with in- and out-asymptotes:

$$\mathbf{U}(t)|\psi\rangle \xrightarrow[t\to-\infty]{} \mathbf{U}^0(t)|\psi\rangle_{in}$$

$$\xrightarrow[t\to+\infty]{} \mathbf{U}^0(t)|\psi\rangle_{out}$$

Every $|\psi>_{in}$ (or $|\psi>_{out}$) in H labels the in- (or out-) asymptote of a unique actual orbit $\mathbf{U}(t)|\psi>$ and the operators $\Omega_\pm$ map each $|\psi>_{in}$ (or $|\psi>_{out}$) in H onto the corresponding scattering state $|\psi>$ in R.

$$|\psi\rangle = \mathbf{\Omega}_+|\psi\rangle_{in} = \mathbf{\Omega}_-|\psi\rangle_{out}$$

The operators are isometric. This means that for each normalized $|\psi>_{in}$ or $|\psi>_{out}$ in H, there is a unique, corresponding normalized $|\psi>$ in R; and conversely, for each normalized $|\psi>$ in R, there are unique normalized asymptotes $|\psi>_{in}$ and $|\psi>_{out}$. The operators $\Omega_\pm$ are known as the "Møller"[22-23,25] wave operators. They describe the connections between the asymptotic states and the state at t = 0. However, in an actual scattering experiment, what is observed (either in a bubble or in a cloud chamber) is not the "uninteresting" actual scattering "orbits" in the scattering region, but the asymptotes. Hence, in the scattering theory, it is more useful to be able to describe the evolution of the in-asymptote into the out-asymptote directly. Since the Møller operators are isometric, such a description is not difficult to obtain, at least formally. Indeed, since $\Omega_\pm^+ \Omega_\pm = I$ by virtue of isometry of the Møller operators, then

$$|\psi\rangle_{out} = \mathbf{\Omega}_-^+\mathbf{\Omega}_-|\psi\rangle_{out} = \mathbf{\Omega}_-^+|\psi\rangle_0$$
$$= \mathbf{\Omega}_-^+\mathbf{\Omega}_+|\psi\rangle_{in} = \mathbf{S}|\psi\rangle_{in}$$

where $\mathbf{S} = \Omega_-^{+}\,\Omega_+$ is the "scattering operator". In the absence of the bound states, S becomes a "unitary normalization operator" exactly as described before. In the following section the actual form of the scattering matrix **S** will be derived, the discrete spectrum, of course, being absent.

XVI. "DERIVATION" OF THE SCATTERING OPERATOR S

Equation 9.92 shows that the Møller operators act in two steps. The first stage involves operating on a state vector $|\psi\rangle \subset H$ by the "free" evolution operator $\mathbf{U}^0(t)$. The resultant is then operated by the adjoint $\mathbf{U}^{+}(t)$ of the evolution operator $\mathbf{U}(t)$ corresponding to the full Hamiltonian. A diagrammatic representation of this two-stage operation is given below.

$$\underset{\subset H}{|\psi\rangle} \xrightarrow{U^0(t)} \mathbf{U}^0(t)|\psi\rangle \xrightarrow{U^+(t)} \underset{\subset H}{|\psi\rangle_\rho}$$

Let us now translate all this from the quantum domain and Dirac notation to the scattering problem of this monograph, which is essentially a Fourier transform of the time-dependent Schrödinger equation

$$i|\dot{\psi}\rangle = \mathbf{H}|\psi\rangle$$

The "free" evolution operator $\mathbf{U}^0(t)$ corresponds to the scattering "free" Schrödinger operator $\mathbf{L}_0 = \mathbf{L}(q = 0) = \mathbf{D}'' + \lambda$. The operator $\mathbf{U}(t)$ corresponds to the full Schrödinger operator $\mathbf{L}$ in our case. The arbitrary state vector $|\psi\rangle \subset H$ is equivalent to an arbitrary square-integrable function $u(x) \subset L_2(R)$. Thus, $\mathbf{U}^0(t)|\psi\rangle$ of the above diagram is equivalent to an $\mathbf{L}_0$ Fourier transform of u(x), i.e., transforming u(x) by the Krein functionals involving $\mathbf{L}_0$ eigenfunctions. The second operation $\mathbf{U}^{+}(t)$ corresponds to back Fourier transforming with kernels as Krein functionals corresponding to $\mathbf{L}$. Thus, $\mathbf{U}^0(t)|\psi\rangle$ corresponds to the $\mathbf{K}_0$ operation on u(x), $\mathbf{K}_0$ being the Krein functional mapping from $\mathbf{L}_0$. Similarly, operating with $\mathbf{U}^{+}(t)$ parallels the actions of $\mathbf{K}^*$, the $\mathbf{K}$ operation being with the Krein functionals having $\mathbf{L}$ eigenfunctions. Hence, in the present context, the Møller operator will be diagrammed as

$$\underset{\subset L_x}{u(x)} \xrightarrow{\mathbf{K}_0} \underset{\subset L_0}{u_0(k)} \xrightarrow{\mathbf{K}^*} \underset{\subset L_x}{v(x)}$$

$$\hat{\mathbf{u}}_0(k) = \int_{-\infty}^{\infty} \mathbf{u}(x)\xi_0^*(x,k)dx$$

where

$$\xi_0^* = \left[\exp(+ikx), \exp(-ikx)\right]/\bar{t}(k)$$

and

$$\mathbf{K}_0\mathbf{u}(x) = \hat{\mathbf{u}}_0(k)$$

where $\hat{\mathbf{u}}_0(k) = (\hat{u}_{01}, \hat{u}_{02})^T$. $\mathbf{K}^* = \mathbf{K}^{-1}$ acts only on the range of $\mathbf{K}$, whereas $\hat{\mathbf{u}}_0(k) \subset L_{\rho 0}$ for the free operator $\mathbf{L}_0$. It is then necessary that $\mathbf{L}$ be spectrally represented on the same space as $L_{\rho 0}$.

It was already demonstrated that

$$\|u\|_x^2 = \|\hat{u}\|_\rho^2$$

Generalizing, we have

$$(u, v)_x = (\hat{u}, \hat{v})_\rho = (\mathbf{K}u, \mathbf{K}v)\rho$$

Then

$$\lambda\|\hat{u}\|_\rho = (\lambda u, u)_x = (\mathbf{L}u, u)_x = (\mathbf{KL}u, \mathbf{K}u)_x = \left(\mathbf{KLK}^+\hat{u}, \hat{u}\right)_\rho$$

Consider a unitary "intertwining" operator **U** such that

$$\mathbf{LU} = \mathbf{UL}_0$$

Then

$$(\mathbf{L}_0 v, u)_x = \left(\mathbf{K}_0(\mathbf{L}_0 v),\ \mathbf{K}_0 u\right)_{\rho_0} = \left(\left(\mathbf{K}_0\mathbf{U}^+\right)\mathbf{UL}_0 v,\ \left(\mathbf{K}_0\mathbf{U}^+\right)(\mathbf{U}u)\right)_{\rho_0}$$

$$= \left(\left(\mathbf{K}_0\mathbf{U}^+\mathbf{L}\left(\mathbf{K}_0\mathbf{U}^+\right)\left(\mathbf{K}_0\mathbf{U}^+\right)\mathbf{U}v,\ \left(\mathbf{K}_0\mathbf{U}^+\right)(\mathbf{U}u)\right)\right)_{\rho_0}$$

On the other hand,

$$(\mathbf{L}_0 v, u)_x = \left(\mathbf{U}^+\mathbf{LU}v,\ \mathbf{U}^+\mathbf{U}u\right)_x = (\mathbf{LU}v,\ \mathbf{U}u)_x = (\mathbf{KLU}v,\ \mathbf{KU}u)_\rho$$

$$= \left(\left(\mathbf{KLK}^+\right)\mathbf{K}(\mathbf{U}v),\ \mathbf{K}(\mathbf{U}u)\right)_\rho$$

Comparing, it follows that

$$\mathbf{K} = \mathbf{K}_0\mathbf{U}^+ \text{ and } L_\rho = L_{\rho 0}$$

Defining $\mathbf{K}^+\mathbf{K}_0 \equiv \mathbf{U}_-$ and from the earlier result $\mathbf{KLK}^+ = \mathbf{k}^2\mathbf{I}$, it follows that

$$\mathbf{LU}\_u = \mathbf{k}^2\mathbf{U}\_u = \mathbf{U}\_k^2 u = \mathbf{U}\_\mathbf{L}_0 u$$

This yields the intertwining operator relationship $\mathbf{LU}_- = \mathbf{U}_-\mathbf{L}_0$, showing that $\mathbf{U}_- = \mathbf{K}^+\mathbf{K}_0$ is an intertwining operator. Consider the **K** operation with the Krein functionals with $\xi(x,k)$ of Equation 9.91 as eigenfunctions. Then

$$\mathbf{K}^+\mathbf{K}_0 u(x) = \mathbf{U}\_u(x) = \int_{-\infty}^{\infty}\xi(x,k)\hat{\mathbf{u}}_0(k)d\rho(k)$$

From Equation 9.7, we have

$$\overline{\psi} = \bar{t}\phi + \bar{r}_{-}\overline{\phi}$$

Also, from Equation 9.8,

$$\overline{\phi} = \bar{t}\psi + \bar{r}_{+}\overline{\psi}$$

Rearranging and expressing ϕ and ψ in terms of $\overline{\phi}$ and $\overline{\psi}$, and from Equation 9.12a we obtain

$$\xi(x,k) = \overline{\xi}(x,k)\mathbf{s}(k)$$

where

$$\mathbf{s}(k) = \begin{pmatrix} 1/t & -\bar{r}_{+}/t \\ r_{+}/t & 1/t \end{pmatrix}(k)$$

and

$$\overline{\xi}(x,k) = \left(\hat{\overline{\psi}}, \hat{\overline{\phi}}\right)$$

Therefore,

$$\mathbf{U}_{-}u(x) = \int_{-\infty}^{\infty} \overline{\xi}(x,k)\mathbf{s}(k)\hat{\mathbf{u}}_0(k)d\rho(k)$$

$$= \int_{-\infty}^{\infty} \overline{\xi}(x,k)\hat{\mathbf{w}}_0(k)d\rho(k) = \mathbf{U}_{+}w(x)$$

where

$$\hat{\mathbf{w}}_0(k) = \mathbf{s}\hat{\mathbf{u}}_0$$

$\mathbf{U}_{+}$ represents the K-operation corresponding to the Krein functionals $\xi(k)$ and $w(x) \subset L_x$ whose $\mathbf{L}_0$ transformation is the vector $\mathbf{s}(k)\ \hat{\mathbf{u}}_0(k)$. Thus,

$$\mathbf{U}_{+}w(x) = \mathbf{U}_{-}u(x)$$

from which we obtain

$$\mathbf{U}_{+}^{+}\mathbf{U}_{+}w(x) = w(x) = \mathbf{U}_{+}^{+}\mathbf{U}_{-}u(x)$$

Writing $\mathbf{S}$ for $\mathbf{U}_{+}^{+}\,\mathbf{U}_{-}$, we have

$$\mathbf{S}u(x) = w(x)$$

From this we obtain

$$\mathbf{S}\mathbf{K}_0^{*}\hat{\mathbf{u}}_0(k) = \mathbf{K}_0^{*}\hat{\mathbf{w}}_0(K) = \mathbf{K}_0^{*}\mathbf{s}\hat{\mathbf{u}}_0$$

Comparing,

$$\mathbf{SK}_0^* = \mathbf{K}_0^*\mathbf{s}$$

or

$$\mathbf{K}_0\mathbf{SK}_0^* = \tilde{\mathbf{S}} = \mathbf{s}$$

where $\tilde{\mathbf{S}}$ is the $\mathbf{K}_0$-transformed $\mathbf{S}$. That is, $\tilde{\mathbf{S}}$ is the $\mathbf{L}_0$ representative of $\mathbf{S}$. Then

$$\tilde{\mathbf{S}} = \begin{pmatrix} 1/t & -\bar{r}_+/t \\ r_+/t & 1/t \end{pmatrix} = \begin{pmatrix} 1/t & -r_-/t \\ r_+/t & 1/t \end{pmatrix}(k) = \begin{pmatrix} T_+(k) & R_-(k) \\ R_+(k) & T_-(k) \end{pmatrix}$$

This is the "scattering matrix" encountered earlier.

XVII. ORTHOGONALITY RELATIONS FOR THE JOST SOLUTIONS ON R

Some orthogonality relations satisfied by the Jost solutions of the Schrödinger problem on the entire real axis will be established below. Let $y_+(x,k)$ resp. $y_-(x,k)$ denote the Jost solutions $\psi(x,k)$ resp. $\phi(x,k)$. $\tilde{y}_\pm$, $\tilde{r}_\pm$, $\tilde{t}$ will denote the Jost solutions $y_\pm$, the reflection and transmission coefficients, respectively, at $k = \tilde{k}$. From the earlier results (Section VI.A),

$$\int_{-\infty}^{\infty} y_\pm \tilde{y}_\pm dx = \left(k^2 - \tilde{k}^2\right)^{-1} W\left[y_\pm, \tilde{y}_\pm\right]\Big|_{-\infty}^{\infty}$$

where

$$W\left[y_\pm, \tilde{y}_\pm\right] = y_\pm \tilde{y}'_\pm - y'_\pm \tilde{y}_\pm$$

In order to evaluate these Wronskians, the representations given by Equations 9.5 and 9.6 will be required along with the asymptotic expressions of Equations 9.9a and 9.9b. Furthermore, it will also be necessary to use the delta sequence

$$\lim_{z\to\infty}\left[\exp(\pm ipz)/p\right] = \pm i\pi\delta(p)$$

This is the well known shifting property of Dirichlet kernel.* This is another way of expressing the more familiar δ-sequence given by

$$\lim_{z\to\infty}(\sin pz)/\pi p \to i\delta(p)$$

Evaluating the Wronskians W at both limits, using Equations 9.5, 9.6, 9.9, and the above δ-sequence, and after some straightforward algebra, we obtain

$$\pi^{-1}\int_{-\infty}^{\infty} y_\pm(x,k)y_\pm\left(x,\tilde{k}\right)dx = \delta\left(k + \tilde{k}\right)\left[1 + r\tilde{r} + t\tilde{t}\right] \mp \delta\left(k - \tilde{k}\right)\left[r\tilde{t} + t\tilde{r}\right]$$

* *Generalized Functions*, R. P. Kanwal, Academic Press, New York, 1983.

Replacing $\tilde{k}$ by –k in the coefficient of $\delta(k + \tilde{k})$ and $\tilde{k}$ by k in that of $\delta(k - \tilde{k})$, using the definitions T = 1/t and R = r/t, and, finally, using Equations 9.5, 9.6, and 9.9, we obtain after some straightforward algebra

$$(2\pi)^{-1}\int_{-\infty}^{\infty} y_{\pm}\tilde{y}_{\pm}dx = |T(k)|^{-2}\delta\left(k+\tilde{k}\right) \mp R_{\pm}(k)[T(k)]^{-2}\delta\left(k-\tilde{k}\right)$$

Similarly,

$$(2\pi)^{-1}\int_{-\infty}^{\infty} y_{\pm}\tilde{y}_{\mp}dx = [1/T(k)]\delta\left(k-\tilde{k}\right)$$

These orthogonality relations hold for the scattering states, that is, for the continuous spectrum where Im k = 0. The potential q(x) is not required to be real.

Other orthogonality relations are

$$\int_{-\infty}^{\infty} y_{\pm}(x,k_i)y_{\pm}\left(x,k_j\right)dx = \left(y_{\pm i}, y_{\pm j}\right) = \delta_{ij}$$

where $y_{\pm i}(x)$, $y_{\pm j}(x)$ are the normalized bound states. Also,

$$\int_{-\infty}^{\infty} dx\ y_{\pm}(x,k)y_{\pm n}(x) = 0$$

The eigenfunctions of the continuum are, therefore, orthogonal to the corresponding eigenfunctions of the discrete spectrum. This again proves what was said in Section XV.A that the scattering and the bound states together span the Hilbert space.

XVIII. SCATTERING DATA AND SPECTRAL TRANSFORM

The collection of data

$$S_{\pm}(k) = \left\{R_{\pm}(k),\ k \subset R;\ D_{\pm j},\ k = ik_j,\ k_j > 0,\ j = 1,\ 2,\ldots.N\right\}$$

are called the "scattering data". Since the reflection coefficients, the point spectrum $\sigma(\mathbf{L})$, and the normalizing constants all depend on the potential q(x), then so does S. This fact is emphasized by writing S as S[q], which is then

$$S_{\pm}[q] = \left\{R_{\pm}(k),\ k \subset R;\ D_{\pm j},\ k = ik_j,\ k_j > 0,\ = 1,\ 2,\ldots.N\right\} \tag{9.93}$$

S[q] resembles a transform between the x- and the k-domain. Hence, S [q(x)] is also variously called the "spectral transform" or the "scattering transform" of the function q(x). Let us assume that the potential q(x) is weak, that is, the contribution from the continuum is small. Let it be further assumed that the number of discrete eigenvalues is at most one. Then $R_{\pm}(K)$ and q(x) are related exactly by a Fourier transform. In the context of scattering, this is the well-known "Born approximation", which was encountered in Section V of this chapter. This will be demonstrated later. The "scattering data" or the "spectral transform" is uniquely determined by the quantities specified in Equation 9.93. The potential q(x) and its spectral transform S have a one-to-one correspondence between them:

$$q(x) \leftrightarrow S[q] \tag{9.94}$$

The forward part q --→ S[q] constitutes the "direct problem", whereas the back transformation q ← S[q] gives rise to the "inverse problem". Specifying S is equivalent to specifying the spectral matrix. The transform given by Equation 9.94 is "nonlinear"* except in the Born approximation, where it reduces to the linear Fourier transform. The same also holds true for the inverse part of the transform. S[q] is perhaps the most important physical quantity in solving inverse problems and, what amounts to the same thing, in solving a certain class of highly important nonlinear partial differential equations. Its importance and significance in the inverse scattering problems can hardly be overemphasized. Let us, therefore, digress here to elucidate this point.

A. RELATION BETWEEN THE FOURIER AND THE SPECTRAL TRANSFORM

Let us consider a Cauchy boundary-value problem given by

$$\dot{u}(x,t) = -iw(z)u'(x,t)$$

$$u(x,0) = u_0(x), \ x \subset R \tag{9.95}$$

u is also assumed to vanish asymptotically as $|x| \to \infty$ at a rate such that the integrals arising in the course of the analysis converge. z is given by $-i\mathbf{D}$, $\mathbf{D} = d/dx$, and w(z) is a polynomial in z. Since z does not contain either u or its derivatives, the Cauchy problem is linear. For a general problem of this sort, it is only rarely possible to obtain an analytical solution by attacking the problem directly, that is, by trying to obtain a solution by quadrature. It is this difficulty that motivates a solution by the Fourier transform.

Consider the Fourier transform pair:

$$u(x,t) = \int_{-\infty}^{\infty} dk \ \exp(ikx)\hat{u}(k,t)$$

$$\hat{u}(k,t) = \int_{-\infty}^{\infty} dx \ \exp(-ikx)u(x,t)$$

The normalization constants are omitted in the Fourier transform expressions. If u^(k,0) is the Fourier transform of $u_0(x)$, then the Fourier-transformed Equation 9.95 is

$$\dot{u}^\wedge(k,t) = -iw(k)u^\wedge(k,t) \tag{9.96}$$

yielding

$$u^\wedge(k,t) = u^\wedge(k,0)\exp[-iw(k)t] \tag{9.97}$$

The solution is then

$$u(x,t) = \int_{-\infty}^{\infty} dk \exp[i\{kx - w(k)t\}]u^\wedge(k,0) \tag{9.98a}$$

where w(k) is the "dispersion". If, in addition, w contains t explicitly, i.e., w = w(z,t), then

* The spectral transform has also been called "nonlinear Fourier transform" by Ablowitz et al.: *Stud. Appl. Math.*, 53, 1974. This reference contains probably the most comprehensive discussion on the subject.

$$u(x,t) = \int_{-\infty}^{\infty} dk \exp\left[i\left\{kx - \int_0^t w(k,t')dt'\right\}\right]u\hat{\ }(k,0) \tag{9.98b}$$

In the event that w also contains x explicitly, then the Fourier transform solutions become difficult unless the dependence on x is linear. However, it will not be necessary to elaborate on this further.

A solution by the Fourier transform is thus achieved in three steps:

1. Write the Fourier transform u^(k,0) of $u_0(x)$, namely,

$$u\hat{\ }(k,0) = \int_{-\infty}^{\infty} dx \exp(-ikx)u_0(x) \tag{9.99}$$

2. Solve the Fourier-transformed time evolution Equation 9.96.
3. Obtain the solution as given by Equations 9.98a and/or 9.98b.

It may be recalled that these steps were discussed in Example 7 of Chapter 2. First and foremost is the fact that the time evolution operator in the k-space is simple. From Equation 9.96, it is just –iw(k) instead of –iw(–i**D**). In Equation 9.98, the operator is only exp{–iw(k)t]. Contrast this to the corresponding operator in the configuration or the x-space which is formally given by the "resolvent operator":

$$u(x,t) = \exp\{-iw(-i\mathbf{D})\}u_0(x) \tag{9.100}$$

Thus, the evolution operator in the x-space is relatively more complex and is amenable to analytical solutions if w(z) = ± z, resulting in a "translation" solution given by

$$u(x,t) = u(x+t) = \exp(t\mathbf{D})u_0(x)$$

The entire operation is schematized diagrammatically in Figure 4.

The "spectral" or "scattering" transforms perform a function for a certain class of nonlinear evolution equations analogous to that performed by the Fourier transforms for linear evolution equations such as Equation 9.95 (see Figure 5). This is true at least in a one-space dimension and for evolution equations of the nonintegral type. Such an evolution equation in one-dimension x is primarily of the form[5]

$$u(x,t) = w(\mathbf{L})u'(x,t) \tag{9.101}$$

where w(z) is, as before, a polynomial in z. However, the operator **L** is now an integro-differential operator containing u and its derivatives. The problem is thus nonlinear. The form of the operator **L** specifying this class of evolution equations is given by[5]

$$\mathbf{L} = \mathbf{D}^2 - 4u(x,t) + 2\mathbf{D}u(x,t)F$$

where

$$\mathbf{D} = \partial/\partial x \text{ and } F = \int_x^{\infty} dy$$

Highly interesting is the fact that a nonlinear equation such as Equation 9.101 is surprisingly closely related to a linear scattering equation such as the Schrödinger equation.

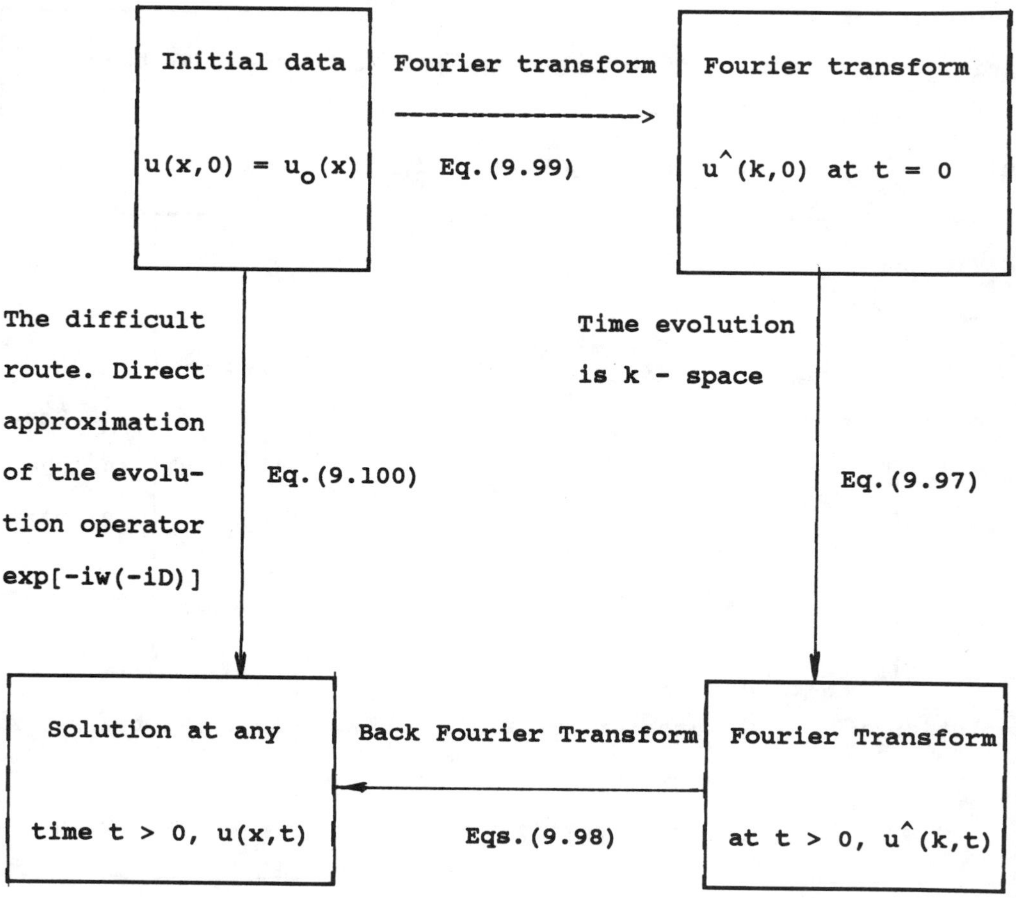

FIGURE 4. A schematic diagram illustrating the Fourier transform technique of solving linear evolution equations.

Assume that the potential in the Schrödinger equation depends on a parameter t, i.e., q = q(x,t). It turns out that if the Schrödinger equation is "iso-spectral", i.e., if the parameter λ in that equation does not depend on t, then the potential q satisfies an evolution equation such as that given by Equation 9.101. It is worth mentioning that although, in keeping with the terminology of quantum mechanics, we are considering evolution in time t, the physical interpretation is immaterial and t could be replaced by any other relevant parameter. Obtaining solutions of Equation 9.101 is, therefore, doubly useful. First, it solves a certain class of difficult nonlinear equations which are highly relevant in the physical sciences. Second, it yields the solution of the inverse problem involving a potential such as the Schrödinger equation, a problem which is extremely important in numerous scientific and engineering applications.

At the moment it will be assumed that for the class of evolution equations given by Equation 9.101, the spectral transform evolves according to the ordinary linear differential equations.

$$R(k,t) = 2ikw\left(-4k^2\right)R(k,t)$$

$$\dot{D}_j(k) = -2k_j w\left(4k_j^2\right)D_j(t), \quad = 1, 2, \ldots . N$$

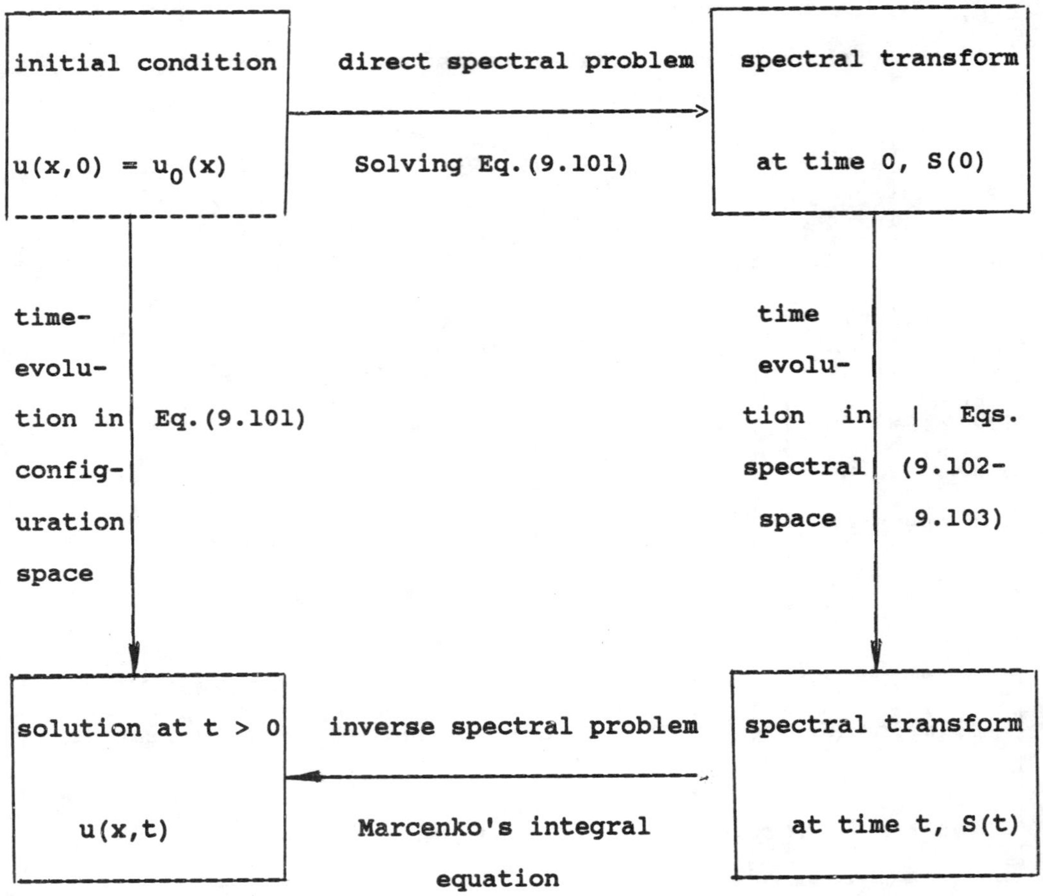

FIGURE 5. A schematic diagram illustrating the scattering transform technique of solving nonlinear evolution equations.

$$\dot{k}_j(t) = 0, \quad = 1,2,\ldots N \tag{9.102}$$

Straightforward integration of Equation(s) 9.95 then yields the time evolution:

$$R(k,t) = R(k,0)\exp\left[2ikw\left(-4k^2\right)t\right]$$

$$D_j(t) = D_j(0)\exp\left[-2k_j w\left(4k_j^2\right)t\right], \quad = 1,2,\ldots N$$

$$k_j(t) = k_j(0) = k_j, \; j = 1,2,\ldots N \tag{9.103}$$

The last of Equation(s) 9.103 implies that the Schrödinger equation being considered is isospectral, i.e., if q(x,t) evolves in the parameter t, then the discrete eigenvalues remain independent of t. This feature is crucial in the solution of the inverse scattering problems.

Like its linear counterpart, the solution of the Cauchy problem for the nonlinear evolution equation (9.101) also proceeds in three stages. The first step is the "direct" problem and can be described as follows. The Cauchy problem for the nonlinear evolution equation is

$$\dot{u} = w(\mathbf{L})u'$$

$$u(x, t = 0) = u_0(x), \ x \subset R$$

As already mentioned, u(x,t) in the above Cauchy problem is the potential of the time-independent Schrödinger equation, i.e.,

$$y'' + [\lambda - u(x,t)]y = 0 \tag{9.104}$$

with t as the parameter. The direct problem consists of replacing u by u_0 in the Schrödinger equation and then calculating the spectral transform corresponding to u_0:

$$S_{\pm}[0] = \left\{R_{\pm}(k,0), \ k \subset R; \ D_{\pm j0}, \ k = ik_{j0}, \ k_{j0} > 0, \ j = 1, 2, \ldots N\right\} \tag{9.105}$$

The second stage comprises in going from $S(0) \rightarrow S(t)$ according to the evolution equations (9.102 and 9.103) for the scattering data. The final stage consists of back spectral transform, i.e., going from $S(t) \rightarrow u(x,t)$. This stage is prescribed by the celebrated Marcenko integral equation and will form the subject matter for the next chapter. We will accept momentarily that, given certain restrictions on the potential, this final stage can, indeed, be performed.

To summarize, solving a nonlinear evolution equation such as that given by Equation 9.101 involves the following three stages.

1. Solving the direct problem: $u_0(x) \rightarrow S_{\pm}(0)$. This involves solving the Schrödinger equation (9.104) with $u_0(x)$ as the potential.
2. The evolution of $S_{\pm}(0)$: $S_{\pm}(0) \rightarrow S_{\pm}(t)$ according to Equations 9.102 and 9.103.
3. The final solution: $S_{\pm}(t) \rightarrow u(x,t)$ via Marcenko's equation.

Consider the solution of a linear system by the Fourier transform. The principal advantage of the technique is not in an easy prescription for analytically reducing the quadratures (as a matter of fact, the integrals appearing in Equation 9.100 can be reduced only very rarely). Rather, the advantage lies in a simplified and transparent solution procedure. In other words, the problem is significantly simplified working in the k-space instead of the x-space. This is easily seen by comparing the forms of the corresponding time evolution operators of Equations 9.97 and 9.100. Moreover, the Fourier transform solution lends some insight into the behavior of the solution u(x,t). For example, Equation 9.100 shows that the solution is completely determined by the "dispersion function" w(k). Also, the solution clearly demonstrates the dependence of the evolution of u(x,t) on the prescribed initial condition $u_0(x)$. Specifically, the solution illustrates vividly the asymptotic dispersion of the localized initial disturbance $u_0(x)$, the Fourier transform of which is also localized around some $k = k_0$. These standard features are not trivially evident from the differential equation itself.

As in the case of the linear Fourier transform, the three-stage solution procedure for the nonlinear evolution equation in a one-space dimension and of the nonintegral type is schematized below.

The analogy between the spectral transform procedure and its linear Fourier transform counterpart continues to hold even in interpreting their significance. In both cases, the three steps involved in obtaining a solution only rarely yield an analytical expression. However, working in k-space substantially simplifies the problem. Also, both transforms lend insight into the behavior of the solution. Moreover, if the nonlinearities in the integro-differential evolution operator **L** are summarily neglected, then the analogy becomes exact and the spectral transform reduces to the Fourier transform.

Consider the nonlinear evolution equation

$$\dot{u}(x,t) = \alpha(\mathbf{L})u'(x,t) = u'(x,t) + \beta(\mathbf{L})u'(x,t)$$

where it has been assumed that α is a polynomial. Now the linear term of **L** is $\mathbf{D}^2$, the second derivative. Then the linearized version of the above equation becomes

$$\dot{u}(x,t) = \alpha\left(\mathbf{D}^2\right)u'(x,t) = \left\{-iw(-i\mathbf{D})\right\}u(x,t)$$

$$w(z) = -z\alpha\left(-z^2\right)$$

yielding

$$\hat{u}(k,t) = \hat{u}(k,0)\exp\left[-iw(k)t\right]$$

as before. Moreover, from Equation 9.103, the reflection coefficient $R_{\pm}(k,t)$ evolves in this approximation according to

$$R_{\pm}(k,t) = R_{\pm}(k,0)\exp\left[2ik\alpha\left(-4k^2\right)t\right]$$

or

$$R_{\pm}(k/2,t) = R_{\pm}(k/2,0)\exp\left[-iw(k)t\right]$$

It will be found later that in this linearized approximation, the inverse solution according to Marcenko's equation is

$$u(x,t) = \int_{-\infty}^{\infty} dk \exp(ikx)(ik)R_{\pm}(k/2,t)$$

identifying

$$ikR_{\pm}(k/2,t) = \hat{u}(k,t)$$

Hence, the inversion of the spectral transform becomes identical to the back Fourier transform. This is the case of the Born approximation mentioned before.

A few comments are in order here. First, the main point in applying either the Fourier transform (linear case) or the spectral transform (nonlinear evolution) is that it is simpler to work in the k-space than in the x-space. However, for the spectral transform, the k-space contains both the continuum and the discrete spectrum. These two parts evolve independently and can be investigated separately. The evolution of the discrete part may give rise to "solitons", the localized disturbances that can propagate without change in shape even after colliding nonlinearly with other solitons. The contribution from the continuum, on the other hand, does not remain localized, but, rather, spreads in space as it evolves, thereby displaying dispersive behavior. The evolution of the continuum is called "radiation" or the "background". Since the discrete spectrum is absent in the case of the radiation, its evolution can be treated approximately by considering the linearized version of the nonlinear evolution equation. Second, the analogy between the Fourier and the spectral transform technique breaks down in two situations. If the space is more than one-dimensional, the spectral transform technique becomes more difficult although some progress has been made for three-space dimensions.[26-29] Extension to multidimensional space, on the other hand, poses no conceptual difficulty in the Fourier transform. The analogy also breaks down if the operator **L** is of an integral type, that is, if the evolution equation is of the form

$$\dot{u}(x,t) = \int_{-\infty}^{\infty} dx'\, K(x,x',t)u(x',t)$$

or symbolically

$$\dot{u}(x,t) = \mathbf{K}(x,x',t)u(x',t)$$

If the operator **K** is convolutional in form, that is, if

$$\mathbf{K}(x,x',t) = \mathbf{K}(x - x',t)$$

then the Fourier transform technique is applicable. In that case,

$$\dot{\hat{u}}(k,t) = \mathbf{K}(k,t)\hat{u}(k,t)$$

$$K(x,t) = \int_{-\infty}^{\infty} K(k,t)\exp(ikx)dk$$

and

$$K(k,t) = \int_{-\infty}^{\infty} K(x,t)\exp(-ikx)dx$$

No simple counterpart exists in the nonlinear context.

B. TIME EVOLUTION OF THE SCATTERING DATA

In the previous section, the mathematical formulas describing the time evolution of the scattering data were mentioned in passing in Equations 9.102 and 9.103. In this section, these formulas will be derived. The derivation of these relations requires that we establish the connection between the linear Schrödinger problem and the nonlinear K-dV equation. This is usually established through Lax's generalization.[30] Let the potential in the Schrödinger equation depend on a parameter which we will specifically consider to be the time t. Thus,

$$y'' + [\lambda - q(x,t)]y = 0 \tag{9.106}$$

Let us determine the condition under which Equation 9.106 is isospectral, that is, the eigenvalue λ is independent of the parameter t. Equation 9.106 can be rewritten as

$$\mathbf{L}y = \lambda y \tag{9.107}$$

where, as usual, **L** is the Schrödinger operator

$$\mathbf{L} = \mathbf{D}^2 - q(x,t),\ \mathbf{D} = d/dx$$

Let it be further assumed that there is a differential time evolution operator **B** which, when acting upon y, yields $\dot{y}$. Therefore,

$$\dot{y} = \mathbf{B}y \tag{9.108}$$

Differentiating Equation 9.107 with respect to time yields

$$\begin{aligned}(\mathbf{L}y)_t &= \mathbf{L}\dot{y} + \dot{\mathbf{L}}y = \dot{\lambda}y + \dot{y}\lambda \\ &= \dot{y}'' - q\dot{y} - \dot{q}y = \dot{\mathbf{L}}y - q\dot{y} + \dot{y}''\end{aligned} \tag{9.109}$$

This then shows that $\dot{\mathbf{L}} = -\dot{\mathbf{q}}$. Combining this result with Equations 9.107, 9.108, and 9.109, one obtains

$$(-\dot{q} + [\mathbf{L},\mathbf{B}])y = \dot{\lambda}y \tag{9.110}$$

where $[\mathbf{L},\mathbf{B}] = \mathbf{LB} - \mathbf{BL}$, the commutator of **L** and **B**. The condition of isospectrality is obtained from Equation 9.110 by setting the RHS to be zero.

$$(-\dot{q} + [\mathbf{L},\mathbf{B}]) = 0 \tag{9.111}$$

Let us assume that the form of the operator **B** is such that

$$\mathbf{B} = a\mathbf{D}^3 + f\mathbf{D} + g$$

This then results in

$$\{\mathbf{L},\mathbf{B}\}y = (2f' + 3aq')\mathbf{D}^2 y + (f'' + 2g' + 3aq'')\mathbf{D}y + (g'' + aq''' + fq')y$$

Let us further demand that the coefficients of $\mathbf{D}^2y$ and $\mathbf{D}y$ vanish. This results in the following functional forms for f and g:

$$f = -(3/2)aq + c_1 \tag{9.112}$$

and

$$g = -(3/4)aq' + c_2$$

c_1 and c_2 being the constants of integration, and functions of time only. Then

$$[\mathbf{L},\mathbf{B}]y = \left[(1/4)a(q''' - 6qq') + c_1q'\right]y \tag{9.113}$$

Since the differential equation for q is given by Equation 9.111, we have

$$\dot{q} - 6qq' + q''' - c_1q' = 0 \tag{9.114}$$

In deriving Equation 9.114, the constant a in Equation 9.113 was set to be equal to – 4.

Since the functions f and g are known from Equation 9.112, the time evolution of y is obtained from Equation 9.108 as

$$\dot{y} = \mathbf{B}y = \left(-4\mathbf{D}^3 + 6q\mathbf{D} + 3q' + c_1 + c_2\right)y \tag{9.115}$$

The functions c_1 and c_2 can be taken to be zero without any loss of generality as they can be eliminated by proper changes of variables. Thus, by changing $x \to x'$ such that

$$dx' = dx + c_1(t)dt$$

the constant c_1 can be eliminated. Similarly, by going over to the variable

$$\tilde{y} = y\exp\left(\int c_2(t)dt\right)$$

the function c_2 can also be eliminated. Finally, therefore, we have the following:

$$\text{equation for the potential: } \dot{q} - 6qq' + q''' = 0 \tag{9.116}$$

$$\text{time evolution of y: } \dot{y} = \mathbf{B}y = \left(-4\mathbf{D}^3 + 6q\mathbf{D} + 3q'\right)y \tag{9.117}$$

Equation 9.116 is the most familiar form of the K-dV equation. This nonlinear evolution equation is the condition to be imposed upon the potential so that the Schrödinger equation becomes isospectral.

The K-dV equation (9.116) determines the time evolution of the potential. Since the spectral transform or the scattering data are associated with the potential, these will also evolve in time. Let us, therefore, determine how the scattering data evolve in time if the potential evolves according to the K-dV equation. Specifically, we are seeking the time evolution of the reflection coefficient R(k,t) and of the bound state normalization constants $D_{\pm j}(t)$. Consider Equation 9.5 for the Jost solution with added t-dependence. Then

$$\psi(x,k;t) = t(k,t)\bar{\phi}(x,k;t) + r_{-}(k,t)\phi(x,k;t) \tag{9.118}$$

The time t and the transmission coefficient t(k,t) are not to be confused. In order to determine the reflection and transmission coefficients, it is sufficient to consider only the asymptotic limit $|x| \to \infty$. In this limit, Equation 9.117 reduces to

$$\dot{y} = -4y''' \tag{9.119}$$

Let y(x,k;t) be written as

$$y(x,k;t) = h(k,t)\psi(x,k;t) \tag{9.120}$$

the proportionality factor being considered to have time dependence. In this limit, $\psi(x,k;t) \to \exp(ikx)$, and using Equation 9.120 in Equation 9.119, one obtains

$$\dot{h}(k,t) = 4ik^3h$$

from which

$$h(k,t) = h(k,0)\exp\left(4ik^3t\right) \tag{9.121}$$

In the other limit, $x \to -\infty$, the Jost solution is

$$\psi(x,k;t) = t(k,t)\exp(ikx) + r_-(k,t)\exp(-ikx)$$

Then

$$y(x,k;t) = h(k,0)\exp\left(4ik^3t\right)\left[t(k,t)\exp(ikx) + r_-(k,t)\exp(-ikx)\right]$$

Substituting this equation in Equation 9.119 and equating the coefficients of the exponentials separately yields

$$t(k,t) = t(k,0) \tag{9.122}$$

and

$$r_-(k,t) = r_-(k,0)\exp\left(-8ik^3t\right)$$

From this, the normalization constant is calculated as

$$D_{+j}(t) = D_{+j}(0)\exp\left(-8ik^3t\right)$$

$$= D_{+j}(0)\exp\left(8k_j^3t\right) \tag{9.123}$$

where $k = ik_j$ and $k_j > 0$ is the pole of the transmission coefficient. The reflection coefficient $R_+(k,t)$, being the ratio $[r_+(k,t)/t(k,t)]$, is then given by

$$R_-(k,t) = R_-(k,0)\exp\left(-8ik^3t\right) \tag{9.124}$$

Similar expressions are obtained for $D_{-j}(t)$ and R_+ with – sign changed to + sign. For readable details, the reader is referred to References 33 to 36.

XIX. A LIST OF SOME IMPORTANT NONLINEAR EVOLUTION EQUATIONS SOLVABLE BY INVERSE SCATTERING TRANSFORM

A list of nonlinear evolution equations and their applications is given below. These equations occur in diverse disciplines of the physical sciences and govern an astonishingly large number of highly important physical phenomena. Hence, the importance of these equations can hardly be overemphasized. If not for the inverse scattering techniques, these otherwise difficult equations would have remained unsolved. This list, therefore, emphasizes the importance and significance of the spectral or scattering transform technique.

1. The Korteweg-de Vries (K-dv) equation

$$\dot{u} - u''' - 6uu' = 0$$

Applications: fluid dynamics and plasma physics

2. The modified K-dv equation

$$\dot{u} + u''' \pm 6u'u^2 = 0$$

Application: acoustic wave propagation in anharmonic lattices and Alfvén waves in collisionless plasma

3. The translational K-dv equation

$$\dot{u} + u''' - 6f(t)uu' = 0$$

This equation is solvable for $f(t) = t^{-1/2}$ or f(t) relatively well behaved in a wavelength. Application: surface-wave propagation between two shallow layers of a fluid.

4. The cylindrical K-dv equation

$$\dot{u} + u''' - 6uu' + (u/2t) = 0$$

Application: plasma physics. This is a special case of the translational K-dv when f(t) $= t^{-1/2}$.

5. The Benjamin-Ono equation

$$\dot{u} + Hu'' - 6uu' = 0$$

$$Hf(x) = \pi^{-1} ⨍_{-\infty}^{\infty} [f(x')/(x - x')]dx'$$

Application: internal waves in stratified fluids.

6. The Burger equation

$$\dot{u} - u'' - 2uu' = 0$$

Application: fluid dynamics and nonlinear acoustics.

7. The Boussinesq equation

$$\ddot{u} - u'' - u'''' + 3(u^2)'' = 0$$

Application: propagation of long waves in shallow water.

8. The cubic or nonlinear Schrödinger equation (NLS)

$$i\dot{u} + u'' \pm 2|u|^2 u = 0, \text{ u complex}$$

Application: wave propagation in weakly nonlinear dispersive systems and resonant wave interactions.

9. The derivative NLS equation

$$i\dot{u} + u'' + i\left(|u|^2 u\right)' = 0, \text{ u complex}$$

Application: plasma physics.

10. The Wadati-Konno-Ichikawa-Schimizu equation

$$i\dot{u} + \left[\left(1 + |u^2|\right)^{-\frac{1}{2}} u\right]'' = 0$$

11. The Harry-Dym equation

$$\dot{u} + 2\left(u^{-\frac{1}{2}}\right)'' = 0$$

12. The Liouville equation

$$\dot{u}' = \exp(\eta u), \ \eta \text{ an arbitrary constant}$$

Applications: field theory and parametric wave amplification.

13. The sine-Gordon equation

$$\dot{u}' \pm \sin u = 0$$

$$\ddot{\phi} - \phi'' \pm \sin \phi = 0$$

Applications: differential geometry, nonlinear optics, and relativistic field theory

14. The double sine-Gordon equation

$$\dot{u}' \pm \left[\sin u + \eta \sin(u/2)\right] = 0, \ \eta \text{ constant}$$

Applications: nonlinear optics and low-temperature physics

The other important evolution equations in the fields of solid-state physics, particle theory, and relativistic physics solvable by the inverse scattering transform are

15. The O(n) sigma model

$$\dot{\mathbf{V}}' + \left(\mathbf{V}' \cdot \dot{\mathbf{V}}\right)\mathbf{V} = 0, \ \mathbf{V} = \text{an n-dimensional unit vector}$$

16. The Chiral field equation

$$\left(\mathbf{U}^{+}\dot{\mathbf{U}}\right)_t + \left(\mathbf{U}^{+}\mathbf{U}'\right)' = 0$$

$$\mathbf{U} = \mathbf{U}(x,t), \text{ a unitary matrix}$$

17. The Yang-Mills equation

$$\left(\mathbf{U}^{+}\dot{\mathbf{U}}\right)_t - \left(\mathbf{U}^{+}\mathbf{U}'\right)_{\bar{z}} = 0$$

$$\mathbf{U} = \mathbf{U}(x,y,t), \text{ a matrix}$$

$$\partial/\partial z = \partial_x - i\partial_y$$

$$\partial/\partial \bar{z} = \partial_x + i\partial_y$$

18. The Einstein equation

$$\left(g^{1/2}\mathbf{G}'\mathbf{G}^{-1}\right)_t + \left(g^{1/2}\dot{\mathbf{G}}\mathbf{G}^{-1}\right)' = 0$$

$$\mathbf{G} = \mathbf{G}(x,t), \text{ a } (2\times 2) \text{ symmetrical matrix}, \; g = \det \mathbf{G}$$

An exhaustive list and a detailed account of these and many other evolution equations solvable by inverse scattering transform and given in the excellent monograph of Calogero and Degasperis.[5]

REFERENCES

1. **E. A. Hylleraas:** Calculation of a perturbing central field of force from the elastic scattering phase shift, *Phys. Rev.*, 74, 1948.
2. **C. Froberg:** Calculation of the interaction between two particles from the asymptotic phase, *Phys. Rev.*, 72, 1948.
3. **V. Bargmann:** On the connection between phase shifts and scattering potential, *Rev. Mod. Phys.*, 21, 1949.
4. **M. Verde:** Asymptotic expansion of phase shifts at high energies, *Nuovo Cimento*, 2, 1955.
5. **F. Calogero and A. Degasperis:** *Spectral Transform and Solitons*, North-Holland, Amsterdam, 1982.
6. **G. L. Lamb:** *Elements Of Soliton Theory*, John Wiley & Sons, New York, 1980.
7. **S. M. Rytov:** Correlation theory for Rayleigh scattering of light, *Sov. Phys. JETP*, 6, 1958.
8. **A. B. Migdal:** *Qualitative Methods in Quantum Theory*, W. A. Benjamin, Reading, MA, 1977.
9. **A. Messiah:** *Quantum Mechanics*, North-Holland, Amsterdam, 1958.
10. **C. J. Joachain:** *Quantum Collision Theory*, North-Holland, Amsterdam, 1975.
11. **I. Schiff:** *Quantum Mechanics*, McGraw-Hill, New York, 1968.
12. **E. C. Titchmarsh:** *Theory of Functions*, Oxford University Press, New York, 1939.
13. **E. A. Coddington and N. Levinson:** *Theory of Ordinary Differential Equations*, McGraw-Hill, New York, 1955.
14. **W. Eckhaus and A. van Harten:** *The Inverse Scattering Transformation and the Theory of Solitons*, North-Holland, Amsterdam, 1981.
15. **J. B. Conway:** *Functions of One Complex Variable*, Springer Verlag, 1973.
16. **R. G. Newton:** *Scattering of Waves and Particles*, McGraw-Hill, New York, 1966.
17. **L. D. Fadeev:** Properties of the S-matrix of the one-dimensional Schrodinger equation, *Am. Math. Soc.*, 65, 1964.

18. **H. Dym and H. P. McKean:** *Gaussian Processes, Function Theory, and the Inverse Spectral Problem*, Academic Press, New York, 1976.
19. **V. De Alfaro and T. Regge:** *Potential Scattering*, John Wiley & Sons, New York, 1965.
20. **M. J. Lighthill:** *Introduction to Fourier Analysis and Generalized Functions*, Cambridge University Press, 1958.
21. **S. Tanaka:** KdV equation: construction of solutions in terms of scattering data, *Osaka J. Math.*, 11, 1974.
22. **J. R. Taylor:** *Scattering Theory*, John Wiley & Sons, New York, 1972.
23. **D. B. Pearson:** *Quantum Scattering and Spectral Theory*, Academic Press, London, 1988.
24. **R. K. Dodd, J. C. Eilbeck, J. D. Gibbon, and H. C. Morris:** *Solitons and Nonlinear Wave Equations*, Academic Press, London, 1982.
25. **C. H. Wilcox:** Scattering theory and wave operators in the abstract theory of scattering, *J. Functional Anal.*, 12, 1973.
26. **L. D. Fadeev:** *Sov. Phys. Dokl.*, 10, 1966.
27. **L. D. Fadeev:** *Sov. Phys. Dokl.*, 11, 1966.
28. **R. G. Newton:** The Gel'fand-Levitan method in the inverse scattering problem, in *Scattering Theory in Mathematical Physics*, J. A. LaVita and J. P. Marchand, Eds., D. Reidel, Dordrecht, Holland, 1974.
29. **K. Chadan and P. S. Sabatier:** *Inverse Problems In Quantum Scattering Theory*, Springer-Verlag, New York, 1977.
30. **P. D. Lax:** Integrals of nonlinear equations of evolution and solitary waves, *Comm. Pure Appl. Math.*, 21, 1968.
31. **P Deift and E. Trubowitz:** *Comm. Pure Appl. Math.*, 32, 1979.
32. **M. Reed:** in *Schrödinger Equation,* W. Thirring and P. Urban, Eds., Springer-Verlag, New York, 1977.
33. **G. L. Lamb, Jr.:** Elements of Soliton Theory, Wiley-Interscience, New York, 1980.
34. **P. G. Drazin and R. S. Johnson:** *Solitons: An Introduction,* Cambridge University Press. Cambridge, 1989.
35. **M. J. Ablowitz and H. Segur:** *Solitons and the Inverse Scattering Transform, SIAM Studies in Applied Mathematics,* Vol. 4, Philadelphia, 1981.
36. **S. Novikov, S. Manakov, L. P. Pitaevskii, and V. E. Zakharov:** *Theory of Solitons: The Inverse Scattering Method,* Consultants Bureau, New York, 1984.

Chapter 10

THE MARCENKO INTEGRAL EQUATION

I. INTRODUCTION

The correspondence between the potential q(x) and its spectral transform S[q] was established in the previous chapter. S[q] was interpreted as a nonlinear forward transform, the spectral transform of q(x). This, in effect, is a mapping between the potential and its associated scattering data. It was also pointed out that the potential q(x) could be reconstructed by essentially "back transforming" the "forward" spectral transform S[q]. Since the transform is nonlinear, its inverse is not as straightforward as that of the linear Fourier transform, in which case the transform operators in both directions are unitary. The spectral transform is inverted indirectly through the construction of an inhomogeneous Fredholm integral equation of the second kind. As is already known from Chapters 1 and 5, such an equation expresses the unknown function to be determined as a sum of a known function and an integral which contains the unknown function itself. The known, given function constitutes the inhomogeneous term. In the Fredholm equation expressing the inverted spectral transform, the spectral data appear in the inhomogeneous term and also in the kernel. The solution of this equation along a characteristic is then related to the potential to be reconstructed. The integral equation referred to is the all-important "Marcenko" equation of inverse scattering, named after the Russian mathematician Marcenko.[1] It is the most fundamental equation for solving inverse scattering problems on the entire real axis.

The Levin transforms were discussed in Chapter 8. The potential was found to be directly related to the derivative of the Levin transform prescribed along a characteristic. The determination of the potential thus involves the determination of the Levin transform. Marcenko's equation is the integral equation for the Levin transform which constitutes the unknown function to be determined by this Fredholm equation. The inhomogeneous term of this equation, which is a function of the spectral data, is assumed to be obtained from the direct solution of the problem or else known from experiments.

Marcenko's equation is often referred to as the Gel'fand-Levitan-Marcenko equation, and for good reasons. It is identical in form to that of the Gel'fand-Levitan equation on the semiaxis R^+. Just as the Gel'fand-Levitan equation is a Fredholm integral equation of the second kind in which the Povzner-Levitan kernel is the unknown, so is the Marcenko equation, in which it is the Levin transform for which the equation is to be solved. The Marcenko equation, however, uses the irregular boundary conditions of the Schrödinger equation, that is, conditions at infinity. The Gel'fand-Levitan equation, on the other hand, concerns itself with the regular condition at the origin. The Marcenko equation is, therefore, based on the Jost solutions of the Schrödinger equation, as opposed to the radial solution used in the Gel'fand-Levitan equation. The condition at the origin is, therefore, not particularly significant for Marcenko's equation. Even on the half-line, Marcenko's equation uses the Jost solutions satisfying the condition at ∞. There is also a deeper difference. The Gel'fand-Levitan equation uses the "triangular" property of the Fourier transforms of a class of solutions of the Schrödinger equation and the closure relation for these. The Marcenko method, on the other hand, uses the relation (on R) between two classes of solutions of the Schrödinger equation which have different regions of analyticity (upper and lower half-plane) and with prescribed asymptotic behavior. The Marcenko method is then essentially the problem of Riemann-Hilbert. For details, see R. G. Newton, *Inverse Schrödinger Scattering in Three Dimensions,* Springer, New York, 1989.

II. MARCENKO'S EQUATION ON THE ENTIRE LINE

The Marcenko equation can be derived in both time and frequency domain, as was done for the Gel'fand-Levitan equation, and there is more than one way of deriving Marcenko's equation for $x \subset R$. Two derivations will be presented in each domain. The standard derivation starts with the scattering equation for the Jost solutions given by Equations 9.5 through 9.8 and then applying the Levin transform to it. The resultant equation is then Fourier transformed and various relevant quantities are evaluated by applying Cauchy's integral theorem. The analyticity and continuity properties of the Jost functions, Jost solutions, and the Levin transforms are required in order to perform these operations. This was the main reason for discussing these properties rigorously in Chapters 7, 8, and 9. The Fourier transform derivation is presented first.

A. THE FOURIER TRANSFORM DERIVATION

The definition of the Fourier transform pair is that of Chapter 8. Consider the scattering equation for the Jost function f(x,k) given by Equation 9.48 and reproduced below for convenience.

$$T_+(k)f_-(x,k) = R_+(k)f_+(x,k)\exp(2ikx) + \bar{f}_+(x,k)$$

The Levin representation of $f_+(x,k)$ is given by Theorem 8.7,

$$f_+(x,k) = 1 + (2\pi)^{1/2} J_+(x,k)$$

where $J_+(x,k)$ is the inverse Fourier transform of $I_+^\wedge(x,s)$, which was defined in Chapter 8 as

$$I_+^\wedge(x,s) = H(s)B_+^\wedge(x,s)$$

H being the Heaviside function. The scattering equation for $f_-(x,k)$ is then

$$\begin{aligned} T_+(k)f_-(k) - 1 &= R_+(k)\exp(2ikx) + (2\pi)^{1/2} R_+(k)\exp(2ikx)J_+(x,k) + \\ &\quad + (2\pi)^{1/2}(J_+(x,k))^* \\ &= R_+(k)\exp(2ikx) + (2\pi)^{1/2} R_+(k)\exp(2ikx)J_+(x,k) + \\ &\quad + (2\pi)^{1/2} J_+^\wedge(x,k) \end{aligned} \tag{10.1}$$

where $J_+^\wedge(x,k) = J_+(x,-k)$. The last step follows from the fact that $J_+(x,k)$ is a real quantity. The quantity $J_+^\wedge$ is to be interpreted as the direct Fourier transform of $I_+^\wedge(x,s)$, the conjugate variable being k.

Let us rewrite the LHS of Equation 10.1 as

$$T_+(k)f_-(x,k) - 1 = T_+(k)[f_-(x,k) - 1] + [T_+(k) - 1]$$

From Section V.B of Chapter 9, we know that for $|k| \to \infty$ and k real,

$$R_+(k) \to O(1/k)$$

and

$$T_+(k) \to 1 + O(1/k)$$

Then $R_+(k) \subset L_2(R)$. Also, from Chapter 7 we have the bound

$$|f_-(x,k) - 1| \leq [\eta_-(x)/|k|] \exp[\eta_-(x)/|k|]$$

On the basis of these estimates we conclude that the quantity $\{T_+(k)f_-(x,k) - 1\} \sim O(1/|k|)$. Since the Levin transforms were already shown to belong to $L_2(R)$, all the terms of Equation 10.1 are $L_2(R)$. The operations of the Fourier transformation can thus be performed.

Next, we inverse Fourier transform both sides of Equation 10.1, that is, multiply both sides of this equation by $(2\pi)^{-1/2} \exp(iks)$ and then, integrating in k from $-\infty$ to $+\infty$, we obtain

$$(2\pi)^{-1/2}[T_+(k)f_-(k) - 1]^\vee(x,s) = (2\pi)^{-1/2}[R_+(k)\exp(2ikx)]^\vee(x,s) + \\ +[R_+(k)\exp(2ikx)J_+(x,k)]^\vee(x,s) + I_+^\wedge(x,s) \quad (10.2)$$

The first term on the right-hand side of this equation can be written as

$$(2\pi)^{-1/2}[R_+(k)\exp(2ikx)]^\vee(x,s) = [T_{-2x}R^\vee](s) \quad (10.3)$$

where T_y is the translation operator defined by

$$\{T_y f\}(x) = f(x - y)$$

In other words, Equation 10.3 is the inverse Fourier transform of $R_+(k)$ in the variable k, the corresponding conjugate variable being $(2x + s)$.

Similarly, the second term on the right-hand side can be evaluated as follows.

$$[R_+(k)\exp(2ikx)J_+(x,k)]^\vee(x,s) \\ = \int_{-\infty}^{\infty} R_+(k)\exp(2ikx)\left[(2\pi)^{-1/2}\int_{-\infty}^{\infty} I_+^\wedge(x,s')\exp(iks')ds'\right]\exp(iks)dk$$

The integrals are absolutely convergent and the order of integration can be interchanged, yielding

$$(2\pi)^{1/2}[R_+(k)\exp(2ikx)J_+(x,k)]^\vee(x,s) = \int_{-\infty}^{\infty} I_+^\wedge(x,s')\{T_{-(2x+s)}R^\vee\}(s')ds' \quad (10.4)$$

The transform of the left-hand side of Equation 10.2 is relatively more complicated and can be calculated by complex integration. The contour will consist of the real axis between –R and +R and a semicircle of radius R in the upper half-plane. Because of the decay of the integrand discussed above, the contribution from the semicircle disappears as $R \to \infty$. The only contribution is then from the poles of the transmission coefficient, which were shown in Chapter 9 to be finite in number and located on the imaginary axis. The application of Cauchy's integral theorem then results in

$$(1/2\pi)\int_{-\infty}^{\infty}\left[T_+(k)f_-(x,k)-1\right]\exp(iks)\,dk = i\sum_{j=1}^{N}\left(f_{-j}/\dot{t}_{+j}\right)\exp\left(-k_j s\right)$$

where $f_{-j} = f_-(x,k = ik_j)$. Using the normalization relation given by Equation 9.47, and the fact that $c_{-j}c_{+j} = 1$, we have

$$\begin{aligned} f_{-j}/\dot{t}_{+j} &= ic_{+j}D_{+j}f_{-j} = iD_{+j}\left(c_{+j}\phi_j\right)\exp\left(-k_j x\right) = iD_{+j}\psi_j\exp\left(-k_j x\right) \\ &= iD_{+j}f_{+j}\exp\left(-2k_j x\right) \end{aligned}$$

Defining

$$e_j^x(y) = \exp\left[-k_j(y+x)\right]$$

finally yields

$$\begin{aligned} (2\pi)^{-1/2}\left[T_+(k)f_-(x,k)-1\right]^{\vee}(x,s) &= -(2\pi)^{1/2}\sum_{j=1}^{N} D_{+j}f_{+j}e_j^{2x}(s) \\ &= -(2\pi)^{1/2}\sum_{j=1}^{N} D_{+j}e_j^{2x}(s)\left[1+(2\pi)^{1/2}J_+\left(x,k_j\right)\right] \end{aligned} \tag{10.5}$$

Therefore, from Equation 10.2,

$$\text{Eq. } 10.5 = \text{Eq. } 10.3 + \text{Eq. } 10.4 + I_+^{\wedge}(x,s)$$

Rearranging terms and defining

$$\Omega_+(x+z) = \left[T_{-x}R^{\vee}\right](z) + \sum_{j=1}^{N} D_{+j}e_j^x(z)$$

the above equation can be written as

$$0 = \Omega_+(s+2x) + I_+^{\wedge}(x,s) + \int_{-\infty}^{\infty} I_+^{\wedge}(x,s)\Omega_+(s+s'+2x)\,ds'$$

Since $I_+^{\wedge}(x,s) = H(s)B_+^{\wedge}(x,s)$, the equation can be further reduced to

$$0 = \Omega_+(s+2x) + B_+^{\wedge}(x,s) + \int_0^{\infty} B_+^{\wedge}(x,s')\Omega_+(s'+s+2x)\,ds'$$

The full expression for $\Omega_+(z)$ is

$$\Omega_+(z) = \hat{R}_+(z) + \sum_{j=1}^{N} D_{+j} \exp(-k_j z)$$

where

$$\hat{R}_+(z) = (1/2\pi)\int_{-\infty}^{\infty} R_+(k)\exp(ikz)dk$$

The equation derived above pertains to the Jost function $f_+(x,k)$. The integral equation for the corresponding Jost solution $\psi(x,s)$ can be written as

$$\Omega_+(s+x) + K_+(x,s) + \int_+ K_+(x,s')\Omega_+(s'+s)ds' = 0$$

where $K_+(x,s) = B_+^{\wedge}(x,s-x)$ is the Levin transform for the Jost solution $\psi(x,k)$.

Similarly, the corresponding equation for the other Jost solution $\phi(x,s)$ is given by

$$\Omega_-(s+x) + K_-(x,s) + \int_- K_-(x,s')\Omega_-(s'+s)ds' = 0$$

Combining, we obtain

$$\Omega_\pm(s+x) + K_\pm(x,s) + \int_\pm K_\pm(x,s')\Omega_\pm(s'+s)ds' = 0 \tag{10.6}$$

and

$$\Omega_\pm(z) = \hat{R}_\pm(z) + \sum_{j=1}^{N} D_{\pm j} \exp(\mp k_j z) \tag{10.7}$$

Equation(s) 10.6 are the celebrated Marcenko integral equations.

For the above derivation, it was necessary that the potential satisfied a growth condition of order 2 since the integrals involving T(k) required that T(k) be continuous at the origin. Summarizing, the following theorem is established.

Theorem 10.1

Let the potential q(x) satisfy a growth condition of order 2. Then the Levin transforms $B_\pm^{\wedge}(x,s)$ for the Jost functions f(x,k) are given by the integral equations

$$\Omega_\pm(s+2x) + B_\pm^{\wedge}(x,s) + \int_\pm B_\pm^{\wedge}(x,s')(s'+s+2x)ds' = 0$$

with

$$\Omega_\pm(z) = \hat{R}_\pm(z) + \sum_{j=1}^{N} D_{\pm j} \exp(\mp k_j z)$$

For the Jost solutions $\psi(x,s)$ resp. $\phi(x,s)$, the integral equations for the corresponding Levin transforms K_+ resp. K_- are given by

$$\Omega_\pm(s+x) + K_\pm(x,s) + \int_\pm K_\pm(x,s')\Omega_\pm(s'+s)ds' = 0$$

and

$$\Omega_\pm(z) = \hat{R}_\pm(z) + \sum_{j=1}^{N} D_{\pm j} \exp(\mp k_j z)$$

B. DERIVATION USING THE COMPLETENESS RELATION

An elegant and simpler derivation[2,3] of Marcenko's equation can be given using the generalized Levin transform involving two potentials. Before proceeding further with this derivation, let us introduce the generalized Levin representations.

1. The Generalized Levin Representations

In Chapter 8, the Levin representations were given in terms of the scattering free solution. Let us now generalize the Levin representations in terms of the solution of the Schrödinger equation with a known potential other than zero. This means that the plane-wave functions exp($\pm$ikx) are now to be replaced by the Jost solutions corresponding to the known potential. Consider, therefore, two potentials $q^{(j)}(x)$, j = 1,2 and an operator **A** such that

$$\mathbf{L}^{(1)}\mathbf{A} = \mathbf{A}\mathbf{L}^{(2)},\ \mathbf{L}^{(j)} = -\partial^2/\partial x^2 + q^{(j)}(x) \tag{10.8}$$

A is an operator that transforms a solution $y^{(2)}$ of the Schrödinger operator $\mathbf{L}^{(2)}$ to a solution $y^{(1)}$ of $\mathbf{L}^{(1)}$, that is,

$$y^{(1)} = \mathbf{A}y^{(2)}$$

In addition, it is assumed that **A** is of the form **A** = **I** + **Z** such that

$$\psi^{(1)}(x,k) = A\left[\psi^{(2)}(x,k)\right] = \psi^{(2)}(x,k) + \int_{-\infty}^{\infty} Z(x,x')\psi^{(2)}(k,x')dx' \tag{10.9}$$

where Z(x,x′) is the kernel. Clearly, Z(x,x′) = 0 if $q^{(1)}(x)$ equals $q^{(2)}(x)$. Moreover, the solution $\psi^{(1)}(x)$ and $\psi^{(2)}(x)$ must be identical to each other in the absence of the potentials. Now, the potentials to be considered here vanish as $|x| \to \infty$. Since the Jost solutions cannot vanish in this asymptotic limit, Z(x,x′) must vanish as $|x'| \to \infty$. Furthermore, in order to ensure that the solutions are identical in the asymptotic limit when the potential disappears, we will require the derivative of the kernel to also vanish as x′ goes to infinity, that is, $Z(x,x') \to 0$ as $|x'| \to \infty$. Collecting these,

$$Z = 0,\ q^{(1)} = q^{(2)} \tag{10.10}$$

and

$$Z(x,\pm\infty) = \dot{Z}(x,\pm\infty) = 0$$

The derivative with respect to x′ is denoted by "·". Equation 10.8 can then be rewritten as

$$\mathbf{q}^{(1)} - \mathbf{q}^{(1)} = \left[\mathbf{D}^2, \mathbf{Z}\right] - \mathbf{q}^{(1)}\mathbf{Z} + \mathbf{Z}\mathbf{q}^{(2)}, \qquad \mathbf{D} = \partial/\partial x$$

$[\mathbf{P},\mathbf{Q}]$ denotes the commutation between two operators $\mathbf{P}$ and $\mathbf{Q}$. Writing $\mathbf{q}$ instead of q implies that the above is an operator equation and becomes a functional identity if both sides are made to operate on a function f(x). The functional identity reads

$$\left[q^{(1)}(x) - q^{(2)}(x)\right]\delta(x - x') = Z''(x,x') - \ddot{Z}(x,x') - \left[q^{(1)}(x) - q^{(2)}(x')\right]Z(x,x') \quad (10.11)$$

The Jost solution must satisfy the asymptotic boundary condition

$$\lim_{x\to\infty} \psi^{(j)}(x,k) \to \exp(ikx)$$

The representation (10.9) can be made to satisfy this asymptotic condition if the lower limit of integration is set equal to x. Moreover, in view of the fact that Equation 10.11 contains Dirac's delta function, Z must contain a Heaviside function. In other words, the kernel Z(x,x′) must be "triangular". Let us, therefore, write

$$Z(x,x') = K(x,x')H(x' - x) \quad (10.11a)$$

where H is the Heaviside function. Equation 10.9 then takes the form

$$\psi^{(1)}(x,k) = A\left[\psi^{(2)}(x,k)\right] = \psi^{(2)}(x,k) + \int_+ K(x,x')\psi^{(2)}(k,x')dx' \quad (10.12)$$

Replacing Equation 10.12 and the triangular representation for Z(x,x′) of Equation 10.11a into Equation 10.11 and then equating the coefficients of the singularities H(x′ – x) and δ(x′ – x) separately, we obtain the following boundary-value problem:

$$K''(x,x') - \ddot{K}(x,x') - \left[q^{(1)}(x) - q^{(2)}(x)\right]K(x,x') = 0, \; x' > x \quad (10.13)$$

and

$$K'(x,x) + \dot{K}(x,x) = -(1/2)\left[q^{(1)} - q^{(2)}\right](x) \quad (10.14)$$

respectively. Evaluating Equation 10.14 along the characteristic x = x′ results in the condition

$$K(x,x') = (1/2)\int_+ \left[q^{(1)} - q^{(2)}\right](x')dx' \quad (10.15)$$

We have used the fact that along the characteristic

$$K'(x,x)\big|_{x=x'} + \dot{K}(x,x')\big|_{x=x'} = \left[K(x,x)\right]' = -q/2$$

As before, K(x,x) = K(x,x+0). If the second potential $q^{(2)}(x)$ is set to zero, then the corresponding Jost solution $\psi^{(2)}(x,k)$ reduces simply to exp(ikx). The boundary-value problem (10.13) and Equation 10.14 then become identical to that of the Levin transform for the Jost solution ψ(x,k) and Equation 10.12 also reduces to the Levin transform for ψ(x,k), as were established in Chapter 8 (Section V.A). The representation of Equation 10.12 and the corresponding boundary-value problem (Equations 10.13 and 10.14) are, therefore, generalizations of the Levin problems of Chapter 8 to the case when the second potential $q^{(2)}(x)$ is other than zero.

If the potentials $q^{(j)}$, j = 1,2, are real, then K(x,x′) is also real and K(x,x′) = [K(x,x′)]*, where * denotes a complex conjugate. Moreover, if the potentials are assumed to vanish after some arbitrarily large x, that is, $q^{(j)}(x) = 0$, $|x| > x_\infty$, then for $|x| > x_\infty$, Equation 10.13 reduces to a simple wave propagation problem with δ(x ± x′) as solutions. In view of the triangularity of the kernel expressed by the condition x′ ≥ x, we conclude that the kernel also has compact support. Thus,

$$K(x,x') = 0, \qquad x_\infty < x \le x'$$

In addition, due to the asymptotic boundary condition imposed on the Jost solution, we must have

$$\int_+ K(x,x')\exp(ikx')dx' = 0, \qquad x_\infty < x \le x'$$

Equation 10.15 specifies the boundary condition along one of the characteristics. The problem is, therefore, a Goursat problem. It is known[4,5] that a Goursat problem such as that of K above possesses a unique solution in the domain x′ ≥ x and $x' > 2x_\infty - x$. Analogous results can also be obtained for the second Jost function ϕ(x,k).

Summarizing, the following theorem is obtained.

Theorem 10.2

Let $\psi^{(j)}(x,k)$ be the Jost solution of the Schrödinger problem

$$\psi^{(j)''}(x,k) + \left[k^{(2)} - q^{(j)}(x)\right]\psi^{(j)}(x,k) = 0, \qquad j = 1,\ 2$$

with the boundary condition

$$\lim_{x\to\infty}\left[\exp(-ikx)\psi^{(j)}(x,k)\right] = 1, \qquad j = 1,2$$

Let K(x,x′) satisfy the Goursat boundary-value problem

$$K''(x,x') - \ddot{K}(x,x') - \left[q^{(1)}(x) - q^{(2)}(x)\right]K(x,x') = 0, \qquad x \le x'$$

$$K(x,x+0) = K(x,x) = (1/2)\int_+\left[q^{(1)}(y) - q^{(2)}(y)\right]dy$$

such that

$$K(x,\pm\infty) = \dot{K}(x,\pm\infty) = 0$$

$$K(x,x') = [K(x,x')]^*$$

$$K(x,x') = 0, \qquad x_\infty < x \le x'$$

and

$$\int_+ K(x,x')\exp(ikx')dx' = 0, \qquad x_\infty < x \le x'$$

Then K(x,x′) is the unique solution to the Goursat problem in the half-plane $x' - x > 0$, and forms the kernel in the representation

$$\psi^{(1)}(x,k) = \psi^{(2)}(x,k) + \int_+ K(x,x')\psi^{(2)}(x',k)dx'$$

2. Derivation of the Marcenko Equation

Let us return to the derivation of Marcenko's integral equation and begin by rearranging the completeness relation of Equation 9.86. Using the relations

$$\overline{\psi}(x,k) = \psi\left(x,-k^*\right) = \psi^*\left(x,k^*\right), \quad k \subset C_+$$

$$\overline{\psi}(x,k) = \psi(x,-k) = \psi^*(x,k), \qquad k \subset R$$

$$\bar{t}(k) = t(-k) \text{ and } \bar{r}(k) = r(-k)$$

the continuous part of the completeness relation is written as

$$\left[\psi(x,k)\overline{\psi}(x',k) + \overline{\psi}(x,k)\overline{\psi}(x',k)\left\{\bar{r}_+(k)/\bar{t}(k)\right\}\right](1/4\pi) +$$

$$\left[\overline{\psi}(x,k)\psi(x',k) + \psi(x,k)\psi(x',k)\left\{r_+(k)/t(k)\right\}\right](1/4\pi)$$

Integrating the quantity within the upper braces with respect to k from $-\infty$ to $+\infty$ yields

$$\int_{-\infty}^{\infty}\left[\psi(x,k)\overline{\psi}(x',k) + \overline{\psi}(x,k)\overline{\psi}(x',k)\left\{\bar{r}_+(k)/\bar{t}(k)\right\}\right]dk$$

$$= \int_{-\infty}^{\infty}\left[\overline{\psi}(x,k)\psi(x',k) + \psi(x,k)\psi(x',k)\left\{r_+(k)/t(k)\right\}\right]dk$$

Hence, the completeness relation can be reformulated as

$$\delta(x-x') = (1/2\pi)\int_{-\infty}^{\infty}\left[\overline{\psi}(x,k)\psi(x',k) + \psi(x,k)\psi(x',k)\left\{r_+(k)/t(k)\right\}\right]dk + \\ + \sum_{j=1}^{N} D_{+j}\psi_j(x)\psi_j(x') \tag{10.16}$$

Consider two potentials $q^{(j)}(x)$, j = 1,2, and the generalized Levin transforms for the corresponding Jost solutions, namely,

$$\psi_{\pm}(x,k) = \psi_{\mp}(x,k) + \int_{\pm} K_{\pm}(x,x')\psi_{\mp}(x',k)dx' \tag{10.17}$$

The + sign is for j = 1 and the minus sign for j = 2. Multiplying $\psi^{(1)}(x',k)$ by the quantity $[R_+^{(2)}\psi^{(2)}(x,k) + \overline{\psi}^{(2)}(x,k)]$, integrating with respect to k over R, using Equation 10.17 for j = 1 and the completeness relation given by Equation 10.16, we obtain the identity

$$(1/2\pi)\int_{-\infty}^{\infty} dk\psi^{(1)}(x',k)\left[R_+^{(2)}(k)\psi^{(2)}(x,k) + \overline{\psi}^{(2)}(x,k)\right]$$

$$= \int_{+x'} dx''\delta(x - x'')K^{(1)}(x',x'') + \delta(x - x') - \sum_{j=1}^{N^{(2)}} D_{+j}^{(2)}\psi_j^{(2)}(x)\psi_j^{(1)}(x') \tag{10.18}$$

In Equation 10.18, $N^{(2)}$ and $D_{+j}^{(2)}$ correspond to $q^{(2)}(x)$.

Similarly, multiplying $\psi^{(2)}(x,k)$ by $[R_+^{(1)}\psi^{(1)}(x',k) + \overline{\psi}^{(1)}(x',k)]$, integrating with respect to k over R, using Equation 10.17 for j = 2 and the completeness relation given by Equation 10.16, another identity is obtained, namely,

$$(1/2\pi)\int_{-\infty}^{\infty} dk\psi^{(2)}(x,k)\left[R_+^{(1)}(k)\psi^{(1)}(x',k) + \overline{\psi}^{(1)}(x',k)\right]$$

$$= \int_{+} dx''\delta(x' - x'')K^{(2)}(x,x'') + \delta(x - x') - \sum_{j=1}^{N^{(1)}} D_{+j}^{(1)}\psi_j^{(1)}(x')\psi_j^{(2)}(x) \tag{10.19}$$

Subtracting Equation 10.19 from Equation 10.18 yields

$$(1/2\pi)\int_{-\infty}^{\infty} \psi^{(2)}(x,k)\psi^{(1)}(x',k)\left[R_+^{(2)}(k) - R_+^{(1)}(k)\right]dk +$$

$$+ (1/2\pi)\int_{-\infty}^{\infty}\left[\psi^{(1)}(x',k)\overline{\psi}^{(2)}(x,k) - \overline{\psi}^{(1)}(x',k)\psi^{(2)}(x,k)\right]dk$$

$$= \sum_{j=1}^{N^{(1)}} D_{+j}^{(1)}\psi_j^{(1)}(x')\psi_j^{(2)}(x) - \sum_{j=1}^{N^{(2)}} D_{+j}^{(2)}\psi_j^{(2)}(x)\psi_j^{(1)}(x') - \int_{+}\delta(x' - x'')K^{(2)}(x,x'')dx'' +$$

$$+ \int_{+x'}\delta(x - x'')K^{(1)}(x',x'')dx''$$

Upon changing k to –k and by virtue of the reality conditions cited at the beginning of this section, the second integral on the L.H.S. disappears. For x′ > x, the last integral on the R.H.S. vanishes and the remaining integral reduces to $-K^{(2)}(x,x')$ This results in the following expression for the generalized Levin representations:

$$K_+^{(2)}(x,x') = -(1/2\pi)\int_{-\infty}^{\infty} dk\psi^{(1)}(x',k)\psi^{(2)}(x,k)\left[R_+^{(2)}(k) - R_+^{(1)}(k)\right] +$$

$$+\sum_{j=1}^{N^{(1)}} D_{+j}^{(1)}\psi_j^{(1)}(x')\psi_j^{(2)}(x)-\sum_{j=1}^{N^{(2)}} D_{+j}^{(2)}\psi_j^{(1)}(x')\psi_j^{(2)}(x), \qquad x<x' \tag{10.20}$$

The role of K(x,x′) in the solution of the inverse problem can now be appreciated. The direct problem consisted of the unique determination of K(x,x′) through the Goursat problem (Equations 10.13 through 10.15) from the specification of the potentials. Equation 10.20, on the other hand, shows the dependence of K(x,x′) on the reflection coefficient $R_+^{(j)}(k)$, the eigenvalues $k_i^{(j)}$, and the normalization constants $D_+^{(j)}$. In other words, the kernel K(x,x′) is to be determined from the specification of the spectral data $S[q^{(j)}]$. Equation 10.20 for K(x,x′) is not the desired expression yet as it involves the Jost solutions for both potentials. One of the potentials, say, $q^{(1)}$, being given, the unknown is eliminated through the application of the generalized Levin representation given by Equation 10.17 for j = 2. Therefore, substituting this equation in Equation 10.20, the following integral representation is obtained

$$K(x,x')+\Omega_+(x,x')+\int_+ dx''\,K(x,x'')\Omega_+(x',x'')=0, \qquad x<x' \tag{10.21}$$

where the kernel $\Omega_+(x,x')$ is

$$\Omega_+(x,x')=(1/2\pi)\int_{-\infty}^{\infty} dk\;\psi^{(1)}(x,k)\psi^{(1)}(x',k)\left[R_+^{(2)}(k)-R_+^{(1)}(k)\right]+$$

$$+\sum_{j=1}^{N^{(1)}} D_{+j}^{(1)}\psi_j^{(1)}(x)\psi_j^{(1)}(x')-\sum_{j=1}^{N^{(2)}} D_{+j}^{(2)}\psi_j^{(1)}(x)\psi_j^{(1)}(x') \tag{10.22}$$

The superscript (2) has been omitted in Equation 10.21. Equation 10.21 is the Marcenko integral equation and Equation 10.22 is the expression for the kernel function. It is a linear Fredholm integral equation of the second kind, the inhomogeneous term of which contains the spectral data which must either be given or determined from the known potential. The variable x plays the role of a parameter.

As a special case, let $q^{(1)}(x)$ be zero and $q^{(2)}(x)$ be q(x). Then Marcenko's equation reduces to

$$K(x,x')+\Omega(x+x')+\int_+ dx''K(x,x'')\Omega(x'+x'')=0, \qquad x<x'$$

The kernel is now

$$\Omega(z)=(1/2\pi)\int_{-\infty}^{\infty} dk\,\exp(ikz)R(k)-\sum_{j=1}^{N} D_{+j}\exp(-k_j z)$$

From Equation 10.14, the potential is reconstructed as

$$q(x)=-2[K(x,x)]'$$

The results of the first derivation are then completely recovered.

III. DERIVATION IN THE TIME DOMAIN

Marcenko's equation can be derived entirely in the time domain. In order to do this, let us consider the time-dependent wave equation

$$u'' - \ddot{u} - qu = 0$$

where, $u = u(x,t)$ is the wave amplitude, $q = q(x)$ is the potential, which may be an inhomogeneity, as in the case of an elastically braced string, or may be related to electron number density if the medium is a plasma. We assume that $x \subset R$ and that $q(x) \subset R^+$. The medium is initially assumed to be quiescent. Thus, if it is an elastically braced string, the initial displacement of the elastic medium is assumed to be zero. If it is a plasma medium, then the quiescent condition corresponds to the absence of any bias, with no initial electric and magnetic fields. The inhomogeneity is assumed to be nonnegative, bounded, and continuous with continuous derivatives.

Consider that a δ-function pulse is incident from the left and is impingent on the free space-medium interface at $x = 0$, $q(x)$ occupying the space $x \geq 0$. Thus, the incident field is

$$u_{inc}(x,t) = \delta(x - t)$$

The wave reflected at the interface then propagates to the left through the potential free space. That is,

$$u_{refl}(x,t) = R(x + t)$$

As in the case of the inverse scattering on the half-line (Chapter 6), the objective is to show that the depth-dependent information about the medium or, in other words, the potential can be determined uniquely from the time record of the reflection coefficient R at the interface $x = 0$.

A. METHOD OF "CHARACTERISTIC FIELDS"

We consider the following boundary-value problem

$$u'' - \ddot{u} - qu = 0, \quad x \geq 0$$

$$u(0,t) = \delta(-t) + R(t)$$

$$u'(0,t) = \delta'(-t) + R'(t)$$

$$u(x,t) = 0, \qquad x > t \tag{10.23}$$

The two boundary conditions represent continuity conditions at the interface, while the third is the causality condition, indicating that the disturbance cannot be ahead of the wavefront. Following Balanis,[6] we write down the following two boundary-value problems:

$$u_1'' - \ddot{u}_1 - qu_1 = 0, \qquad x \geq 0$$

$$u_1(0,t) = \delta(-t)$$

$$u_1'(0,t) = \delta'(-t) \tag{10.24}$$

and

$$u_2'' - \ddot{u}_2 - qu_2 = 0, \qquad x \geq 0$$

$$u_2(0,t) = \delta(+t)$$

$$u_2'(0,t) = \delta'(+t) \tag{10.25}$$

Fourier transforming with respect to time, we obtain

$$u_1^{\wedge}{}'' + \left[k^2 - q\right]u_1^{\wedge} = 0, \qquad x \geq 0$$

$$u_1^{\wedge}(0,k) = 1$$

$$u_1^{\wedge}{}'(0,k) = ik \tag{10.26}$$

and

$$u_2^{\wedge}{}'' + \left[k^2 - q\right]u_2^{\wedge} = 0, \qquad x \geq 0$$

$$u_2^{\wedge}(0,k) = 1$$

$$u_2^{\wedge}{}'(0,k) = -ik \tag{10.27}$$

Differentiating the Wronskian

$$W\left(u_1^{\wedge}, u_2^{\wedge}\right) = u_1^{\wedge}u_2^{\wedge}{}' - u_2^{\wedge}u_1^{\wedge}{}'$$

with respect to x and using Equations 10.26 and 10.27 yields

$$W\left(u_1^{\wedge}, u_2^{\wedge}\right) = -2ik$$

Therefore, for $k \neq 0$, $u_1^{\wedge}$ and $u_2^{\wedge}$ are two linearly independent solutions and form a fundamental set. The Fourier-transformed solution $u^{\wedge}(x,k)$ of the original wave equation can then be expressed as a linear combination of these two fundamental solutions. Therefore,

$$u^{\wedge}(x,k) = C_1(k)u_1^{\wedge}(x,k) + C_2(k)u_2^{\wedge}(x,k), \quad k \neq 0$$

where $C_1(k)$ and $C_2(k)$ are functions of k to be determined. From the boundary conditions of Equation 10.24, it is easily seen that the following relations hold:

$$1 + R(k) = C_1(k) + C_2(k)$$

and

$$ik\{1 - R(k)\} = ik\{C_1(k) - C_2(k)\}$$

This yields $C_1(k)$ to be unity and $C_2 = R(k)$. Then

$$u^{\wedge}(x,k) = u_1^{\wedge}(x,k) + R(k)u_2^{\wedge}(x,k)$$

Inverse Fourier transforming, we obtain

$$u(x,t) = u_1(x,t) + \int_{-\infty}^{\infty} R(t - t')u_2(x,t')dt' \qquad (10.28)$$

In the medium (plasma or string), the high frequencies are propagated with little or no attenuation, that is, the medium acts as a high-pass filter. The characteristic fields $u_1^{\wedge}$ and $u_2^{\wedge}$ can then be written as

$$u_1^{\wedge}(x,k) = \exp(ikx) + K_1^{\wedge}(x,k)$$

and

$$u_2^{\wedge}(x,k) = \exp(-ikx) + K_2^{\wedge}(x,k)$$

The functions $K_1^{\wedge}$ and $K_2^{\wedge}$ contain all the information about the medium and consist primarily of low-frequency components. That these functions contain information about the medium is evident from the fact that as $q \to 0$, these functions also vanish since the solutions of the wave equations in the absence of q are the plane waves $\exp(\pm ikx)$. Back transforming yields

$$u_1(x,t) = \delta(x - t) + K_1(x,t)$$

and

$$u_2(x,t) = \delta(x + t) + K_2(x,t)$$

where the δ-functions represent the sharp leading edges and K_1 (x,t), K_2(x,t), the scattering. The fields u_1(x,t) and u_2(x,t) thus consist of sharp leading edges, followed by wakes trailing behind. It may be recalled that a similar situation was encountered in Chapter 6. The fields u_1 and u_2 will be called the "characteristic fields". Balanis reserves this term for the K_1 and K_2 functions.

Replacing the above representations of the characteristic fields to Equation 10.28, we obtain

$$u(x,t) = \delta(x - t) + K_1(x,t) + R(x + t) + \int_{-\infty}^{\infty} K_1(x,t')R(t + t')dt' \qquad (10.29)$$

In deriving Equation 10.29, we have used the fact that

$$u_2(x,t) = u_1(x,-t)$$

This follows from the time invariance of the wave equation and will be shown later.

It is clear that the information about the medium is contained in the wake $K_1(x,t)$, which must, therefore, bear some relation to the potential $q(x)$. Hence, the objective is to determine $K_1(x,t)$ and express it in terms of the potential $q(x)$. Using the causality condition given earlier, namely, $u(x,t) = 0$, $x > t$, an equation for $K_1(x,t)$ is obtained in terms of the measured data $R(t)$ in the form

$$0 = K_1(x,t) + R(x+t) + \int_{-\infty}^{\infty} K_1(x,t')R(t+t')dt' \tag{10.30}$$

In the boundary-value problem (10.24) for u_1, the Cauchy data are prescribed on the line $x = 0$. In addition, by virtue of the δ-functions, these data are concentrated at $t = 0$. The function $K_1(x,t)$ must, therefore, have the region of influence $x = \pm t$ as its support. It then follows that $K_1 = 0$, $|t| < x$. Using this result in Equation 10.30 yields

$$0 = K_1(x,t) + R(x+t) + \int_{-x}^{x} K_1(x,t')R(t+t')dt' \tag{10.31}$$

From the causality condition of the field at the interface $x = 0$, we can write that

$$R(t) = 0 \text{ if } t \le 0$$

This means that $R(t + t') = 0$ if $t' \le -t$. This then allows the final reduction of the equation (10.31) for $K_1(x,t)$ to the form

$$0 = K_1(x,t) + R(x+t) + \int_{-t}^{x} K_1(x,t')R(t+t')dt' \tag{10.32}$$

Equation 10.32 is then Marcenko's equation in the time domain.

It remains to demonstrate the relation between $K_1(x,t)$ and the potential $q(x)$. In order to do this, let us substitute

$$u_1(x,t) = \delta(x-t) + K_1(x,t)$$

in Equation 10.24. The following boundary-value problem is obtained for $K_1(x,t)$:

$$K_1'' - \ddot{K}_1 - qK_1 = q\delta(x-t)$$

$$K_1(0,t) = 0$$

$$K_1'(0,t) = 0 \tag{10.33}$$

From Equation 10.33, it is clear that $K_1(x,t)$ undergoes a jump discontinuity across the characteristic $x = t$ while, for t such that $|t| > x$, $K = K_0$, the solution of the homogeneous equation.

$$K_0'' - \ddot{K}_0 - qK_0 = 0$$

Across the characteristic x = –t, there is no discontinuity and we have $K_1(x,-x) = 0$.

As in Chapter 6, we represent $K_1(x,t)$ at the characteristic x = t as

$$K_1(x,t) = H(x-t)K_0(x,t)$$

where H is the Heaviside function. Replacing this representation in Equation 10.33 and equating the coefficients of the δ-function singularity in the resulting equation, we obtain

$$K_1'(x,t)\big|_{x=t} + \dot{K}_1(x,t)\big|_{x=t} = q(x)/2$$

yielding

$$dK_1(x,x)/dx = q(x)/2 \tag{10.34}$$

As before, we have followed the standard practice of writing $K_1(x,x)$ for $K_1(x,x+0)$. Grouping all the results together, the boundary-value problem for the kernel $K_1(x,t)$ is

$$\begin{aligned} &K_1'' - \ddot{K}_1 - qK_1 = 0, && x \ge 0, \quad |t| \le x \\ &dK_1(x,n)/dx = q(x)/2, && x \ge 0 \\ &K_1(x,-x) = 0, && x \ge 0 \end{aligned} \tag{10.35}$$

and

$$K_1(x,t) = 0, \qquad |t| > x, \quad x \ge 0$$

Since the value of the function is specified along a characteristic, it is again a Goursat problem and, as discussed several times before, has a unique, bounded, and continuous solution for every finite x. Hence, once the data R(t) are specified, $K_1(x,t)$ can be determined uniquely as a solution of the Goursat problem (10.35) and, hence, the potential q(x) through the relation (10.34).

Similar analysis for the other characteristic field u_2 results in the boundary-value problem

$$\begin{aligned} &K_2'' - \ddot{K}_2 - qK_2 = 0, && x \ge 0, \quad |t| \le x \\ &dK_2(x,-x)/dx = q(x)/2, && x \ge 0 \\ &K_2(x,-x) = 0, && x \ge 0 \end{aligned} \tag{10.36}$$

and

$$K_2(x,t) = 0, \qquad |t| > x, \quad x \ge 0$$

Comparing Equations 10.35 and 10.36 yields the relation

$$u_2(x,t) = u_1(x,-t)$$

as was assumed previously from the time invariance of the wave equation.

As already mentioned, as $|k| \to \infty$, the solutions $\hat{u_1}(x,k)$ and $\hat{u_2}(x,k)$ become the plane progressing waves $\exp(\pm ikx)$. This indicates that the wakes $K_1(x,t)$ and $K_2(x,t)$ tend toward zero as $|k| \to \infty$. This means that the high frequencies of the δ-function pulse are propagated without attenuation, whereas the low-frequency components are reflected. It is thus reasonable to assume that the reflected wave contains primarily low frequency components. Hence, R(t) will be continuous in time. On the basis of this continuity argument, Balanis[7] shows that Marcenko's integral equation is also valid if $-x \leq t \leq x$. The analysis presented here applies for any q(x) which is real and integrable.

B. FOURIER TRANSFORM DERIVATION

Finally, we present yet another derivation in the time domain, which is due to Lamb.[8] The derivation is based on the scattering equations of various Jost solutions, the Levin representations, and the representations of the relevant Jost solutions in terms of the characteristic fields of the previous section. Let us consider, for example, the scattering equation for the Jost solution $\psi(x,k)$, which is reproduced here for convenience as Equation 10.37.

$$T_{-}(k)\psi(x,k) = R_{-}(k)\phi(x,k) + \bar{\phi}(x,k) \tag{10.37}$$

Defining

$$\Gamma_{-}(k) = T_{-}(k) - 1$$

Equation 10.37 can be written as

$$\Gamma_{-}(k)\psi(x,k) + \psi(x,k) = R_{-}(k)\phi(x,k) + \bar{\phi}(x,k) \tag{10.38}$$

Let us consider the transform from k to t as the direct and from t to k as the inverse Fourier transform. According to the convention adopted in this text,

$$\begin{aligned} f(x,k) &= (2\pi)^{1/2}[y(x,t)]^{\vee}(k) \\ y(x,t) &= (2\pi)^{1/2}[f(x,k)]^{\wedge}(t) \end{aligned} \tag{10.39}$$

is the Fourier transform pair. Then from the integral representation

$$\psi(x,k) = \exp(ikx) + \int_{+} K(x,x')\exp(ikx')dx'$$

where $x' = ct = t$, the speed of propagation set to unity, we obtain using Equation 10.39

$$y_1(x,t) = \delta(t-x) + H(t-x)K_1(x,t) \tag{10.40}$$

$y_1(x,t)$ is the Fourier transform of $\psi(x,k)$. Similarly,

$$y_2(x,t) = (2\pi)^{1/2}[\overline{\psi}(x,k)]^{\wedge}(t)$$

$$= \delta(t + x) + H(-t - x)K_1(x,-t) \tag{10.41}$$

$$y_3(x,t) = (2\pi)^{1/2}[\phi(x,k)]^{\wedge}(t)$$

$$= \delta(t + x) + H(t + x)K_2(x,-t) \tag{10.42}$$

and

$$y_4(x,t) = (2\pi)^{1/2}[\overline{\phi}(x,k)]^{\wedge}(t)$$

$$= \delta(t - x) + H(x - t)K_2(x,t) \tag{10.43}$$

We next calculate the Fourier transform of the LHS of Equation 10.38. The transform of the first term on the LHS is a convolution integral and, hence, the transform can be written down immediately

$$[\mathrm{LHS}](x,t) = (2\pi)^{-(1/2)}\int_{-\infty}^{\infty}\Gamma_-(t - t')[\psi(x,k)]^{\wedge}(t')dt' + [\psi(x,k)]^{\wedge}(t)$$

$$= \int_{-\infty}^{\infty}\Gamma_-(t - t')y_1(x,t')dt' + [\psi(x,k)]^{\wedge}(t)$$

$$= \Gamma_-(t - x) + \int_+\Gamma_-(t - t')K_1(x,t')dt' + y_1(x,t)(2\pi)^{1/2}$$

by virtue of the representation (10.40).

Now,

$$\Gamma_-(z) = (2\pi)^{-(1/2)}\int_{-\infty}^{\infty}\Gamma_-(k)\exp(-ikz)dk$$
$$= (2\pi)^{-(1/2)}\oint_C\Gamma_-(k)\exp(-ikz)dk, \quad k \subset C_+ \tag{10.44}$$

The contour C is a line segment on Re k from –R to +R and a semicircle of radius R in the upper half-plane. The transmission coefficient is holomorphic in the upper half-plane and if $t < x$, the contour in the integral of Equation 10.44 can be closed in the upper half-plane. Let the presence of the bound states be ignored at the moment. The integral of Equation 10.44 is then zero, having no singularity in C_+. If $x > t$, then y_1 also disappears by virtue of its representation (10.40). Hence, in the absence of bound states and with $x > t$, we conclude that the entire LHS is zero.

The RHS can be calculated entirely analogously and yields

$$[\mathrm{RHS}](x,t) = (2\pi)^{-(1/2)}\int_{-\infty}^{\infty}R_-(t - t')[\phi(x,k)]^{\wedge}(t')dt' + [\overline{\phi}(x,k)]^{\wedge}(t')$$

Again using the representations (10.42 and 10.43) and recalling that x > t, the RHS is obtained as

$$[\text{RHS}](x,t) = R_-(t+x) + \int_- R_-(t+t')K_2(x,t')dt' + (2\pi)^{(1/2)}K_2(x,t) \qquad (10.45)$$

Equating to the LHS, which is zero, yields the Marcenko equation

$$\hat{R}_-(t+x) + K_2(x,t) + \int_- dt'\,\hat{R}_-(t+t')K_2(x,t')dt' = 0, \qquad x > t \qquad (10.46)$$

where

$$\hat{R}_-(z) = (1/2\pi)\int_{-\infty}^{\infty} R_-(k)\exp(-ikz)dz$$

Starting with the scattering equation

$$T_-(k)\phi(x,k) = R_+(k)\psi(x,k) + \overline{\psi}(x,k)$$

and proceeding exactly analogously, the Marcenko equation is obtained for $K_1(x,t)$ as

$$\hat{R}_+(t+x) + K_1(x,t) + \int_+ dt'\,R_+(t+t')K_2(x,t') = 0, \quad x < t$$

Equations 10.46 and 10.47 can be combined into a single equation

$$\hat{R}_\pm(t+x) + K_\pm(x,k) + \int_\pm dt''\,R_\pm(t+t'')K_\pm(x,t'') = 0, \qquad x < t$$

with

$$\hat{R}_\pm(z) = (1/2\pi)\int_{-\infty}^{\infty} R_\pm(k)\exp(\pm ikz)dz$$

The presence of the bound states has been neglected thus far. It is, however, straightforward to incorporate the bound states, if any, into the formalism. If the bound states are present, then the Fourier transforms of the left-hand sides will be nonzero because the transmission coefficients will now have poles on the positive imaginary axis and the residues in the Fourier-transformed transmission coefficients will be nonzero. Consider again scattering Equation 10.38 with the wave incident from the left. The Fourier transform of Γ_- is now given by

$$\Gamma_-(z) = (2\pi)^{-(1/2)}\int_{-\infty}^{\infty}\Gamma_-(k)\exp(-ikz)dk$$

$$= -(2\pi)^{1/2}\sum_{j=1}^{N} c_{-j}D_{-j}\exp(k_j z), \qquad k_j > 0$$

The Fourier transform of the LHS of Equation 10.38 now becomes $[\text{LHS}](x,t)$

$$= \Gamma_-(t-x) + \int_+ dt'K_1(x,t')\Gamma_-(t-t') + y_1(x,t)(2\pi)^{1/2}$$

$$= -(2\pi)^{(1/2)} \sum_{j=1}^{N} c_{-j} D_{-j} \exp\{k_j(t-x)\} -$$

$$-(2\pi)^{(1/2)} \sum_{j=1}^{N} c_{-j} D_{-j} \int_+ dt'K_1(x,t') \exp\left[k_j(t-t')\right] + y_1(x,t)(2\pi)^{1/2}$$

$$= -(2\pi)^{(1/2)} \sum_{j=1}^{N} c_{-j} D_{-j} \exp\left(k_j t\right)\left[\exp\left(-k_j x\right)\right] +$$

$$+\int_+ dt'K_1(x,t') \exp\left[-k_j t'\right)\right] + y_1(x,t)(2\pi)^{1/2}$$

$$= -(2\pi)^{(1/2)} \sum_{j=1}^{N} c_{-j} D_{-j} \exp\left(k_j t\right)\psi_j(x) + y_1(x,t)(2\pi)^{1/2}$$

$$= -(2\pi)^{(1/2)} \sum_{j=1}^{N} D_{-j} \exp\left(k_j t\right)\phi_j(x) + y_1(x,t)(2\pi)^{1/2}$$

$$= -(2\pi)^{(1/2)} \sum_{j=1}^{N} D_{-j} \exp\left(k_j t\right)\left[\exp\left(k_j x\right) + \int_- dt'K_2(x,t') \exp\left(k_j t'\right)\right] + y_1(x,t)(2\pi)^{1/2}$$

$$= -(2\pi)^{(1/2)} \sum_{j=1}^{N} D_{-j} \exp\left[k_j(t+x)\right] -$$

$$-(2\pi)^{(1/2)} \sum_{j=1}^{N} D_{-j} \int_- dt'K_2(x,t') \exp\left[k_j(t+t')\right] + y_1(x,t)(2\pi)^{1/2}$$

We now use Equation 10.40 for $y_1(x,t)$, equate the above expression for the LHS to that of the RHS given by Equation(s) 10.45, and obtain the following identity:

$$-\sum_{j=1}^{N} D_{-j} \exp\left[k_j(t+x)\right] - \sum_{j=1}^{N} D_{-j} \int_- dt'K_2(x,t') \exp\left[k_j(t+t')\right]$$

$$= \hat{R}_-(t+x) + K_2(x,t) + \int_- dt'\hat{R}_-(t+t')K_2(x,t')$$

Rearranging,

$$\left[\hat{R}_-(t+x) + \sum_{j=1}^{N} D_{-j} \exp\left\{k_j(t+x)\right\}\right] + K_2(x,t) +$$

$$+\int_{-} dt' \left[\hat{R}_{-}(t+t') + \sum_{j=1}^{N} D_{-j} \exp\{k_j(t+t')\} \right] K_2(x,t') = 0$$

Defining

$$\Omega_{-}(t+x) = \hat{R}_{-}(t+x) + \sum_{j=1}^{N} D_{-j} \exp\{k_j(t+x)\}$$

we finally obtain the Marcenko equation.

$$\Omega_{-}(t+x) + K_2(x,t) + \int_{-} dt' \Omega_{-}(t+t') + K_2(x,t') = 0, \qquad x > t$$

For the right propagating scattering problem, the expression is similarly

$$\Omega_{+}(t+x) + K_1(x,t) + \int_{+} dt' \Omega_{+}(t+t') + K_1(x,t') = 0, \qquad x < t$$

Combining the two into a single expression, we obtain

$$\Omega_{\pm}(t+x) + K_{\pm}(x,t) + \int_{+} dt' \Omega_{\pm}(t+t') K_{\pm}(x,t') = 0, \; x \gtrless t \qquad (10.47)$$

This, then, is the Marcenko integral equation with the bound states taken into account. As can be seen from Equation 10.47, the presence of the bound states is to modify the reflection coefficients $R_{\pm}$ to the modified functions $\Omega_{\pm}$, which are given by

$$\Omega_{\pm}(z) = \sum_{j=1}^{N} D_{\pm j} \exp(\mp k_j z) + (1/2\pi) \int_{-\infty}^{\infty} dk R_{\pm}(k) \exp(\pm ikz)$$

By considering scattering from a delta function potential at the origin, Lamb has given interesting physical interpretations of various time-domain representations of the Jost functions. Consider, for example, a potential of the form

$$q(x) = q_0 \delta(x)$$

q_0 being the strength of the potential. Also consider the Jost function $\psi(x,k)$ given by

$$\psi(x,k) = \exp(ikx) - k^{-1} \int_{+} dx' \sin k(x-x') q(x') \psi(x',k)$$

For a delta function potential, we have

$$\psi(x,k) = \exp(ikx) \qquad x \geq 0$$

$$= \exp(ikx) - (q_0/k) \sin kx, \qquad x \leq 0$$

Since from Equation 10.39

$$\psi(x,k) = (2\pi)^{1/2}\left[y_1(x,t)\right]^{\vee}(x,k)$$

then $y_1(x,t)$ must be of the form

$$y(x,t) = \delta(t-x) + (q_0/2)[H(t-x) - H(t+x)] \tag{10.48}$$

if $\psi(x,k)$ is to correspond to the Jost solution for $x \le 0$. We then have

$$\psi(x,t) = \delta(t-x), \qquad x \ge 0$$

$$= \delta(t-x) + (q_0/2)[H(t-x) - H(t+x)], \qquad x \le 0$$

Now, for $x \le 0$, one can write

$$H(t-x) - H(t+x) = H(t-x)H[-(t+x)]$$

For $x \le 0$, therefore, one obtains

$$y(x,t) = \delta(t-x) + H(t-x)[(q_0/2)H\{-(t+x)\}]$$

$$= \delta(t-x) + H(t-x)K_1(x,t)$$

This, then, is exactly Equation 10.39, with $K_1(x,t)$ given by the quantity within the braces above. Inspection of Equation 10.48 reveals that $y(x,t)$ represents the following scattering process. A delta pulse along with a positive step of amplitude $(q_0/2)$ is incident on the delta function potential at the origin from the left, that is, from $x \to -\infty$. After being scattered, the delta function pulse propagates to the positive infinity, while a reflected step of negative amplitude $(-q_0/2)$ returns to the negative infinity and annihilates the positive-going step as it sweeps backward.

The corresponding noncausal scattering is represented analogously as

$$y(x,-t) = \delta(t+x) + (q_0/2)[H\{-(t+x)\} - H(x-t)]$$

This represents a scattering process in which a delta function pulse and a step function pulse are incident on the delta function potential simultaneously from the right and from the left, respectively. After being scattered, the delta function pulse propagates toward negative infinity, while a step of negative amplitude $(-q_0/2)$ propagates to the right and annihilates the incident left-propagating step as it encounters it in its path. Analogous descriptions can be given for the other transforms.

IV. SOME PROPERTIES OF Ω

We intend to show that Ω is an absolutely continuous function, absolutely integrable on $]a,x[$, and that its derivatives satisfy a growth condition of order 2 if the potential does so. We will proceed by first establishing that the derivative Ω' of Ω exists. In order to demonstrate this, we differentiate Marcenko's equation for $K_+(x,y)$ in x and then set $x = x'$ in the resulting equation. This gives

$$\Omega'_+(2x) - (1/2)q(x) - K_+(x,x)\Omega_+(2x) + \int_+ K'_+(x,x')\Omega_+(x+x')dx' +$$

$$+ \int_+ K_+(x,x')\Omega'_+(x+x')dx' = 0 \tag{10.49}$$

The subscript + will be dropped in the rest of the discussion. Differentiating the first integral on the RHS with respect to x yields

$$\left[\int_+ K'(x,x')\Omega(x+x')dx'\right]' = q(x)\int_+ K(x,x')\Omega(x+x')dx' +$$

$$+\int_+ K'(x,x')\Omega'(x+x')dx' + \int_+ \ddot{K}(x,x')\Omega(x+x')dx' - K'(x,x)\Omega(2x) \tag{10.50}$$

In deriving Equation 10.50, the hyperbolic propagation equation for K(x,x′), namely,

$$K''(x,x') - \ddot{K}(x,x') = q(x)K(x,x')$$

was used. Using the fact that

$$K'(x,x')\big|_{x=x'} + \dot{K}(x,x')\big|_{x=x'} = \left[K(x,x)\right]' = -q(x)/2$$

Equation 10.50 is rewritten as

$$\text{LHS} = q(x)\left[K(x,x) + \Omega(2x) + \int_+ K(x,x')\Omega(x+x')dx'\right] -$$

$$-q(x)K(x,x) - q(x)\Omega(2x)/2 - \int_+ \dot{K}(x,x')\dot{\Omega}(x+x')dx' \tag{10.51}$$

The first term on the RHS of Equation 10.51 disappears by virtue of the Marcenko equation itself at x = x′. Since $\Omega'(x + x') = \dot{\Omega}(x + x')$, we obtain

$$\text{LHS} = -q(x)\left[K(x,x) + \Omega(2x)/2\right] + J(x) \tag{10.52}$$

where J(x) is given by

$$J(x) = \int_+ \left(K' - \dot{K}\right)(x,x')\Omega'(x+x')dx'$$

It is not difficult to show that

$$J(x) - q(x)\Omega(2x)/2 = -\left[\int_+ \dot{K}(x,x')\Omega(x+x')dx'\right]' \tag{10.53}$$

Using the identity (10.53) and the relation

$$q(x) = -2\left[K(x,x)\right]'$$

in Equation 10.52, we obtain

$$\text{LHS} = \left[K^2(x,x) - \int_+ \dot{K}(x,x')\Omega(x+x')dx'\right]'$$

Finally, then

$$\int_+ K'(x,x')\Omega(x+x')dx' = K^2(x,x) - \int_+ \dot{K}(x,x')\Omega(x+x')dx' \tag{10.54}$$

Substitution of Equation 10.54 in Equation 10.49 yields

$$\Omega'(2x) = [q(x)/2] - K^2(x,x) - 2\int_+ K(x,x')\Omega'(x+x')dx' \tag{10.55}$$

The objective is to obtain an equation for $\Omega'(2x)$ suitable for an iterative solution. Equation 10.55 then requires further reduction. Let us write

$$K^2(x,x) = -\int_+ \left(\dot{K}^2\right)(x,x')dx' = -2\int_+ K(x,x')\dot{K}(x,x')dx'$$

$(\dot{K}^2)$ is the differentiation of K^2 with respect to x'. Differentiating the Marcenko equation

$$K(x,x') + \Omega(x+x') + \int_+ K(x,x'')\Omega(x'+x'')dx'' = 0$$

with respect to x' and substituting in the above identity for $K^2(x,x)$ yields

$$K^2(x,x) = 2\int_+ K(x,x')\dot{\Omega}(x+x')dx' +$$

$$+2\int_+ K(x,x')\left[\int_+ K(x,x'')\dot{\Omega}(x'+x'')dx''\right]dx'$$

Replacing $x'' + x'$ by $2x''$ and after some straightforward algebra, we obtain

$$K^2(x,x) = 2\int_+ K(x,x')\dot{\Omega}(x+x')dx' +$$

$$+2\int_+ dx' K(x,x')\int_{(x+x')/2}^{\infty} K(x,2x''-x')\Omega_{x''}(2x'')dx'' \tag{10.56}$$

The differentiation with respect to x'' has been written explicitly as subscript x'' in order to distinguish it from the derivatives in x and x'. Combining Equations 10.55 and 10.56 with the definitions

$$-Q(x,x') = 4\left[K(x,2x'-x) + (1/2)(K*K)(x')\right]$$

$$(K*K)(x) = \int_+ dx' K(x,x')\int_{(x+x')/2}^{\infty} K(x,2x''-x')dx''$$

$*$ denoting the operation of convolution, an expression for $\Omega'(2x)$ is obtained in the iteration format:

$$\Omega'(2x) = [q(x)/2] + \int_+ Q(x,x')\dot{\Omega}(2x')dx' \tag{10.57}$$

Equation 10.57 is a Volterra integral equation and relates the potential to the reflection coefficients through Ω.

Further analysis will require the estimates of the bound of $[\Omega'(2x) - q(x)/2]$. Iterating Equation 10.57 yields

$$\Omega'(2x) = \sum_{n=0}^{\infty} \Omega'^{(n)}(2x) = \sum_{n=1}^{\infty} \Omega'^{(n)}(2x) + q(x)/2$$

where

$$\Omega'^{(0)} = q(x)/2$$

The n-th iterate is

$$\Omega'^{(n+1)}(2x) = \prod_0^n \int_{x_i} Q(x_{i-1}, x_i) q(x_i) / 2dx_i$$

where $x_0 = x$, $x \le x_1 \le x_2 \le \ldots \le x_n$, and

$$Q(x_i, x_j) = -4\left[K(x_i, 2x_j - x_i) + (1/2)(K * K)(x_i)\right]$$

Consider the first iterate $\Omega'^{(1)}$. Taking the absolute value, we have

$$\left|\Omega'^{(1)}\right| \le \int_+ \left|K(x, 2x' - x)\right|\left|q(x')\right|dx' +$$

$$+ \int_+ \left|K(x,x')\right|\left[\int_{(x+x')/2}^{\infty} \left|K(x, 2x'' - x')\right|\left|q(x'')\right|dx''\right]dx'$$

From Chapter 8 (Equations 8.29 and 8.30),

$$\left|K(x,y)\right| \le \eta(x)\exp\{\gamma(x)\}, \qquad y \ge x$$

and

$$\int_+ dx'\left|K(x,x')\right| \le \gamma(x)\exp[\gamma(x)]$$

From these

$$\left|\Omega'^{(1)}\right| \le \eta^2(x)\exp\{\gamma(x)\} + \eta^2(x)\exp\{\gamma(x)\}\int_+ \left|K(x,x')\right|dx'$$

$$\le \eta^2(x)\exp\{\gamma(x)\} + \eta^2(x)\gamma(x)\exp\{2\gamma(x)\}$$

$$\le C_1(x)\eta^2(x)$$

Therefore,

$$\left|\int_+ Q(x,x')q(x')dx'\right| \le C_1(x)\eta^2(x), \ ]a,x[$$

where]a,x[is any half-line $-\infty < a \le x < \infty$.

In the same fashion, the second iterate can be estimated as follows:

$$\left|\Omega'^{(2)}\right| = \left|\int_+ Q(x,x')\int_{+x'}^{\infty} Q(x',x'')q(x'')dx''dx'\right|$$

$$\le \int_+ \left|Q(x,x')\right| \left|\int_{+x'}^{\infty} \left|Q(x',x'')q(x'')\right| dx''\right| dx', \quad x' \ge x$$

$$\le C_1(x)\eta^2(x)\int_+ \left|Q(x,x')\right| dx'$$

$$\le C_2(x)\eta^2(x), \quad]a,x[$$

Proceeding in this manner, the bound for the n-th iterate is obtained:

$$\left|\Omega'^{(n)}\right| \le C_n(x)\eta^2(x), \qquad n > 0, \quad]a,x[$$

where $C_n(x)$ is monotonically decreasing in x. Then,

$$\left|\Omega'(2x) - q(x)/2\right| \le C(x)\eta^2(x), \qquad]a,x[$$

Since the iterates of a Volterra equation converge, the bound exists and we finally have

$$\left|\Omega'(2x) - q(x)/2\right| \le C(x)\eta^2(x) \le C(a)\eta^2(x), \qquad]a,x[\tag{10.58}$$

again by virtue of a monotonic decrease of C(x) with x.

Let us now show that the function $|\Omega'(2x)|(1+|x|^2)$ exists. Toward this, let us write

$$\left|\Omega'(2x)\right| = \left|\Omega'(2x) - q(x)/2 + q(x)/2\right|$$

$$\le \left|\Omega'(2x) - q(x)/2\right| + \left|q(x)/2\right|$$

Then,

$$\int_+ \left|\Omega'(2x')\right|\left(1+|x'|^2\right)dx'$$

$$\le (1/2)\int_+ \left|q(x')\right|\left(1+|x'|^2\right)dx' + C(a)\int_+ \eta^2(x')\left|\left(1+|x'|^2\right)dx'\right.$$

The second term on the RHS is

$$\le C(a)\int_+ \eta(x')\left[\int_{+x'}|q(x'')|dx''|\right]\left(1+|x'|^2\right)dx'$$

$$\le C(a)\int_+ \eta(x')dx'\left[\int_+|q(x'')|\left(1+|x''|^2\right)dx''\right]$$

$$= C(a)\left[\int_+(x'-a)|q(x')|dx'\right]\left[\int_+|q(x'')|\left(1+|x''|^2\right)dx''\right]$$

Collecting these results together, we finally obtain

$$\int_+|\Omega'(2x')|\left(1+|x'|^2\right)dx'$$

$$\le \int_+|q(x')|\left(1+|x'|^2\right)[\cdot]dx'$$

where

$$[\cdot] = (1/2) + C(a)\int_+(x'-a)|q(x')|dx'$$

Provided that $q(x) \subset L_2^1$,

$$\int_+|\Omega'(2x)|\left(1+|x'|^2\right)dx' < \infty, \qquad]a,x[$$

From this it follows that if the potential satisfies a growth condition of order 2, then so does the function $\Omega'(x)$. It also follows that this derivative is L_1 on $]a,x[$. Hence, the function Ω is absolutely continuous on any half-line. Since $\Omega(z)$ is essentially $R(z)$, the discrete sum over the point spectrum vanishing as $|x| \to \infty$, it can be said that the derivative of the reflection coefficient, that is, $\hat{R}'(z)$, is also of growth of order 2 if the growth of the potential is of the same order.

The absolute integrability of $\Omega'(x)$ on any half-line can also be shown. Now, because of the absolute continuity

$$\Omega(x') = -\int_{+x'}\Omega_{x''}(x'')dx''$$

Then

$$\int_+ dx'\,|\Omega(x')| \le \int_+ dx'\int_{+x'}|\Omega_{x''}(x'')|dx'' = \int_+ dx'\,(x'-x)|\Omega_{x'}(x)|dx$$

$$\le C_1(x) \le C_1(a) < \infty, \qquad]a,x[\tag{10.59}$$

These results can be summarized in the following Lemma.

Lemma 10.1

Let the potential q(x) be of the growth of order 2. Then the Fourier transforms $\hat{R}_\pm(z)$ of the reflection coefficients $R_\pm(k)$ in the scattering data are also of the growth of order 2. That is,

$$\int_{\pm} \hat{R}'_{\pm}(z)\left[1+z^2\right]dz < \infty, \qquad]a,z[$$

where the prime means differentiation with respect to z. Moreover, $\hat{R}_{\pm}'(z)$ is locally L_1 and, hence, the Fourier transforms $\hat{R}_{\pm}(z)$ are absolutely continuous on any half-line.

V. ESTIMATES OF K(x,x′) AND ITS FIRST DERIVATIVES

The estimates for the kernel K(x,x′) and its first derivatives will be obtained next. These bounds were already established in Chapter 8 in terms of the potential and the functions $\gamma_{\pm}(x)$ and $\eta_{\pm}(x)$. It will be shown here that analogous expressions hold if the potential in these estimates is replaced by the derivatives of $\Omega_{\pm}$ and $\gamma_{\pm}(x)$ and $\eta_{\pm}(x)$ by the analogous quantities with $\Omega_{\pm}'$ in place of the potential q(x). In the previous section, Ω was found to be absolutely continuous on any half-line. Hence,

$$\left|\Omega(x)\right| \le \int_{+}\left|\Omega'(x')\right|dx' = w(x) \tag{10.60}$$

since $\Omega \to 0$ as $x \to \infty$ and the function w(x) is defined by Equation 10.60.

Let **K** be an integral operator such that

$$(\mathbf{K}f)(x) = \int_{+}\Omega(x+x')f(x')dx', \qquad]a,x[$$

where f(x) is an L_1 function on this line. Let us calculate the L_1 norm $\|\mathbf{K}\|_1$ of **K** and show that the norm exists on]a,x[. Calculating the norm, we have

$$\begin{aligned}\|\mathbf{K}f\|_1 &= \int_{+}|\mathbf{K}f|(x')dx' \\ &\le \int_{+}dx'\int_{+x'}dx''\left|f(x'')\right|\left|\Omega(x'+x'')\right. \\ &\le \int_{+}dx'\int_{+}dx''\left|f(x'')\right|\left|\Omega(x'+x'')\right| \\ &\le \left[\int_{+}\left|\Omega(x+x')\right|dx'\right]\left[\int_{+}\left|f(x')\right|dx'\right] \\ &\le \left[\int_{+}\left|\Omega(x')\right|dx'\right]\left[\int_{+}\left|f(x')\right|dx'\right] \\ &\le \|f\|_1\left[\int_{+}\left|\Omega(x')dx'\right|\right]\end{aligned}$$

Hence,

$$\|\mathbf{K}\|_1 \le \int_{+}dx'\left|\Omega(x')\right|dx' \le C_1(x) < \infty \tag{10.61}$$

from Equation 10.59. The norm $\|\mathbf{K}\|_1$, therefore, exists.

The Marcenko equation for the kernel function K(x,x′) can be written in the form

$$K(x,x') + [\mathbf{I} + \mathbf{K}]^{-1}\Omega(x + x') = 0 \tag{10.62}$$

It will be shown in the subsequent section that Equation 10.62 has a unique solution on any half-line. The inverse $[\mathbf{I} + \mathbf{K}]^{-1}$ must, then, exist and be bounded on any half-line. Moreover, since $\|\mathbf{K}\|_1 \to 0$ as $|x| \to \infty$, it follows that

$$\max_{]a,x[}\|\mathbf{I} + \mathbf{K}\|^{-1} \le C_2(a) \tag{10.63}$$

With this in mind, let us first establish the bound for K(x,x′). From Marcenko's equation,

$$|K(x,x')| \le w(x + x')\left[1 + \int_+ |K(x,x')|dx' = w(x + x')\left[1 + \|K\|_1\right]\right. \tag{10.64}$$

Using Equation 10.62, we also obtain

$$\|K(x,x')\|_1 \le \left\|(\mathbf{I} + \mathbf{K})^{-1}\right\|_1 \|\Omega\|_1 \le C_2(a)\|\Omega\|_1$$

by virtue of Equation 10.63. It then remains to determine the norm $\|\Omega(x + x')\|_1$. From Equation 10.59,

$$\|\Omega(x + x')\|_1 = \int_+ |\Omega(x + x')|dx' \le C_1(2x) \tag{10.65}$$

Therefore,

$$\|K(x,x')\|_1 \le C_2(a)C_1(2x)$$

Combining this with Equation 10.64 yields the bound

$$|K(x,x')| \le C_3(x)w(x + x') \tag{10.66}$$

The estimates of the bounds for the derivatives of K(x,x′) can also be established similarly. Let us differentiate the Marcenko equation with respect to x and obtain

$$K'(x,x') + \Omega'(x + x') - K(x,x)\Omega(x + x') + \int_+ K'(x,x'')\Omega(x' + x'')dx'' = 0$$

This can be rewritten as

$$[\mathbf{I} + \mathbf{K}]K'(x,x') = K(x,x)\Omega(x + x') - \Omega'(x + x')$$

From this, we have

$$K'(x,x') = [\mathbf{I} + \mathbf{K}]^{-1}\left[K(x,x)\Omega(x + x') - \Omega'(x + x')\right]$$

Taking norms of both sides yields

$$\left\|K'(x,x')\right\|_1 \le \left\|[\mathbf{I}+\mathbf{K}]^{-1}\right\|_1 \left[\left|K(x,x)\right|_1 \left\|\Omega(x+x')\right\|_1 + \left\|\Omega'(x+x')\right\|_1\right] \tag{10.67}$$

From Equations 10.63, 10.65, and 10.66, we have for]a,x[,

$$\left\|[\mathbf{I}+\mathbf{K}]^{-1}\right\|_1 \le C_2(a)$$

$$\left\|\Omega(x+x')\right\|_1 \le C_1(2x)$$

and

$$\left|K(x,x)\right| \le C_3(x)w(2x)$$

Now

$$\left\|\Omega'(x+x')\right\|_1 \le \left|\Omega'(2x)\right| = w(2x)$$

by Equation 10.60. Combining these results, it follows from Equation 10.67 that

$$\left\|K'(x,x')\right\|_1 \le C_2(a)w(2x)\left[1+C_1(x)C_3(2x)\right]$$

Taking the x' derivative of Marcenko's equation,

$$\dot{K}(x,x') + \dot{\Omega}(x+x') + \int_+ K(x,x'')\dot{\Omega}(x'+x'')dx'' = 0$$

This leads to the estimate

$$\left|\dot{K}(x,x') + \dot{\Omega}(x+x')\right| \le \int_+ \left|K(x,x'')\right|\left|\Omega(x'+x'')\right|dx''$$

$$\le C_3(x)\int_+ w(x+x'')\left|\Omega(x'+x'')\right|dx''$$

$$\le C_3(x)\int_+ \left[\int_{x+x'}^{\infty}\left|\Omega(z)\right|dz\right]\left|\Omega(x'+x'')\right|dx''$$

Changing the integration variable z to $z - x'$ and using the estimate of Equation 10.65, we obtain

$$\left|\dot{K}(x,x') + \dot{\Omega}(x+x')\right| \le C_3(x)w(2x)w(x+x')$$

The step before the last follows from the fact that $x' > x > x'$.

Similarly,

$$\left|K'(x,x') + \Omega'(x+x')\right| \le C_4(x)w(x+x')w(2x)$$

By following the same procedures, it is also shown that

$$\left|K'(x,x)+\Omega'(2x)\right| \le C_5(x)w^2(2x)$$

Finally, since from the Goursat problem for K(x,x′) we know that

$$q(x) = -2\Big[K(x,x\}'$$

then K′(x,x) satisfies a growth condition of order 2 if the potential does so. This can, however, be established independently.

The estimates of various bounds for the Levin transform K(x,x′) and its partial first derivatives in x and x′ are summarized in the Lemma below.

Lemma 10.2

The Levin transform K(x,x′) and its first partial derivatives satisfy the following bounds:

$$\left|K(x,x')\right| \le C_3(x)w(x+x')$$

$$\left|\dot{K}(x,x')+\dot{\Omega}(x+x')\right| \le C_3(x)w(x+x')w(2x)$$

$$\left|K'(x,x')+\Omega'(x+x')\right| \le C_4(x)w(2x)w(x+x')$$

$$\left|K'(x,x)+\Omega'(x,x)\right| \le C_5(x)w^2(2x)$$

Moreover, the L_1 norm $||K||_1$ of K exists and is given by

$$\left\|K(x,x')\right\|_1 \le C_2(x)C_1(2x)$$

Finally, K′(x,x) satisfies a growth condition of order 2 if the potential does so, that is,

$$\int_+\left[K(x',x')\right]'\left(1+x'^2\right)dx' \le \infty, \qquad]a,x[$$

These inequalities are exactly analogous to those established in Chapter 8 if Ω′ resp. w is replaced by q(x) resp. η(x), and the $C_j(x)$'s by functions of γ(x). Analogous estimates also hold for $K_-(x,x')$. The materials of the two previous sections were based on Reference 2.

VI. THE UNIQUENESS AND THE EXISTENCE OF SOLUTION

The uniqueness of the solution of the Marcenko equation will be proved via the application of Fredholm's alternative, as was done earlier on several occasions. The first step is to show that the homogeneous equation admits of only a trivial solution. Let us consider, therefore, the homogeneous equation

$$[\mathbf{I}+\mathbf{K}]K = 0$$

It is also assumed that K(x,y) = 0, y < x. Then we define an extension K to K such that

$$\begin{aligned} K(x,y) &= K(x,y), & y \geq x \\ &= 0, & y < x \end{aligned}$$

Let K be a solution in $L_2[x,\infty)$. Forming an inner product of the homogeneous equation with K yields for a real potential

$$(K,[\mathbf{I}+\mathbf{K}]K) = (K,K) + (K,\mathbf{K}K)$$

$$= (K,K) + \sum_{j=1}^{N} D_j\left(K,e_j^x\right)^2 + \int_{-\infty}^{\infty} dx' K(x,x') \int_{-\infty}^{\infty} dx'' \hat{R}(x'+x'') K(x,x'')$$

Recognizing that

$$\hat{R}(x+y) = (1/2\pi)\int_{-\infty}^{\infty} dk R(k) \exp\left[ik(x+y)\right]$$

and using Parseval's relation we have the identity

$$\int_{-\infty}^{\infty} dx' K(x,x') \int_{-\infty}^{\infty} dx'' \hat{R}(x'+x'') K(x,x'') + (K,K)$$

$$= (2\pi)\left[\int_{-\infty}^{\infty}\left[K(k)K(-k) + R(k)\{K(k)\}^2\right]dk\right.$$

$$= (2\pi)\left[\int_{-\infty}^{\infty}\left[K(k)\{K(-k) + R(k)K(k)\}\right]dk\right.$$

$$= (2\pi)\int_{-\infty}^{\infty} K(k)F(k)dk$$

where

$$F(k) = K(-k) + R(k)K(k)$$

The argument x in K was suppressed. We then obtain

$$(2\pi)\int_{-\infty}^{\infty} dk K(k)F(k) + \sum_{j=1}^{N} D_j\left(K,e_j^x\right)^2 = 0$$

Taking the absolute value of both sides,

$$(2\pi)\int_{-\infty}^{\infty} dk\left|K(k)\right|\left|F(k)\right| + \sum_{j=1}^{N} D_j\left|\left(K,e_j^x\right)^2\right| = 0$$

Defining

$$G(k) = \left|T(k)\right|K(k)$$

and recalling the unitarity relation

$$|T(k)|^2 + |R(k)|^2 = 1$$

the above inequality reduces to

$$0 = \pi\int_{-\infty}^{\infty} dk\left[|F(k)|^2 + |G(k)|^2\right] + \sum_{j=1}^{N} D_j\left|\left(K, e_j^x\right)^2\right|$$

This at once yields

$$F(k) = G(k) = 0$$

We recall from Chapter 8 that either (1) $T(k) = \alpha k + O(|k|)$ as $|k| \to 0$, Im $k \geq 0$, or (2) $T(k)$ = constant > 0 as $|k| \to 0$, Im $k \geq 0$. If case (2) holds, then it follows from the definition of $G(k)$ that $K(k) = 0$ and, hence, the triviality of the solution to the homogeneous equation. In the event that case (1) is true, then $K = 0$ may be problematic. However, for $|k| \to 0$, $R(k) = -1$, and, hence, from the definition of $F(k)$, $K(k)$ is again zero, establishing the existence of only a trivial solution. Marcenko's equation thus admits of a unique solution.

VII. ILLUSTRATIONS

The derivation of Marcenko's integral equation and the proof of the existence and uniqueness of its solution completes the essential formalism of the inverse scattering technique. The rest of the chapter will be occupied basically with illustrating the methodology with a few simple, well-known, but important examples. Before proceeding with the illustrations, let us summarize the main steps involved in the inverse scattering solution of the Schrödinger equation for a potential evolving in time and thereby solving a class of nonlinear partial differential equations. The essential steps are described as follows.

1. The Direct Problem

This involves solving the equation

$$y''(x,\lambda) + [\lambda - q(x,t)]y(x,\lambda) = 0, \quad x \subset R$$

at time $t = 0$, that is, solving the above equation with $q(x,t) = q(x,0)$, which is presumably known. The final results of Step 1 are the scattering data $S[q(x,0)]$.

2. Time Evolution of the Scattering Data

These were established in the final section of Chapter 9 and consist of the following equations:

$$R_{\pm}(k,t) = R_{\pm}(k,0)\exp\left(\pm 8ik^3 t\right) \tag{10.68}$$

and

$$D_{\pm j}(t) = D_{\pm j}(0)\exp\left(\pm 8k_j^3 t\right) \tag{10.69}$$

3. The Inverse Solution

Solve the Marcenko equation

$$K_{\pm}(x,x',t)+\Omega_{\pm}(x+x',t)+\int_{\pm}K_{\pm}(x,x',t)\Omega_{\pm}(x'+x'',t)dx''=0 \tag{10.70}$$

where

$$\Omega_{\pm}(z,t)=\sum_{j=1}^{N}D_{\pm j}(0)\exp\left\{\pm\left(8k_j^3t-k_jz\right)\right\}+$$

$$(1/2\pi)\int_{-\infty}^{\infty}R_{\pm}(k,0)\exp\left\{\pm i\left(8k^3t+kz\right)\right\} \tag{10.71}$$

Let us now present a few examples. Our first example will be a delta function initial condition. This will be discussed at some length for the following reasons. With this potential, it is possible to perform all the calculations (at least for t = 0) explicitly in each of the three steps of the inverse scattering procedure, including the direct solution, the determination of the spectral transform, and the solution of the Marcenko equation. For most potentials, including even such well-known cases as those of the rectangular barrier or an exponential potential, performing all these steps explicitly may be at least difficult. Moreover, their direct solutions may be too involved to be presented as illustrative examples. Since the main objective here is to solve the inverse problem, it will be assumed that the direct solutions or the spectral transforms have been calculated beforehand if these turn out to be too involved.

Example 1. Delta Function Potential

Consider a delta function potential

$$q(x)=2q_0\delta(x)$$

where q_0 is the strength of the potential. Depending upon the sign of q_0, the potential is either attractive or repulsive. If q_0 is positive, it is a repulsive potential and if q_0 is negative, the potential is attractive. The two cases will be considered separately.

a. Attractive Potential: $q(x) = -2q_0\,\delta(x)$

The potential is at time t = 0. According to the inverse scattering technique, the first step is to solve the direct problem. In this case, the direct problem satisfies the equation

$$y''+\{\lambda+2q_0\delta(x)\}y=0$$

It is reasonable to assume that y″ will behave like a delta function at x = 0. Therefore, the first derivative must undergo a jump discontinuity at that point. Integrating over an interval [ε,ε] and then letting ε → 0, we obtain

$$y'|_{-\varepsilon}^{\varepsilon}=-\int_{-\varepsilon}^{\varepsilon}\{\lambda+2q_0\delta(x)\}ydx=-2q_0y(0)\ \text{as}\ \varepsilon\to 0 \tag{10.72}$$

Let us next consider the continuous states. For this, we set $\lambda = -k^2$, where k is real. Let a plane wave exp(–ikx) be incident on the delta function potential from the right. For the potential under consideration, the fundamental solutions of the scattering equations are

$$\phi = \exp(--ikx) - (2q_0/k)\sin kx, \qquad x \geq 0$$

$$= \exp(-ikx), \qquad x \leq 0$$

$$\psi = \exp(ikx) + (2q_0/k)\sin kx, \qquad x \leq 0$$

$$= \exp(ikx), \qquad x \geq 0 \tag{10.73}$$

One of these fundamental solutions was used in the last section. Substituting Equation(s) 10.69 in the scattering Equations 9.5 and 9.6, and then using the definition $R_{\pm}(k) = r_{\pm}(k)/t_{\pm}(k)$, the reflection coefficient $R_+(k)$ is found to be

$$R_+(k) = iq_0/(k - iq_0) \tag{10.74}$$

The same expression also holds for $R_-(k)$. Equation 10.70 is applicable for both repulsive and attractive potentials.

In order to obtain the bound states, we set $\lambda = -k^2$, k real. This yields the following solution for the wave function:

$$y(x) = \exp(-kx), \quad x > 0$$

and

$$y(x) = \exp(kx), \quad x < 0$$

Again, the continuity condition at the origin is satisfied, as is the square integrability over R. From the jump condition, Equation 10.72, we then obtain

$$-k - k = -2q_0 y(0) = -2q_0$$

yielding

$$k = q_0 \tag{10.75}$$

The potential has, therefore, only one bound state. The normalization constant is calculated as follows.

$$D_+(0) = \left[\int_{-\infty}^{\infty} y^2(x)dx\right]^{-1}$$

$$= \left[2\int_0^{\infty} \exp(-2kx)dx\right]^{-1}$$

$$= k = q_0$$

The spectral transform is then

$$S[q(x)] = \{R_+(k,0), \quad k = ik_j, \quad k_j > 0, \quad j = 1, \quad D_{+j}(0)\}$$

where

$$R_+(k,0) = -q_0/[q_0 + ik]$$

$$k_j = k_1 = q_0$$

and

$$D_{+j}(0) = D_{+1}(0) = q_0$$

We have, therefore, performed step 1 of the inverse scattering transform, namely, the direct solution and the determination of the spectral transform.

In order to inverse solve it for the potential, the Marcenko equation (10.70) is to be solved and the kernel Ω determined according to Equation 10.71. The Fourier transform of the reflection coefficient of Equation 10.74 can be evaluated from the integral

$$R_+(z,0) = (iq_0/2\pi)\int_{-\infty}^{\infty} \exp(ikz)[k - iq_0]^{-1} dk$$

The pole of the integrand is at $k = iq_0$. We apply Cauchy's theorem for evaluating the integral. If $z > 0$, the contour can be closed in the upper half-plane and, for this location of the pole, Cauchy's theorem shows that the integral is $-q_0 \exp(-q_0 z)$. If, on the other hand, $z < 0$, then the contour will be closed in the lower half-plane and since there is no singularity there, the integral will be zero. Therefore,

$$R(z,0) = -q_0 \exp(-q_0 z)H(z)$$

H being Heaviside's function. The kernel $\Omega_+(z,0)$ is then

$$\Omega_+(z,0) = D_{+1}(0)\exp(-k_1 z) + R_+(z,0)$$

Using Equations 10.74 and 10.75, we obtain

$$\Omega_+(z,0) = 0 \qquad \text{if } z > 0$$

$$= q_0 \exp(-q_0 z) \qquad \text{if } z > 0$$

That is, $$\Omega_+(z,0) = q_0 \exp(-q_0 z)H(-z)$$

Substituting in the Marcenko equation yields

$$K_+(x,x',0) + q_0 \exp[-q_0(x + x')]H[-(x + x')] +$$

$$+q_0\int_x^{-x'} dx'' \exp[-q_0(x'+x'')]K_+(x,x'',0)=0, \qquad x'>x \tag{10.76}$$

If $x > -x'$, then $K_+(x,x',0) = 0$ identically. In the opposite case $(x + x') < 0$, the equation can be satisfied by taking K_+ to be a constant which equals $-q_0$. This then gives

$$K_+(x,x',0) = -q_0 H[-(x+x')]$$

Setting $x = x'$ and using the relation

$$q(x,0) = -2[K(x,x',0)]'$$

yields

$$q(x,0) = 2q_0[H(-2x)]' = -4q_0\delta(2x) = -2q_0\delta(x)$$

where the identity

$$\delta(ax) = (1/|a|)\delta(x)$$

was used.

b. Repulsive Potential: $q(x) = 2q_0\,\delta(x)$, $q_0 > 0$

Following the same procedures as above, we have from Equation 10.74

$$R_+(k,0) = -\,iq_0/[k+iq_0], \qquad q_0>0$$

The Fourier transform is

$$R_+(z,0) = -q_0 \exp(q_0 z)H(-z)$$

The reflection coefficient expressed in this form, the Marcenko equation that results in this case is the same as that given by Equation 10.76. We, therefore, again obtain its solution as $K_+(x,x';0) = q_0\, H[-(x + x')]$, leading to the potential

$$q(x,0) = 2q_0\delta(x)$$

However, in this case, $q_0 > 0$.

The calculations for time evolution can also be performed. The time variation of the spectral transform is given by

$$k_1(t) = k_1(0) = q_0$$

$$D_\pm(t) = D_\pm(0)\exp\left(\pm 8q_0^3 t\right)$$

and

$$R_{\pm}(k,t) = R_{\pm}(k,0)\exp\left(\pm 8ik^3 t\right)$$

The kernel is then

$$\Omega_{\pm}(z,t) = q_0 \exp(\pm 8q_0 t \mp q_0 z) + (1/2\pi)\int_{-\infty}^{\infty} R_{\pm}(k,0)\exp\left(\pm i\left(8q_0^3 t + kz\right)\right\}dk$$

The solution of the Marcenko equation is, however, difficult to obtain in terms of elementary functions.

A qualitative description of the evolving potential based on numerical solution gives the following picture. The initial pulse eventually breaks up into two parts: a solitary wave moving to the right and a dispersive wave train moving to the left and decaying with time. In the asymptotic limit in time, the left-going dispersive wave decays as $t^{-1/3}$ as $t \to \infty$. The solitary wave, on the other hand, takes the form

$$q(x,t) \approx -2q_0^2 \sec h^2\left\{q_0\left(x - 4q_0^2 t - \theta\right)\right\}$$

where θ is a phase factor. These are for the case of the attractive potential. If the potential is repulsive, then the solitary wave is absent.

The delta function potential discussed in this example illustrates an interesting point. The potential clearly belongs to those for which the reflection coefficient is a rational function of k. If a potential is infinitely differentiable, that is, $q(x) \subset C_{\infty}$, then its corresponding reflection coefficient $R \to 0$ as $|x| > \infty$ faster than any power of k (see Chapter 9). If R is a rational function of k, this is clearly not possible. The potential cannot, therefore, be infinitely differentiable and the point where this differentiability fails is the origin.

Example 2. Initial Condition $= \delta'(x)$

A second example[8,9] involving a delta function where the reflection coefficients are found to be the simplest is that of an initial pulse profile given by $\delta'(x)$, that is, $q(x,0) = \delta'(x)$. We follow Lamb[8] here. The direct solutions[8] for this pulse profile show that the reflection coefficient $R_-(k,0)$ resp. $R_+(k,0)$ are + 1 resp. – 1. Let us consider the right reflection R_+, for example. If the discrete spectrum is assumed to be absent, then the kernel in the Marcenko equation can be calculated to be

$$\Omega_+(x,t) = -(1/2\pi)\int_{-\infty}^{\infty} dk \exp\left\{i\left(kx + 8k^3 t\right)\right\} \tag{10.77}$$

since $R_+ = -1$. This integral can be recognized as the integral representation of the Airy function.[10] Writing $k = \alpha\theta$, and $8\alpha^3 t = (1/3)$, the Airy integral becomes

$$\Omega_+(x,t) = -(\alpha/2\pi)\int_{-\infty}^{\infty} d\theta \exp\left\{i\left[\theta\alpha x + (1/3)\theta^3\right]\right.$$

$$= -\alpha \mathrm{Ai}(\alpha x)$$

Since $\alpha x = x(24t)^{-1/3}$, the kernel has a space and time dependence of the form $\Omega_+(x,t) = t^{-1/3}\, w(\zeta)$, where $\zeta = xt^{-1/3}$. Let us assume that $K_+(x,x',t) = t^{-1/3}\, A(\zeta,\eta)$ where $\eta = yt^{-1/3}$. Marcenko's equation then reduces to

$$w(\zeta+\eta) + A(\zeta,\eta) + \int_{+\zeta} d\zeta' w(\zeta'+\eta)A(\zeta,\zeta') = 0$$

and the pulse evolution is given by

$$q(x,t) = -2[K_+(x,x;t)]' = -2t^{-1/3}[A(\zeta,\zeta)]', \qquad ' = \partial/\partial x$$

$$= -2t^{-2/3}A_\zeta(\zeta,\zeta) = t^{-2/3}f(\zeta)$$

The differentiation with respect to ζ is explicitly written as subscript ζ and the function f is defined by the equation.

Now, since

$$[\partial/\partial t] = -[\zeta/3t][\partial/\partial\zeta]$$

we have

$$[\partial q/\partial t] = -(1/3)t^{-5/3}\left(2f + \zeta f_\zeta\right)$$

Let us substitute $g(\zeta) = 6f(\zeta)$ and consider the K-dV equation

$$q'' - 6qq' + q''' = 0$$

The result is

$$3g''' - 3gg' - \left(2g + \zeta g_\zeta\right) = 0 \tag{10.77a}$$

Set $g = h'$ and make the small amplitude approximation, that is, assume that g is sufficiently small so that the nonlinear term is neglected. This results in the following equation for h:

$$h'' - (1/3)\zeta h = 0$$

Its solution that vanishes as $\zeta \to \infty$ is given by the Bessel function K of order of one third and can be written as

$$\zeta^{1/2}K_{1/3}\left[2(-\zeta/3)^{3/2}\right]$$

Using the asymptotic form for the Bessel function, we have

$$g \approx \zeta^{1/4}e^{-2(\zeta/3)^{3/2}} \text{ as } \zeta \to \infty$$

Equation 10.77a was numerically solved by Karpman.[9] The solution that reduces to the small amplitude form given above shows that the pulse is unstable as $\zeta \to -\infty$. The potential grows in oscillation in this limit. The instability is due to the fact that the pulse profile $\delta'(x)$ is too singular a function to be a physically meaningful initial condition. Such an initial condition is stronger than any of the growth conditions encountered so far and, hence, is not strictly amenable to a solution by Marcenko's equation, which postulates a growth condition of order 2.

In order to recover the $\delta'(x)$ initial condition, let us set t = 0 and obtain from Equation 10.77

$$\Omega_+(x,0) = -\delta(x)$$

The corresponding Marcenko equation is

$$K_+(x,x',0) + \Omega_+(x+x',0) + \int_+ dx'' K_+(x,x'',0)\Omega_+(x'+x'',0) = 0, \qquad x' > x$$

This equation can be satisfied upon choosing $K_+(x,x',0) = \delta(x+x')$ and using the identity

$$\int dy\,\delta(x-y)\delta(y-a) = \delta(x-a)$$

Therefore,

$$q(x,0) = -2[K_+(x,x,0)]'$$

$$= -2[\delta(2x)]' = -\delta'(x)$$

Example 3: Reflectionless Potentials

If the potential is reflectionless, then R(k,t) = 0, and the inverse solution can be obtained explicitly. We illustrate the solution for K_+ since the K_- case is similar. For a reflectionless potential, the Marcenko equation (10.70) becomes

$$K_+(x,x',t) + \sum_{j=1}^{N} D_{+j}(t)\exp\{-k_j(x+x')\} +$$

$$+\sum_{j=1}^{N} D_{+j}(t)\exp(-k_j x')\int_+ dx''\exp(-k_j x'')K_+(x,x'',t) = 0 \tag{10.78}$$

where $D_{+j}(t)$ is given by Equation 10.69.

Equation 10.78 can be solved by assuming a solution of the form

$$K_+(x,x',t) = -\sum_{j=1}^{N} d_{+j}h_j(x,t)\exp(-k_j x') \tag{10.79}$$

where $d_{+j}(t) = [D_{+j}(t)]^{1/2}$. Substituting Equation 10.79 in Equation 10.78, the following set of algebraic equations are obtained for the h_j's:

$$h_j(x,t) + \sum_{m=1}^{N} d_{+j} d_{+m} \left[\exp\left\{ -\left(k_j + k_m\right)x \right\} / \left(k_j + k_m\right) \right] h_m(x,t) = d_{+j} \exp\left(-k_j x\right) \quad (10.80)$$

Equation 10.80 can be written in a matrix form as

$$(\mathbf{I} + \mathbf{C})\mathbf{H} = \mathbf{E} \quad (10.81)$$

where

$$\mathbf{I}_{jm} \equiv \delta_{jm},$$

$$\mathbf{C}_{jm} \equiv d_{+j} d_{+m} \left[\exp\left\{ -\left(k_j + k_m\right)x \right\} \right] / \left(k_j + k_m\right) \quad (10.82)$$

$$\mathbf{H} \equiv \begin{pmatrix} h_1 \\ h_2 \\ \vdots \\ h_N \end{pmatrix}, \qquad \mathbf{E} = \begin{pmatrix} d_{+1} \exp(-k_1 x) \\ \vdots \\ d_{+N} \exp(-k_N x) \end{pmatrix}$$

If Q_{jm} denotes the cofactors of the determinant Δ of $(\mathbf{I} + \mathbf{C})$, then

$$\Delta \equiv \sum_{m=1}^{N} \left[\delta_{jm} + d_{+j} d_{+m} \left\{ \exp\left[-\left(k_j + k_m\right)x \right] / \left(k_j + k_m\right) \right\} Q_{jm} \right.$$

The application of Cramer's rule[11] results in

$$h_j(x,t) = \Delta^{-1} \sum_{m=1}^{N} D_{+1} \exp(-k_m x) Q_{jm} \quad (10.83)$$

Replacing Equation 10.83 in Equation 10.79 and then setting $x' = x$, we obtain

$$K_+(x,x,t) = -\Delta^{-1} \sum_{j=1}^{N} \sum_{m=1}^{N} d_{+j} d_{+m} \exp\left[-\left(k_j + k_m\right)x \right] Q_{jm} = \Delta^{-1}\Delta' = \left[\log \Delta\right]' \quad (10.84)$$

The potential is then determined as

$$q(x,t) = -2\left[K_+(x,x,t)\right]' = -2\left[\log \det(\mathbf{I} + \mathbf{C})\right]'' \quad (10.85)$$

In order to prove that Equation 10.81 possesses a unique solution, it is necessary to show that the quadratic form of the matrix $\mathbf{C}$ is positive definite. It may be recalled that the quadratic form Q of a matrix $\mathbf{A}$ is given by[11]

$$Q = \mathbf{z}^T \mathbf{A} \mathbf{z}$$

where the superscript T denotes transpose. For the matrix **C** then,

$$Q = \sum_{i=1}^{N} \sum_{j=1}^{N} d_{+i} d_{+j} \left[k_i + k_j\right]^{-1} \exp\left[-\left(k_i + k_j\right)x\right] y_i y_j$$

$$= \int_+ dx' \left[\sum_{n=1}^{N} d_{+n} \exp\left(-k_n x'\right) y_n\right]^2 > 0$$

The matrix **C** is, therefore, positive definite and Equation 10.81 admits, therefore, of a unique solution.

Moreover, det **C** also has a simple form, which can be seen as follows.

$$\det \mathbf{C} = \left[\prod_{1}^{N} d_{+i}^2\right] \exp\left[-2\left(\sum_{j=1}^{N} k_j\right)x\right] \det\left[k_m + k_n\right]^{-1}$$

$$= a \exp(-bx)$$

where a and b are both positive, the latter because of the fact that det **C** > 0.

Let us next determine the physical meaning of the h_i functions in Equation 10.79. In order to do this, let us operate both sides of Equation 10.79 by the Schrödinger operator

$$\mathbf{L}_j = \left(d^2/dx^2\right) - \left[k_j^2 - q(x)\right]$$

The reduction is complicated and involves considerable algebraic manipulation. However, the net result[12] is a simple expression:

$$\mathbf{L}_j h_j + \sum d_{+j} d_{+m} \left[\exp\left\{-\left(k_j + k_m\right)x\right\} / \left(k_j + k_m\right)\right] \mathbf{L}_m h_m = 0$$

This means that the matrix equation (10.81) now becomes

$$(\mathbf{I} + \mathbf{C})\mathbf{LH} = 0 \tag{10.86}$$

where $\mathbf{L}\ \mathbf{H} = [\mathbf{L}_1\ h_1, \mathbf{L}_2\ h_2, \ldots, \mathbf{L}_N\ h_N]^T$. Since $\mathbf{I} + \mathbf{C}$ is nonsingular, the homogeneous Equation 10.86 possesses a unique solution and according to Fredholm's alternative, this solution must be zero. Therefore,

$$\mathbf{LH} = 0$$

implying that the functions h_j are the eigenfunction of the Schrödinger equation corresponding to the k_j eigenvalues.

Furthermore, the functions h_j are also the normalized eigenfunctions and can be seen as follows. Let us rewrite Equation 10.79 as

$$h_j + \sum_{m=1}^{N} d_{+j} d_{+m} h_m \exp\left\{-2k_m x\right) = d_{+m}$$

where $h_j = h_j \exp(k_j x)$. Taking the limit $x \to \infty$ yields

$$\lim_{x \to \infty} h_j \exp(k_j x) = d_{+j}$$

The functions h_j are, therefore, the normalized eigenfunctions.

As a result of the inverse scattering theory,[13-16] the time asymptotic behavior of the potential is described by

$$q(x,t) = -2 \sum_{j=1}^{N} k_j^2 \operatorname{sech}^2\left[k_j\left(x - 4k_j^2 t - \theta_j\right) \text{ as } t \to \infty\right.$$

where the phase term θ_j is given by

$$\theta_j = \left(1/2k_j\right)\left[\ln\left\{D_{+j}(0)/2k_j\right\} + \sum_{m=1}^{N} \ln\left\{\left(k_j - k_m\right)/\left(k_j + k_m\right)\right\}\right] \tag{10.87}$$

Similar expressions are also obtained for $t \to -\infty$. These results will be derived in the appendix.

The reason for discussing this example at such length is the following. Equation 10.86 recovers the reflectionless potential q(x,t) via the inverse scattering transform. According to the discussion of this methodology at the end of the previous chapter, this means that we have obtained the solution of the K-dV equation. It turns out that the solution q with N-discrete eigenvalues is exactly the N-soliton solution of the K-dV equation and is one of the most important problems of the inverse scattering theory.

In the next two examples, we specialize the N-soliton solution to one soliton and two soliton solutions.

Example 4. $q(x,0) = -2 \operatorname{sech}^2 x$: One-Soliton Solution of the K-dVE

This is an example of a reflectionless potential that results in a single soliton as the solution of the K-dV equation. The direct problem for this potential was solved in the final example of Chapter 2. The potential was found to possess only one bound state at the location $k = i$ on the positive imaginary axis. This, therefore, gives $k_i = 1$. Furthermore, the normalization constant D_{+1} was determined to be 2. With these data, the kernel of the Marcenko equation becomes

$$\Omega_+(z,t) = 2\exp(8t - z)$$

and the Marcenko equation becomes

$$K_+(x,x',t) + 2\exp(8t - x - x') + 2\int_+ K_+(x,x'',t)\exp(8t - x' - x'')dx'' = 0 \tag{10.88}$$

We now solve the Marcenko equation following the procedure outlined in Example 3 and write

$$K_+(x,x',t) = w(x,t)\exp(-x') \tag{10.89}$$

where w(x,t) is written in place of d(t)h(x,t) of Equation 10.79. Substitution of Equation 10.89 in the Marcenko equation (10.88) leads to the following equation for w(x,t):

$$w(x,t)\exp(-x') + 2\exp(8t - x - x') + 2\exp(8t - x')w(x,t)\int_+ \exp(-2x'')dx'' = 0$$

Calculating the simple integral in the above equation and after some straightforward algebra, the solution is obtained as

$$w(x,t) = -\exp(4t)\,\text{sech}(x - 4t)$$

Replacing this in Equation 10.89 and setting $x' = x$, we obtain

$$K_+(x,x,t) = -\exp(-x + 4t)\,\text{sech}(x - 4t)$$

Then the inverse solution for the potential is given by

$$q(x,t) = -2\left[-\exp(-x + 4t)\,\text{sech}(x - 4t)\right]'$$

Upon differentiating and using the trigonometric identity

$$\text{sech}x = 2\left[\exp(x) + \exp(-x)\right]^{-1}$$

we obtain for the potential

$$q(x,t) = -2\,\text{sech}^2(x - 4t)$$

This is a propagating pulse. Its velocity is 4 and from Equation 10.87, its phase lag can be calculated as

$$\theta = (2k_1)^{-1}\ln\{D_{+1}(0)/2k_1\} = 0$$

Example 5. $q(x,0) = -6\,\text{sech}^2(x)$: Two-Soliton Solution Of K-dVE

This is an example of a two-soliton solution of the K-dV equation. The potential evolving in time is also known as the quadratic Bargmann potential. Before proceeding with the calculations, let us summarize some results about the direct solutions for the potentials of the form

$$q(x) = q_0\,\text{sech}^2(x)$$

Define r by

$$r = \{(1/4) - q_0\}^{1/2}$$

Some general results of the direct solution are

$$T(k) = \Gamma\left[(1/2) + r - ik\right]\Gamma\left[(1/2) - r - ik\right]\Big/\left[\Gamma(1 - ik)\Gamma(-ik)\right] \qquad (10.90)$$

$$R(k) = \pi^{-1}\cos(\pi r)\Gamma\left[(1/2) + r - ik\right]\Gamma\left[(1/2) - r - ik\right]\Gamma(ik)/\Gamma(-ik) \qquad (10.91)$$

If the potential is attractive, then the poles of T(k) are located at

$$k_j = (1/2) + \left[(1/4) + V^2\right]^{1/2} - j, \qquad j = 1, 2, \ldots, N$$

where $q_0 = -V^2$. For a repulsive potential, that is, $q_0 = V^2$, the poles of T(k), as follows from Equation 10.90, are all in the lower half-plane, implying that there are no bound states. These conclusions follow from Equation 10.90 since $\Gamma(1/z)$ is entire. The normalization constants are given by

$$D_j = \Gamma\left(2k_j + j\right)/\left[\Gamma(j)\Gamma\left(k_j\right)\Gamma\left(1 + k_j\right)\right], \qquad j = 1, 2, \ldots, N \qquad (10.92)$$

If the potential is reflectionless, then R(k) must be zero. The factor of cos (πr) in Equation 10.91 implies that for a reflectionless potential q_0 is of the form m(m + 1), m $\subset$ Z^+. m is also the number of eigenvalues which assume the simple form $k_j = j$, j = 1,2,...,m in this case. Moreover, the normalization constants given by Equation 10.92 can be rewritten for a reflectionless potential as

$$D_j = \Gamma(m + j + 1)/\left[\Gamma(j)\Gamma(j + 1)\Gamma(m - j + 1)\right] \qquad (10.93)$$

Let us now return to the potential of this example, namely, $q = -6\,\mathrm{sech}^2\,x$. For this potential, $V^2 = 6$ and m = 2. Using Equation 10.93 and the fact that $k_j = j$, the poles in the upper half-plane are $k_1 = 1$ and $k_2 = 2$, and the normalization constants are $D_{+1} = 6$ and $D_{+2} = 12$. The kernel of the Marcenko equation is then

$$\Omega_+(z,t) = 6\exp[(8t - z) + 12\exp(64t - 2z)$$

and the Marcenko equation becomes

$$K_+(x,x';t) + \left[6\exp(8t - x - x') + 12\exp(64t - 2x - 2x')\right] +$$

$$\int_+ K_+(x,x'',t)\left[6\exp(8t - x' - x'') + 12\exp(64t - 2x' - 2x'')\right]dx'' = 0$$

In order to recover the potential, we use the inversion formula (10.87) at time t = 0. The determinant of $(\mathbf{I} + \mathbf{C})$ becomes

$$\Delta = |\mathbf{I} + \mathbf{C}| = \begin{vmatrix} 1 + (D_{+1}/2)\exp(-2x) & (D_{+2}/3)\exp(-3x) \\ (D_{+1}/3)\exp(-3x) & 1 + (D_{+2}/4)\exp(-4x) \end{vmatrix}$$

From this we obtain

$$K_+(x,x;0) = \left[\ln(\Delta)\right]' = -P/Q \qquad (10.94)$$

where

$$P = D_{+1}\exp(-2x) + D_{+2}\exp(-4x) + [D_{+1}D_{+2}/72]\exp(-6x) \tag{10.95}$$

and

$$Q = 1 + (D_{+1}/2)\exp(-2x) + (D_{+2}/4)\exp(-4x) + (D_{+1}D_{+2}/72)\exp(-6x)$$

Using the trigonometric identity

$$\cosh x = (1/2)[\exp(x) + \exp(-x)]$$

and after some straightforward reduction, we obtain, upon setting $x' = x$, the following simple expression:

$$K(x, x, 0) = -3\exp(-x)\operatorname{sech} x$$

where the numerical values have been substituted for the various quantities involved. Differentiating with respect to x yields the potential

$$q(x, 0) = -6[\exp(-x)\operatorname{sech} x]' = -6\operatorname{sech}^2 x$$

Let us now determine the time evolution of the potential. In this case, Equation 10.94 with P and Q expressed by Equation 10.95 are still valid if the D_+ is replaced by $D_+(t)$. K_+ is now $K_+(x,x',t)$. Define

$$\exp(\lambda) = [D_{+2}/2D_{+1}]^{1/2}$$

and

$$\exp(\mu) = [D_{+2}D_{+1}/72]^{1/2}$$

In terms of these quantities, $K_+(x,x',t)$ becomes

$$K_+(x, x', t) = -p/s$$

where

$$p = 3\exp(x - \lambda) + 6\exp\{-(x - \lambda)\} + 3\exp\{-(3x - \mu)\}$$

and

$$s = \cosh(3x - \mu) + 3\cosh(x - \lambda)$$

Upon differentiating, setting $x' = x$, and after a lengthy manipulation, the potential is obtained as

$$q(x,t) = -12\,U/V \tag{10.96}$$

where

$$U = 3 + 4\cosh(2x + \lambda - \mu) + \cosh(4x - \lambda - \mu) \tag{10.97}$$

and

$$V = [\cosh(3x - \mu) + 3\cosh(x - \lambda)]^2 \tag{10.98}$$

Now, from Equation 10.69, we obtain

$$D_{+1}(t) = 6\exp(8t)$$

and

$$D_{+2}(t) = 12\exp(64t)$$

Then,

$$\exp(\lambda) = [D_{+2}/2D_{+1}]^{1/2} = \exp(28t)$$

and

$$\exp(\mu) = [D_{+2}D_{+1}/72]^{1/2} = \exp(36t)$$

yielding $\lambda = 28$ and $\mu = 36$. With these parameters and from Equations 10.96 through 10.98, the time-dependent potential is obtained as

$$q(x,t) = -12[3 + 4\cosh(2x - 8t) + \cosh(4x - 64t)]\big/[\cosh(3x - 36t) + 3\cosh(x - 28t)]^2 \tag{10.99}$$

Interesting conclusions follow from Equation 10.99. In order to see the remarkable nature of the motion expressed by this equation, let us define two variables:

$$X_1 = x - 16t$$

and

$$X_2 = x - 4t$$

In terms of X_1, Equation 10.99 reduces to

$$q(x,t) = -12\left[3 + 4\cosh(2X_1 + 24t) + \cosh 4X_1\right] / \left[3\cosh(X_1 - 12t) + \cosh(3X_1 + 12t)\right]^2$$

Let X_1 be fixed and consider that $t \to \infty$. Then

$$q(x,t) \approx -12\left[2\exp(2X_1 + 24t) / \left[(3/2)\exp(-X_1 + 12t) + (1/2)\exp(3X_1 + 12t)\right]^2\right.$$

The terms vanishing with t as $t \to \infty$ were neglected. This can be further reduced to the form

$$q(x,t) \approx -32 / \left[3^{-1/2}\exp(2x_1) + 3^{1/2}\exp(-2X_1)\}\right]^2$$

A final reduction results in

$$q(x,t) \approx -8\,\mathrm{sech}^2\left(x_1 - \ln 3^{1/2}\right) \tag{10.100}$$

In terms of the second variable X_2 fixed, the analogous calculations are

$$q(x,t) = -12\left[3 + 4\cosh 2X_2 + \cosh(4x_2 - 48t)\right] / \left[3\cosh(X_2 - 24t) + \cosh(3X_2 - 24t)\right]^2$$

$$\approx -24\left[\exp(-4x_2 + 48t)\right] / \left[3\exp(-X_2 + 24t) + \exp(24t - 3X_2)\right]^2$$

$$= -2\mathrm{sech}^2\left(X_2 + \ln 3^{1/2}\right) \tag{10.101}$$

Again, the final step is in the limit $t \to \infty$. Performing similar calculations, we can show that Equation 10.99 admits of the asymptotic form given by

$$q(x,t) \approx -2\,\mathrm{sech}^2\left(X_2 \pm \ln 3^{1/2}\right) - 8\,\mathrm{sech}^2\left(2X_1 \mp \ln 3^{1/2}\right) \tag{10.102}$$

as $t \to +\infty$.

These equations show that the motion is composed of two individual solitary waves given by Equations 10.100 and 10.101. It is interesting to observe that the wave with the greater speed is also deeper (Equation 10.100), whereas the wave with the slower speed is shallower (Equation 10.101). Equation 10.102 indicates that at time $t \to -\infty$, the deeper wave was on the left and was moving toward the shallower wave in front of it. Since it is faster, it overtook the shallower wave and at $t = 0$, they interacted, resulting in the sech^2 (x) potential. Subsequent to the interaction, as $t \to \infty$, the two waves are again separated and proceed without any alteration in their shapes. All that remains is a phase shift. The waves just described are called the "solitons". They are not merely solitary waves or pulses. In order to qualify as solitons, they must interact in the particular way just described. A good account of this behavior can be found in Drazin.[17]

APPENDIX A

At the end of Example 4, the mathematical forms of the time asymptotic N-soliton solution of the K-dV equation and the associated phase lag were presented. This represented the interaction between N number of solutions as they journey to $t \rightarrow \infty$. N is the number of eigenvalues of the Schrödinger equation, the potential in which is the N-soliton solution of the K-dV equation. This also implies that a superposition of N number of solutions (each solitary wave solution is a solution of the K-dV equation) is also a solution of the K-dV equation. If one remembers that the K-dV equation is a nonlinear partial differential equation, the result is certainly most remarkable. Considering the importance of this result, we present a brief derivation of the N-soliton interactions. The detailed results can be found in References 13 through 16. For the presentation here, we follow Reference 12.

Our starting point is Equation 10.79 and its differentiation with respect to x. Equation 10.79 will be recast in terms of the following variables:

$$z_m(x, x', t) = d_{+m} h_m(x, t) \exp(-k_m x')$$

$$\zeta = x - 4k_i^2 t$$

$$\Delta_m(\zeta) = D_{+m}^{-1}(0) \exp(2k_m \zeta) \tag{A.1}$$

In terms of these variables, we have from Equation 10.85

$$q(x, t) = -2K'(x, x, t) = 2 \sum_{m=1}^{N} z_m'(x) \tag{A.2}$$

In terms of the variable z_m, Equation 10.79 can be rewritten as

$$d_{+m}^{-2} \exp(2k_m x) z_m + \sum_{n=1}^{N} z_n / (k_m + k_n) = 1, \quad m = 1, 2, \ldots, N \tag{A.3}$$

Differentiating Equation A.3 yields

$$d_{+m}^{-2} \exp(2k_m x) z_m' + \sum_{n=1}^{N} z_n' / (k_m + k_n) = 1, \quad m = 1, 2, \ldots, N \tag{A.4}$$

The objective here is to solve Equation A.4 for z_m', substitute the solution into Equation A.2, and sum over the eigenvalues to recover the potential. The computations will, of course, be carried out in the time asymptotic limits $t \rightarrow \pm\infty$. It is convenient to work in the Lagrangian frame of reference, that is, in the coordinate system moving with the soliton the natural speed of which is given by the variable ζ of Equation A.1. Switching to the Lagrangian coordinate system and in terms of the variables ζ and Δ_m, Equations A.3 and A.4 become

$$\Delta_m(\zeta) z_m \exp\left\{-8k_m\left(k_m^2 - k_i^2\right)t\right\} + \sum_{n=1}^{N} z_n / (k_m + k_n) = 1, \quad m = 1, 2, \ldots, N \tag{A.5}$$

and

$$\Delta_m(\zeta)z'_m \exp\left\{-8k_m\left(k_m^2 - k_i^2\right)\right\}t\Big\} + \sum_{n=1}^{N} z'_n/\left(k_m + k_n\right)$$

$$= -2k_m z_n \Delta_m(\zeta)\exp\left\{-8k_m\left(k_m^2 - k_i^2\right)\right\}t\Big\} \quad i, m = 1, 2, \ldots, N \tag{A.6}$$

Let us now consider the asymptotic limits $t \to \pm\infty$. The limit $t \to +\infty$ will be considered first.

Case 1. Asymptotic Behavior as $t \to \infty$

In the limit that $t \to \infty$, Equation A.5 reduces to the following identities, depending upon whether m <, =, or > i. If $m < i$, then the time-dependent term vanishes in the asymptotic limit and we obtain

$$\sum_{n=1}^{N} z_n/\left(k_m + k_n\right) = 1, \qquad m < i \tag{A.7}$$

$$\Delta_i(\zeta)z_i + \sum_{n=1}^{N} z_n/\left(k_m + k_n\right) = 1, \qquad m = i \tag{A.8}$$

and

$$z_m = 0, \ \ m > i \tag{A.9}$$

In deriving the above result, it was tacitly assumed that the eigenvalues were ordered as $k_i > k_{i+1}$, $i = 1,2,\ldots,N-1$. Combining Equations A.7 through A.9 into a single expression, we have

$$\sum_{n=1}^{N} z_n/\left(k_m + k_n\right) = 1 - \delta_{mi}\Delta_i(\zeta)z_i, \qquad m \le i \tag{A.10}$$

and

$$z_m = 0, \ \ m > i \tag{A.11}$$

δ is the Kronecker delta.

Similar results for z_m' are

$$\sum_{n=1}^{N} z'_n/\left(k_m + k_n\right) = -\delta_{mi}\Delta_i(\zeta)\left(2k_i z_i + z'_i\right), \ m \le i \tag{A.12}$$

and

$$z'_m = -2k_m z_m = 0, \qquad m > i \tag{A.13}$$

We want to solve Equations A.12 and A.13 for z_m'. However, the expression for z_m' contains z_i and, hence, it is necessary to solve Equations A.10 and A.11 for z_m also. Let us consider these equations. Equations A.10 and A.11 can be written in a matrix form as

$$\mathbf{MZ} = \mathbf{B} \tag{A.14}$$

where

$(\mathbf{M})_{mn} = [k_m + k_n]^{-1}$ is a $(i \times i)$ matrix,

$\mathbf{Z} = [z_1, z_2, \ldots\ldots, z_n]^T$, a column vector

$\mathbf{B} = [1, 1, \ldots\ldots, 1 - \Delta_i(\zeta)z_i]^T$, also a column vector.

Applying Cramer's rule to Equation A.14 yields

$$\det \mathbf{M} = z_m \sum_{n=1}^{i} M_{mn}\Delta_i(\zeta)M_{mi}z_i, \quad m \le i \tag{A.15}$$

where M_{ij} are the cofactors of det $\mathbf{M}$. Setting m = i in Equation A.15 and denoting by $\tilde{\mathbf{M}}$ the $\mathbf{M}$ matrix with its last column replaced by a unit vector, we obtain

$$z_i = \det \tilde{\mathbf{M}} \Big/ \left[\det \mathbf{M} + \Delta_i(\zeta)\det \tilde{\tilde{\mathbf{M}}}\right] \tag{A.16}$$

where $\tilde{\tilde{\mathbf{M}}}$ is a $(i{-}1 \times i{-}1)$ matrix formed from $\mathbf{M}$ by deleting the final column.

The results for z_m', that is, the solutions of Equations A.12 and A.13 are

$$z_m' \det \mathbf{M} = -\Delta_i(\zeta)\mathbf{M}_{mi}(2k_i z_i + z_i'), \qquad m \le i$$

and

$$z_i' = -2\,\Delta_i(\zeta)k_i z_i \det \tilde{\mathbf{M}} \Big/ \left[\det \mathbf{M} + \Delta_i(\zeta)\det \tilde{\tilde{\mathbf{M}}}\right] \tag{A.17}$$

Substituting Equation A.16 into Equation A.17 and then summing up yields

$$\lim_{t\to\infty} \sum_{m=1}^{i} z_m' = -2\,\Delta_i(\zeta)k_i \Big/ \left[a + \Delta_i(\zeta)b\right]^2, \qquad \zeta \text{ constant} \tag{A.18}$$

where a = det $\mathbf{M}$/det $\tilde{\mathbf{M}}$ and b = det $\tilde{\tilde{\mathbf{M}}}$/ det $\tilde{\mathbf{M}}$.

Equation A.18 formally describes the time asymptotic (+ ∞) behavior of the potential or the solution of the K-dV equation for a soliton propagating at a free speed of $4k_i^2$ corresponding to the eigenvalue k_i. All that remains is to evaluate Equation A.18 explicitly. In order to do that, the determinants must be evaluated. It is not difficult to evaluate these determinants. The expression for det $\mathbf{M}$ is given by

$$\det \mathbf{M} = \det \tilde{\mathbf{M}}\left[\prod_{m=1}^{i-1}(k_i - k_m) \Big/ \prod_{m=1}^{i}(k_i + k_m)\right] \tag{A.19}$$

Similarly,

$$\det \tilde{\mathbf{M}} = \det \tilde{\tilde{\mathbf{M}}}\left[\prod_{m=1}^{i-1}(k_i - k_m) \Big/ \prod_{m=1}^{i-1}(k_i + k_m)\right] \tag{A.20}$$

Substituting Equations A.19 and A.20 into Equations A.18 and using the definition of Δ_m given in Equation A.1, we finally obtain for ζ constant

$$\begin{aligned}
\lim_{t\to\infty} q(x,t) &= \lim_{t\to\infty} 2\sum_{m=1}^{i} z'_m \\
&= -8k_i^2 \exp\left[2k_i(\theta - \theta_i)\right] \Big/ \left[1 + \exp\{2k_i(\theta - \theta_i)\}\right]^2 \\
&= -2k_i^2 \operatorname{sech}^2\left[k_i(\theta - \theta_i)\right] \\
&= -2k_i^2 \operatorname{sech}^2\left[k_i\left(x - 4k_i^2 t - \theta_i\right]\right.
\end{aligned} \tag{A.21}$$

The phase θ_i is given by the expression

$$\theta_i = (1/2k_i)\ln\left[\{d_{+i}^2(0)/2k_i\}\prod_{m=1}^{i-1}\{(k_i - k_m)/(k_i + k_m)\}^2\right] \tag{A.22}$$

Equation A.21 is a solitary wave which is propagating in the positive x-direction with a speed of $4k_i^2$ and an amplitude of $2k_i^2$. This is the soliton solution of the K-dV equation corresponding to the eigenvalue $k = k_i$.

Let us now consider the asymptotic form for $t \to -\infty$.

Case 2. Asymptotic Behavior as $t \to -\infty$

Recalling that the eigenvalues were ordered as $k_i > k_{i+1}$, $i = 1,2,\ldots,N-1$, it can be seen easily that in the limit $t \to -\infty$, Equations A.10 through A.13 remain the same, but with the following modifications:

$$\sum_{n=1}^{N} z_n/(k_m + k_n) = 1 - \delta_{mi}\Delta_i(\zeta)z_i, \quad m \ge i$$

and

$$z_m = 0, \quad m \le i - 1$$

Similar results for z_m' are

$$\sum_{n=1}^{N} z'_n/(k_m + k_n) = -\delta_{mi}\Delta_i(\zeta)(2k_i z_i + z'_i), \quad m \ge i$$

and

$$z'_m = -2k_m z_m = 0, \qquad m \le i-1$$

The rest of the mathematics remains the same and yields the asymptotic solutions as (ζ constant)

$$\begin{aligned} \lim_{t\to\infty} q(x,t) &= \lim_{t\to\infty} 2\sum_{m=1}^{i} z'_m \\ &= -2k_i^2 \operatorname{sech}^2\left[k_i(\theta - \theta'_i)\right] \end{aligned} \tag{A.23}$$

The phase term is now given by

$$\theta'_i = (1/2k_i)\ln\left[\left\{d_{+i}^2(0)/2k_i\right\}\prod_{m=i+1}^{i-1}\left\{(k_i - k_m)/(k_i + k_m)\right\}^2\right] \tag{A.24}$$

Calculating the difference in phase yields

$$\theta_i - \theta'_i = (1/k_i)\left[\ln\left\{\prod_{m=i}^{i-1} p_{mi} \Big/ \prod_{m=i+1}^{N} Q_{mi}\right)\right. \tag{A.25}$$

where

$$P_{mi} = (k_m - k_i)/(k_m + k_i)$$

and

$$Q_{mi} = (k_i - k_m)/(k_i + k_m)$$

In summary, then, let q(x,t) be a solution of the K-dV equation such that it is a reflectionless potential of the isospectral Schrödinger equation with t as the parameter. Then each eigenvalue $\lambda_i = -k_i^2$ corresponds to a soliton which at time $t \to -\infty$ was the solitary wave of Equation A.21 with the phase given by Equation A.22. At the other asymptotic limit $t \to \infty$, the solution is again a solitary wave given by Equation A.23, whose phase is now described by Equation A.24. From Equations A.21 and A.23, it is seen that the amplitude of the wave remains the same at $2k_i^2$ and the speed also remains unchanged at $4k_i^2$. The only difference in its entire travel from $t \to -\infty$ to $t \to \infty$ is just the phase change given by Equation A.25.

REFERENCES

1. **A. K. Agranovich and V. A. Marcenko:** *The Inverse Problem of Scattering Theory*, Gordon and Breach, London, 1963.
2. **R. K. Dodd, J. C. Eilbeck, J. D. Gibbon and H. C. Morris:** *Solitons and Nonlinear Wave Equations*, Academic Press, New York, 1984.

3. **F. Calogero and A. Degasperis:** *Spectral Transform and Solitons I*, North-Holland, Amsterdam, 1982.
4. **R. Courant and K. O. Friedrichs:** Methods of Mathematical Physics, Vol. 2, New York University, New York, 1943.
5. **P. Garabadien:** *Partial Differential Equations*, John Wiley & Sons, New York, 1964.
6. **G. N. Balanis:** The plasma inverse problem, *J. Math. Phys.*, 13, 1972.
7. **G. N. Balanis:** *Plasma Inverse Scattering Theory*, Ph.D. thesis, California Institute of Technology, 1972.
8. **G. L. Lamb Jr.:** *Elements of Soliton Theory*, John Wiley & Sons, New York, 1980.
9. **V. I. Karpman:** *Nonlinear Waves in Dispersive Media*, Pitman, London, 1975.
10. **M. Abramowitz and I. Stegun:** *Handbook of Mathematical Functions*, NBS Applied Mathematics Series 55, 1964.
11. **F. R. Gantmacher:** *Theory of Matrices*, Chelsea, New York, 1959.
12. **P. L. Bhatnagar:** *Nonlinear Waves in One-dimensional Dispersive Systems*, Clarendon Press, Oxford, 1979.
13. **C. S. Gardner, J. M. Greene, M. D. Kruskal and R. M. Miura:** Kortweg-de Vries equation and generalizations. VI. Methods for exact solution, *Comm. Pure Appl. Math.*, 27, 1974.
14. **V. E. Zakharov:** Kinetic equation for solitons, *Sov. Phys. JETP*, 33, 1971.
15. **M. Wadati and M. Toda:** The exact N-soliton solution of the Kortweg-de Vries equation, *J. Phys. Soc. Jpn.*, 32, 1972.
16. **S. Tanaka:** On the N-tuple wave solutions of the Kortweg-de Vries equation, *Publ. Res. Inst. Math. Sci. Kyoto Univ.*, 8, 1972.
17. **P. G. Drazin:** *Solitons*, Cambridge University Press, Cambridge, 1983.
18. **I. W. Kay:** *Inverse Scattering Papers (1955-1962)*, Math Science Press, Brooklyn, MA, 1982.

Chapter 11

DISCRETE SYSTEMS AND TRANSMISSION LINES

I. INTRODUCTION

Up until now we have considered continuous systems, that is, those systems the characteristics of which are continuous functions of the independent variables. For example, the potential or the inhomogeneity q considered so far depended continuously on the variable x. These continuous systems were governed by differential equations and possessed a continuous spectrum. In addition, their eigenfunctions varied continuously with space and/or time. On the other hand, the systems which are discrete have their characteristics as discrete functions of the independent variables. Thus a flexible string is a continuous system, its mass density varying continuously with length, whereas a body consisting of rigid masses, rods, and weightless springs is a discrete system. However, the concept of a discrete system such as the one consisting of a finite number of masses connected by springs is unnecessarily restrictive. The more general concept of a discrete system is any system the motion of which can be described in terms of a finite number of generalized coordinates. Such a concept of a discrete system applies also to the systems treated by the finite element technique. The discrete systems are governed not by partial differential but by ordinary matrix differential equations. Their spectra consist solely of a discrete number of eigenvalues and, instead of eigenfunctions, they possess "eigenvectors" having discrete components, that is, their normal modes are vector functions of the independent variables. An exhaustive discussion of discrete systems in general is the well-known work of Atkinson.[1]

In this chapter, attention will be focused on two types of discrete systems: (1) discrete second-order mechanical systems and (2) lossless, nonuniform transmission lines. The mechanical systems, composed of rigid masses connected by weightless springs, will be subjected to vibrational motion, whereas the transmission line dynamics will be due to wave propagation. The transmission line problem is a prototype of wave propagations problems in elastic media. The matrix formulation of the transmission line problems will be seen to mimic the continuous integral equations of Gel'fand-Levitan, Marcenko, and Krein.

It was mentioned earlier that the discrete systems are described by ordinary matrix instead of partial differential equations. It is interesting that the inverse problems in vibration and in transmission line involve inversions of two distinct matrices. The objective of inverting the mechanical systems in vibration is to recover the distributions of masses and spring constants from the spectral data which consist of the normal modes and their frequencies. The problem reduces generally to inverting a "Jacobi" matrix. On the other hand, the inverse problems of wave propagations in elastic media, the objective of which is to determine the impedance distribution, generally involve inversions of a "Hankel" and/or a "Toeplitz" matrix. Once the inversion of a Jacobi matrix is performed, the important inverse problems of second-order discrete mechanical systems in vibration are solved. Similarly, once a Hankel and/or a Toeplitz matrix is inverted, so are the problems of wave propagation in discrete elastic media.

The present chapter will be divided into two parts: (1) inverse problems for a Jacobi matrix and their connections with the inverse solutions of discrete second-order vibrational systems and (2) inverse problems of wave propagation in a lossless, nonuniform transmission line. The transmission line problem will be treated both in discrete and continuous formulation, whereas the vibrational systems will be treated only as discrete systems. We will also illustrate (Appendix D) the interesting connection between a Toeplitz matrix problem (direct and inverse) with that of the transmission line (inverse and direct).

For the inverse problem of a continuous transmission line, a theory developed by Brown and Wilcox will be presented. The mathematical background relevant to this chapter and the properties of various matrices required are discussed in detail in the appendices at the end.

II. JACOBI MATRICES AND THEIR INVERSIONS

A. JACOBI MATRICES IN INVERSE PROBLEMS IN VIBRATION

The Jacobi matrices and their properties are discussed briefly in Appendix A. In this section, we discuss the relevance of a Jacobi matrix in mechanical vibrations and, consequently, the significance of its inversion in solving the inverse problems in vibration. The discussions of this section draws heavily from Gladwell.[2]

Consider a mechanical system having N degrees of freedom characterized by the inertial parameters m_i, stiffness constants k_i, and displacements u_i from equilibrium. Three examples are shown in Figure 1. In Figure 1(a), the m_i's are the masses, whereas the m_i's of Figure 1(b) correspond to the moments of inertia. The stiffness constants k_i's correspond to the spring stiffness of (a), torsional stiffness of (b), and the parameter (T/l_i) of the transverse vibration of (c). These examples are important since they describe the simplest possible discrete models of longitudinal (Figure 1(a)) and torsional (Figure 1(b)) vibrations of a rod and the transverse vibration of a taut spring (Figure 1(c)). Let us consider "free" end conditions as shown in Figure 1(a) to (c). In Figure 1(c) this means that the end of the string is attached to a massless ring which can slide up and down on a frictionless vertical rod, whereas in the other two cases this means that there is no restriction on the displacement of the end point. Assuming small departures from equilibrium, such systems are described by the matrix equation[1]

$$\begin{pmatrix} m_1 & & & \\ & m_2 & \mathbf{0} & \\ & & \cdot & \\ & \mathbf{0} & \cdot & \\ & & & m_n \end{pmatrix}\begin{pmatrix} \ddot{u}_1 \\ \ddot{u}_2 \\ \cdot \\ \cdot \\ \ddot{u}_n \end{pmatrix} + \begin{pmatrix} \mathbf{k_1} + k_2 & -k_2 & & & \mathbf{0} \\ -k_2 & k_2 + k_3 & -k_3 & & \\ & & & \cdot & \\ \mathbf{0} & & & \cdot & \\ & & & -k_n & k_n \end{pmatrix}\begin{pmatrix} u_1 \\ u_2 \\ \cdot \\ \cdot \\ u_n \end{pmatrix} = \begin{pmatrix} F_1 \\ F_2 \\ \cdot \\ \cdot \\ F_n \end{pmatrix}$$

or equivalently,

$$\mathbf{M}\ddot{\mathbf{u}} + \mathbf{K}\mathbf{u} = \mathbf{F} \tag{11.1}$$

M is the "inertia" or the "mass" matrix and **K** the "stiffness" matrix. **u** and **F** are the displacement (from equilibrium) and the applied force vector, respectively. The vector **0** indicates the null section of the matrix. **M** is diagonal, that is, $(\mathbf{M})_{ij} = M_{ij}\delta_{ij}$ and **K** is tridiagonal, that is, $K_{i,i+1} = K_{i+1,i}$ and the only nonzero elements are on the principal and two codiagonals. Both **M** and **K** are positive definite. Moreover, **K** is symmetrical. This is characteristic of conservative systems such as that described by Equation 11.1.

Define

$$C = \mathrm{diag}\left[k_1, k_2, \ldots, k_N\right]$$

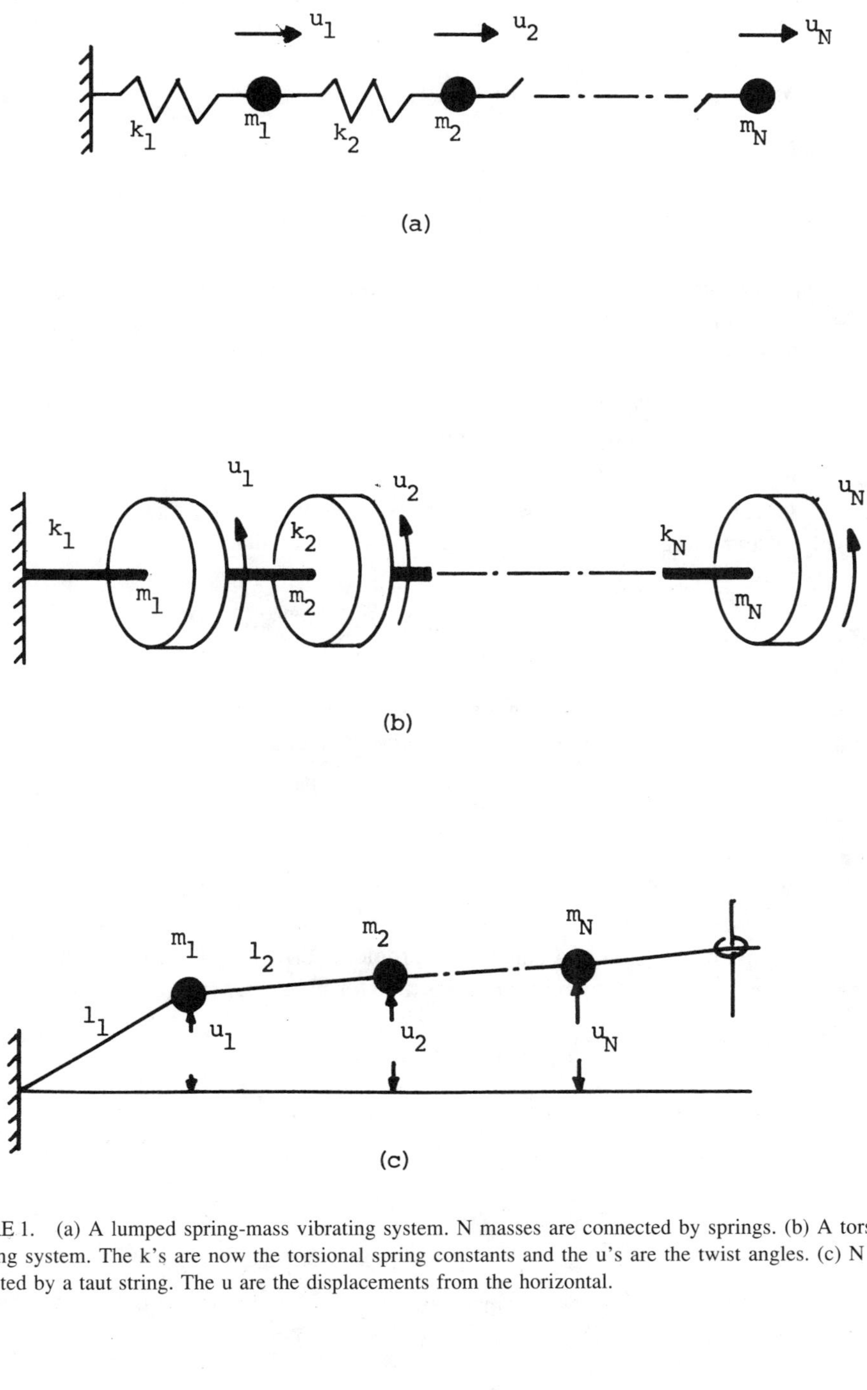

FIGURE 1. (a) A lumped spring-mass vibrating system. N masses are connected by springs. (b) A torsionally vibrating system. The k's are now the torsional spring constants and the u's are the twist angles. (c) N masses connected by a taut string. The u are the displacements from the horizontal.

and

$$\mathbf{E} = \begin{pmatrix} 1 & -1 & 0 & \dots & 0 \\ 0 & 1 & -1 & \dots & 0 \\ . & . & . & . & . \\ 0 & \dots & & 1 & -1 \\ 0 & \dots & & 0 & 1 \end{pmatrix}$$

Using **C** and **E**, Equation 11.1 can be rewritten as

$$\mathbf{M\ddot{u}} + \mathbf{ECE}^{\mathrm{T}}\mathbf{u} = \mathbf{F} \tag{11.1a}$$

where

$$\mathbf{K} = \mathbf{ECE}^{\mathrm{T}}$$

For a more general system with the generalized coordinates $\{q_i\}_1^N$, the governing equation is that of Lagrange written as

$$\left[\frac{d}{dt}\left(\frac{\partial T}{\partial \dot{q}_r}\right)\right] - \frac{\partial T}{\partial q_r} + \frac{\partial V}{\partial q_r} = Q_r, \quad (r = 1, 2, \ldots, N)$$

T, V, and Q represent kinetic energy, potential energy, and the generalized force, respectively. T and V are given by the matrices T and V such that

$$T = \left(\frac{1}{2}\right)\dot{\mathbf{q}}^{\mathrm{T}}\mathbf{T}\dot{\mathbf{q}} \text{ and } V = \left(\frac{1}{2}\right)\mathbf{q}^{\mathrm{T}}\mathbf{V}\mathbf{q}$$

In terms of these matrices and the generalized coordinate vector **q**, the Lagrange equation takes for free vibration the following form.

$$\mathbf{T\ddot{q}} + \mathbf{Vq} = \mathbf{Q} \tag{11.1b}$$

T, **V** are symmetric and **Q** is the matrix for the generalized force. Recall that in the Lagrange equation, q_r and $\dot{q}_r$ are independent variables and the kinetic energy is a function of $\dot{q}_r$ only.

We next formulate the standard eigenvalue problem corresponding to Equation 11.1. Let us consider the homogeneous equation. Fourier transforming we write

$$\mathbf{u} = \mathbf{U}\sin(\omega t + \phi)$$

U is the amplitude vector (constant), ω resp. ϕ is the frequency resp. the phase angle. Equation 11.1 then reduces to the form

$$(\mathbf{K} - \lambda\mathbf{M})\mathbf{U} = 0, \ \lambda = \omega^2 \tag{11.2}$$

For an (N × N) system, the eigenvalue problem has N eigenvalues (not necessarily distinct) $\{\lambda_i\}_1^N$ and N eigenvectors $\{\mathbf{x}^{(i)}\}_1^N$ such that*

$$0 \le \lambda_1 \le \lambda_2 \ldots \le \lambda_N$$

and

$$(\mathbf{K} - \lambda_i\mathbf{M})\mathbf{x}^{(i)} = 0$$

* The matrices are assumed to be positive definite. If **K** is only positive semi-definite, at least one eigenvalue will be zero.

$\omega_i = \lambda_i^{1/2}$ is the i-th "natural frequency" and the eigenvector $\mathbf{x}^{(i)}$ is an eigen- or a "normal" mode of the system.

Equation 11.2 can be further reduced to a more standard form through the following substitutions. Define

$$\mathbf{m} = \text{diag.}\left(m_i^{\frac{1}{2}}, i = 1, 2, \ldots . N \right)$$

Then $\mathbf{m} = \mathbf{m}^T$ and $\mathbf{M} = \mathbf{m}\mathbf{m}^T$. Furthermore, let us define a vector $\mathbf{X}$ such that

$$\mathbf{X} = \mathbf{m}^T\mathbf{U} \text{ or } \mathbf{U} = \left(\mathbf{m}^T\right)^{-1}\mathbf{X}$$

Then replacing $\mathbf{M}$ by $\mathbf{m}\mathbf{m}^T$, $\mathbf{U}$ by $(\mathbf{m}^T)^{-1}\mathbf{X}$, and premultiplying Equation 11.2 by $\mathbf{m}^{-1}$ yields

$$(\mathbf{J} - \lambda\mathbf{I})\mathbf{X} = 0, \quad \lambda = \omega^2 \tag{11.2a}$$

where,

$$\mathbf{J} = \mathbf{m}^{-1}\mathbf{K}\left(\mathbf{m}^T\right)^{-1} = \mathbf{m}^{-1}\mathbf{K}\mathbf{m}^{-1} \tag{11.3}$$

The matrix $\mathbf{J}$, therefore, has the same eigenvalues as those of $\mathbf{K}$. $\mathbf{J}$ is symmetric, tridiagonal, and positive definite or semidefinite depending on whether $\mathbf{K}$ is positive definite or semidefinite. $\mathbf{J}$ is thus a Jacobi matrix and can be shown[2] to be sign-oscillatory. This means that the sign reverse $\mathbf{B}^*$ of $\mathbf{B}$, namely, $(b^*)_{ij} = (-1)^{i+j}b_{ij}$ is "oscillatory" (Appendix A).

The lumped-mass-spring discrete systems in vibration are, therefore, represented by sign-oscillatory Jacobi matrices. Moreover, Equation 11.3 shows that the Jacobi matrix $\mathbf{J}$ contains information regarding inertial and spring constants. The inversion of a discrete second-order vibrational system then consists of two steps: inversion of $\mathbf{J}$ on the basis of spectral data (see Section II.B) and unraveling this reconstructed $\mathbf{J}$ to obtain the mass and stiffness distributions.

Let us illustrate this by considering a specific example. Consider the unconstrained system of Figure 1(a) and assume that the associated Jacobi matrix $\mathbf{J}$ has been inverse solved. Various techniques for inverting a Jacobi matrix will be discussed momentarily. Step 1 of the inverse solution is thus assumed. Step 2, namely, the unraveling of the reconstructed $\mathbf{J}$ matrix for recovering the mass and stiffness information, will be discussed now. This unraveling procedure requires the knowledge of the impulse response of the system and one other parameter, namely, the total mass.

Let only the mass m_1 be given a unit displacement by a force concentrated only at this mass. Since the other masses are unconstrained, they will undergo rigid body displacements and the displacement vector $\mathbf{U}$ will be given by

$$\mathbf{U} = [1,\ 1,\ 1, \ldots ,\ 1]$$

From Equations 11.2, 11.3, and the relation $\mathbf{U} = \mathbf{m}^{-1}\mathbf{X}$, it follows that

$$\mathbf{J}\mathbf{X} = \mathbf{m}^{-1}\mathbf{K}\mathbf{m}^{-1}\mathbf{X} = \mathbf{m}^{-1}\mathbf{K}\mathbf{U} = \left[m_1^{-\frac{1}{2}}k_1, 0, 0, \ldots, 0 \right] \tag{11.4}$$

and

$$\mathbf{X} = \mathbf{mU} = \left[m_i^{\frac{1}{2}};\ i = 1,\ 2,\ \ldots\ N \right]$$

Let us now make a guess of the value of the component $X_N = m_N^{1/2}$. Since the inversion procedure is assumed to have been completed, the matrix **J** is known. The tridiagonal matrix equation (11.4) can now be solved without knowing the nonzero component $m_1^{-1/2}\, k_1$. The masses m_i will be obtained with respect to the assumed mass m_N. Then from the knowledge of the total mass m, all the individual masses can be computed. This is, of course, an iterative procedure. Total mass must be specified if uniqueness is to be achieved. See discussion in the next section. Total stiffness constant might have done as well.

The tridiagonal stiffness matrix is known since

$$\mathbf{K} = \mathbf{M}^{\frac{1}{2}}\mathbf{J}\mathbf{M}^{\frac{1}{2}}$$

and the inversion of the Jacobi matrix **J** is assumed. From Equation 11.1a the diagonal stiffness matrix **C** is known by virtue of the relation that $\mathbf{E}^{-1}\mathbf{K}\mathbf{E}^{-T} = \mathbf{C}$. The reconstruction is thus completed.

This completes the demonstration that by inverting the associated Jacobi matrix, a second-order vibrational system can be inverted. It remains for us to discuss the procedures by which a Jacobi matrix can be inverted. The classical methods for inverting a Jacobi matrix **J** requires two spectra, one for the matrix itself and the other for its principal submatrix, that is, the matrix obtained from **J** by deleting one row and one column. These may be either the first or the last ones. The physical reason for these two spectra will be discussed and the way in which these eigenvalues interlace with each other will be explained below.

B. INVERSION OF A JACOBI MATRIX

Let us now consider the problem of inverting a Jacobi matrix. The data input for the inverse problems in vibration consist of the spectra, that is, the eigenvalues and the eigenvectors of vibration are prescribed. From this one constructs a Jacobi matrix which has the specified data as its spectra.

There are basically four variants of this inverse problem. The first and the more traditional formulation is to construct a Jacobi matrix **J** from two sets of data — the eigenvalues $\{\lambda_i\}^N{}_1$ of **J** and the set of eigenvalues $\{\mu_i\}_1{}^{N-1}$ of a principal submatrix (either left or right) of order N – 1. The second variant assumes only one set of eigenvalues $\{\lambda_i\}^N{}_1$ as the data and constructs a "persymmetric" Jacobi matrix corresponding to this set of eigenvalues. Once the persymmetric matrix is known, the original Jacobi matrix can be constructed. There is a third inverse problem which may be considered as a variation of the first version and will be enunciated later. Finally, in the last variant, the eigenvalues and the absolute value of one of the eigenvectors are specified. The inverse problem consists of determining a Jacobi matrix that possesses the given spectra.

Theorem 11.1. Inverse Problem I

Let $\{\lambda_i\}^N{}_1$ and $\{\mu_i\}_1{}^{N-1}$ be two strictly increasing interlacing sequences such that

$$\lambda_i < \mu_i < \lambda_{i+1},\ i = 1,\ldots,N-1$$

Let $\sigma(\mathbf{J}(N;a,b))$ denote the set of eigenvalues of a Jacobi matrix **J** of order N with a_i's and b_i's as the diagonal and the codiagonal elements, respectively. Then there exists a unique Jacobi matrix **J** such that $\sigma(\mathbf{J}(N,a,b)) = \{\lambda_i\}^N{}_1$ and $\sigma(\mathbf{J}(N-1;a,b)) = \{\mu_i\}_1{}^{N-1}$.

Theorem 11.2. Inverse Problem II

Given a strictly increasing sequence $\{\lambda_i\}^N{}_1$, construct a persymmetric Jacobi matrix of order N having $\lambda_1, \lambda_2, \ldots, \lambda_N$ as its eigenvalues.

Theorem 11.3. Inverse Problem III

Given two strictly increasing sequences $\{\lambda_i\}^N{}_1$ and $\{\lambda^*{}_i\}^N{}_1$ with $\lambda_i < \lambda^*{}_i < \lambda_{i+1}\ \forall$ i, determine an N-th order Jacobi matrix **J** which has $\lambda_1, \lambda_2, \ldots, \lambda_N$ as its eigenvalues and for which the matrix $\mathbf{J}^*$, obtained from **J** by changing a_N to $a^*{}_N$, has $\lambda^*{}_1, \lambda^*{}_2, \ldots, \lambda^*{}_N$ as its eigenvalues.

Theorem 11.4. Inverse Problem IV

Given the eigenvalues $(\lambda_i)^N{}_1$

$$\lambda_1 < \lambda_2 < \ldots < \lambda_N$$

and the normalized eigenvectors $\mathbf{u}^{(i)}$ such that the absolute values of either $(u_1{}^{(i)})^N{}_1$ or $(u_N{}^{(i)})^N{}_1$ are known, construct a Jacobi matrix **J** having the specified data as its spectra.

Theorems 11.1 and 11.2 appear to have been investigated first by Hochstadt.[3] A simpler inductive proof was given by Gray and Wilson[4] for Theorem 11.1. Hald[5] presented a more detailed analysis of Theorems 11.1 and 11.2 along with discussions of their well-posedness and also gave illustrative examples. Relatively more recent stable reconstructions of Jacobi matrices are due to de Boor and Golub.[6] Theorem 11.3 is due to de Boor and Golub.[6] Theorem 11.4 is based on the Lanczos[7] algorithm. In proving these theorems, two facts will be important: the interlacing properties of $\sigma(\mathbf{J}(N;a,b))$ and $\sigma(\mathbf{J}(N-1;a,b))$ and the properties of the so-called "monic" polynomials. The monic polynomials, their properties, and their relations with the Jacobi matrices are discussed in Appendix B.

Let us now consider the physical meaning of the two spectra and the strict interlacing of their eigenvalues as in Theorems 11.1 and 11.3. First, consider a simple system with only one mass m attached to rigid supports by springs of stiffness k_1 and k_2. The natural frequency of this system is, of course, $\omega^2 = \lambda_2 = (k_1 + k_2)/m$ which is supposedly known. Since both numerator and denominator are homogeneous of degree 1, then obtaining uniqueness of the solution requires that we specify either the total mass or the resultant spring constant. In other words, ω^2 is invariant to multiplication of k_1, k_2 and m by any constant. A third condition is also required if all three parameters, namely, m, k_1, and k_2 are to be recovered. This third condition can be obtained by making the end condition, for example, that of the spring k_2, free. From this system with free end condition we obtain the ratio (k_1/m), knowing the eigenfrequency $\omega_1{}^2 = \lambda_1 = k_1/m$. Thus, given the knowledge of either the total mass or the total spring constant, m, k_1, and k_2 can be obtained uniquely if ω_1 and ω_2, that is, λ_1, λ_2, are given such that $0 < \omega_1 < \omega_2$, or, which is identical, $0 < \lambda_1 < \lambda_2$, that is, the eigenvalues strictly interlace. Also, $m:k_1:k_2 = 1:\lambda_1:\lambda_2 - \lambda_1$.

The problem discussed above is thus closely connected to imposing a constraint condition on the system. Let us, therefore, carry the analysis a little further. Equation 11.1 does not contain any constraint conditions. This means that the generalized coordinates $\{q_i\}^N{}_1$ are not related to each other. If a constraint is imposed, then $\{q_i\}^N{}_1$ must satisfy a constraint relation of the form

$$f(\mathbf{q}) = f(q_1, q_2, \ldots, q_n) = 0$$

If $\mathbf{d} = \{f_i'\}^N{}_1$, $f_i' = (\partial f/\partial q_i|_{q=0}$, then the condition of the constraint is (for small oscillations)

$$\mathbf{q}^T\mathbf{d} = 0$$

The spectrum of a vibrating system subjected to constraint can be obtained through the application of the "Rayleigh Principle".[8] Rayleigh's principle is that of the variational calculus and can be explained as follows. One defines what is called the Rayleigh quotient λ_R. For systems described by any of the Equations 11.1, 11.1a, or 11.1b, the Rayleigh quotient is given by

$$\lambda_R = \frac{\mathbf{v}^T \mathbf{D} \mathbf{v}}{\mathbf{v}^T \mathbf{A} \mathbf{v}}$$

where $\mathbf{v}$ stands for the Fourier amplitude of either $\mathbf{u}$ or $\mathbf{q}$, the matrix $\mathbf{D}$ for $\mathbf{K}$ or $\mathbf{V}$ and the matrix $\mathbf{A}$ for $\mathbf{M}$ or $\mathbf{T}$. At the stationary values of the Rayleigh quotient, $\lambda_R = \lambda_i$ and $\mathbf{v} = \mathbf{v}_i$. That is, the stationary values of λ_R are the eigenvalues of the system and the corresponding $\mathbf{v}$'s the eigenvectors. Since $\mathbf{A}$ is positive definite, the denominator is strictly greater than zero and, hence, obtaining the stationary values of λ_R is equivalent to that of the quantity $\mathbf{v}^T\mathbf{D}\mathbf{v} - \lambda\mathbf{v}^T\mathbf{A}\mathbf{v}$, with λ acting as a Lagrange multiplier with some constraint $\mathbf{v}^T\mathbf{A}\mathbf{v}$.

In order to apply Rayleigh's principle with the constraint condition as given above, the Rayleigh quotient Z is constructed as (considering 11.2)

$$\mathbf{Z} = \mathbf{U}^T\mathbf{K}\mathbf{U} - \lambda\mathbf{U}^T\mathbf{M}\mathbf{U} - 2\Lambda\mathbf{U}^T\mathbf{d}$$

with λ and Λ as the Lagrange multipliers where the Fourier amplitude U has been used. The spectra are then determined from the conditions $(\mathbf{Z}')_i = 0,\ \forall\ i$, that is, the stationary values of the function $\mathbf{Z}$. It can be shown[2] that application of Rayleigh's principle results in the spectral equation

$$B(\lambda) = \sum_{i=1}^{N} \frac{\left\{\left(\mathbf{P}\mathbf{U}^T\mathbf{d}\right)_i\right\}^2}{m_i} (\lambda_i - \lambda)^{-1} = 0$$

$\mathbf{P} = (\Pi^T)^{-1}$, where Π is the diagonal matrix of the "principal coordinates". In other words, Π transforms the unconstrained eigenvectors to those of the constrained system. Now, $\mathbf{M}$ is positive definite and thus $m_i^{-1}\{(\mathbf{P}\mathbf{U}^T\mathbf{d})_i\}^2 > 0$ for $(\mathbf{P}\mathbf{U}^T\mathbf{d})_i \neq 0,\ \forall\ i$. Then the spectral condition $B(\lambda)$ will have the form shown in Figure 2. This can be seen as follows. If $(\mathbf{P}\mathbf{U}^T\mathbf{d})_i > 0$, then the B function steadily increases between a large negative to a large positive with a zero crossing in between, which must correspond to the constrained system. Therefore, if a linear system is subjected to a linear constraint, each of its natural frequencies increases in such a way as not to exceed the next higher frequency of the original system. Stated differently, if λ^* corresponds to the constrained system, then $\lambda_i < \lambda_i^* < \lambda_{i+1}$, $i=1,2,\ldots,N-1$.

We now turn to the problem of inverting a Jacobi matrix when its spectrum is given. We first consider the variant as formulated in Theorem 11.1. The problem of inverting a Jacobi matrix is more traditional in this formulation and occurs often in the literature. Three different proofs will be presented. The first is due to Hochstadt,[3] who pioneered this work. This is followed by the proof supplied by Gray and Wilson,[4] which is by far the simplest and is based on the principle of induction. A proof which is relatively more recent and stable from the viewpoint of machine computation is due to de Boor and Golub[6] and uses the properties of the monic polynomials. There is also the well-known proof due to Hald.[5] However, from the point of view of numerical inversion, Hald's construction is unstable.[6] We do not give Hald's proof, for which his original article must be consulted.

We start with Hochstadt's proof. Hochstadt assumes that the given spectrum cannot be

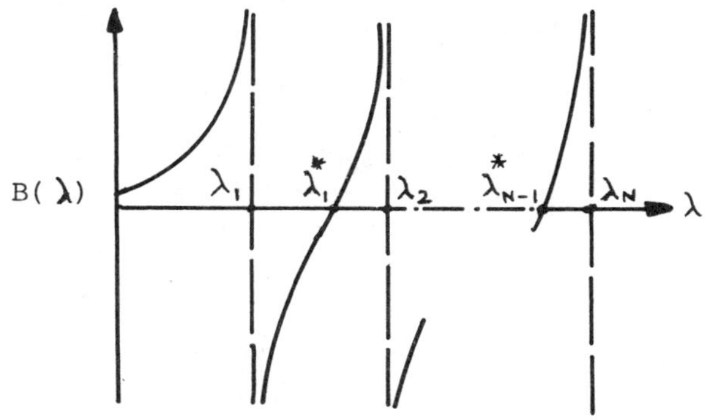

FIGURE 2. A plot illustrating that a vibrating system, when subjected to a linear constraint, vibrates with eigenvalues which interlace those of the unconstrained system.

arbitrary, but, rather, must *a priori* correspond to at least one Jacobi matrix. Let **J**(N;a,b) denote a Jacobi matrix of order N with diagonal elements a_i's and codiagonal elements b_i's, where the a_i's and b_i's are real and all b_i's are positive. Let $\{\lambda_i\}^N{}_1$ correspond to its spectrum. Let this also be true for the right principal submatrix **J**(N–1;a,b) and its spectrum $\{\mu_i\}_1{}^{N-1}$. The problem is to reconstruct a Jacobi matrix **J** (N;a,b) such that it has the spectrum $\{\lambda_i\}^N{}_1$ and its principal minor **J**(N – 1;a,b), i.e., the matrix **J** with the first column and row deleted has the spectrum$\{\mu_i\}_1{}^{N-1}$. For brevity, **J**(N) and **J**(N – 1) will be written instead of **J**(N;a,b) and **J**(N – 1;a,b), respectively. Let us define the following polynomials:

$$J_N(\lambda) = \sum_{i=0}^{N}(\lambda - \lambda_i) = B(\lambda) \tag{11.5}$$

and

$$J_{N-1}(\lambda) = \prod_{i=1}^{N}(\lambda - \mu_i) = A(\lambda) \tag{11.6}$$

Every eigenvalue of a Jacobi matrix is distinct (Appendix B), and the following strict inequalities hold:

$$\lambda_i > \lambda_{i+1}, \quad i = 0, 1, \ldots., N$$

$$\mu_i > \mu_{i+1}, \quad i = 1, 2, \ldots., N$$

The interlacing condition is, of course,

$$\lambda_i > \mu_{i+1} > \lambda_{i+1}, \forall\, i$$

B(λ) in Equation 11.5 is not to be confused with the function shown in Figure 2.

Let $\mathbf{u}^{(i)}$ be the i-th eigenvector of **J**(N) so that

$$\mathbf{J}(N)\mathbf{u}^{(i)} = \lambda_i \mathbf{u}^{(i)}$$

Hochstadt assumes that the eigenvector $\mathbf{u}^{(i)}$ has been so normalized that its first component $u_1{}^{(i)}$ is unity. It is not difficult to see that if this component were zero, then $\mathbf{u}^{(i)}$ itself would vanish. Let us define the permutation matrix carrying the set (1, 2, ... N) to the set (N, N – 1, ... 1) by **S**. Then

$$\mathbf{S} = \left(\delta_{i+j,N+1}\right)_{i,j=1}^{N} = \mathbf{S}^{-1} \tag{11.7}$$

In Hochstadt's notation,

$$\mathbf{S} = \left(\delta_{i+j,N}\right)_{i,j=0}^{N} = \mathbf{S}^{-1}$$

Thus, $\mathbf{S} = \delta = \{\delta_K\}$, $\{\delta_K\}$ being a set of canonical unit vectors. $\{\delta\}_i = \delta_{ik}$ is the Kronecker delta. In Hochstadt's notation,

$$\mathbf{S} = \begin{pmatrix} 0 & 0 & . & . & . & 0 & 1 \\ 0 & 0 & . & . & . & 1 & 0 \\ . & . & . & . & . & . & . \\ . & . & . & . & . & . & . \\ . & . & . & . & . & . & . \\ 0 & 1 & . & . & . & 0 & 0 \\ 1 & 0 & . & . & . & 0 & 0 \end{pmatrix}$$

and $\mathbf{S}\,\mathbf{S} = \mathbf{S}^2 = \mathbf{I}$, the identity matrix. $\delta_0 = (1,0,0, \ldots 0,0)$. The inner product $<\mathbf{u}^{(i)}, \delta_0>$ is then unity and is the first component of the eigenvector $\mathbf{u}^{(i)}$. The eigenvectors are normalized. The matrix **S** will play an important role later in connection with the inversion of a persymmetric Jacobi matrix. We now come to the proof.

1. Proof of Inverse Problem 1 (Hochstadt[3])

Consider the equation

$$(\lambda\mathbf{I} - \mathbf{J})\mathbf{F} = \delta_0 \tag{11.8}$$

λ is not an eigenvalue. A simple application of Cramer's rule shows that

$$\langle \mathbf{F}, \delta_0 \rangle = \frac{\begin{pmatrix} 1 & -b_0 & 0 & \ldots & 0 \\ 0 & \lambda - a_1 & -b_1 & \ldots & 0 \\ 0 & -b_1 & \lambda - a_2 & \ldots & 0 \\ . & . & . & . & . \\ . & . & . & . & . \\ . & . & . & . & . \\ 0 & 0 & 0 & . & \lambda - a_N \end{pmatrix}}{B(\lambda)} = \frac{A(\lambda)}{B(\lambda)} \tag{11.9}$$

$<\mathbf{F},\delta_0>$ is the first component of the vector **F**. From Equation 11.8, we can write

$$\mathbf{F} = (\lambda\mathbf{I} - \mathbf{J})^{-1}\delta_0 \tag{11.10}$$

If $|\lambda| > \max|\lambda_i|$, then the Neumann expansion of Equation 11.10 results in

$$\mathbf{F} = \sum_{k=0}^{\infty} \frac{\left(\mathbf{J}^k \delta_0\right)}{\lambda^{k+1}}$$

Hence,

$$\left\langle \mathbf{F}, \delta_0 \right\rangle = \sum_{k=0}^{\infty} \frac{\left(\mathbf{J}^k \delta_0, \delta_0\right)}{\lambda^{k+1}} \tag{11.11}$$

A partial fraction decomposition of the RHS of Equation 11.9 yields

$$\frac{A(\lambda)}{B(\lambda)} = \sum_{i=0}^{N} A(\lambda_i)\left[B'(\lambda_i)(\lambda - \lambda_i)\right] \tag{11.12}$$

If $|\lambda| > \max|\lambda_i|$, we have

$$\frac{A(\lambda)}{B(\lambda)} = \sum_{i=0}^{N} \sum_{k=0}^{\infty} \lambda_i^k \frac{A(\lambda_i)}{\left[B'(\lambda_i)\lambda^{k+1}\right]} \tag{11.13}$$

Comparing Equations 11.9, 11.11, and 11.13, we obtain

$$\left(\mathbf{J}^k \delta_0, \delta_0\right) = \sum_{i=0}^{N} \lambda_i^k \frac{A(\lambda_i)}{B'(\lambda_i)} \qquad k = 0,1,2\ldots \tag{11.14}$$

Equation 11.14 is the basic equation from which the Jacobi matrix J(N) can be reconstructed. Thus, for k = 1

$$\left(\mathbf{J}\delta_0, \delta_0\right) = a_0 = \sum_{i=0}^{N} \frac{\lambda_i A(\lambda_i)}{B'(\lambda_i)}$$

and for k = 2

$$\left(\mathbf{J}^2 \delta_0, \delta_0\right) = a_0^2 + b_0^2 = \sum_{i=0}^{N} \frac{\lambda_i^2 A(\lambda_i)}{B'(\lambda i)}$$

Since a_0 is already known, b_0 can be determined. Recall that b_0 was chosen positive. Thus J_{11} and J_{12} are determined.

In general, the k-th term can be expressed as

$$\mathbf{J}^k \delta_0 = \sum_{i=0}^{k} d_i \delta_i \tag{11.15}$$

where the coefficients d_i depend on the a_i and b_i, with i running from 0 through k – 1. Also,

$$\mathbf{J}\delta_k = b_{k-1}\delta_{k-1} + a_k \delta_k + b_k \delta_{k+1} \tag{11.16}$$

Combining Equations 11.15 and 11.16 yields

$$\mathbf{J}^{k+1}\delta_0 = d_k\left[a_k\delta_k + b_k\delta_{k+1}\right] + d_k b_{k-1}\delta_k + R$$

where the remainder R involves the δ_i, a_i, and b_i, with i running from 0 up to (k – 1). Using Equations 11.14 and 11.16, we obtain

$$\left(\mathbf{J}^{2k+1}\delta_0,\delta_0\right) = \left(\mathbf{J}^{k+1}\delta_0,\mathbf{J}^{k}\delta_0\right)$$

$$= d_k^2 a_k + R$$

$$= \sum_{i=0}^{N} \lambda_i^{2k+1}\frac{A(\lambda_i)}{B'(\lambda_i)} \tag{11.17}$$

Again, the remainder term R depends on various a_i and b_i, with i running up to (k – 1). Equation 11.17 then determines the a_k. Exactly analogously, we have

$$\left(\mathbf{J}^{2k+2}\delta_0,\delta_0\right) = \left(\mathbf{J}^{k+1}\delta_0,\mathbf{J}^{k+1}\delta_0\right)$$

$$= d_k^2\left[a_k^2 + b_k^2\right] + R$$

$$= \sum_{i=0}^{N} \lambda_i^{2k+2}\frac{A(\lambda_i)}{B'(\lambda_i)} \tag{11.18}$$

Equation 11.18 determines the b_k. With this, the first (k + 1) rows having been determined the rest can be found successively.

The procedure leads to a unique Jacobi matrix. Otherwise, if there were another matrix $\tilde{\mathbf{J}}$ with the same spectra, we would have

$$\left(\mathbf{J}^{k}\delta_0,\delta_0\right) = \left(\tilde{\mathbf{J}}^{k}\delta_0,\delta_0\right), \quad k = 0,1,2,\ldots N$$

This is impossible unless $\mathbf{J} = \tilde{\mathbf{J}}$.

2. Proof of Inverse Problem 1 (Gray and Wilson[4])

Gray and Wilson[4] have given a simpler, inductive proof of Hochstadt's reconstruction of a unique Jacobi matrix from its specified spectrum and that of its principal subminor. Their proof is based on establishing the following inequality.

Lemma 11.1

Let

$$0 < \lambda_i < \mu_i < \lambda_{i+1}, \quad i = 1,\ldots N-1$$

Then

$$F(N,\lambda,\mu)=\sum_{i=1}^{N}\lambda_i-\sum_{i=1}^{N-1}\mu_i-\frac{Q}{S}>0$$

where

$$Q=\prod_{i-1}^{N}\lambda_i \text{ and } S=\prod_{i-1}^{N-1}\mu_i$$

Proof

For N = 2, $0<\lambda_1<\lambda_1<\lambda_2$

$$F(2,\lambda,\mu)=\lambda_1+\lambda_2-\mu_1-\lambda_1\frac{\lambda_2}{\mu_1}$$

$$=\frac{(\mu_1-\lambda_1)(\lambda_2-\mu_1)}{\mu_1}>0$$

For N = 3,

$$F(3,\lambda,\mu)=F(2,\lambda,\mu)+(\lambda_3-\mu_2)+\lambda_1\lambda_2\frac{\left[1-\frac{\lambda_3}{\mu_2}\right]}{\mu_1}$$

Let $\lambda_3=\mu_2+\delta$. Then $\lambda_3-\mu_2=\delta$ and

$$1-\frac{\lambda_3}{\mu_2}=1-1-\frac{\delta}{\mu_2}=-\frac{\delta}{\mu_2}$$

Then

$$\lambda_1\lambda_2\frac{\left[1-\frac{\lambda_3}{\mu_2}\right]}{\mu_1}=-\left(\frac{\lambda_1\lambda_2}{\mu_1\mu_2}\right)\delta$$

Therefore,

$$(\lambda_3-\mu_2)+\lambda_1\lambda_2\frac{\left[1-\frac{\lambda_3}{\mu_2}\right]}{\mu_1}=\delta\left[1-\frac{\lambda_1\lambda_2}{\mu_1\mu_2}\right]>0$$

by the hypothesis of Lemma 11.1. $F(m,\lambda,\mu)$ depends linearly on λ_m, the coefficient of which, $[1-Q(m-1)/S(m-1)]$, is > 0. Hence, $F(m,\lambda,\mu)$ decreases as λ_m decreases. Then by induction

$$F(m,\lambda,\mu)>F(m,\lambda,\mu)\Big|_{\lambda_m=\mu_{m-1}}=F(m-1,\lambda,\mu)>0$$

Proof of Inverse Problem 1:

Let N = 2. Then **J**(2) is

$$\mathbf{J}(2;a,b) = \begin{pmatrix} a_1 & b_1 \\ b_1 & a_2 \end{pmatrix}$$

Then by the hypothesis of the problem, a_1 must equal to μ_1, i.e.,

$$\mathbf{J}(2;a,b) = \begin{pmatrix} \mu_1 & b_1 \\ b_1 & a_2 \end{pmatrix} \tag{11.19}$$

If

$$a_2 = \lambda_1 + \lambda_2 - \mu_1$$

and

$$b_1 = \left[(\mu_1 - \lambda_1)(\lambda_2 - \mu_1)\right]^{\frac{1}{2}}$$

then **J** given by Equation 11.19 is the unique matrix having the required properties.

Let us assume that the theorem is true for $N \le m - 1$. Since

$$a_m = \text{Trace}\{\mathbf{J}(m;a,b)\} - \text{Trace}\{\mathbf{J}(m-1;a,b)\},$$

we set

$$a_m = \sum_{i=1}^{m} \lambda_i - \sum_{i=1}^{m-1} \mu_i \tag{11.20}$$

This is certainly true for N = 2 as above. We now define a polynomial $q(\lambda)$ of degree $(m - 2)$ as

$$q(\lambda) = \prod_{k=1}^{m}(\lambda_k - \lambda) - (a_m - \lambda)\prod_{k=1}^{m-1}(\mu_k - \lambda)$$

The roots of $q(\lambda)$, $\{\upsilon_i\}_1^{m-2}$, are real and interlace the μ_i's. This is a consequence of the fact that $q(\mu_j)$ is opposite in sign from $q(\mu_{j+1})$ for $j = 1, \ldots, m - 2$.

The assumption of induction guarantees that a unique $\mathbf{J}(m-1)$ exists such that its spectrum is $\{\mu_i\}_1^{m-1}$ and $\mathbf{J}(m-2)$ has spectrum $\{\upsilon_i\}_1^{m-2}$. a_m is known from the previous construction (11.20) and $\{a_i\}_1^{m-1}$ and $\{b_i\}_1^{m-2}$ can be recovered from $\mathbf{J}(m-1;a,b)$. In order to determine b_{m-1}, we consider the three-term recurrence formula (Appendix A, Section B) for the Jacobi matrix and set $\lambda = 0$. This results in the following expression for b_{m-1}:

$$b_{m-1}^2 = \frac{\left[\left(\sum_{k=1}^{m}\lambda_k - \sum_{k=1}^{m-1}\mu_k\right)\prod_{j=1}^{m-1}\mu_j - \prod_{k=1}^{m}\lambda_k\right]}{\prod_{j=1}^{m-2}\upsilon_j}$$

By virtue of the assumption that $0 < \lambda_1$, the denominator is positive, and Lemma 11.1 guarantees that the numerator is also greater than zero. b_{m-1} is thus uniquely determined.

3. Construction Due To de Boor and Golub[6]

Consider the polynomials given by Equations 11.5 and 11.6. Let the index i run from 1 to N. The weight function is then (see Appendix B)

$$w_i = \left[P(N-1;\xi_i)P'(N;\xi_i)\right]^{-1}, \quad i = 1,\ldots N$$

The polynomials $P_i(\lambda)$ are then constructed in their natural order according to (Appendix A, Section B)

$$P(-1;\lambda) = 0$$

$$P(0;\lambda) = 1$$

$$P(i;\lambda) = (\lambda - a_i)P(i-1;\lambda) - b_{i-1}^2 P(i-2;\lambda), \qquad i = 1,2,\ldots,N \tag{11.21}$$

The elements a_i and b_i are computed according to the formulas (Equations B.5 and B.6, Appendix A)

$$a_i = \frac{(P(i-1;\lambda), \lambda P(i-1;\lambda))}{\|P(i-1;\lambda)\|^2}, \quad i = 1,\ldots N$$

and

$$b_i = \frac{\|P(i;\lambda)\|}{\|P(i-1;\lambda)\|}, \quad i = 1, 2,\ldots, N-1$$

The uniqueness of the construction follows from that of the monic polynomials.

In the weights w_i computed according to Equation B.15, both $P(N-1;\xi_i)$ and $P'(N;\xi_i)$ occur in the denominator. Computationally, this results in the w_i being susceptible to over- and underflow. This can be avoided by considering the logarithms of these polynomials. For large matrices, this can be expensive. If, on the other hand, one of these polynomials appears in the numerator, then the problem of under- or overflow is less severe. This can be seen from the inequality

$$\frac{(\lambda_i - \mu_{i-1})(\mu_i - \lambda_i)}{(\lambda_i - \lambda_1)(\lambda_N - \lambda_i)} < \frac{P(N-1;\lambda_i)}{P'(N;\lambda_i)} < 1$$

This is a consequence of the interweaving condition

$$\lambda_i < \mu_i < \lambda_{i+1}, \forall\, i$$

The factor containing $(\lambda_i - \lambda_i)$ resp. $(\lambda_N - \lambda_i)$ is to be omitted if i = 1 resp. i = N. This means that an over or an underflow is highly unlikely to occur. Let us, therefore, consider weights of the form (B.15a)

$$w_i = \frac{P(N-1;\xi_i)}{P'(N;\xi_i)}, \forall i$$

and see what the consequences are. This will prove to be highly interesting. Consider a "persymmetric" Jacobi matrix $\bar{\mathbf{J}}(N;a,b)$ defined by

$$\bar{\mathbf{J}}(N;a,b) \equiv \mathbf{S}^{-1}\mathbf{J}(N;a,b)\mathbf{S}$$

where the orthogonal permutation matrix **S** was defined earlier in Equation 11.7. Therefore, **J**(N;a,b) is of the form

$$\bar{\mathbf{J}}(N;a,b) = \begin{pmatrix} \bar{a}_1 & \bar{b}_1 & \cdot & & & \\ \bar{b}_1 & \bar{a}_2 & \bar{b}_2 & \cdot & 0 & \\ \cdot & \cdot & \cdot & \cdot & \cdot & \\ \cdot & \cdot & \cdot & \cdot & \cdot & \cdot \\ & 0 & & \bar{b}_{N-2} & \bar{a}_{N-1} & \bar{b}_{N-1} \\ & & & \cdot & \bar{b}_{N-1} & \bar{a}_N \end{pmatrix}$$

where

$$\bar{a}_i = a_{N+1-i} \text{ and } \bar{b}_i = b_{N-i}, \forall i$$

The persymmetric matrix is thus a reflection of the original matrix **J** about its second diagonal. Analogous to Equation 11.21, we construct

$$\bar{P}(-1;\lambda) = 0, \ \bar{P}(0;\lambda) = 1$$

and

$$\bar{P}(i;\lambda) = \prod_{j=1}^{i}(\lambda - \lambda_j)$$

Lemma 11.2

$$P(N-1;\lambda_i)\bar{P}(N-1;\lambda_i) = \left[\prod_{i=1}^{N-1} b_i\right]^2$$

for i = 1, 2, ..., N.

Proof

Let **A** be the matrix $(\lambda_i - \mathbf{J})$. Then (Appendices A and B)

$$P(N-1;\lambda_i) = \det \mathbf{A}\begin{pmatrix} 1,\ldots,N-1 \\ 1,\ldots,N-1 \end{pmatrix}$$

and

$$\bar{P}(N-1;\lambda_i) = \det \mathbf{A}\begin{pmatrix} 2,\ldots,N \\ 2,\ldots,N \end{pmatrix}$$

Moreover, det$\mathbf{A}$ = 0 since $\mathbf{A} = (\lambda_i - \mathbf{J})$ is singular. Application of Sylvester's identity (Appendix A) using

$$\mathbf{A}\begin{pmatrix} 2,\ldots,N-1 \\ 2,\ldots,N-1 \end{pmatrix}$$

as the pivotal block yields

$$0 = \det \mathbf{A}\begin{pmatrix} 2,\ldots,N-1 \\ 2,\ldots,N-1 \end{pmatrix} \det \mathbf{A} = \det \begin{vmatrix} \det \mathbf{A}\begin{pmatrix} 1,\ldots,N-1 \\ 1,\ldots,N-1 \end{pmatrix} & \det \mathbf{A}\begin{pmatrix} 1,\ldots,N-1 \\ 2,\ldots,N \end{pmatrix} \\ \det \mathbf{A}\begin{pmatrix} 2,\ldots,N \\ 1,\ldots,N-1 \end{pmatrix} & \det \mathbf{A}\begin{pmatrix} 2,\ldots,N \\ 2,\ldots,N \end{pmatrix} \end{vmatrix}$$

$$= P(N-1;\lambda_i)\bar{P}(N-1;\lambda_i) - \left[\prod_{i=1}^{N-1} (-)^{N-1} b_i\right]^2$$

and the Lemma is proved.

The above result brings us to the solution of inverse problem 2. Lemma 11.2 and Equation B.15 show that if the reconstruction is done according toB.15a, then the matrix reconstructed is not the Jacobi matrix $\mathbf{J}$, but the persymmetric matrix $\bar{\mathbf{J}}$. Also, the Jacobi matrix is persymmetric, then $\mathbf{J} = \bar{\mathbf{J}}$ and so $P(i) = \bar{P}(i)$, $\forall$ i. Lemma 11.2 then implies that

$$\{P(N-1;\lambda_i)\}^2 = \left[\prod_{i=1}^{N-1} b_i\right]^2, \forall i$$

Since the weight factor w_i is determined to within a scalar multiple, it follows that we need to know only P(N) in order to reconstruct a persymmetric matrix $\bar{\mathbf{J}}$. Once $\bar{\mathbf{J}}$ is known, $\mathbf{J}$ is also reconstructed, but in reverse order. Inverse problem 2 is thus solved. This is the argument due to de Boor and Golub. A more rigorous proof of this inverse problem is presented below.

4. Inversion of a Persymmetric Jacobi Matrix (Hochstadt[3])

The method is similar to the first variant considered in Section III.A. However, it cannot be applied directly without modifications. This is because the data $\{\mu_i\}_1^{N-1}$ are not specified so that $B(\lambda)$ is given, but not $A(\lambda)$, whereas both $A(\lambda)$ and $B(\lambda)$ must be known for the procedure of Section III.A to apply. The polynomial $A(\lambda)$ is thus to be constructed. Hochstadt does this as follows. The first step is to expand the vector δ_0 in terms of the eigenvectors of $\mathbf{J}$:

$$\delta_0 = \sum_{i=0}^{n} \frac{\left(\mathbf{u}^{(i)}, \delta_0\right)\mathbf{u}^{(i)}}{\left\|\mathbf{u}^{(i)}\right\|^2} = \sum_{i=0}^{n} \frac{\mathbf{u}^{(i)}}{\left\|\mathbf{u}^{(i)}\right\|^2}$$

recalling that $(\mathbf{u}^{(i)},\delta_0) = 1$. Now, using Neumann's expansion again, we obtain

$$(\lambda - \mathbf{J})^{-1}\delta_0 = \sum_{i=0}^{N} \frac{\mathbf{u}^{(i)}}{\left[\left\|\mathbf{u}^{(i)}\right\|^2(\lambda - \lambda_i)\right]} \tag{11.22}$$

and

$$\left((\lambda - \mathbf{J})^{-1}\delta_0, \delta_0\right) = \sum_{i=0}^{N} \frac{\left(\mathbf{u}^{(i)}, \delta_0\right)}{\left[\left\|\mathbf{u}^{(i)}\right\|^2(\lambda - \lambda_i)\right]} = \sum_{i=0}^{N} \frac{1}{\left[\left\|\mathbf{u}^{(i)}\right\|^2(\lambda - \lambda_i)\right]} \tag{11.23}$$

Now, from Section III.A, we know that

$$\left[(\lambda - \mathbf{J})^{-1}\delta_0, \delta_0)\right] = \frac{A(\lambda)}{B(\lambda)}$$

Comparing this relation with Equation 11.23 yields

$$A(\lambda) = B(\lambda) = \sum_{i=0}^{N} \frac{1}{\left[\left\|\mathbf{u}^{(i)}\right\|^2(\lambda - \lambda_i)\right]} \tag{11.24}$$

Since $\mathbf{J}$ is persymmetric, it follows that

$$\mathbf{SJS} = \mathbf{J} \tag{11.25}$$

$$\mathbf{J}\mathbf{u}^{(i)} = \lambda_i \mathbf{u}^{(i)} \tag{11.26}$$

and hence

$$\mathbf{SJSS}\mathrm{u}^{(i)} = \lambda_i \mathbf{S}\mathbf{u}^{(i)}$$

since $\mathbf{S}^2 = \mathbf{I}$. Then by Equation 11.25 we have

$$\mathbf{JS}\mathbf{u}^{(i)} = \lambda_i \mathbf{S}\mathbf{u}^{(i)} \tag{11.27}$$

Since all eigenspaces of $\mathbf{J}$ are one dimensional, Equations 11.26 and 11.27 show that $\mathbf{S}\,\mathbf{u}^{(i)}$ and $\mathbf{u}^{(i)}$ are linearly dependent so that

$$\mathbf{S}\mathbf{u}^{(i)} = k_i \mathbf{u}^{(i)}, \qquad i = 0, 1, \ldots, N \tag{11.28}$$

where $\{k_i\}$ are constants. Premultiplying Equation 11.28 by $\mathbf{S}$ and using the fact that $\mathbf{S}\,\mathbf{S} = \mathbf{I}$, we have $k_i = \pm 1$.

Hochstadt next computes the component $((\lambda - \mathbf{J})^{-1}\,\delta_0, \delta_N)$ as

$$\left((\lambda-\mathbf{J})^{-1}\delta_0,\delta_N\right)=\frac{\begin{pmatrix}\lambda-a_0 & -b_0 & 0 & \ldots 1\\ -b_0 & \lambda-a_1 & -b_1 & \ldots 0\\ 0 & -b_1 & \lambda-a_2 & \ldots 0\\ \cdot & \cdot & \cdot & \cdot\\ \cdot & \cdot & \cdot & \cdot\\ \cdot & \cdot & \cdot & \cdot\\ 0 & 0 & 0 & 0\end{pmatrix}}{B(\lambda)}=\prod_{i=0}^{N-1}\frac{b_i}{B(\lambda)} \tag{11.29}$$

Using Equation 11.22 and expressing Equation 11.29 as partial fractions, we have

$$\sum_{i=0}^{N}\frac{\left(\mathbf{u}^{(i)},\delta_N\right)}{\left[\left\|\mathbf{u}^{(i)}\right\|^2(\lambda-\lambda_i)\right]}=\sum_{i=0}^{N}\frac{\left[\prod_{i=0}^{N-1}b_i\right]}{B'(\lambda_i)(\lambda-\lambda_i)} \tag{11.30}$$

Now

$$\left(\mathbf{u}^{(i)},\delta_N\right)=\left(\mathbf{S}\mathbf{u}^{(i)},\mathbf{S}\delta_N\right)=k_i\left(\mathbf{u}^{(i)},\delta_0\right)=k_i$$

This yields

$$\left|\left(\mathbf{u}^{(i)},\delta_n\right)\right|=1$$

Using this result in Equation 11.30, we obtain

$$\left\|\mathbf{u}^{(i)}\right\|^2=\frac{|B'(\lambda_i)|}{\left[\prod_{i=0}^{N-1}b_i\right]} \tag{11.31}$$

Now from the given data, $B'(\lambda_i)$ is known. Hence, from Equations 11.24 and 11.31, it follows that

$$\frac{A(\lambda)}{B(\lambda)}=\sum_{i=0}^{N}\frac{\left[\prod_{i=0}^{N-1}b_i\right]}{|B'(\lambda_i)|(\lambda-\lambda_i)} \tag{11.32}$$

The product term containing the b_i can be determined from the limit

$$\lim_{\lambda\to\infty}\frac{\lambda A(\lambda)}{B(\lambda)}=1$$

This follows from the polynomial definitions of A(λ) and B(λ) given in Section III.A. Hence, from Equation 11.32, we obtain

$$\prod_{i=0}^{N-1} b_i = \left[\sum_{i=0}^{N} \left\{ \frac{1}{|B'(\lambda_i)|} \right\} \right]^{-1}$$

The right-hand side of Equation 11.32 and, hence, A(λ) are known. The procedures of Section III.A can then be applied in order to reconstruct the desired Jacobi matrix.

5. Proof of the Third Inverse Problem[6]

We now come to the proof of the third variant of the inversion of a Jacobi matrix, which is closely related to problem 1. From the hypothesis of the problem, $P(N-1;\lambda) = P^*(N-1;\lambda)$ and $b_i = b_i^*$ since only a_N has been changed to a^*_N. Then using Equation 11.21 in order to calculate $P(N,\lambda)$ and $P^*(N;\lambda)$ and subtracting yields

$$\prod_{i=1}^{N}\left(\lambda - \lambda_i^*\right) - \prod_{i=1}^{N}(\lambda - \lambda_i) = \left(a_N - a_N^*\right)P(N-1;\lambda) \tag{11.33}$$

Replacing $P(N-1;\lambda)$ by its polynomial expansion (Equation 11.5) and comparing the coefficients of on both sides of the resulting equation obtains

$$\sum_{i=1}^{N}\left(\lambda - \lambda_i^*\right) = a_N - a_N^* \tag{11.34}$$

from which a^*_N can be calculated if a_N is known. Using Equations 11.33 and 11.34, $P(N-1;\lambda)$ can be obtained in terms of the known data $\{\lambda_i\}^N_1$ and $\{\lambda^*_i\}^N_1$. Setting $\xi_i = \lambda_i$ (Appendix B) and calculating $P(N-1;\lambda)$ from Equations 11.33 and 11.34, the weight functions are given by

$$w_i = -\frac{1}{\left[P'(N;\lambda_i)\prod_{i=1}^{N}\left(\lambda_i - \lambda_i^*\right)\right]}$$

We have put a factor of –1 in front arbitrarily since, as shown earlier, w_i only need be known up to a scalar multiple. By the hypotheses of the problem, namely,

$$\lambda_i < \lambda_i^* < \lambda_{i+1} \forall i$$

it is obvious that the weights are positive. The weight functions being known, the desired Jacobi matrix can be reconstructed.

6. Well-Posedness of the Inverse Jacobi Problem

The last topic to be discussed in connection with the inversion of a Jacobi matrix is the question of its well-posedness, that is, does the reconstructed matrix depend continuously on the data? In other words, if there is a perturbation in the data, does it cause a corresponding perturbation in the reconstructed matrix or does it cause the inversion to be unstable?

That the reconstruction is stable can be seen as follows. Consider Forsythe's three-term recurrence relation (Equation B.4). The parameters $a = \alpha$ and $b = \beta$ there depend continuously on ξ and w. In view of the connection between this recurrence relation and that of the principal minors of the Jacobi matrix (Appendix B), it follows that a,b of **J** also depend continuously on ξ and w. The weights w_i and the polynomials $P'(N;\xi_i)$, $P(N-1;\xi_i)$, depend continuously on ξ_i and the specified spectra. Hence, setting $\xi_i = \lambda_i$, it follows that the reconstructed Jacobi matrix **J** also depends continuously on the data.

Theorem 11.5

Let $\mathbf{J}(N)$ and $\mathbf{J}(N-1)$ be the Jacobi matrices with given spectra $\{\lambda_i\}$ and $\{\mu_i\}$, respectively, as in Theorem 11.1. Then **J** depends continuously on the given spectra.

A simple formal proof of the continuous dependence of **J** on λ and μ is due to Hochstadt[3] and is presented below.

Proof

Let $\{\tilde{\lambda}_i\}$ and $\{\tilde{\mu}_i\}$ denote the spectra corresponding to $\tilde{\mathbf{J}}(N)$ and $\tilde{\mathbf{J}}(N-1)$, respectively, such that

$$\left|\lambda_i - \tilde{\lambda}_i\right| < \varepsilon, \qquad i = 0,1,2,\ldots,N$$

$$\left|\mu_i - \tilde{\mu}_i\right| < \varepsilon, \qquad i = 1,2,\ldots,N$$

Define $A(\lambda)$, $\tilde{A}(\lambda)$, $B(\lambda)$, and $\tilde{B}(\lambda)$ for **J** and $\tilde{\mathbf{J}}$ as before. Since these are polynomials, we have

$$\left|B'(\lambda_i) - \tilde{B}'(\tilde{\lambda}_i)\right| < \beta_i\varepsilon, \qquad i = 0,1,\ldots,N$$

and

$$\left|A(\lambda_i) - \tilde{A}(\tilde{\lambda}_i)\right| < \alpha_i\varepsilon, \qquad i = 1,2,\ldots,N$$

$\{\beta_i\}$ and $\{\alpha_i\}$ are suitable constants. From Equation 11.14, it follows that

$$\left|\left(\mathbf{J}^k\delta_0,\delta_0\right) - \left(\tilde{\mathbf{J}}^k\delta_0,\delta_0\right)\right| < K\varepsilon, \qquad k = 1,2,\ldots,N+1$$

K being a constant. From the various relations derived in Section III.A, it is not difficult to see that

$$\left|a_i - \tilde{a}_i\right| < M\varepsilon, \qquad i = 0,1,\ldots,N$$

and

$$\left|b_i - \tilde{b}_i\right| < M\varepsilon, \qquad i = 1,2,\ldots,N$$

M being another suitable constant. This proves that the reconstructed Jacobi matrix **J** depends continuously on the data.

Perhaps the most rigorous proof is due to Hald.[4] In the course of his proof, Hald also establishes the boundedness of the elements of the Jacobi matrix in terms of the data. Hald's proof is lengthy and will not be discussed in detail; reference is made to his original article.

III. INVERSE PROBLEMS FOR NONUNIFORM, LOSSLESS TRANSMISSION LINES

We now discuss lossless, nonuniform transmission lines and their associated inverse problems. Both discrete and continuous versions of transmission line problems will be analyzed. The discrete inverse solutions will be discussed after Bruckstein, Kailath and Levy[8,9] who present the most general and elegant formalism of the problem. We will call this the Bruckstein-Kailath-Levy formalism. The continuous version of the inverse transmission line problem will be presented according to Brown[29] and Wilcox.[30,31] We begin with the discrete transmission line.

A. INVERSE SOLUTION OF DISCRETE LINES: BRUCKSTEIN-KAILATH-LEVY FORMALISM

1. The Basic Transmission Line

We will now consider the inverse problem of wave propagation in a nonuniform, lossless transmission line. The problem consists of determining the distribution of the line's impedance from suitably defined scattering data. What the scattering data are will be pointed out as we proceed. The line is modeled as a layered, cascaded system which is discretized into N sections, over each of which the line impedance is assumed to remain constant, changing only at the interfaces between the layers. For waves propagating along the line, reflection thus occurs only at these interfaces where the impedance of the line changes. Otherwise, the waves propagate "freely", that is, they are only either time delayed (causal propagation) or time advanced (noncausal propagation). The transmission line is, therefore, also conceptualized as a cascaded system in which waves propagate freely over any section (delay), being reflected only locally at the interfaces (gain). This is what is known as the "gain-delay" model in which the line is layered into "gain" sections separated by "delays". Each gain section is specified completely by a wave interaction matrix that depends on a local parameter, which can be either the reflection coefficient or the local impedance. The gain consists of propagating waves interacting with each other, part of each wave being "backscattered" and added to the counterpropagating one. This characterization suggests a direct, time-domain recursive procedure for solving the inverse problem for a lossless line. Moreover, the same inversion method is also applicable to more general wave propagation models having similar gain-delay structures. The details of the basic physical structure of this gain-delay transmission line will be considered first.

Let us consider the vector

$$\mathbf{w}(x,t) = \left[v(x,t), i(x,t)\right]^T$$

where v(x,t), i(x,t) are the voltage and current in the line at a point x and a time t, respectively. If Z(x) is the line impedance at x, then a nonuniform, lossless transmission line is governed by what is known as the "telegrapher's equations",[10] which are written as

$$\mathbf{w}'(x,t) = \begin{pmatrix} 0 & -Z(x) \\ -Z^{-1}(x) & 0 \end{pmatrix} \dot{\mathbf{w}}(x,t) \tag{11.35}$$

where, as before, $\cdot = \partial/\partial t$. Introducing the transformations

$$\tilde{\mathbf{w}}(x,t) = \left[\tilde{v}(x,t), \tilde{i}(x,t)\right]^T = \begin{pmatrix} Z^{-\frac{1}{2}}(x) & 0 \\ 0 & Z^{\frac{1}{2}}(x) \end{pmatrix} \mathbf{w}(x,t) \tag{11.35a}$$

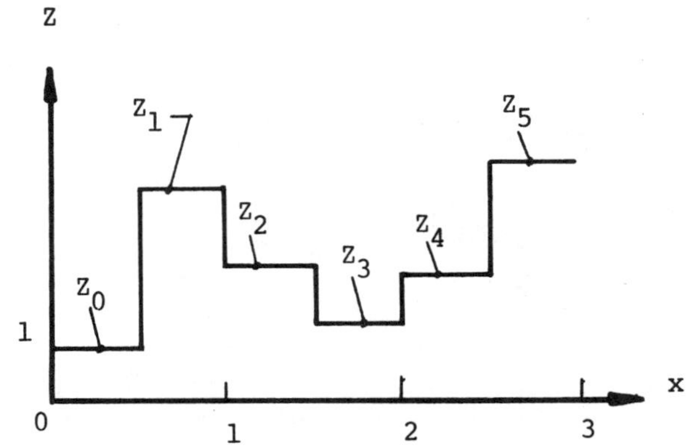

FIGURE 3. The impedence "taper" of a lossless, nonuniform transmission line, that is, the variation of the impedence as a function of the distance x from the origin. The segments into which the line is discretized are of length 1/2.

and

$$r(x) = \left(\tfrac{1}{2}\right)\left[\ln Z(x)\right]' \tag{11.36}$$

we have the following telegrapher's equation:

$$\mathbf{W}'(x,t) = \begin{pmatrix} -\dfrac{\partial}{\partial t} & -r(x) \\ -r(x) & \dfrac{\partial}{\partial t} \end{pmatrix} \mathbf{W}(x,t) \tag{11.37}$$

where

$$\mathbf{W}(x,t) = \left[W_+(x,t), W_-(x,t)\right]^T$$

and

$$W_\pm(x,t) = \left(\frac{1}{2}\right)\left[\tilde{v}(x,t) \pm \tilde{i}(x,t)\right] \tag{11.37a}$$

By virtue of Equation 11.36 and according to the gain-delay characterization of the transmission line in which r(x) is constant over any cascade, Equation 11.37 reduces to

$$\mathbf{W}'(x,t) = \begin{pmatrix} -1 & 0 \\ 0 & 1 \end{pmatrix} \dot{\mathbf{W}}(x,t), \quad x \subset \text{any cascade}$$

This is just the equation for freely propagating waves, W_+ resp. W_- corresponding to right resp. left propagation. Therefore,

$$W_\pm(x,t) = W_\pm(x \mp t)$$

Consider the gain-delay modeled transmission line shown in Figure 3. The line is discretized into units of length 1/2. The n-th section is characterized by the fact that

$$Z(x) = Z(n-1), \quad x \subset \left[\frac{(n-1)}{2}, \frac{n}{2}\right)$$

By assumption, $Z(x) \geq 0$ and $Z(0) = 1$. Defining a time-shift operator by the matrix

$$\Delta = \begin{pmatrix} \Delta & 0 \\ 0 & \Delta^{-1} \end{pmatrix}$$

where Δ_+, Δ_- are the time-delay and time-advance operators, respectively, that is,

$$\Delta_\pm f(t) = f\left(t \mp \frac{1}{2}\right)$$

We then have

$$\mathbf{W}\big|_{x=\left(\frac{n}{2}\right)-} = \Delta\mathbf{W}\big|_{x=\left[\frac{(n-1)}{2}\right]+} \tag{11.38}$$

Δ_+ will be written as just Δ, omitting the + subscript. Equation 11.38 means that the value of the vector **W** at the interface x = (n/2)– is obtained from its value at the interface x = [(n–1)/2]+ by replacing the time t in W_+ resp. W_- by (t – 1/2) resp. (t + 1/2). The time-shift operator Δ resp. Δ_- is thus applied to W_+ resp. W_-. The operator Δ_- when applied to W_- is unphysical since it is noncausal because it advances time.

The vector **w**(x,t) is continuous all along the transmission line. This means that at any interface (n/2):

$$\mathbf{w}(x,t)\big|_{\left(\frac{n}{2}\right)+} = \mathbf{w}(x,t)\big|_{\left(\frac{n}{2}\right)-}$$

Then by defining

$$\left(\frac{1}{2}\right)\begin{pmatrix} Z^{-\frac{1}{2}} & Z^{\frac{1}{2}} \\ Z^{-\frac{1}{2}} & -Z^{\frac{1}{2}} \end{pmatrix}$$

by $\wp(Z_n)$, we have from Equation (11.38)

$$\mathbf{W}(x,t)\big|_{\left(\frac{n}{2}\right)+} = \wp(Z_n)\mathbf{w}(x,t)\big|_{\left(\frac{n}{2}\right)+}$$

$$= \wp(Z_n)\mathbf{w}(x,t)\big|_{\left(\frac{n}{2}\right)-}$$

$$= \wp(Z_n)\wp^{-1}(Z_{n-1})\mathbf{W}(x,t)\big|_{\left(\frac{n}{2}\right)-}$$

$$= \mathbf{T}(n)\mathbf{W}(x,t)\big|_{\left(\frac{n}{2}\right)-}$$

$$= \mathbf{T}(n)\Delta\mathbf{W}(x,t)\big|_{\left[\frac{(n-1)}{2}\right]+} \tag{11.39}$$

The matrix $\mathbf{T}(n)$ is called the "transfer matrix" and can be expanded as

$$\mathbf{T}(n) = \left(1 - r_n^2\right)^{-\frac{1}{2}} \begin{pmatrix} 1 & -r_n \\ -r_n & 1 \end{pmatrix}$$

where

$$r_n = \frac{(Z_n - Z_{n-1})}{(Z_n + Z_{n+1})} \tag{11.40}$$

Since $Z_n > 0$, n, it follows that $|r_n| \le 1$.

In Equation 11.39, W_- is obtained at x = [(n/2)–] at a time t from its value of x = {[(n–1)/2]+} at a time (t + 1/2). This is clearly unphysical and is a consequence of operating by the noncausal time-advance operator Δ_-. The actual physical situation is, of course, otherwise: $W_-(x,t)$ at x = [(n/2)+] is the incident wave at a time t, from which is obtained its value at x = {[(n–1)/2+]}, which occurs at a later time (t + 1/2). On the basis of signal flow path reversal, Bruckstein and Kailath[8] show that the proper causal equation can be written as

$$\begin{pmatrix} W_+\left(\left\{\frac{n}{2}\right\}+,t\right) \\ W_-\left(\left\{\frac{(n-1)}{2}\right\}+,t\right) \end{pmatrix} = \begin{pmatrix} 1 & 0 \\ 0 & \Delta \end{pmatrix} \mathbf{S}_n \begin{pmatrix} \Delta & 0 \\ 0 & 1 \end{pmatrix} \begin{pmatrix} W_+\left(\left\{\frac{(n-1)}{2}\right\}+,t\right) \\ W_-\left\{\left(\frac{n}{2}\right)+,t\right\} \end{pmatrix} \tag{11.41}$$

where the "scattering matrix" $\mathbf{S}_n$ is of the form

$$\mathbf{S}_n = \left(1 - r_n^2\right)^{\frac{1}{2}} \begin{pmatrix} 1 & \dfrac{-r_n}{\left(1 - r_n^2\right)^{\frac{1}{2}}} \\ \dfrac{r_n}{\left(1 - r_n^2\right)^{\frac{1}{2}}} & 1 \end{pmatrix} \tag{11.42}$$

If Equation 11.41 is written out fully with the scattering matrix $\mathbf{S}_n$ given by Equation 11.42

and then compared to Equation 11.39, it will be seen that the algebraic relationship between W_+ and W_- is not changed. The causal scattering Equations 11.41 and 11.42 are compatible with the unphysical Equation 11.39 insofar as the algebraic relationship between the right- and left-propagating waves is concerned. The transfer matrix **T**(n) is, however, modified. Notice that only the causal time-delay operator Δ appears in Equation 11.41.

Writing out Equation 11.41 in full, we obtain

$$W_+\left[\left(\frac{n}{2}\right)+, t\right] = \left(1 - r_n^2\right)^{\frac{1}{2}} W_+\left[\left\{\frac{(n-1)}{2}\right\}+, t - \frac{1}{2}\right] - r_n W_-\left[\left(\frac{n}{2}\right)+, t\right] \tag{11.43}$$

and

$$W_-\left[\left\{\frac{(n-1)}{2}\right\}+, t\right] = \left(1 - r_n^2\right)^{\frac{1}{2}} W_-\left[\left(\frac{n}{2}\right)+, t - \frac{1}{2}\right] + r_n W_+\left[\left\{\frac{(n-1)}{2}\right\}+, t - 1\right] \tag{11.44}$$

Let us assume that for time < t, no left-propagating wave reaches the interface (n/2)+. Then from Equation 11.44, we obtain

$$r_n = \frac{W_-\left[\left\{\frac{(n-1)}{2}\right\}+, t\right]}{W_+\left[\left\{\frac{(n-1)}{2}\right\}+, t - 1\right]}$$

This shows that r_n is a local "reflection coefficient" for the n-th section. This then obtains $(1 - r_n^2)^{1/2}$ as the amplitude transmitted. Using Equation 11.40, the local impedance Z_n of the n-th section of the line can be expressed in terms of the local reflection coefficients r_n as

$$\begin{aligned} Z_n &= Z_{n-1}\frac{(1 + r_n)}{(1 - r_n)} \\ &= \prod_1^n \frac{(1 + r_i)}{(1 - r_i)} \end{aligned}$$

Again, $Z_0 = 1$.

From Equations 11.43 and 11.44, the following relation can be obtained:

$$\left\|W_+\left[\left\{\frac{(n-1)}{2}\right\}+, \cdot\right]\right\|^2 + \left\|W_-\left[\left(\frac{n}{2}\right)+, \cdot\right]\right\|^2 = \left\|W_+\left[\left(\frac{n}{2}\right)+, \cdot\right]\right\|^2 + \left\|W_-\left[\left\{\frac{(n-1)}{2}\right\}+, \cdot\right]\right\|^2 \tag{11.45}$$

where $\|\cdot\|$ is the L_2 norm. This expresses the physical property of conservation of energy for a lossless transmission line. Equation 11.45 is a consequence of the unitarity of the scattering matrix $\mathbf{S}_n$. As a matter of fact,

$$\mathbf{S}_n \mathbf{S}_n^T = \mathbf{S}_n^T \mathbf{S}_n = \mathbf{I}$$

A simple rearrangement of the terms of Equation 11.45 yields

$$\left\| W_+\left[\left(\frac{n}{2}\right)+,\cdot\right]\right\|^2 - \left\| W_-\left[\left(\frac{n}{2}\right)+,\cdot\right]\right\|^2 = \left\| W_+\left[\left\{\frac{(n-1)}{2}\right\}+,\cdot\right]\right\|^2 - \left\| W_-\left[\left\{\frac{(n-1)}{2}\right\}+,\cdot\right]\right\|^2 \quad (11.46)$$

This corresponds to the **J** orthogonality of the transfer matrix $\mathbf{T}_n$, namely,

$$\mathbf{T}_n\mathbf{J}\mathbf{T}_n^T = \mathbf{J} = \begin{pmatrix} 1 & 0 \\ 0 & -1 \end{pmatrix}$$

Therefore, both Equations 11.45 and 11.46 or the unitarity and the **J** orthogonality of $\mathbf{S}_n$ express energy conservation. Equation 11.41 or Equations 11.43 and 11.44 represent a physical scattering interaction process in which $W_+[\{(n-1)/2\}+,\cdot]$ resp. $W_-[(n/2)+,\cdot]$ are the incident waves and $W_+[(n/2)+,\cdot]$ resp. $W_-[\{(n-1)/2\}+,\cdot]$ are the reflected waves.

The evolution equation for $\mathbf{w}(x,t)$ similar to Equation 11.39 can be obtained as

$$\mathbf{w}\Big|_{\left(\frac{n}{2}\right)+} = \wp^{-1}(Z_{n-1})\Delta\wp(Z_{n-1})\mathbf{w}\Big|_{\left[\frac{(n-1)}{2}\right]+} \quad (11.47)$$

The counterpart of the $\mathbf{S}_n$ matrix for the vector **w** can also be derived as

$$\mathbf{S}_n^{\mathbf{w}} = \begin{pmatrix} 2Z^{-\frac{1}{2}} & -1 \\ Z^{-1} & -Z^{-\frac{1}{2}} \end{pmatrix}$$

2. Inverse Solution of the Transmission Line Problem

The direct or forward problem for wave propagation in a nonuniform, lossless transmission line was considered in the previous section, where the line impedance, or what is the same thing, the local reflection coefficients were known. We will now discuss the inverse problem where the line impedance or the distribution of the local reflection coefficients is to be determined from the scattering data $W_\pm(0,t)$. $W_+(0,t)$ is the excitation at $x = 0$ and $W_-(0,t)$ is the response of the line at $x = 0$ corresponding to the excitation $W_+(0,t)$. It is necessary to assume that the line was quiescent prior to the application of $W_+(0,t)$ so that the response recorded is necessarily due to the causal propagation of waves in the line.

A natural way to inverse solve the nonuniform gain-delay-structured transmission line is to invert the line in a recursive manner, layer by layer. Thus, the reflection coefficient of the first section is determined (discussed below) from the excitation response. This is then used to determine the reflection coefficient of the next, second section and so on. In this way, successively determining the reflection coefficient of a section and then peeling off the associated layer, the transmission line can be determined to any desired depth. This procedure, called the "layer-peeling" or "direct" method, was applied by the geophysicists[11-14] in inverse solving the geophysical structures of a layered earth model. In geophysics, the direct method is known as the "dynamic deconvolution" or "downward continuation" method. The method invokes only causality and needs only to recursively compute the discrete waveforms at increasing depths of the layered medium, for which rules are usually simple. This direct or layer-peeling technique will be described first.

At $t = 0-$, the line is quiescent, i.e., $W_\pm(x,0-) = w_\pm(x,0-) = 0$. At $t = 0+$, a current input is applied at $x = 0$ and the voltage response is recorded at $x = 0$ as a function of time. Let us assume that the input and the response are continuous in time. Then for $W_{\pm}$, we can write

$$W_{-}(0,t) = 0, \forall t \subset [0,1) \tag{11.48}$$

and

$$W_{-}(0,t) = r_1 W_{+}(0,t-1) \; \forall \; t \subset [1,2] \tag{11.49}$$

These are the obvious consequences of causality. Equation 11.48 simply says that we cannot expect a response at $x = 0$ for times less than one round-trip time in the first section. Equation 11.49, on the other hand, implies that it will take at least two round-trip times for the effect of any left-propagating input at the first interface to be felt at $x = 0$. From Equations 11.48 and 11.49, the determination of r_1 is imminent and we have

$$r_1 = \frac{W_{-}(0,t)}{W_{+}(0,t)}, \quad \forall \, t \subset [1,2)$$

By the exact same reasoning, we can also obtain

$$r_2 = \frac{W_{-}(1,t)}{W_{+}(1,t)}, \quad \forall \, t \subset [2,3)$$

Although we do not have $W_{\pm}|_{(n/2)+}$ from direct measurement, we can create "artificial scattering data" by using the transfer matrix $\mathbf{T}_n$ and Equation 11.49. Then

$$\mathbf{W}(1,t) = \mathbf{T}_1 \Delta \mathbf{W}(0,t)$$

This then yields $\mathbf{W}(1,t)$ in terms of the measured data $\mathbf{W}(0,t)$. Essentially, then, once the reflection coefficient of the first section is determined, its associated matrix can be used to "peel off" the effect of the first layer of the medium. The problem then becomes that of a line parameterized by $\{r_1, r_2, ...\}$. In other words, using r_1 and the original scattering data, "artificial scattering data" are generated for the line extending over $[2,\infty)$. This can be continued in this way, successively determining a reflection coefficient and peeling off the associated layer, thereby determining as much of the line as desired. This procedure is, therefore, appropriately termed the "layer-peeling" algorithm.

Once the local reflection coefficients $\{r_i\}^n{}_1$ are obtained, the impedance profile can be recovered by appealing to the relation previously derived, namely,

$$Z_n = \prod_{i=1}^{n} \frac{(1 + r_i)}{(1 - r_i)}, \; Z_0 = 1$$

However, direct recovery of $\{Z_i\}^n{}_1$ is also possible.

$$Z_1 = \frac{v(1,t)}{i(1,t)} \forall \, t \subset [1,2)$$

$w(1,t)$ can be obtained from Equation 11.47 and the fact that $v(0,t) = i(0,t)$ for $t < 1$ since $Z_0 = 1$.

Similarly, by induction

$$Z_n = \frac{v(n,t)}{i(n,t)} \forall\, t \subset [n, n+1)$$

The signal at a depth n is again obtained from Equation 11.47 by knowing it at a depth of (n – 1).

The layer-peeling method described so far works directly on the scattering data and needs only one division in order to determine the reflection coefficient without the necessity to calculate convolution integrals or inner products. These are, therefore, efficient computationally. As a rule, these methods require $O(N^2)$ computations in order to retrieve the transmission line up to a depth of N. However, parallel processing with an array of N processors can result in a speed up time of O(N). With only a single processor, the speed up is $O(N \log^2 N)$ through the use of the "doubling technique". Further details will not be discussed here; the reader may consult References 15 and 18.

The effect of noise on transmission line synthesis using the direct method has also been investigated.[19-21] The error accumulation with depth due to noisy data, that is, the rate of error growth with depth, is determined by the running product of the form $(1 + |r_n|)/(1 - |r_n|)$. This implies that if the reflection coefficients are away from unity, the error in medium identification due to noisy data will not grow explosively. Moreover, if the reflection coefficients remain small most of the time, then the line can be reconstructed to a significant depth before the error due to noisy data accumulates to a destructive level.

A most interesting observation is the close connection between the layer-peeling method and an algorithm due to Schur[22] when formulated in terms of the z-transform (Appendix C). Schur's algorithm concerns a recursive procedure for testing if a power series in z^{-1} is bounded outside a unit circle in the complex plane. The close relation between the layer-peeling algorithm of inverse scattering on the transmission line and Schur's recursive procedure was first pointed out by Dewilde and Dym.[23] This interesting point will not be discussed in detail here; only a brief sketch will be presented.

The inverse solution discussed in this section shows that the reflection coefficient of the n-th section of the line is given by

$$r_n(t) = \frac{W_-(n-1,t)}{W_+(n-1,t)}, \quad t \subset [n+1, n+2)$$

Taking the z-transform of both sides results in

$$\tilde{r}_n(z) = \frac{\tilde{W}_-(n-1,z)}{\tilde{W}_+(n-1,z)}$$

the symbol ~ denoting the z-transformed quantities. Using Equation 11.39 in the transform domain in the equation above for $\tilde{r}_n(z)$ yields

$$\tilde{r}_n(z) = \frac{[z\tilde{r}_{n-1}(z) - \tilde{r}_n]}{[1 - \tilde{r}_n z\tilde{r}_{n-1}(z)]}$$

This is a discrete version of Riccati's equation.[23] The reflection coefficient $\tilde{r}_n$ is determined as

$$\tilde{r}_n = \lim_{z\to\infty} z\tilde{r}_{n-1}(z)$$

For example,

$$\tilde{r}_1 = \lim_{z\to\infty} z\tilde{r}_0(z) = \frac{\tilde{W}_-(0,1)}{\tilde{W}_+(0,0)}$$

We can also rewrite the expression for $\tilde{r}_n(z)$ as

$$\tilde{s}_n(z) = z\frac{[\tilde{s}_{n-1}(z) - \tilde{r}_n]}{[1 - \tilde{r}_n z\tilde{s}_{n-1}(z)]}$$

where $\tilde{s}_n(z) = z\tilde{r}_n(z)$. This turns out to be exactly identical to the recursion relation derived by Schur for checking if a power series $\tilde{S}(z)$ in z^{-1} is bounded by one outside a unit circle in a complex plane, that is, if

$$\left|\tilde{S}(z)\right| < 1 \text{ if } |z| > 1$$

Schur derived the result that for this to happen,

$$\lim_{z\to\infty} \tilde{S}_{n-1}(z) = \tilde{r}_n$$

must satisfy the inequality $|\tilde{r}_n| < 1$.

3. Linear Matrix Equations of Discrete Inverse Scattering

The direct, layer-peeling, fast inversion algorithm was described in the previous section. The scattering data were used directly without involving any special computations of convolution integrals, inner products, etc. However, since a discretized, layered model (whether of a transmission line or of geostructures) was invoked, the direct methods are not considered to yield solutions to a general inverse scattering problem. Traditionally, the inverse scattering solutions are obtained in terms of linear integral equations, which reduce to linear matrix equations in the case of discrete media. The present section discusses these "indirect" inversion procedures. The matrix equations of discrete inverse scattering will be derived first and the connection between these equations and their linear integral equation counterparts will be demonstrated next. Of the discrete formalisms of continuous inverse scattering, the derivations due to Case and Kac[27] and Sharpe[28] for potential scattering, Ware and Aki[24] for acoustical scattering in a layered elastic medium, and Berryman and Greene[25] in geophysical inversion are well known. Recently, Bruckstein and Kailath[8] and Bruckstein, Levy, and Kailath[9] have presented a general formulation of the problem from which the discrete versions of the well-known formalisms of continuous inverse scattering can be obtained by suitably tailoring the scattering data for the line. We present this generalized procedure and then specialize it to appropriate scattering data in order to recover the integral equations of Marcenko, Gel'fand-Levitan, and Krein.

Let us recall the fundamental elements of inverse scattering. The medium is perturbed in a predetermined way by a known input. One then measures the output or the response of the system to the given input. The output is essentially a convolution of the input with the system's impulse response. In other words,

$$\text{output} = \text{input} * \text{impulse response}$$

where $*$ means the operation of convolution. In all of the integral equations considered thus far, this convolutional form is apparent. The case of the discrete systems is also no exception. In order to formulate the problem in this direction, let us consider the z-transform (see Appendix C) of Equation 11.39, which is

$$\tilde{\mathbf{W}}(z,n) = T_n\tilde{\Delta}(z)\tilde{\mathbf{W}}(z,n-1) \tag{11.50}$$

The caret ~ denotes the z-transform, and n resp. n – 1 denotes x = (n/2)+ resp. x = [(n – 1)/2]+. Then from Equation 11.50, one obtains by iteration

$$\tilde{W}(z,n) = \tilde{M}(z,n)\tilde{W}(z,0) \tag{11.51}$$

where

$$\tilde{\mathbf{M}}(z,n) = \begin{pmatrix} M_{11} & M_{12} \\ M_{21} & M_{22} \end{pmatrix}(n,z) = \prod_{i=n}^{1} \mathbf{T}_i\tilde{\Delta}(z)$$

The time-shift operator $\tilde{\Delta}$ is given by

$$\tilde{\Delta} = \begin{pmatrix} z^{-\frac{1}{2}} & 0 \\ 0 & z^{\frac{1}{2}} \end{pmatrix}$$

It is readily seen from Equation 11.51 that the equations of the ± components of W and, hence, of its z-transforms are completely symmetric. This implies that the equations remain unaltered if $W_+(W_-)$ is interchanged with $W_-(W_+)$ and if the time-shift operation is also reversed, that is, if $\Delta(\Delta_-)$ is replaced by $\Delta_-(\Delta)$. In the language of the z-transform, this means replacing $z^{1/2}(z^{-1/2})$ by $z^{-1/2}$ $(z^{1/2})$. It then follows that

$$M_{12}(n,z) = M_{21}\left(n,z^{-1}\right)$$

and

$$M_{22}(n,z) = M_{11}\left(n,z^{-1}\right) \tag{11.52}$$

From Equation 11.51 and using the convolution property of the z-transforms, we obtain

$$W(n,t) = W(0,t) * M(n,t) \tag{11.53}$$

Written out fully, the equations for the components are

$$W_+(n,t) = W_+(0,t) * M_{11}(n,t) + W_-(0,t) * M_{12}(n,t)$$

and (11.54)

$$W_-(n,t) = W_+(0,t) * M_{21}(n,t) + W_-(0,t) * M_{22}(n,t)$$

Equation 11.53 relates the waves at a depth of x to those at the boundary. **M** is then a "transition matrix" of the medium over the interval [0,x]. It is also seen that the transition matrix satisfies the same equation of motion as that of **W**, namely,

$$\mathbf{M}'(n,t)=\begin{pmatrix} -\frac{\partial}{\partial t} & -r(n) \\ -r(n) & \frac{\partial}{\partial t} \end{pmatrix}\mathbf{M}(n,t) \tag{11.55}$$

From Equation 11.54, it follows that **M** satisfies the boundary condition

$$\mathbf{M}(0,t)=\begin{pmatrix} \delta(t) & 0 \\ 0 & \delta(t) \end{pmatrix} \tag{11.56}$$

The symmetry conditions of Equation 11.52 take the following forms in the time domain:

$$M_{11}(n,t)=M_{22}(n,-t)$$

and

$$M_{21}(n,t)=M_{12}(n,-t)$$

The equations in terms of **M** provide a transmission description of the medium. One can also consider a "scattering" description by writing

$$\begin{pmatrix} W_+(x,t) \\ W_-(x,t) \end{pmatrix}=\begin{pmatrix} T_-(x,t) & R_+(x,t) \\ R_-(x,t) & T_+(x,t) \end{pmatrix}*\begin{pmatrix} W_+(0,t) \\ W_-(0,t) \end{pmatrix} \tag{11.57}$$

where $T_\pm$, and $R_\pm$ denote the transmission and reflection coefficients, respectively, and the subscripts – resp. + correspond to a left-incident resp. a right-incident wave. The Fourier-transformed (in time) matrix of transformation in Equation 11.57 is known as the "scattering matrix" associated with the medium over the length [0,x]. The **S** matrix is related to the transition matrix **M** through the Fourier transform

$$\hat{\mathbf{S}}(x,\omega)=\left\{\begin{pmatrix} 1 & 0 \\ 0 & 0 \end{pmatrix}\hat{\mathbf{M}}(x,\omega)+\begin{pmatrix} 0 & 0 \\ 0 & 1 \end{pmatrix}\right\}\left\{\begin{pmatrix} 0 & 0 \\ 0 & 1 \end{pmatrix}\hat{\mathbf{M}}(x,\omega)+\begin{pmatrix} 1 & 0 \\ 0 & 0 \end{pmatrix}\right\}^{-1}$$

The solutions of Equation 11.55 and the boundary condition of Equation 11.56 can be written as[9]

$$M_{11}(x,t)=\delta(x-t)+m_{11}(x,t)\overline{H}(x,t)$$

and (11.58)

$$M_{21}(x,t)=m_{21}(x,t)\overline{H}(x,t)$$

where $\overline{H}(x,t) = H(t + x) - H(t - x)$, and the vector $\mathbf{m} = (m_{11}, m_{21})^T$ satisfies the equation

$$\mathbf{m}'(x,t)=\mathbf{A}\mathbf{m}(x,t),\ -x\le t\le x$$

The region $-x \le t \le x$ follows from the region of nonzero $\overline{H}(x,t)$, **A** being the matrix in Equation 11.55. Only the components m_{11} and m_{21} have been considered since the other two, m_{12} and m_{22}, are related to **m** via the symmetry properties of **M**. Equations 11.58 imply that the **m** vector has a support over $t \subset [-x,x]$ and, hence, is noncausal. Equation 11.54 then shows that the matrix **M** has support over $t \subset [-x,x]$. This, in turn, implies that the elements of the **S** matrix follow the causality conditions below.

$$R_+ = R_- = 0,\ t < 0$$

and

$$T_+ = T_- = 0,\ t < x$$

Let us now return to Equation 11.54 and consider n instead of x. Adding the two equations yields

$$\hat{W}(x,t) = W_+(0,t) * \psi(n,t) + W_-(0,t) * \psi(n,-t) \tag{11.59}$$

where

$$\hat{W} = W_+ + W_-$$

$$\psi(n,t) = M_{11}(n,t) + M_{21}(n,t)$$

and

$$\psi(n,-t) = M_{22}(n,t) + M_{12}(n,t)$$

From the causality conditions, we have

$$W_+(n,t) = 0,\ t < n$$

and

$$W_-(n,t) = 0, \qquad t < n+1 \tag{11.60}$$

Writing out the convolutions in Equation 11.59 explicitly and using the causality condition (11.60), we obtain

$$\begin{aligned} \sum_{-n}^{n} W_+(0,t-i)\psi(n,i) + \sum_{-n}^{n} W_-(0,t-i)\psi(n,-i) &= 0 \qquad && t < n \\ &= W_+(n,n) && t = n \end{aligned}$$

This equation can be written in the matrix notation as

$$\left\{\mathbf{L}_T\left[\mathbf{W}_+^n\right] + \mathbf{L}_T\left[\mathbf{W}_-^n\right]\bar{\mathbf{I}}\right\}\psi^n = \begin{pmatrix} 0 \\ 0 \\ \\ \\ W_+(n,n) \end{pmatrix} \tag{11.61}$$

where $\mathbf{L}_T$ is a lower triangular Toeplitz matrix whose first column is the vector $\mathbf{W}^n_\pm = [W_\pm(0,0), W_\pm(0,1) \ldots W_\pm(0,n)]T$, ψ^n is a column vector with dimension (n + 1) with all nonzero elements of the sequence $\psi(n,t)$ from the span $t \subset [-n,n]$ in the increasing order of the time index, and **I** is a matrix with ones on the antidiagonal and acts as a time-reversal operator. Equation 11.61 is the general equation of inverse scattering in a discrete system, of which the lossless, nonuniform transmission line is a prototype.

At this point, it is interesting to compare the determination of the reflection coefficients with that obtained by the direct methods. Again, from the results of Section III.A.2, we have

$$r_{n+1} = \frac{W_-(n,n+1)}{W_+(n,n)}$$

The time appears as n + 1 in the numerator as a consequence of causality. In terms of the results of this section, this expression is given by (Equation 11.54)

$$r_{n+1} = \frac{\left[W_-(0,n+1)*M_{11}(n,-n-1) + W_-(0,n+1)*M_{11}(n,-n-1)\right]}{W_+(n,n)}$$

where

$$W_+(n,n) = \prod_{n}^{1}\left(1 - r_i^2\right)^{\frac{1}{2}} W_+(0,0)$$

This demonstrates explicitly how the determination of the reflection coefficients by the indirect methods involves computations of convolution integrals. The number of computations involved in retrieving a transmission line up to a depth of N which require linear matrix equations of size $(N \times N)$ is of the order of $O(N^3)$ if solutions are obtained by a straightforward way such as by Gaussian elimination. However, as shown in Reference 8, the number of computations is speeded to $O(N^2)$ by using the relation derived earlier, namely,

$$\tilde{\mathbf{M}}(n,z) = \mathbf{T}(n)\tilde{\Delta}(z)\tilde{\mathbf{M}}(n-1,z)$$

For further details of the numerical aspects, we refer the reader to the reference cited above.

4. Special Forms of Linear Matrix Equations

Equation 11.61 is a general linear matrix equation that describes the inverse scattering in the transmission line. Bruckstein and Kailath have shown how to specialize this equation to variously tailored scattering data pairs $W_\pm(0,t)$. Let us note that due to the causality conditions, namely, $W_+(n,t) = 0$, $t < n$ and $W_-(n,t)$, $t < n+1$,

$$W_+(n,n) = \prod_{i=n}^{1}\left(1 - r_i^2\right)^{\frac{1}{2}} W_+(0,0), \qquad t = n$$

and the general formula (11.61) then reduces to

$$\left\{\mathbf{L}_T\left[\mathbf{W}_+^n\right]+\mathbf{L}_T\left[\mathbf{W}_-^n\right]\bar{\mathbf{I}}\right\}\psi^n=\begin{pmatrix}0\\0\\\cdot\\\cdot\\\prod_{i=n}^{1}\left(1-r_i^2\right)^{\frac{1}{2}}W_+(0,0)\end{pmatrix} \tag{11.62}$$

The transmission line under consideration is a discrete, time-invariant linear system and, hence, operates on the propagating signals linearly. It is, therefore, to be expected that the voltage response v(0,t) must be linear in the current input i(0,t). This implies that if i(0,t) is piecewise constant, so will be v(0,t), and if i(0,t) is an impulsive forcing current, the response v(0,t) will also consist of impulses. With this in mind, let us now consider some specially designed scattering data pairs.

Case 1: The Discrete Gel'fand-Levitan Equation

Consider a delta function as the current input. This leads to the scattering data

$$i(0,t)=\delta(t) \text{ and } v(0,t)=\delta(t)+2\sum_1^\infty h_i\delta(t-i)$$

where the h_i are the weights of the impulse sequences. Correspondingly, the right- and left-propagating wave variables are

$$W_+(0,t)=\delta(t)+\sum_1^\infty h_i\delta(t-i)$$

and

$$W_-(0,t)=\sum_1^\infty h_i\delta(t-i)$$

respectively. This follows from the definitions (Equations 11.35a and 11.37a) of $W_\pm$ and from the fact that Z(x=0) = 1. Substituting these variables in Equation 11.62, we obtain

$$\left\{\mathbf{L}_T\left[\mathbf{E}+\mathbf{H}^n\right]+\mathbf{L}_T\left[\mathbf{H}^n\right]\bar{\mathbf{I}}\right\}\psi^n=\begin{pmatrix}0\\0\\\cdot\\\cdot\\\prod_{i=n}^{1}\left(1-r_i^2\right)^{\frac{1}{2}}\end{pmatrix} \tag{11.63}$$

where

$$\mathbf{H}^n=\left[0,h_1,h_2,\ldots,h_n\right]^T$$

and

$$\mathbf{E}=\left[1,0,0,\ldots,0\right]^T$$

The coefficient matrix $\mathbf{L}_T[\mathbf{E}+\mathbf{H}^n]$ is a lower triangular Toeplitz matrix, whereas $\mathbf{L}_T[\mathbf{H}^n]\bar{\mathbf{I}}$ is a right triangular Hankel matrix. Thus, the coefficient matrix of Equation 11.63 is Toeplitz + Hankel. Equation 11.63 is the discrete version of the Gel'fand-Levitan equation.

It was seen earlier that the continuous Gel'fand-Levitan integral equation had a completely symmetric kernel, whereas the discrete version (Equation 11.63) does not. However, a slight modification of this equation will lead to a completely symmetric equation with a completely symmetric kernel. Toward this, we discretize the Gel'fand-Levitan equation not for ψ^n,but, rather, for the linear combination $\psi^n + \bar{\mathrm{I}}\psi^n$ which, in turn, is the same as $\{\psi(n,t) + \psi(n,-t)\}$. We also note the identity

$$\left(\mathbf{I}+\bar{\mathbf{I}}\right)\left(\mathbf{I}+\bar{\mathbf{I}}\right)=2\left(\mathbf{I}+\bar{\mathbf{I}}\right) \tag{11.64}$$

Multiplying both sides of Equation 11.64 by $\bar{\mathbf{I}}$, adding the direct and the time-reversed equations, and then using the above identity, we obtain the completely symmetric Gel'fand-Levitan equation for the vector $\psi^n + \bar{\mathbf{I}}\psi^n$.

$$\left\{\mathbf{I}+\left(\tfrac{1}{2}\right)\left(\mathbf{I}+\bar{\mathbf{I}}\right)\mathbf{L}_T\left(\mathbf{H}^n\right)\left(\mathbf{I}+\bar{\mathbf{I}}\right)\right\}\left(\psi^n+\bar{\mathbf{I}}\psi^n\right)=\begin{pmatrix}\prod\limits_{i=n}^{1}\left(1-r_i^2\right)^{\frac{1}{2}}\\ 0\\ 0\\ \cdot\\ \cdot\\ \prod\limits_{i=n}^{1}\left(1-r_i^2\right)^{\frac{1}{2}}\end{pmatrix} \tag{11.65}$$

Case 2: The Discrete Marcenko Equation

Let us now deal directly with the wave variables $W_\pm$ instead of v and i. Consider that $W_+(0,t) = \delta(t)$. In the geophysical context, such a source can be generated through explosive means. Then we have as the scattering data pair

$$W_+(0,t)=\delta(t)$$

and

$$W_-(0,t)=\sum_1^\infty h_i\delta(t-i)$$

With these data, Equation 11.61 results in the discrete Marcenko equation:

$$\left\{\mathbf{L}_T\left[\mathbf{H}^n\right]\bar{\mathbf{I}}+\mathbf{I}\right\}\psi^n=\begin{pmatrix}0\\ 0\\ \cdot\\ \cdot\\ \prod\limits_{i=n}^{1}\left(1-r_i^2\right)^{\frac{1}{2}}\end{pmatrix} \tag{11.66}$$

The coefficient matrix is of Hankel structure.

Case 3: The Krein Equation

Still another well-known equation of inverse scattering can be obtained which possesses a purely symmetric Toeplitz structure. It uses the same scattering data as that in Case 1. However, we write Equation(s) 11.55 in the form

$$W_+(n,t) = M_{11}(n,t) + \left[\Sigma h_i \delta(t-i)\right] * \left[M_{11}(n,t) + M_{12}(n,t)\right]$$

and

$$W_-(n,t) = M_{21}(n,t) + \left[\Sigma h_i \delta(t-i)\right] * \left[M_{11}(n,t) + M_{12}(n,t)\right]$$

Define:

$$\phi(n,t) = M_{11}(n,t) + M_{12}(n,t)$$

$$= M_{22}(n,-t) + M_{21}(n,-t)$$

Changing t to –t in the second of the above equations and adding the Krein equation results in

$$\left\{\mathbf{I} + \mathbf{L}_T\left[\mathbf{H}^n\right] + \bar{\mathbf{I}}\mathbf{L}_T\left[\mathbf{H}^n\right]\bar{\mathbf{I}}\right\}\phi^n = \begin{pmatrix} 0 \\ 0 \\ \cdot \\ \cdot \\ \prod_{i=n}^{1}\left(1-r_i^2\right)^{\frac{1}{2}} \end{pmatrix}$$

The matrix is symmetric Toeplitz.

5. Connection With Continuous Inverse Scattering

Let $W_+(0,t)$ contain a leading impulse. Then we have the following probing waves:

$$W_+(0,t) = \delta(t) + w_+(0,t)H(t)$$

$$W_-(0,t) = w_-(0,t)H(t) \tag{11.67}$$

$W_\pm$ are some piecewise continuous formations. These forms follow from the causality conditions. This means that we can write in general:

$$W_\pm(x,t) = \delta(t-x) + w_\pm(x,t)H(t-x), \;\; \delta(t-x)$$

only for W_+ and does not appear in W_-. Considering that $\mathbf{M}(x,t)$ has support over $[-x,x]$ and using the causality for $W_\pm(x,t)$, namely, $W_\pm(x,t) = 0$, $t < x$, the general Equations 11.54 can be written as

$$\int_{-x}^{t} W_+(0,t-t')M_{11}(x,t')dt' + \int_{-x}^{t} W_-(0,t-t')M_{12}(x,t')dt' = 0 \tag{11.68}$$

and

$$\int_{-x}^{t} W_{+}(0,t-t')M_{21}(x,t')dt' + \int_{-x}^{t} W_{-}(0,t-t')M_{12}(x,t')dt' = 0$$

Using the expressions for M_{11} and M_{21} given by Equation 11.58 and from the symmetry relation for **M** (see equations following Equation 11.56, Equation 11.68 can be reduced to

$$\int_{-x}^{t} W_{+}(0,t-t')m_{11}(x,t')dt' + \int_{-t}^{x} W_{-}(0,t+t')m_{21}(x,t')dt' = 0$$

and (11.69)

$$W_{-}(0,t+x) + \int_{-t}^{x} W_{-}(0,t+t')m_{11}(x,t')dt' + \int_{-x}^{t} W_{+}(0,t-t')m_{21}(x,t')dt' = 0$$

Finally, from the probing wave expressions of Equation 11.67, we obtain two coupled Fredholm integral equations of the second kind for $m_{11}(x,t)$ and $m_{21}(x,t)$:

$$m_{11}(x,t) + \int_{-x}^{t} w_{+}(0,t-t')m_{11}(x,t')dt' + \int_{-t}^{x} w_{-}(0,t+t')m_{21}(x,t')dt' = 0$$

and (11.70)

$$w_{-}(0,t+x) + m_{21}(x,t) + \int_{-t}^{x} w_{-}(0,t+t')m_{11}(x,t')dt' + \int_{-x}^{t} w_{+}(0,t-t')m_{21}(x,t')dt' = 0$$

These equations are the continuous forms of the discrete Equation 11.61 and must be solved for $m_{11}(x,t)$ and $m_{21}(x,t)$.

Define:

$$K(x,t) = m_{11}(x,t) + m_{21}(x,t)$$

Adding the two integral equations (11.70) and again using the causality requirements that $w_{\pm}(x, t) = 0,\ t < x$, we obtain

$$w_{-}(0,t+x) + K(x,t) + \int_{-x}^{x}\left[w_{+}(0,t-t') + w_{-}(0,t+t')\right]K(x,t')dt = 0 \qquad (11.71)$$

Consider the boundary condition

$$w_{+}(0,t) = w_{-}(0,t) = h(t)$$

and the symmetrized kernel

$$K_s(x,t) = \left(\frac{1}{2}\right)\left[K(x,t) + K(x,-t)\right]$$

Bruckstein, Levy, and Kailath[8] show that Equation 11.71 reduces to

$$K_s(x,t) + \overline{H}(x \pm t) + \int_{0}^{x} H(|t \pm t'|)K_s(x,t')dt' = 0 \qquad (11.72)$$

where $\overline{H}(y \pm z) = (1/2)[h(z + y) + h(z - y)]$, and $0 < t < x$. Equation 11.72 is the equation of Gel'fand and Levitan. The boundary condition above implies that the end x = 0 is terminated by a perfect reflector. This is equivalent to considering the problem on the half-line, which is what the equation of Gel'fand and Levitan describes.

If, on the other hand, the boundary conditions are those of Case 2 of Section III.A.4, namely,

$$w_+(0,t) = 0$$

and

$$w_-(0,t) = R(t)H(t)$$

R(t) being the coefficient of reflection, then Equation 11.71 is the classical equation of Marcenko. This is not the condition of perfect reflection and, hence, allows the propagation of the left-going wave. The problem is, therefore, on the entire line which constitutes the domain of Marcenko's equation.

Let L(x,t) be the continuous version of ψ(n,t) of Case 3 of the previous section. Define:

$$L(x,t) = m_{11}(x,t) + m_{12}(x,t) \tag{11.73}$$

Let the boundary conditions be that of Case 1. Then replacing t by – t in the second of Equation(s) 11.70 and adding it to the first yields the equation of Krein:

$$h(x-t) + L(x,t) + \int_{-x}^{x} h(|t-t'|)L(x,t')dt' = 0$$

Also,

$$K_s(x,t) = L(x,t) + L(x,-t)$$

6. Inverse Solutions

In this section, we show how the inverse problem is solved. Substituting Equation 11.58 into Equation 11.55 and after some algebra, we find that

$$-r(x)m_{21}(x,x-) = [m_{11}(x,x-)]'$$

$$r(x) = -2m_{21}(x,x-) \tag{11.74}$$

and

$$r^2(x) = 2[m_{11}(x,x-)]'$$

The solutions of Equation(s) 11.70 then reconstruct the medium by exploiting the Equation(s) 11.74. The inverse problem is then solved.

The equation of motion (11.36) describes the transmission of waves through an infinitesimally small section of the medium. Equation 11.53, which is the basic equation for indirect inversion, relates the waves at a depth x to the waves at the boundary and is a description of aggregated layers over the interval [0,x]. Since Equation 11.36 is the point of departure for the direct method, it is clear that the two methods are complementary to each other. The direct method is a layer-peeling method in which one layer is determined at a time and subsequently peeled off. The indirect methods, on the other hand, can be said to be "layer adjoining" methods in which the medium is built up recursively for increasing depth by adjoining the successively identified layers. These two methods can be considered to be duals of each other.

B. BROWN-WILCOX THEORY OF INVERSE PROBLEM OF NONUNIFORM TRANSMISSION LINES

1. Eigenvalue Problem for the Transmission Line and Formulation of the Inverse Problem

The final topic of this chapter concerns an inverse reflection problem of electric wave propagation on a lossless, nonuniform, continuous transmission line. The theory to be presented was developed by Brown[29] and Wilcox[30,31] as early as 1965. This rigorous and sophisticated theory has, however, received little attention. The theory is elaborate and can only be given a summary presentation here. The interested reader, however, may easily fill up the blanks from the references cited above.

The transmission line equation is

$$\mathbf{A}(x,t)\mathbf{a}(x,t) = 0 \tag{11.76}$$

where the matrix $\mathbf{A}(x,t)$ is

$$\mathbf{A}(x,t) = \begin{pmatrix} C(x)\dfrac{d}{dt} & \dfrac{\partial}{\partial x} \\ \dfrac{\partial}{\partial x} & L(x)\dfrac{d}{dt} \end{pmatrix}$$

and the column vector $\mathbf{a}(x,t)$ is $[E(x,t),I(x,t)]^T$. L and C are the inductance and the capacitance, respectively, and E and I the voltage and the current in the line, respectively. The line is considered to be inhomogeneous on a compact subinterval [0,X], the rest being homogeneous. Quantities pertaining to the homogeneous portion of the line will be indicated by a subscript 0. Thus

$$C(x) = C_0(x) + q_1(x)$$

$$L(x) = L_0(x) + q_2(x)$$

the inhomogeneities $q_1(x)$ and $q_2(x)$ being identically zero outside the interval [0,X]. The capacitance and the inductance are considered to be real valued, positive functions which are twice continuously differentiable, that is, $L,C \subset C_2(R^+)$, and are bounded away from zero and infinity for all finite x. The impedence of the line is given by $z(x) = [L(x)/C(x)]^{1/2}$ and similarly for z_0. A function j(x) called the "velocity function" is defined as

$$j(x) = \int_0^x \left[C(x')L(x')\right]dx'$$

The reciprocal of this function [1/j(x)] is a measure of the local characteristic velocity.

The boundary condition I(0,t) = 0 is imposed. Moreover, the initial state $\mathbf{a}(x,t = 0)$ is required to satisfy the energy requirement

$$\int_0^\infty \left[C(x)\left|E(x,0)\right|^2 + L(x)\left|I(x,0)\right|^2\right]dx = \left\|\mathbf{a}(x,0)\right\|_1^2 < \infty$$

This is the definition of the norm of a Hilbert space H_1. That is, if $\mathbf{f} = (f_1,f_2) \subset H_1$, then

$$\int_0^\infty \left[L(x)|f_1|^2 + C(x)|f_2|^2\right]dx = \|f\|_1^2$$

An inner product is defined on H_1 as

$$(f,g)_1 = \int_0^\infty \left(Lf_1g_1^* + Cf_2g_2^*\right)dx \tag{11.77}$$

the * denoting, as usual, the complex conjugate.

Fourier transforming Equation 11.76 yields a spectral problem, which can be written as

$$\mathbf{T}(x,\lambda)\mathbf{y}(x,\lambda) = 0 \tag{11.78}$$

$$y_1(0) = 0 \tag{11.78a}$$

where the matrix $\mathbf{T}$ is

$$\mathbf{T}(x,\lambda) = \begin{pmatrix} \dfrac{\partial}{\partial x} & i\lambda C(x) \\ i\lambda L(x) & \dfrac{\partial}{\partial x} \end{pmatrix}$$

and $\mathbf{y}(x,\lambda) = (y_1,y_2)^T = [I(\text{Re}\exp(i\lambda t))^{-1}, E(\text{Re}\exp(i\lambda t))^{-1}]^T$. Attention will be focused on this spectral problem. $\lambda = \omega$ will be considered to be real unless otherwise specified. Equation 11.78 can be rewritten alternately as

$$\mathbf{T}(x,\lambda)\mathbf{y}(x,\lambda) = \lambda\mathbf{y}(x,\lambda) \tag{11.79}$$

where

$$\mathbf{T} = \begin{pmatrix} 0 & \left(\dfrac{i}{L}\right)\dfrac{\partial}{\partial x} \\ \left(\dfrac{i}{C}\right)\dfrac{\partial}{\partial x} & 0 \end{pmatrix} \tag{11.79a}$$

Equation 11.79 is in the form of an eigenvalue equation. It will be shown momentarily that the operator $\mathbf{T}$ possesses only a continuous spectrum. Its eigenfunctions are, therefore, generalized eigenfunctions (see Chapter 9) which do not belong to the Hilbert space H_1. However, as in Chapter 9, these can be used in expanding a function in H_1. This results in an expansion theorem.

Let $R(\omega)$ be the reflection coefficient. The scattering data consist of $R(\omega)$, the velocity function $j(x)$, and the parameters C_0 and L_0 of the uniform part of the line. The inverse problem consists of determining the capacitance and the inductance functions $C(x)$ and $L(x)$ from this data. It should be mentioned that if $j(x)$ is not known then it is not possible to determine C and L separately. In that case only the line impedence $z(x)$ can be recovered.

The basic idea behind the Brown-Wilcox theory can be described as follows. Based on the expansion theorem, a transformation operator $\mathbf{U}$ is obtained such that $\mathbf{TU} = \mathbf{UT}_0$, $\mathbf{T}_0$ being the unperturbed operator corresponding to $\mathbf{T}$. The operators $\mathbf{T}$ and $\mathbf{U}$ are then used to derive a differential equation connecting the line impedence function $z(x)$ to certain kernel functions.

The latter are then shown to possess integral representations solving which the kernel functions are obtained in terms of the data prescribed. Interestingly enough, the integral equations for the kernel functions turn out to be none other than the equations of Gel'fand and Levitan.

Let us begin by establishing two basic properties of the operator **T**.

2. The operator T for the Transmission Line

Definition

Let **h** be an element of H_1 such that it is absolutely continuous, possesses a derivative z′ which is also in H_1, and, in addition, $f_1(0) = 0$. This class of functions will be denoted by D_T.

D_T is a dense set in H_1.

Theorem 11.6

The transformation **T** is self-adjoint in H_1.

Proof

The proof consists of two steps. First, the self-adjointness is proved on the elements $\subset D_T$. The proof is then extended to the elements in H_1. Let $\mathbf{h} \subset D_T$. It then follows that limit $x \to \infty$, $\mathbf{h} \to 0$. Furthermore, by virtue of the conditions imposed on the functions L and C, it also follows that $\mathbf{Th} \subset H_1$, $\mathbf{h} \subset D_T$.

Now consider two functions **f** and **g** in D_T. If b is an arbitrary positive number, then using the definition of the inner product (11.77) on H_1 we obtain

$$i\int_0^b \left(f_2' g_1^* + f_1' g_2^*\right)dx = -i\int_0^b \left(f_2 g_1'^* + f_1 g_2'^*\right)dx + i\left(f_1 g_2^* + f_2 g_1^*\right)\Big|_0^b$$

Since $\mathbf{f},\mathbf{g} \subset D_T$, then taking the limit $b \to \infty$ yields

$$(\mathbf{Tf},\mathbf{g})_1 = (\mathbf{f},\mathbf{Tg})_1 \tag{11.80}$$

Let $\mathbf{g},\mathbf{G} \subset H_1$. Assume that for $\mathbf{f} \subset D_T$ and for all $\mathbf{g},\mathbf{G} \subset D_T$, the following identity is satisfied, namely

$$(\mathbf{Tf},\mathbf{g})_1 = (\mathbf{f},\mathbf{g})_1 \tag{11.81}$$

Let $\mathbf{f} \subset D_T$ is such that $f_1 = 0$ and $f_2 = 0$ if $x > b$, b arbitrary. Equation 11.81 with the inner product as defined by Equation 11.77 then leads to the fact that g_1 is absolutely continuous, $g_1(0) = 0$ and that $G_2 = i\, g_1'(x)/C(x)$ for $0 < x < \infty$. Similarly, applying Equation 11.81 to the elements $\mathbf{f} \subset D_T$ such that $\mathbf{f} = 0$, $x > b$, one obtains that g_2 is absolutely continuous and $G_1 = i g_2'(x)/L(x)$, $0 < x < \infty$. Combining these results then yields that $\mathbf{G} = \mathbf{T}^+\mathbf{g}$, $\mathbf{g} \subset D_T$.

This result combined with Equation 11.81 shows that $\mathbf{T}^+ \subset \mathbf{T}$. On the other hand, Equation 11.80 implies that $\mathbf{T} \subset \mathbf{T}^+$. Hence, $\mathbf{T} \subset \mathbf{T}^+$ and also $\mathbf{T}^+ \subset \mathbf{T}$. Therefore, $\mathbf{T} = \mathbf{T}^+$ and the theorem is proved.

Next it is shown that **T** possesses only a continuous spectrum for $\lambda = \omega$. Let $f_+(x,\lambda)$ and $f_-(\lambda)$ be two solutions of Equation 11.78 such that for $x > X$,

$$f_\pm(x,\lambda) = \left(f_{\pm 1}(x,\lambda), f_{\pm 2}(x,\lambda)\right)$$

$$= \exp\left(\frac{\pm i\lambda x}{c_0}\right), \mp z_0 \exp\left(\frac{\pm i\lambda x}{c_0}\right)$$

It is not difficult to see that the Wronskian $W[f_+,f_-] = 2z_0 \neq 0$. Hence f_+ and f_- are two linearly independent solutions of Equation 11.78. Any other solution can, therefore, be represented as a linear combination of f_+ and f_-. Let $\psi(x,\lambda) = (\psi_1,\psi_2)$ be the solution vector of the system 11.78 and 11.78a. By definition, $\psi_1(0,\lambda) = 0$. Let $\psi_2(0,\lambda) = i[z(0)]^{1/2}$. By making use of the Wronskians of various sets of solutions, it can be shown that

$$2iz_0\left(\frac{\psi(x,\lambda)}{z(0)^{\frac{1}{2}}}\right) = f_{-1}(0,\lambda)f_+(x,\lambda) - f_{+1}(0,\lambda)f_-(x,\lambda) \tag{11.82}$$

The solutions f are plane waves. Then for $\lambda = \omega$, and by virtue of the conditions imposed on the functions L and C, it is seen that f cannot vanish for $\lambda = \omega$. In the absence of any loss mechanism, the waves cannot also vanish in the interior. Hence is the following

Lemma 11.3

$f(0,\lambda)$ cannot vanish for $\lambda = \omega$.

Theorem 11.7

The operator **T** has no eigenvalues.

Proof

Using this result and Equation 11.82, one obtains

$$L_0|\psi_1|^2 + C_0|\psi_2|^2 = C_0 z(0)\frac{\left[|f_{-1}(0,\omega)|^2 + |f_{+1}(0,\omega)|^2\right]}{2}, \quad x > X$$

By virtue of Lemma 11.3 it follows that $||\psi_1||_1 = \infty$. Hence for $\lambda = \omega$, no solution of the system (11.78-11.78a) lies in H_1.

Theorem 11.8

All ω are in the continuous spectrum of the operator **T**.

Proof

The objective is to show that for $\lambda = \omega$, the inverse operator $(\mathbf{T} - \lambda\mathbf{I})^{-1}$ cannot exist as a bounded transformation if $\lambda = \omega$. Let $(\mathbf{T} - \omega\mathbf{I})^{-1}$ exist as a bounded transformation in H_1. Then there exists an element $\mathbf{Y} \subset H_1$ for each element $\mathbf{g} \subset H_1$ such that

$$(\mathbf{T} - \omega\mathbf{I})\mathbf{Y} = \mathbf{g} \tag{11.83}$$

It is, therefore, sufficient to show that there exists an element $\mathbf{g} \subset H_1$ for which no $\mathbf{Y} \subset H_1$ can be found that satisfies Equation 11.83.

Let $\mathbf{g}$ be such that $g_1 = 0$ and g_2 is twice continuously differentiable vanishing at $x = 0$ and for $x > x_1 > X$. For this $\mathbf{g}$, Equation 11.83 becomes

$$iY_2' - \omega L Y_1 = 0$$

$$iY_1' - \omega C Y_2 - C g_2 = 0$$

$$Y_1(0) = 0$$

A solution of this system is (using the method of variation of parameters)

$$L|Y_1(x)|^2 + C|Y_2(x)|^2 = C_0\left[|m_1|^2 + |m_2|^2\right]/2, \qquad x > x_1 \tag{11.84}$$

where

$$m_1 = \int_0^{x_1} Cf_{-2}g_2 dx - if_{-1}(0,\omega)Y_2(0)$$

and

$$m_2 = \int_0^{x_1} Cf_{+2}g_2 dx + if_{+2}(0,\omega)Y_2(0)$$

By virtue of Lemma 11.3, there must exist at least one function g_2 in the chosen class for which m_1 and m_2 cannot vanish simultaneously. Otherwise, the Wronskian of $\mathbf{f}_+$ and $\mathbf{f}_-$ will vanish, resulting in a contradiction. Hence from Equation 11.84 it follows that $\|\mathbf{Y}\|_1 = \infty$ for this $\mathbf{g}$ in H_1 so that there is no $\mathbf{Y}$ in H_1 satisfying the identity 11.83. The theorem is, therefore, proved.

3. Expansion Theorem

The eigenfunctions of **T** are then generalized eigenfunctions which can be used in expanding a function in H_1. Similar expansions were encountered earlier in Chapter 9, where it was shown how generalized eigenfunctions could be incorporated into the framework of a Hilbert space. Moreover, in Chapter 3 we discussed Parseval's relation on the half-line and expansions of functions in $L_2(R^+)$ in terms of their generalized Fourier transforms — which were functions in $L_2(R)$ — with respect to the spectral function. Similar analysis is also applicable to the system 11.78-11.78a. However, modifications are to be made on account of three facts: (1) we are dealing with vector functions instead of the scalar functions of Chapter 3, (2) the definition of H_1 is different from that of L_2, and (3) the eigenfunctions are those of the system (11.78-11.78a) instead of a Sturm-Liouville problem. An expansion theorem taking these modifications into account is quite involved and has been extensively discussed by Wilcox[30,31] and Roos and Sangren.[32,33] In related problems of one-dimensional Dirac systems, extensive discussions of expansion theorems are due to Levitan and Sargsjan.[34] We refrain from discussing these details, which can be found in the references cited, and merely summarize the results.

Theorem 11.9 Expansion theorem for f ⊂ H_1 associated with the operator T

If $\mathbf{f} = (f_1,f_2) \subset H_1$, then

$$\lim_{\substack{\omega_1\to\infty \\ \omega_2\to\infty}} \int_0^{\infty}\left[L(x)\left|f_1(x) - \left(\frac{1}{\pi}\right)\int_{\omega_1}^{\omega_2}\psi_1(x,\omega)\left\{\frac{M(\omega)}{N(\omega)}\right\}d\omega\right|^2 + \right.$$

$$\left. +C(x)\left|f_2(x) - \left(\frac{1}{\pi}\right)\int_{\omega_1}^{\omega_2}\psi_2(x,\omega)\left\{\frac{M(\omega)}{N(\omega)}\right\}d\omega\right|^2\right]dx = 0 \tag{11.85}$$

where the functions M(ω) and N(ω) are defined as

$$\lim_{r\to\infty}\int_{-\infty}^{\infty}\left|M(\omega)-\int_0^{x_1}\left[\frac{\{L(x')f_1(x')\psi_1(x',\omega)-C(x')f_2(x')\psi_2(x',\omega)\}}{N(\omega)}\right]dx'\right|^2 d\omega=0,$$

$$\text{as } x_1\to\infty,\qquad N(\omega)=\left[\mu^2+\upsilon^2\right]^{\frac{1}{2}},$$

$$\mu(\omega)=1+\int_0^{\infty}\psi_1(x',\omega)\left[\frac{F(x')}{\omega}\right](\cos\omega j(x'))z^{\frac{1}{2}}(x')dx'$$

and

$$\upsilon(\omega)=-\int_0^{\infty}\psi_1(x',\omega)\left[\frac{F(x')}{\omega}\right](\sin\omega j(x'))z^{\frac{1}{2}}(x')dx'$$

The function F(s) is given by

$$F(s)=\left(\frac{1}{2}\right)\left[z^{\frac{1}{2}}(s)\left(\frac{dj}{ds}\right)\right]^{-1}$$

The corresponding Parseval identity is obtained as

$$\int_0^{\infty}\left[L|f_1|^2+C|f_2|^2\right]dx=\left(\frac{1}{\pi}\right)\int_{-\infty}^{\infty}|M|^2 d\omega$$

The function $M \subset L_2(R)$. This result is similar in form to that derived in Chapter 3. M will be referred to as the generalized transform of **f**.

4. The Transformation Operator

We now come to the transform operator **U**. Before proceeding, some properties of the solution ψ will be summarized, the details of which can be found in References 29 and 32 These are stated in the following lemma.

Lemma 11.4

Let λ = u + iv. Then $\psi_1(x,u)$ is real and an odd function of u while $\psi_2(x,u)$ is imaginary and even in u. The forms of these functions are

$$\psi_1(x,\lambda)=\left[\frac{1}{z(x)^{\frac{1}{2}}}\right]\sin\lambda j(x)+d_1(x,\lambda)$$

and

$$\psi_2(x,\lambda)=\left[iz(x)^{\frac{1}{2}}\right]\cos\lambda j(x)+d_2(x,\lambda)$$

where

$$d_m = O\left(\frac{(\exp(vj(x))}{|\lambda|}\right), \quad \lambda \subset C_+$$

Furthermore,

$$\int_{-\infty}^{\infty}|d_m(x,\lambda)|^2 = O(\exp(2vj(x)), \quad \lambda \subset C_+$$

With this lemma, let us return to Equation 11.85 of Theorem 11.9 and define

$$G(\omega) = \int_0^{\infty}[L(x')g_1(x')\psi_1(x',\omega) - C(x')g_2(x')\psi_2(x',\omega)]dx' \tag{11.86}$$

Since $M(\omega) \subset L_2(R)$ and $N(\omega)$ is bounded, then it follows that for $\mathbf{g} \subset H_1$, $G(\omega) \subset L_2(R)$ being the function $M(\omega)N(\omega)$. Equation 11.86 then defines an operator

$$\mathbf{P}(\mathbf{g}) = G(\omega): \mathbf{g} \subset H_1 \text{ and } G(\omega) \text{ as defined by Equation 11.86}$$

The equality is in the $L_2(R)$ sense. The adjoint operator $\mathbf{P}^+$ is then defined by

$$\left(\mathbf{P}(\mathbf{g}), G^+\right)_{L_2(R)} = \left(\mathbf{g}, \mathbf{P}^+G^+\right)_1$$

such that it holds for all $\mathbf{g} \subset H_1$, G^+ is in $L_2(R)$, and $\mathbf{P}^+G^+ \subset H_1$. It then follows that

$$\mathbf{P}^+G^+ = \mathbf{g}^+$$

$$= \left(\frac{1}{\pi}\right)\left(\int_{-\infty}^{\infty}G^+(\omega)\psi_1^*(x,\omega)d\omega, -\int_{-\infty}^{\infty}G^+(\omega)\psi_2^*(x,\omega)d\omega\right) \tag{11.87}$$

Similarly for the unperturbed quantities

$$\mathbf{P}_0(\mathbf{g}) = G_0: \qquad G_0(\omega) = \int_0^{\infty}[L_0g_1\psi_{10} - C_0g_2\psi_{20}]dx \tag{11.88}$$

The integral converges with respect to L_2 norm.

The determination of the operator **U** will require the following two observations, First, we note that in view of the expression for **P**,$\mathbf{P}^+$ given by Equations 11.86 and 11.87, respectively, the expansion theorem can be expressed in the form $\mathbf{P}^+N^{-2}\mathbf{P} = \mathbf{I}$, the identity operator. The second fact which is proved in Reference 29 is the lemma given below.

Lemma 11.5

Every element $M(\omega)$ in $L_2(R)$ is the transform of some element in H_1 and also of some element in H_{10}.

With these facts, we consider an element $\mathbf{h} \subset H_1$ such that its transform is MN with respect to the solution vector ψ. Then one obtains the following relations

$$\mathbf{Ph} = (MN)N = MN^2$$

$$\mathbf{PTh} = \omega MN^2$$

$$\mathbf{P}^+\omega M = \mathbf{Th}$$

$$\mathbf{P}^+ M = \mathbf{h}$$

and hence the identity

$$\mathbf{P}^+\omega M = \mathbf{TP}^+ M \tag{11.89}$$

Now if $\mathbf{h}_0 \subset D_{T0}$ has the transform $M(\omega)$ with respect to the solution vector ψ_0, then

$$\mathbf{P}_0\mathbf{h}_0 = M(\omega) \text{ and } \mathbf{P}_0\mathbf{T}_0\mathbf{h}_0 = \omega M(\omega) \tag{11.90}$$

Using the relations given by 11.90 in Equation 11.89 yields

$$\mathbf{P}^+\mathbf{P}_0\mathbf{T}_0\mathbf{h}_0 = \mathbf{TP}^+\mathbf{P}_0\mathbf{h}_0$$

Defining $\mathbf{U} = \mathbf{P}^+\mathbf{P}_0$ then results in the identity

$$\mathbf{TU} = \mathbf{UT}_0$$

The operator $\mathbf{U} = \mathbf{P}^+\mathbf{P}_0$ is then the desired operator of transformation. This leads to the following theorem.

Theorem 11.10
The operator $\mathbf{P}^+\mathbf{P}_0$ is the transformation operator.

5. The Functional Form of the Transformation Operator

The functional form of the transformation operator is derived next. Brown and Wilcox obtain the functional form for the transformation operator **U** by first considering its action on a set of elements dense in H_{10}. The extension to H_{10} then follows by completion since the space H_{10} is complete. Let $\mathbf{g} = (g_1, g_2)$ be in the set of elements dense in H_{10}. Then

$$\mathbf{Ug} = \mathbf{P}^+\mathbf{P}_0[g_1, g_2]^T \tag{11.91}$$

Replacing Equations 11.87 and 11.88 for $\mathbf{P}^+$ and $\mathbf{P}_0$, respectively, in Equation 11.91, one obtains formally

$$\mathbf{Ug} = (f_1, f_2) \text{ with } f_1 = (f_{11}, f_{12}) \text{ and } f_2 = (f_{21}, f_{22})$$

Consider the f_{11} term which, after some significant algebra, reduces to the following expression, namely

$$f_{11}(x)=\int_0^{\infty}\left[F_{11}(x,x')g_1(x')+G_{11}(x,x')\right]dx'$$

where

$$F_{11}(x,x')=\left[\frac{z_0}{z(x)}\right]^{\frac{1}{2}}\delta\left[c_0 j(x)-x'\right]$$

and

$$G_{11}(x,x')=\left(\frac{1}{\pi}\right)\int_{-\infty}^{\infty}L_0 z_0^{-\frac{1}{2}}\sin\left(\frac{\omega X'}{c_0}\right)d_1(x,\omega)d\omega \tag{11.92}$$

By virtue of the properties of the solution ψ given in Lemma 11.4, Titchmarsh's theorem (Appendix, Chapter 8) can be applied to the function d_1. This results in the fact that $G_{11}(x,x') = 0$ if $j(x) < x'/c_0$. The expression for G_{11} is just the Fourier transform of the function d_1 which is in $L_2(R)$, again by virtue of Lemma 11.4. Since the Fourier transform of an L_2 function is also an L_2 function (Theorem 8.4, Chapter 8), it then follows that G_{11} is an L_2 function with respect to x'. The close similarity with the Levin transform of Chapter 8 is easily recognized. Indeed, G_{11} can be considered to be the Levin representation of the f_1 part of **Ug** which is in H_1. The function f_{12} involves the integral of the product $[\cos(\omega x'/c_0)\psi_1(x,\omega)]$ over ω. Since by Lemma 11.4 ψ_1 is odd in ω, this term disappears.

Combining all these results, one obtains the following expression for f_1

$$f_1(x)=\left[\frac{z_0}{z(x)}\right]^{\frac{1}{2}}g_1\left(c_0 j(x)\right)+\int_0^{c_0 j(x)}K_{11}(x,x')g_1(x')dx' \tag{11.93}$$

The analogous result for f_2 is

$$f_2(x)=\left[\frac{z(x)}{z_0}\right]^{\frac{1}{2}}g_2\left(c_0 j(x)\right)+\int_0^{c_0 j(x)}K_{22}(x,x')g_2(x')dx' \tag{11.94}$$

In Equations 11.93 and 11.94, G_{11} and G_{22} have been denoted by K_{11} and K_{22}. Finally, therefore, the functional form of the transformation operator is $\mathbf{U}\mathbf{g} = [f_1,f_2]$, with f_1 and f_2 given by Equations 11.93 and 11.94, respectively. These results will be summarized in the following theorem.

Theorem 11.11

If **g** is any element of H_{10}, then the transformation operator $\mathbf{U} = \mathbf{P}^{+}\mathbf{P}_0$ is represented by the integral operators 11.93 and 11.94 with the kernels $K_{11}(x,x')$ and $K_{22}(x,x')$ square integrable in x' for each x.

6. The Differential Equation for the Inverse Solution

The transformation operator **U** is next used in obtaining a differential equation for the impedence function z(x) in terms of functions determinable from the scattering data. This, therefore, solves the inverse problem formally. In order to derive the differential equation, consider $\mathbf{g} = (g_1, g_2) \subset \mathbf{T}_0$ such that $g_1(0) = 0$. It then follows that $K_{11}(0) = 0$ also. The difference $[\mathbf{TUg} - \mathbf{UT}_0\mathbf{g}\}$ is then calculated. After substantial algebra, the following expression is obtained, namely

$$\mathbf{TUg} - \mathbf{UT}_0\mathbf{g} = \Big(\int_0^{c_0 j(x)} iI_{11}(x,x')g_2(x')dx' + iR_{11}(x,x')g_2(c_0 j(x)),$$

$$\int_0^{c_0 j(x)} iI_{22}(x,x')g_1(x')dx' + iR_{22}(x,x')g_1(c_0 j(x))\Big) = 0$$

where

$$I_{11}(x,x') = \left[\frac{1}{L(x)}\right]K'_{22} + \left(\frac{1}{L_0}\right)\dot{K}_{11}$$

$$I_{22}(x,x') = \left[\frac{1}{C(x)}\right]K'_{11} + \left(\frac{1}{C_0}\right)\dot{K}_{22}$$

$$R_{11}(x,x') = \left[\frac{1}{L(x)z_0^{\frac{1}{2}}}\right]\left[z^{\frac{1}{2}}(x)\right]' + c_0\left[\frac{1}{z(x)}\right]K_{22}\{x, c_0 j(x)\} - \left(\frac{1}{L_0}\right)K_{11}\{x, c_0 j(x)\}$$

and

$$R_{22}(x,x') = \left[\frac{1}{C(x)z_0^{\frac{1}{2}}}\right]\left[\frac{1}{z^{\frac{1}{2}}(x)}\right]' + c_0 z(x)K_{11}\{x, c_0 j(x)\} - \left(\frac{1}{C_0}\right)K_{22}\{x, c_0 j(x)\}$$

"." means differentiation with respect to the variable x'.

Brown and Wilcox then show that for $x' < c_0 j(x)$, the functions I_{11} and I_{22} vanish and that $R_{11}g_2 = R_{22}g_1 = 0$. However, since this is true for all $\mathbf{g}$ in the domain of $\mathbf{T}_0$, it follows that $R_{11} = R_{22} = 0$. Further reduction of these functions finally yields the differential equation, which is given by

$$\left[\frac{1}{c_0 L(x)}\right]z'(x) + \left[\frac{z(x)}{z_0}\right]^{\frac{1}{2}} K_{11}(x, c_0 j(x)) - \left[\frac{z(x)}{z_0}\right]^{-\frac{1}{2}} K_{22}(x, c_0 j(x)) = 0 \quad (11.95)$$

Define the following substitutions

$$p = c_0 j(x)$$

$$M_1(p, x') = z(x)^{\frac{1}{2}} K_{11}(x, x')$$

and

$$M_2(p,x') = \left(\frac{1}{z(x)}\right)^{\frac{1}{2}} K_{22}(x,x')$$

In terms of these changes of variables, Equation 11.95 reduces to

$$[\log z(p)]' = \left(\frac{2}{z_0^{\frac{1}{2}}}\right)[M_1(p,p) - M_2(p,p)] \tag{11.96}$$

The prime in Equation 11.96 refers to differentiation in the variable p. Equation 11.95 or 11.96 is the basic differential equation governing the inverse solution. It should be mentioned that the derivation of Equation 11.95 requires that the kernels K_{11} and K_{22} and their first partial derivatives be continuous for $0 < x' < p$. We do not prove this important assertion here, but refer the interested reader to Brown[29] instead.

7. Integral Equations for the Kernels

The final step in the solution of the inverse problem is to express the functions M_{11} and M_{22} in terms of the data. This is achieved through the integral representations of these functions in terms of kernels determinable from the prescribed data which constitute the reflection coefficient $R(\omega)$ and the velocity function $j(x)$. From $R(\omega)$ another function will be formed which will be denoted by $W(\omega)$. The integral equations for M_{11} and M_{22} will be formulated in terms of this W-function. The functions $R(\omega)$, $W(\omega)$, and the connection between the two will be discussed briefly in the next section. For the moment it will be assumed that the reflection coefficient $R(\omega)$ can be expressed as a ratio $B(\omega)/B(-\omega)$, where, $B(\lambda)$ is an analytic function in the upper-half plane. The W-function $W(\omega)$ is $[1/B(\omega)B(-\omega)]$.

Let us consider Equation 11.92. From Lemma 11.4 it is known that $d_1(x,\omega)$ is odd in ω. Equation 11.92 can, therefore, be put in the form of a Fourier transformation, namely

$$-iK_{11}(x,x') = \left(\frac{1}{\pi}\right)\int_{-\infty}^{\infty} z_0^{\frac{1}{2}} d_1(x, c_0\omega')\exp(-i\omega' x')d\omega'$$

where, $\omega' = \omega/c_0$. Now d_1 is real and continuous. Moreover, we know that $K_{11}(x,x')$ vanishes identically for $|x'| > c_0 j(x)$ and that $K_{11}(x,x')$ is real and odd in x'. Using these facts and applying Fourier transformation to the above equation for $K_{11}(x,x')$ then results in the following expression for $\psi_1(x,\omega)$

$$\psi_1(x,\omega) = \left[\frac{1}{z(x)^{\frac{1}{2}}}\right]\sin\omega j(x) + \int_0^{c_0 j(x)}\left[\frac{1}{z_0^{\frac{1}{2}}}\right]K_{11}(x,x')\sin\left(\frac{\omega x'}{c_0}\right)dx' \tag{11.97a}$$

The corresponding expression for $\psi_2(x,\omega)$ is

$$\psi_2(x,\omega) = iz(x)^{\frac{1}{2}}\cos\omega j(x) + \int_0^{c_0 j(x)} iz_0^{\frac{1}{2}} K_{22}(x,x')\cos\left(\frac{\omega x'}{c_0}\right)dx' \tag{11.97b}$$

Consider Equation 11.97a first. Multiply it by $[W(\omega)L_0/\pi z_0^{1/2}]\sin(\omega x'/c_0)$ and integrate with respect to ω over the entire real axis. After a substantial reduction, the following integral representation is obtained for K_{11}

$$K_{11}(x,x') + \left[\frac{z_0}{z(r)}\right]^{\frac{1}{2}} \Omega_1(c_0 j(x), x') + \int_0^{c_0 j(x)} K_{11}(x,x'')\Omega_1(x',x'')dx'' = 0 \qquad (11.98a)$$

The corresponding expression for K_{22} is

$$K_{22}(x,x') + \left[\frac{z(r)}{z_0}\right]^{\frac{1}{2}} \Omega_2(c_0 j(x), x') + \int_0^{c_0 j(x)} K_{22}(x,x'')\Omega_2(x',x'')dx'' = 0 \qquad (11.98b)$$

$x' < c_0 j(x)$ in the above expressions. The kernels Ω_1 and Ω_2 are given by

$$\Omega_1(x',x'') = \left(\frac{2}{c_0 \pi}\right) \int_0^\infty \sin\left(\frac{\omega x''}{c_0}\right)[W(\omega) - 1]\sin\left(\frac{\omega x'}{c_0}\right)d\omega \qquad (11.99a)$$

$$\Omega_2(x',x'') = \left(\frac{2}{c_0 \pi}\right) \int_0^\infty \cos\left(\frac{\omega x''}{c_0}\right)[W(\omega) - 1]\cos\left(\frac{\omega x'}{c_0}\right)d\omega \qquad (11.99b)$$

In terms of the variables p and M_1, M_2, Equations 11.98a and 11.98b take the form

$$M_1(p,x') + z_0^{\frac{1}{2}}\Omega_1(p,x') + \int_0^p M_1(p,x'')\Omega_1(x',x'')dx'' = 0, \quad x' < p \qquad (11.100a)$$

and

$$M_2(p,x') + \left[\frac{1}{z_0^{\frac{1}{2}}}\right]\Omega_2(p,x') + \int_0^p M_2(p,x'')\Omega_2(x',x'')dx'' = 0, \quad x' < p \qquad (11.100b)$$

Equations 11.100a and 11.100b are the desired integral expressions, solving which the functions M_1 and M_2 are obtained in terms of the data through the kernel functions Ω_1 and Ω_2. Replacing the M-functions thus determined in the differential equation (11.96) then formally solves the inverse problem. These equations can be recognized as the Gel'fand-Levitan equations, suitably couched in the variables appropriate to the transmission line problem.

The continuity of the M-functions and their first partial derivatives follow from those of the corresponding K-functions. Brown and Wilcox show that the Ω-kernel functions are in $L_2(R^+)$ both with respect to x' for x'' fixed and vice versa. It is also demonstrated that these functions are continuous in x' and x'' for $x' \neq x''$. The Ω-functions are essentially the Fourier transforms of $[W(\omega) - 1]$ sin(cos) $(\omega x'/c_0)$, which will be shown in the following section to be absolutely integrable. This property is, therefore, shared also by the Ω-kernels. Finally, it is shown that for each fixed value of p, the integral representations 11.100a and 11.100b possess unique solutions. This is established on the basis of Fredholm's alternative which has been encountered many times throughout the text.

8. Properties of R(ω) and W(ω)

We conclude the presentation of the Brown-Wilcox theory by summarizing the properties of R, W, their interconnection, and the conditions required of R(ω) if it is to be the reflection coefficient corresponding to a capacitance and an inductance function C(x) and L(x), respectively. First, an analytical form can be given to R(ω). The solution vector $\psi(x,\lambda)$ was given in terms of the functions $f(x,\lambda)$ in Equation 11.82. Now, the direct solution of the transmission line equations 11.78 and 11.78a is of the form

$$[y_1(x), y_2(x)] = \left[A(\omega)\exp\left(\frac{i\omega x}{c_0}\right) - R(\omega)A(\omega)\exp\left(-\frac{i\omega x}{c_0}\right),\right.$$

$$\left. -A(\omega)z_0\exp\left(\frac{i\omega x}{c_0}\right) - R(\omega)A(\omega)z_0\exp\left(-\frac{i\omega x}{c_0}\right)\right]$$

Comparing this with Equation 11.82 results in the relation

$$R(\omega) = \frac{f_{+1}(0,\omega)}{f_{-1}(0,\omega)}$$

Calculating the Wronskians of **f** and ψ at x = 0 (which is the same as at any x > X) and making use of the expansion theorem yields after some reduction that the reflection coefficient can be expressed as

$$R(\omega) = [Q(\omega)]^2$$

where

$$Q(\omega) = \left[\frac{1}{N(\omega)}\right][\mu(\omega) - i\upsilon(\omega)]\exp(-i\omega J)$$

J being j(X) = constant. Defining

$$B(\omega) = \mu(\omega) - i\upsilon(\omega)$$

R(ω) reduces to

$$R(\omega) = \exp(-2i\omega J)\frac{B(\omega)}{B(-\omega)}$$

from which is defined the H-function, namely

$$H(\omega) = R(\omega)\exp(2i\omega J)$$

The W-function, on the other hand, is defined as

$$W(\omega) = \left[\frac{1}{N(\omega)^2}\right] = \frac{1}{[\mu(\omega)^2 + \upsilon(\omega)^2]} = \frac{1}{[B(\omega)B(-\omega)]}$$

The introduction of these functions is due to Fadeev and they are extensively treated in his pioneering work.[35] Fadeev discusses the necessary properties of these functions. We will merely summarize these properties and refer the interested reader to Fadeev's original work. Fadeev shows that a unique function $B(-\lambda)$ exists which is analytic and bounded in the upper half-plane, has no zeros, behaves asymptotically as $1 + O(1)$ as $|\lambda| \to \infty$, and reduces to the function $H(\omega) = R(\omega)/R(-\omega)$ on the real axis, if $H(\omega)$ possesses certain required properties. These are

(a) $H(\omega) = R(\omega)\exp(2i\omega J)$; $|H(\omega)| = H(0) = H(\infty) = 1$.
$H(-\omega) = H^*(\omega) = 1/H(\omega)$,
(b) $[H(\omega) - 1]$ has an absolutely integrable Fourier transform, and
(c) $\arg H(\omega)|_{-\infty}^{\infty} = 0$.

The function $B(\omega)$ then determines $W(\omega)$ as defined above and if the W-function satisfies the following properties, namely

(a) $W(\omega) > 0$, $W(-\omega) = W(\omega)$ for $\omega \subset R$ and
(b) $[W(\omega) - 1]$ has an absolutely integrable Fourier transform,

then the W-function is uniquely determined by $H(\omega)$, that is, by the prescribed data $j(x)$ and $R(\omega)$. Fadeev also shows that $H(\omega)$ is determined uniquely by $W(\omega)$ and hence by the prescribed data $j(x)$ and $R(\omega)$.

These properties of $R(\omega)$ and $W(\omega)$ are proved by Brown and Wilcox.

APPENDIX A

A. SOME BACKGROUND MATERIALS ON MATRICES:[2,36] DEFINITIONS AND THEOREMS

Only real and square matrices will be considered. Thus, $\mathbf{A} = (a_{ij})$, i, j, $(a_{ij}) = 1,2...N$. If every $a_{ij} > 0$, then the matrix $\mathbf{A}$ is called "positive". If $a_{ij} \geq 0$ for only some i,j, then $\mathbf{A}$ is "nonnegative".

A p-th order "minor" of a matrix $\mathbf{A}$ is the determinant formed from the elements a_{ij} of $\mathbf{A}$ in p different rows and p different columns. Thus, a p-th order minor is written as

$$\mathbf{A}\begin{pmatrix} i_1 & i_2 & \cdots & i_p \\ j_1 & j_2 & \cdots & j_p \end{pmatrix} = \begin{vmatrix} a_{i_1 j_1} & a_{i_1 j_2} & \cdots & a_{i_1 j_p} \\ a_{i_2 j_1} & a_{i_2 j_2} & \cdots & a_{i_2 j_p} \\ \cdot & \cdot & \cdot & \cdot \\ \cdot & \cdot & \cdot & \cdot \\ a_{i_p j_1} & a_{i_p j_2} & \cdots & a_{i_p j_p} \end{vmatrix} = \hat{\mathbf{A}}_p$$

where

$$1 \leq \begin{pmatrix} i_1 < i_2 < \ldots < i_p \\ j_1 < j_2 < \ldots < j_p \end{pmatrix} \leq N$$

If p = N, then $\hat{\mathbf{A}}_{p=N} = |\mathbf{A}|^{N-1}$, the determinant of $\mathbf{A}$. $\hat{\mathbf{A}}_p$ is the shorthand notation for the minor given above.

Theorem A.1

Let $\mathbf{A} = (a_{ij})^N_1$ be an arbitrary square matrix and let $1 \le p < N$. Put

$$b_{ik} = \mathbf{A}\begin{pmatrix} 1 & 2 & . & . & . & p & i \\ 1 & 2 & . & . & . & p & k \end{pmatrix}, \quad (i, k = p+1, \ldots, N)$$

and

$$\mathbf{B} = (b_{ik})^N_{p+1}$$

Then

$$|\mathbf{B}| = B\begin{pmatrix} p+1 & . & . & . & N \\ p+1 & . & . & . & N \end{pmatrix} = \mathbf{A}\begin{pmatrix} 1 & 2 & . & . & . & p \\ 1 & 2 & . & . & . & p \end{pmatrix}^{N-p-1} \mathbf{A}\begin{pmatrix} 1 & 2 & . & . & . & N \\ 1 & 2 & . & . & . & N \end{pmatrix}$$

A corollary of this theorem is the important Sylvester's Identity. Let

$$1 \le i_1 < i_2 \; . \; . \; . \; < i_p \le N, \quad 1 \le k_1 < k_2 \; . \; . \; < k_p \le N$$

$$b_{ik} = \mathbf{A}\begin{pmatrix} i_1 & i_2 & . & . & . & i_p & i \\ k_1 & k_2 & . & . & . & k_p & k \end{pmatrix}; \qquad \begin{matrix} i = i_p + 1, \ldots N \\ k = k_p + 1, \ldots N \end{matrix}$$

$$i_p + 1 \le l_1 < l_2 \ldots < l_q \le N, \quad k_p + 1 \le m_1 < m_2 \ldots \; m_q \le N$$

Then

$$\mathbf{B}\begin{pmatrix} l_1 & l_2 & \cdots & l_q \\ m_1 & m_2 & \cdots & m_q \end{pmatrix} = \mathbf{A}\begin{pmatrix} i_1 & \cdots & i_p \\ k_1 & \cdots & k_p \end{pmatrix}^{q-1} \mathbf{A}\begin{pmatrix} i_1 & i_2 & \cdots & i_p & l_1 & \cdots & l_q \\ k_1 & k_2 & \cdots & k_p & m_1 & \cdots & m_q \end{pmatrix}$$

This is the well-known Sylvester's Identity.

A matrix **A** is "completely positive" (nonnegative) if not only all its elements, but also all the minors of all orders are positive (nonnegative).

A matrix **A** is called "oscillatory" if it is nonsingular, completely nonnegative, and has positive quasiprincipal elements, i.e., $a_{i,j+1} > 0$, $a_{i+1,j} > 0$ for $i = 1, 2, \ldots, N-1$.

The matrix $\mathbf{A}^*$ with elements a^*_{ij} given by $a^*_{ij} = (-)^{i+j} a_{ij}$ is called the "sign reverse" of **A**. **A** is called "sign oscillatory" if its sign reverse is oscillatory.

B. JACOBI MATRICES

A tridiagonal symmetric matrix is called a Jacobi matrix.

A "principal minor" of a Jacobi matrix **J** is formed by deleting rows and columns. Thus, a right (left) principal minor of order r is formed from **J** by deleting the last (first) r number of rows and columns of the matrix **J** – **I**. Thus

$$P_0 = 1, \quad P_1(\lambda) = a_1 - \lambda, \quad P_2(\lambda) = \begin{vmatrix} a_1 & -\lambda & -b_1 \\ -b_1 & a_2 & -\lambda \end{vmatrix}, \quad \text{etc.}$$

and finally, $P_N(\lambda) = |\mathbf{B} - \lambda\mathbf{I}|$. The minors satisfy the recurrence relation

$$P_{r+1}(\lambda) = (a_{r+1} - \lambda)P_r(\lambda) - b_r^2 P_{r-1}(\lambda), \quad (r = 1,2, \ldots, N-1) \tag{A.1}$$

which allows calculation of $P_2, P_3, \ldots, P_N$ successively from P_0, P_1.

The principal minors of a Jacobi matrix form a "Sturm sequence", that is:

1. $P_0(\lambda)$ has everywhere the same sign ($P_0(\lambda) \equiv 1$).
2. When $P_r(\lambda)$ vanishes, $P_{r+1}(\lambda)$ and $P_{r-1}(\lambda)$ are nonzero and have opposite signs.

It is assumed that $b_r^2 > 0$, $r = 1, 2, \ldots, N-1$.

Some important properties of Jacobi matrices are listed below.

Property 1

The zeros of $P_i(\lambda)$ are real and simple.

Property 2

The eigenvalues of a Jacobi matrix are distinct.

Property 3

If λ_0 is a zero of $P_i(\lambda)$, then as λ passes through λ_0, that is, passes from λ_0- to λ_0+, the sign of $P_i(\lambda)\,P_{i-1}(\lambda)$ changes from + to –.

Property 4

Between any two neighboring zeros of $P_i(\lambda)$ there lies one and only one zero of $P_{i-1}(\lambda)$ and one and only one zero of $P_{i+1}(\lambda)$.

Proof

Let u_1, u_2 be two neighboring zeros and let $P_r(u_1-) > 0$. Then $P_r(u_1+) < 0$ and $P_r(u_2-) < 0$. By Property 3 $P_{r-1}(u_1+) > 0$ and $P_{r-1}(u_2-) < 0$, so that $P_{r-1}(\lambda)$ changes sign between u_1+ and u_2-, and therefore has at least one zero in (u_1,u_2).

Now Property 2 of Sturm's sequence shows that $P_{r+1}(u_i)$ and $P_{r-1}(u_i)$ ($i = 1,2$) have opposite signs. Thus, $P_{r+1}(u_1+) < 0$, $P_{r+1}(u_2-) > 0$ so that $P_{r+1}(\lambda)$ has at least one zero in (u_1,u_2). Now suppose, if possible, that $P_{r-1}(\lambda)$ (or $P_{r+1}(\lambda)$) had two (or more) zeros in (u_1,u_2). Then $P_r(\lambda)$ would have a zero in (u_1,u_2), contrary to the hypotheses that u_1,u_2 are two neighboring zeros. This establishes Property 4.

Property 5

The sequence for the j-th eigenvector $\{u_i^{(j)}\}^N_1$ has exactly $(j-1)$ sign reversals.

Property 6

The zeros of two successive eigenvectors alternate.

APPENDIX B

A. THE MONIC POLYNOMIALS

Let P_N be a linear space of polynomials of order N, i.e., polynomials of degree $k < N$, with real coefficients. On this space one introduces an inner product $<\,,\,>$ with respect to a weight function w which is of the form

$$\langle f,g \rangle = \sum_{i=1}^{N} f(\xi_i)g(\xi_i)w_i, \quad \forall f,g \subset P_N \tag{B.1}$$

where

$$\xi_1 < \ldots . < \xi_N \text{ and } w_i > 0, \ \forall \ i$$

It is easy to establish that <, > of (B.1) satisfies all the requirements of an inner product, namely, those of positive definiteness, bilinearity and symmetry.

Definition

A sequence of "monic" polynomials is a sequence $\{q_i(x)\}_0^N$ such that $q_i(x)$ is of degree i with the coefficient of the leading term x^i being unity.

Theorem B.1

Let P_N be the linear space of polynomials of order N. Then there exists exactly one sequence of monic polynomials $\{q_i(x)\}_0^N$, $q_i(x)$ of degree i, $\forall$ i, which is orthogonal with respect to the inner product < , >, that is,

$$\langle q_i, q_j \rangle = 0, \qquad i \neq j \tag{B.2}$$

A simple proof of this theorem can be given in terms of the Gram-Schmidt orthogonalization procedure.[2]

The next theorem is important for the present discussion.

Theorem B.2: Forsythe[37]

The monic polynomials $\{q_i(x)\}_0^N$ satisfy a three-term recurrence relation

$$q_i(x) = (x - \alpha_i) q_{i-1}(x) - \beta_{i-1}^2 q_{i-2}(x), \qquad i = 1, \ldots . N \tag{B.3}$$

$$q_{-1}(x) = 0, \qquad q_0(x) = 1 \tag{B.4}$$

with the numbers α_i and β_i computed concurrently by

$$\alpha_i = \frac{\langle q_{i-1}, x q_{i-1} \rangle}{\| q_{i-1} \|^2} \tag{B.5}$$

$$\beta_i = \frac{\| q_i \|}{\| q_{i-1} \|}, \qquad i = 1, \ldots . N \tag{B.6}$$

Proof

The monic polynomials are orthogonal with respect to each other according to the inner product < , >. These, therefore, are linearly independent. The polynomial $q_i(x) - x q_{i-1}(x)$ is a polynomial of degree (i – 1). Hence,

$$q_i(x) - x q_{i-1}(x) = \sum_{k=0}^{i-1} c_k q_k(x) \tag{B.7}$$

Taking the inner product of Equation B.7 with $q_j(x)$ (j = 0, 1, ... i –1), we have

$$\langle q_i, q_j \rangle - \langle q_{i-1}, x q_j \rangle = \sum_{k=0}^{i-1} c_k \langle q_k, q_j \rangle = c_j \| q_j \|^2 \tag{B.7a}$$

where the symmetry relation <xp,q> = <p,xq> was used. Now j runs up to (i – 1), and $\langle q_i,q_j\rangle = 0$, $i \neq j$. Hence, for all j up to (i – 3), the factor xq_j is of degree (i – 2) and is, therefore, orthogonal to q_{i-1}. This means that the c_j's are all zero if j is other than (i – 1) and (i – 2). Therefore,

$$q_i(x) - xq_{i-1}(x) = \sum_{i-1}^{i-2} c_k q_k(x) \tag{B.8}$$

From Equation B.7a and using the orthogonality relation (B.2), we have

$$-c_{i-1} = \frac{\langle q_{i-1}, xq_{i-1}\rangle}{\|q_{i-1}\|^2} \tag{B.9}$$

and

$$-c_{i-2} = \frac{\langle q_{i-1}, xq_{i-2}\rangle}{\|q_{i-2}\|^2} \tag{B.10}$$

Substituting Equations B.9 and B.10 in Equation B.8 and rearranging results in Equations B.3 through B.6.

Equations B.3 through B.6 provide the way to construct the monic polynomials $\{q_i(x)\}_1^{N-1}$. Using Equation B.4, one first determines α_1 which, when substituted in Equation B.3, yields $q_1(x)$. Next, the parameters α_2 and β_1 can be computed, yielding $q_2(x)$, and so on.

B. THE WEIGHT FUNCTIONS W_i

The inner products involve weight functions which can be determined as follows. We note that the polynomial

$$q_N(x) = \prod_1^N (x - \xi_i) \tag{B.11}$$

is a monic polynomial of degree N in the sequence $\{q_i(x)\}_0^N$ and is orthogonal to all other $\{q_i(x)\}_0^{N-1}$ since $q_N(\xi_i) = 0$, $\forall$ i. Defining

$$m_i = w_i q_{N-1}(\xi_i) \tag{B.12}$$

we have

$$\sum_{i=1}^N m_i f(\xi_i) = 0, \quad f(\xi_i) \subset P_{N-1}$$

yielding

$$\sum_{i=1}^N m_i \xi_i^k = 0, \quad k = 0,\ 1,\ \ldots,\ N-2$$

since each $x^k \subset P_{N-1}$. In matrix form this is

$$\Delta \mathbf{m} = 0 \tag{B.13}$$

where the n-th row of Δ consists of (n–1) power of $\{\xi_i\}_1^N$ and

$$\mathbf{m} = (m_1,\ m_2,\ \ldots,\ m_N)^T$$

The solution of Equation B.13 is given by[2]

$$m_i = \gamma \left[\prod_{\substack{j=1 \\ j \neq i}}^{N} (\xi_i - \xi_j) \right]^{-1} \tag{B.14}$$

where γ is a constant. Now,

$$q_N(\xi) = \prod_{j=1}^{N} (\xi - \xi_j)$$

Differentiating with respect to ξ and setting $\xi = \xi_i$, it follows that

$$q'_N(\xi_i) = \prod_{\substack{j=1 \\ j \neq i}}^{N} (\xi_i - \xi_j)$$

where the prime denotes differentiation. Then from the definition (B.12) for m_i and Equation B.14, we have

$$w_i = \frac{\gamma}{\left[q_{N-1}(\xi_i) q'_N(\xi_i)\right]},\ \forall\ i,\ w_i = w(\xi_i) \tag{B.15}$$

Using Equation B.15 and the inner product of Equation B.1, the factor γ is determined to be

$$\gamma = \frac{\|q_{N-1}\|^2}{\sum_{i=1}^{N} \frac{\left[q'_{N-1}(\xi_i)\right]}{q'_N(\xi_i)}}$$

All these can be summarized into the following lemma.

Lemma B.1

Let $\{q_i(x)\}_0^N$ be the sequence of monic orthogonal polynomials for the inner product (B.1). Then

$$q_N(\xi_i) = \prod_{i=1}^{N} (\xi - \xi_i)$$

and

$$w_i = \frac{\gamma}{\left[q_{N-1}(\xi_i)q'_N(\xi_i)\right]}, \quad i = 1,\ 2,\ \ldots,\ N$$

where

$$\gamma = \frac{\|q_{N-1}\|^2}{\sum_{i=1}^{N} \frac{\left[q'_{N-1}(\xi_i)\right]}{q'_N(\xi_i)}}$$

The monic polynomials $\{q_i\}_0^N$ satisfy the recurrence relation (B.3) which is the same as that given by Equation A.1 for the Jacobi matrices. As a matter of fact the principal minors of a Jacobi matrix form a set of monic polynomials. Then the results of Section B of Appendix A apply to the sequence of the monic polynomials. This yields that q_N and q_{N-1} cannot be zero at the same time and that their zeros must interlace. Moreover, as a consequence of this, the monic polynomials possess the characteristic that[2]

$$q_{N-1}(\xi_i)\, q'_N(\xi_i)\, \operatorname{sgn}\gamma > 0, \quad \forall\ i$$

This means that the weights w_i are positive.

Lemma B.1 is very important. It shows that by knowing the polynomials q_N and q_{N-1} with their zeros interleaving, the weights w_is can be computed. This, in turn, means that the entire sequence of $\{q_i(x)\}_0^N$ orthogonal with respect to the inner product $< , >$ can now be constructed. This is according to Forsythe's Theorem B.2.

APPENDIX C

A. THE Z-TRANSFORM

Let f[n] be a discrete sequence. Then the series

$$f(z) = \sum_{n=-\infty}^{\infty} f[n]z^{-n} \tag{C.1}$$

called the "z-transform" of f[n] is defined for those z, real or complex, for which the series (C.1) converges. The region of convergence is a ring $r_1 < |z| < r_2$, the radii depending on how f[n] behaves as $n \to \pm\infty$ and f(z) is analytic inside the ring. Two most important properties of a z-transform are given below.

Property 1 (Shifting)

If $f[n] \leftrightarrow f(z)$, then for any integer m,

$$f[n-m] \leftrightarrow z^{-m} f(z)$$

Property 2 (Convolution)

If $f_1[n] \leftrightarrow f_1(z)$ and $f_2[n] \leftrightarrow f_2(z)$, then

$$\sum_{k=-\infty}^{\infty} f_1[k]f_2[n-k] \leftrightarrow f_1(z)f_2(z)$$

APPENDIX D*

A. INTRODUCTION: TOEPLITZ MATRIX AND TRANSMISSION LINE INVERSION

This short Appendix demonstrates in a simple way how a problem for a Toeplitz matrix problem is related to the problem of inverting a transmission line for the capacitance taper. More specifically, the true inverse problem for a Toeplitz matrix requires the knowledge of the capacitance taper of the transmission line, whereas, the inverse solution of the capacitance taper provides the direct solution of a Toeplitz matrix. To elucidate. We follow Caflisch[38] here in the main and begin with some definitions.

A Toeplitz matrix is a symmetric matrix whose elements are constant along each diagonal. Let $\mathbf{A}_m$ be a positive definite m×m Toeplitz matrix with (k,l)th component given by

$$(\mathbf{A}_m)_{k,l} = a_{|k-l|+1}$$

Let $\mathbf{A}_k$ be its $k \times k$ principal minor, the upper left hand $k \times k$ block in $\mathbf{A}_m$, for any $k \leq m$. $\mathbf{A}_k$ is also positive definite and Toeplitz. The "central mass sequence" $\{\rho_1,...,\rho_m\}$ for A_m is defined as

$$\rho_k = \sup\left\{\rho : \mathbf{A}_k - \rho\Pi_k > 0\right\}$$

in which Π_k is th $k \times k$ matrix all of whose entries are 1's, and the inequality means that the matrix on the left is positive definite.

The first two elements of the central mass sequence can be identified almost immediately. These are

$$\rho_1 = a_1 \qquad \rho_2 = \frac{a_1 + a_2}{2}$$

As a matter of fact, it can be proved that the sequence $\{\rho_k\}$ is strictly decreasing, i.e., $\rho_k > \rho_{k+1}$ for $k = 1,...,m-1$.

We next show that the k-th element of the central mass sequence is related to the k-th principal minor of the Toeplitz matrix according to the formula

$$\frac{1}{\rho_k} = \mathbf{l}_k^T \mathbf{A}_k^{-1} \mathbf{l}_k$$

where, $\mathbf{l}_k$ is the k-dimensional column vector with all the entries equal to unity. In order to demonstrate this, the following well known[39] result given in the lemma below will be required.

Lemma D.1

Let u and v be k-vectors and $\mathbf{A}$ a $k \times k$ nonsingular matrix. Then

$$\left(\mathbf{A} - \mathbf{u}\mathbf{v}^T\right)^{-1} = \mathbf{A}^{-1} + \frac{\mathbf{A}^{-1}\mathbf{u}\mathbf{v}^T\mathbf{A}^{-1}}{\mathbf{1} - \mathbf{u}^T\mathbf{A}^{-1}\mathbf{v}'}$$

* Sections of this Appendix have been taken with permission from R. E. Caflisch, *Linear Algebra Appl.*, 38, 1981.

and

$$\det\left(\mathbf{A} - \mathbf{u}\mathbf{v}^{\mathrm{T}}\right) = \left(\mathbf{1} - \mathbf{v}^{\mathrm{T}}\mathbf{A}^{-1}\mathbf{u}\right)\det\mathbf{A}$$

Theorem D.1

The term ρ_k of the central mass sequence for $\mathbf{A}_m$ has the simple form

$$\frac{1}{\rho_k} = \mathbf{l}_k^{\mathrm{T}}\mathbf{A}_k^{-1}\mathbf{l}_k$$

Proof: From the definition of the central mass sequence we have

$$\det\left(\mathbf{A}_k - \rho_k\mathbf{\Pi}_k\right) = 0 \tag{D.1}$$

The matrix Π_k can be written as the product

$$\mathbf{\Pi}_k = \mathbf{l}_k\mathbf{l}_k^{\mathrm{T}}$$

Using Lemma D.1 and Equation D.1, we obtain

$$\det\left(\mathbf{A}_k - \rho_k\mathbf{\Pi}_k\right) = \left(1 - \rho_k\mathbf{l}^{\mathrm{T}}\mathbf{A}_k^{-1}\mathbf{l}_k\right)\det\mathbf{A}_k$$

yielding

$$1 - \rho_k\mathbf{l}_k^{\mathrm{T}}\mathbf{A}_k^{-1}\mathbf{l}_k = 0$$

This results in the relation

$$\frac{1}{\rho_k} = \mathbf{1}_k^{\mathrm{T}}\mathbf{A}_k^{-1}\mathbf{l}_k$$

The direct problem for a Toeplitz matrix consists of determining the central mass sequence given the matrix or rather its principal minors. The inverse problem is to reconstruct the Toeplitz matrix from the knowledge of the central mass sequence. We will show that it is the direct problem for the Toeplitz matrix which is important in so far as the inverse problem for the transmission line is concerned. The scheme is as follows. We first show the connection between the impulse response of the line and the principal minors of a Toeplitz matrix which are themselves Toeplitz. Then we show that the central mass sequence constructed from these minors according to Theorem D.1 is related to the capacitance taper of the line. Hence knowing the central mass sequence results in the knowledge of the capacitance taper. Since the central mass sequence is obtained from the principal minors which are in turn determined by the impulse response, this is equivalent to determining the capacitance taper of the line from its impulse response. Or in other words, the direct problem for a Toeplitz matrix results in the inverse solution of the transmission line problem. On the other hand, if the capacitance taper is known, then the central mass sequence can be determined which in turn determines the Toeplitz matrix whose central mass sequence it is. Hence the knowledge of the direct solution of the transmission line problem determines the inverse solution of a Toeplitz matrix.

B. RELATION BETWEEN IMPULSE RESPONSE AND TOEPLITZ MATRIX

Consider the transmission line equation given by Equation 11.76. For simplicity assume that L(x) = 1/C(x) and and a half-infinite transmission line consisting of m segments each of length (1/2). By definition, the capacitance taper is

$$C(x) = C_m \text{ for } \frac{m-1}{2} \le x < \frac{m}{2}, \quad m = 1,2,\ldots.$$

The initial taper C_0 is assumed to be unity.

Let the line be quiescent initially. The impulse response of the line is by definition the voltage observed at x = 0 in response to a forcing current of δ(t) at x = 0 and is given by

$$v(0,t) = \sum_{n=1}^{\infty} \alpha_n \delta(t - n + 1)$$

For a more general piecewise constant current input, namely

$$i(0,t) = i_m \text{ for } m - 1 \le t < m, \quad m = 1,2,\ldots.$$

the impulse response is of the form

$$v(0,t) = v_m \text{ for } m - 1 \le t < m$$

$$v_m = \sum_{n=1}^{m} \alpha_n i_{m-n+1} \qquad \text{(D.2)}$$

The energy E(m) used in driving this current into the line up until a time of t = m can be calculated as

$$E_m = \int_0^m i(0,t)v(0,t)dt$$

$$= \mathbf{i}_m^T \mathbf{A}_m \mathbf{i}_m$$

$\mathbf{A}_m$ is a m × m positive Toeplitz matrix and can be written as

$$(\mathbf{A}_m)_{kl} = \begin{cases} \alpha_1, & k = 1 \\ \frac{1}{2}\alpha_{|k-l|+1}, & k \ne 1 \end{cases} \qquad \text{(D.3)}$$

The current vector is denoted by

$$\mathbf{i}_m = (i_1,\ldots,i_m)^T$$

E(t) is the total amount of energy in the line up to a time t. i(0,t)v(0,t) is the rate at which energy is fed into the line and the matrix $\mathbf{A}_m$ is called the "energy matrix". As shown by Equation D.4, the components of $\mathbf{A}_k$'s are the coefficients in the impulse response of the transmission line. This demonstrates the relation between the impulse response of a transmission line and a Toeplitz matrix.

Next is shown how the central mass sequence is related to the capacitance taper. In order to demonstrate this, Caflisch considers a current vector i_m which is symmetrical about its center, that is, $i_k = i_{m-k+1}$. Adding the impulse responses at various times up to m, a vector V is formed such that $V_m = (V_1, V_2, \ldots V_m)^T = 2\mathbf{A}_m\mathbf{i}_m$. Using the impulse response given by Equation D.2 and the fact that the current vector $\mathbf{i}_m$ is symmetrical, the k-th element V_k can be calculated as follows.

$$V_k = 2\sum_{l=1}^{m} a_{|k-l|+1} i_l$$

$$= \sum_{j=1}^{k} \alpha_j i_{k-j+1} + \sum_{j=1}^{m-k+1} \alpha_j i_{k-1+j}$$

$$= \sum_{j=1}^{k} \alpha_j i_{k-j+1} + \sum_{j=1}^{m-k+1} \alpha_j i_{(m-k+1)-j+1}$$

$$= v_k + v_{m-k+1}$$

Hence is the lemma

Lemma D.2

Let the current vector $\mathbf{i}_m = (i_1, i_2, \ldots i_m)^T$, that is,

$$i_k = i_{m-k+1}$$

Let v_m be the corresponding impulse response as given by Equation D.2. Define

$$V_m \equiv (V_1, \ldots, V_m) = 2\mathbf{A}_m\mathbf{i}_m$$

Then

$$V_k = v_k + v_{m-k+1}$$

Now from

$$V(x,t) = v(x,t) + v(x, m-t)$$

and

$$I(x,t) = i(x,t) + i(x, m-t), \quad 0 \le t \le m$$

From Lemma D.2 and the fact that the current vector is symmetrical, we have

$$V(0,t) = 2 \text{ and } I(0,t) = 0 \tag{D.4}$$

Now V and I satisfy the telegraph equation (since v and i do) with the boundary condition given by Equation D.4. The solution is

$$V(x,t) = 2 \text{ and } I(x,t) = 0, \quad 0 < t < m - x$$

From this and from the causality condition that the current cannot reach the segment $x > m/2$ in a time $t < m/2$, it follows that

$$v\left(x, \frac{m}{2}\right) = \begin{cases} 1 & \text{for } 0 \leq x < \frac{m}{2} \\ 0 & \text{for } x \geq \frac{m}{2} \end{cases} \tag{D.5}$$

Next we consider the charge conservation law which is written as

$$\int_0^t i(0,s)ds = \int_0^\infty C(x)v(x,t)dx$$

Evaluating the conservation relation at time $t = m/2$ yields the following. For the LHS, we have

$$\int_0^{m/2} i(0,s)ds = \frac{1}{2}\int_0^m i(0,s)ds$$

$$= \frac{1}{2}\sum_{k=1}^{m} i_k$$

$$= \frac{1}{2}\mathbf{1}_m^T \mathbf{i}_m$$

Using Equation D.5 for the line response, the RHS can be reduced to

$$\int_0^\infty C(x)v\left(x, \frac{m}{2}\right)dx = \int_0^{m/2} C(x)dx$$

$$= \frac{1}{2}\sum_{k=1}^{m} C_k$$

These results can be summarized in the theorem below.

Theorem D.2

Let $\mathbf{A}_m$ be the energy matrix for the discrete transmission line with capacitance taper $C_1 = 1, C_2,...C_m$. Let $\mathbf{i}_m$ be the m-dimensional column vector solving

$$\mathbf{A}_m \mathbf{i}_m = \mathbf{1}_m$$

in which $\mathbf{I}_m$ is the vector of all 1's. Then

$$\mathbf{1}_m^T \mathbf{i}_m = \sum_{k=1}^{m} C_k$$

From Theorem D.1 we then finally obtain the relation between the central mass sequence and the capacitance taper, namely,

$$\frac{1}{\rho_k} = \sum_{l=1}^{k} C_l$$

Since a positive definite Toeplitz matrix is determined uniquely by its central mass sequence, hence is the uniqueness theorem follows.

Theorem D.3

For every positive definite m×m Toeplitz matrix $\mathbf{A}_m$, there is exactly one set of positive numbers $\{C_1,...C_m\}$, such that $\mathbf{A}_m$ is the energy matrix for a discrete transmission line with the capacitance taper C(x) given by

$$C(x) = C_k,\ \frac{k-1}{2} \le x < \frac{k}{2},\ k = 1, 2, \ldots, m$$

Furthermore, let $\{\rho_1,...,\rho_m\}$ be the central mass sequence for $\mathbf{A}_m$. Then

$$\frac{1}{\rho_k} = \sum_{l=1}^{k} C_l$$

REFERENCES

1. **F. V. Atkinson:** *Discrete and Continuous Boundary Problems*, Academic Press, New York, 1964.
2. **G. M. L. Gladwell:** *Inverse Problems In Vibration*, Martinus Nijhoff, Dordrecht, 1986.
3. **H. Hochstadt:** On the construction of a Jacobi matrix from spectral data, *Linear Algebra Appl.*, 8, 1974.
4. **L. J. Gray and D. G. Wilson:** Construction of a Jacobi matrix from spectral data, *Linear Algebra Appl.*, 14, 1976.
5. **O. H. Hald:** Inverse eigenvalue problems for Jacobi matrices, *Linear Algebra Appl.*, 14, 1976.
6. **C. de Boor and G. H. Golub:** The numerically stable reconstruction of a Jacobi matrix from spectral data, *Linear Algebra Appl.*, 21, 1978.
7. **C. Lanczos:** An iteration method for the solution of the eigenvalue problem of linear differential and integral operators, *J. Res. Nat. Bur. Stand.*, Sect. B 45, 1950.
8. **A. M Bruckstein and T. Kailath:** Inverse scattering for discrete transmission line models, *SIAM Rev.*, 29, 1987.
9. **A. M. Bruckstein, B. C. Levy and T. Kailath:** Differential methods in inverse scattering, *SIAM J. Appl. Math.*, 45, 1985.
10. **L. Wohlers:** *Lumped and Distributed Passive Networks*, Academic Press, New York, 1969.
11. **E. A. Robinson:** Dynamic predictive deconvolution, *Geophys. Prosp.*, 26, 1978.
12. **E. A. Robinson and S. Treitel:** *Geophysical Signal Processing*, Prentice-Hall, Englewood Cliffs, NJ, 1980.
13. **C. L. Pekeris:** Direct methods of interpretation in resistivity prospecting, *Geophysics*, 5, 1940.
14. **V. Bardan:** Comments on dynamic predictive deconvolution, *Geophys. Prosp.*, 25, 1977.
15. **S. Y. Kung and H. Hu:** A highly concurrent algorithm and pipelined architecture for solving Toeplitz systems, *IEEE Trans. Acoust. Speech Signal Process.*, ASSP-31/1, 1983.
16. **T. K. Citron and T. Kailath:** Euclid's algorithm, scattering theory and a VLSI architecture for decoding Reed-Solomon code, *IEEE Trans. Inf. Theory*, 1985.
17. **J. M. Delsome and M. Morf:** A tree classification of algorithms for Toeplitz and related equations including generalized Levinson and doubling type algorithms, Proc. 19th C.D.C., Albuquerque, 1980.
18. **E. Harrier, Ch. Lubich and M. Schlichte:** Fast numerical solution of nonlinear Volterra convolution equations, *SIAM J. Stat. Comput.*, 6, 1985.
19. **A. M. Bruckstein, I. Koltracht and T. Kailath:** Inverse scattering with noisy data, *SIAM J. Stat. Comput.*, 7, 1986.
20. **W. W. Symes and G. Zimmerman:** Experiments in impedence profile inversion using noisy and bandlimited data, *AMOCO Res. Rep.*, 1982.
21. **G. Arsene, Z. Ceausescu, F. Potra and D. Timotin:** A new algorithm for detecting reflection coefficients in layered media, *Ann. Geophys.*, 1, 1983.

22. **N. I. Akhiezer:** *The Classical Moment Problem*, Hafner, New York, 1965.
23. **P. Dewilde and H. Dym:** Schur recursions, error formulas and convergence of rational estimators for stationary stochastic processes, *IEEE Trans. Inf. Theory*, IT-27, 1981.
24. **J. A. Ware and K. Aki:** Continuous and discrete inverse scattering problems in a stratified elstic medium. I. Plane waves at normal incidence, *J. Acoust. Soc. Am.*, 45, 1969.
25. **J. G. Berryman and R. R. Greene:** Discrete inverse methods for elastic waves in layered media, *Geophysics*, 45, 1980.
26. **M. S. Howard:** Inverse scattering for a layered acoustic medium using the first order equations of motion, *Geophysics*, 48, 1983.
27. **K. M. Case and M. Kac:** A discrete version of inverse scattering problem, *J. Math. Phys.*, 14, 1973.
28. **C. B. Sharpe:** An inversion problem associated with discrete Schrödinger equation, *SIAM*, 32, 1977.
29. **G. L. Brown:** The Inverse Reflection Problem for Electric Waves on Non-Uniform Transmission Lines, Ph.D. thesis, University of Wisconsin, Madison, 1965.
30. **C. H. Wilcox:** The domain of dependence inequality and initial boundary value problems for symmetric hyperbolic systems, *Bull. Am. Math. Soc.*, 70, 1964.
31. **C. H. Wilcox:** Electric wave propagation on non-uniform coupled transmission lines, *SIAM Rev.*, 6, 1964.
32. **B. W. Roos and W. C. Sangren:** Expansions associated with a pair of singular first-order differential equations, *J. Math. Phys.*, 4, 1963.
33. **B. W. Roos and W. C. Sangren:** Spectral theory of Dira's radial relativistic wave equation, *J. Math. Phys.*, 3, 1962.
34. **B. M. Levitan and I. S. Saragsjan:** Introduction to Spectral Theory, American Mathematics Society Monograph, 39, 1975.
35. **L. D. Fadeev:** The inverse problems in the quantum theory of scattering, *J. Math. Phys.*, 4, 1963.
36. **F. R. Gantmacher:** *The Theory of Matrices*, Chelsea, New York, 1959.
37. **G. E. Forsythe:** Generation and use of orthogonal polynomials for data fitting with a digital computer, *J. SIAM*, 5, 1957.
38. **R. E. Caflisch:** Problem for Toeplitz matrices and the systhesis of discrete transmission lines, *Linear Algebra Appl.*, 38, 1981.
39. **C. R. Rao:** *Statistical Inferences and Its Applications*, John Wiley & Sons, New York, 1965.

Chapter 12

THE RADON TRANSFORM

I. INTRODUCTION

In this chapter we discuss a problem of integral geometry, that of determining a function f $\subset R^n$ from its integrals over the (n – 1)-dimensional hyperplanes. The solution of this problem is generally credited to J. Radon.[1] The integrals of f over the hyperplanes are referred to as the "Radon transform" and its inversion yielding f is called the "inverse Radon transform". These transforms will be discussed here on Euclidean spaces.

Let us briefly discuss what is involved. The lower-case letters x, u, k, and ξ will denote vectors. Let $p = (\xi, x)$ denote a hyperplane and ξ a unit vector. The integral of f over this hyperplane, denoted by $\check{f}(\xi,(\xi,x))$, is the Radon transform of f. If $\mathfrak{R}$ and $\mathfrak{R}^{-1}$ denote the Radon transform and its inverse, respectively, then the Radon transform pair is $\mathfrak{R}f = \check{f}$ and $\mathfrak{R}^{-1}\check{f} = f$.

The Radon transform is a powerful tool in solving partial differential equations. Its appeal derives from the fact that it reduces a multidimensional partial differential equation to an equation of one space dimension only, namely, in the variable p (Section VIII). Since the transform remains constant over any hyperplane p which is perpendicular to the unit vector ξ, the inverse transform is essentially a decomposition of f into its "plane wave" components, that is, the functions which are constants over the planes normal to ξ. These plane waves will turn out to be not of exponential type, as in the case of the Fourier transform, but more elementary. The meaning, significance, and forms of these plane wave functions will be discussed in Section V.A. Using the solution of the one-dimensional transformed problem, the multidimensional Cauchy problem can be solved via Radon's inversion.

The Radon transform is also a very potent technique in reconstructing an unknown density distribution f from its attenuation/emission measurements. If the distribution f(x) is probed by passing some radiation through it, then its Radon transform consists of the integrated intensities of the probe beam. Thus, for example, in a highly incandescent plasma medium, the line-integrated spectral intensities "seen" by a monochromator form the Radon transforms. In X-radiography, an incident X-ray beam passes through the body and is attenuated in its travel through the body along the hyperplanes. The attenuated output beam then yields the Radon transform for that hyperplane. In all these applications, the Radon transform is of the form

$$\check{f}(\xi,p) = \int_{\xi} f(x)\,dm(x) \tag{12.1}$$

dm(x) being the Euclidean measure. Inversion of this equation yields the unknown density field f(x). In this form, the Radon transform and its inversion have found surprisingly wide appeal in all fields of image reconstructions, X-radiography, nuclear magnetic resonance, plasma diagnostics, and radioastronomy.[2-30] Kirillov's problems, discussed in Chapter 2, were problems of Radon transform.

The Radon transform and its inversion have been treated extensively by several authors, the most noteworthy being Gel'fand, Graev, and Vilenkin,[31] John,[32] Helgason,[33] Ludwig,[34] and Natterer.[35] The last-mentioned monograph contains serious theoretical discussions of various practical aspects of the application of Radon's transform. On a more practical and elementary level, the book by Deans[36] is especially useful and contains perhaps the most extensive bibliography on the subject. In keeping with the spirit of this volume, the presentation of Radon's problem in this chapter is fundamental in nature. Practical problems such as noise, numerical and experimental inversions, and various subtleties in data acquisition and handling have not been discussed here. These can be found in References 35, 36 and 29.

II. THE DIRECT RADON TRANSFORM: DEFINITION

The direct Radon transform and its properties will be discussed first. The emphasis will be primarily on the spaces $\mathcal{L}(R)$ and $\mathcal{D}(R)$. These spaces were encountered before (see Chapter 8). Although the definition of the Radon transform extends to a broader class of functions, consideration of $\mathcal{L}$ and $\mathcal{D}$ results in certain properties of the transform such as continuous differentiability and the flexibility to deal with various integrals involved.

Let us then consider a function $f(x) \subset \mathcal{L}$, $x \subset R^n$ and let ξ designate a hyperplane. If P^n denotes the space of all hyperplanes in R^n, then by definition

$$\xi = \left[x \subset R^n : (x,\xi) = p\right] \subset P^n, \; p \subset R$$

Then the functional $\check{f}(\xi,(\xi,x))$ defined in Equation 12.1 is called the "Radon transform" of f(x). Equation 12.1 can also be written as

$$\check{f}(p,\xi) = \int_R f(x)\,\delta(p - x \cdot \xi)\,dx \qquad (12.2)$$

It is clear that $\check{f}(p,\xi)$ is defined on the cylinder $R \times S^{n-1}$, S^{n-1} being the surface of a generalized unit sphere embedded in R^n. The Radon transform $\check{f}(\xi,(\xi,x))$ is defined as a function on P^n and the mapping $\check{f} \leftrightarrow f$ relates certain function spaces on R^n and P^n. We notice that (p,ξ) and $(-p,-\xi)$ yield the same hyperplane, that is, it is only necessary to define $\check{f}(p,\xi)$ on the semi-infinite cylinder $R^+ \times S^{n-1}$. This double mapping of ξ by both (p,ξ) and $(-p,-\xi)$ is sometimes expressed by saying that the mapping $(p,\xi) \rightarrow$ is a "double covering" of $R \times S^{n-1}$ onto P^n. One then identifies a continuous, differentiable function ϕ on P^n with another continuous, differentiable function ϕ on the cylinder $R^+ \times S^{n-1}$ such that the function ϕ is symmetrical in p and ξ.

III. CONNECTIONS WITH OTHER TRANSFORMS

The Radon transform is closely related to other transforms, the most important being its connection with the Fourier transform. In this chapter, ~ designates the Fourier transform. This change of notation is made in order to avoid any confusion with the $\wedge$ symbol for Radon transfer and also for Fourier transform in the rest of the book. The Fourier transform of f(x) is given by

$$\tilde{f}(k) = (2\pi)^{-\frac{n}{2}} \int_{R^n} f(x) \exp(-ik \cdot x)\,dx, \; k \subset R^n$$

Replace the Fourier transform variable k by $\rho\xi$, $\rho \subset R^n$ and ξ is a unit vector in R^n. Then from Equations 12.1 or 12.2

$$\tilde{f}(\rho\xi) = (2\pi)^{-\frac{n}{2}} \int_R dp' \int_{x\cdot\xi=p'} f(x)\exp(-i\rho\xi\cdot x)\,dx$$

$$= (2\pi)^{-\frac{n}{2}} \int_R dp' \check{f}(p',\xi)\exp(-i\rho p')\,dp' \tag{12.3}$$

Accordingly, the Fourier transform of f on the "radial slice" $k = \rho\xi$ of the Fourier space is just the one-dimensional Fourier transform of the Radon transform in the "radial" coordinate p'. Thus the n-dimensional Fourier transform can be obtained simply by taking the one-dimensional Fourier transforms of the Radon transform. Likewise, inverting the Fourier transform, the Radon transform can be recovered. Equation 12.3 is often referred to as the "central slice theorem".[29] This fact is extensively used in anatomical and other imaging.

The Radon transform is also closely related to the well-known "Gegenbauer transform".[34,36] We will only outline the basic steps. A Gegenbauer polynomial $C_l^{\upsilon}(x)$ is defined through a generating function

$$\sum_{l=0}^{\infty} C_l^{\upsilon}(x)t^l = \left(1 - 2xt + t^2\right)^{-\upsilon}, \upsilon \neq 0$$

These are orthogonal on [–1,1] with the weight function $(1-t^2)^{\nu-1/2}$.

Suppose that the function f(x) defined in R^n can be expanded in spherical harmonics with coefficients depending on $|x| = r$. Thus

$$f(x) = g_l(r)Y_{lm}(\xi) \tag{12.4}$$

where $\xi = x/|x|$ is a unit vector and $y_{lm}(\xi)$ are the real spherical harmonics of degree l.[37] These form a complete orthonormal set over a unit sphere. Y_{lm}, m = 1,2,... N(n,l) is an orthonormal basis of a finite dimensional space on a unit sphere, the dimension of the space being n. These Y_{lm}s are related to the more familiar spherical harmonics usually denoted by Y_{lm} in the literature. Taking the Radon transform of Equation 12.4, writing the Euclidean measure dm(x) explicitly, and finally converting the resultant expression into a spherical coordinate system yields

$$\check{f}(p,\xi) = \int_0^{\infty} dr\, r^{n-1} g_l(r) \int_{\Omega} d\xi' Y_{lm}(\xi')\delta(p - r\xi\cdot\xi')$$

where $x = r\xi'$ and Ω is, as before, the hypersurface area of a generalized unit sphere embedded in Euclidean n-dimensional space. The integral in ξ' is thus over the surface of a unit sphere. Factoring the delta function, namely,

$$\delta(p - r\xi\cdot\xi') = \left(\frac{1}{r}\right)\delta\left[\left(\frac{p}{r}\right) - \xi\cdot\xi'\right]$$

the above equation reduces to

$$\check{f}(p,\xi) = \int_{|p|}^{\infty} dr\, r^{n-2} g_l(r) \int_{\Omega} d\xi'\, Y_{lm}(\xi')\, \delta\left(\frac{p}{r} - \xi\cdot\xi'\right) \tag{12.5}$$

For $r < |p|$, the integral is zero since $|\xi\cdot\xi'| \leq 1$. Thus $r \geq |p|$ so that $|p/r| \leq 1$. The ξ' integral can be written as

$$\int_\Omega Y_{lm}\left(\frac{p}{r}-\xi\cdot\xi'\right)d\xi' = Y_{lm}(\xi)I_l\left(\frac{p}{r}\right) \tag{12.6}$$

where $I_l(p/r)$ is an unknown function and the reflection property of the spherical harmonic[37] was used, namely,

$$Y_{lm}(\xi) = (-)^l Y_{lm}(\xi)$$

The spherical harmonics are related to the Gegenbauer polynomials through the addition formula[37]

$$C_l^\upsilon(\xi\cdot\xi') = \left[\frac{\Omega_n}{N(n,1)}\right]C_l^\upsilon(1)\sum_{m=1}^{N} Y_{lm}(\xi)Y_{lm}(\xi') \tag{12.7}$$

where Ω_n is the surface area of an n-dimensional sphere and N is an integer which depends on n and l. υ in Equation 12.7 is $\{(n-2)/2\}$. If $\upsilon = (n-2)/2$, then the Gegenbauer polynomial in $(\xi\cdot\xi')$ is a spherical harmonic of degree l if $|\xi| = |\xi'| = 1$. Furthermore, we have the well-known identity[32] (also see Section X)

$$\int_\Omega F(\xi\cdot\xi')d\xi' = \Omega_{n-1}\int_{-1}^{+1}\left(1-t^2\right)^{\frac{(n-3)}{2}} F(t)\,dt \tag{12.8}$$

where $t = \xi\cdot\xi'$. Equation 12.7 and the identity 12.8 can be used to evaluate the function I_l in Equation 12.6 as follows. Multiplying the right hand side of Equation 12.6 by $Y_{lm}(\xi)$ yields

$$\int_\Omega \sum_{m=1}^{N} Y_{lm}(\xi)Y_{lm}(\xi')\delta\left[\left(\frac{p}{r}\right)-\xi\cdot\xi'\right]d\xi' = \sum_{m=1}^{N} Y_{lm}(\xi)Y_{lm}(\xi)I_l\left(\frac{p}{r}\right)$$

Setting $\xi = \xi'$ in 12.7 reduces the identity to

$$\left[\frac{N}{\Omega_n}\right]I_l\left(\frac{p}{r}\right) = \left[\frac{N}{\Omega_n C_l^\upsilon(1)}\right]\int_\Omega C_l^\upsilon(\xi\cdot\xi')\delta\left[\left(\frac{p}{r}\right)-\xi\cdot\xi'\right]d\xi'$$

Using the identity 12.8 yields

$$I_l\left(\frac{p}{r}\right) = \left[\frac{\Omega_{n-1}}{C_l^\upsilon(1)}\right]\int_{-1}^{+1}\left(1-t^2\right)^{\frac{(n-3)}{2}} C_l^\upsilon(t)\delta\left[\left(\frac{p}{r}\right)-t\right)dt$$

where $t = \xi\cdot\xi'$. Finally, since $|p/r| \le 1$, we have

$$I_l\left(\frac{p}{r}\right) = \left[\frac{\Omega_{n-1}}{C_l^\upsilon(1)}\right]C_l^\upsilon\left(\frac{p}{r}\right)\left[1-\left(\frac{p}{r}\right)^2\right]^{\upsilon-\frac{1}{2}}$$

where $\upsilon = (n-2)/2$. From this the Radon transform $\check{f}(p,\xi)$ of Equation 12.5 is finally reduced to

$$\check{f}(p,\xi) = \check{g}_l(p)Y_{lm}(\xi)$$

where, $\check{g}_l(p)$ is given by

$$\check{g}_l(p) = \left[\frac{\Omega_{n-1}}{C_l^{\upsilon}(1)}\right]\int_{|p|}^{\infty} r^{2\upsilon} g_l(r)G(p,r)\,dr$$

$$G(p,r) = \left[C_l^{\upsilon}\left(\frac{p}{r}\right)\right]\left[1-\left(\frac{p}{r}\right)^2\right]^{\upsilon-\frac{1}{2}} \tag{12.9}$$

$\check{g}_l(p)$ given by Equation 12.9 is the "Gegenbauer transform" of $g_l(r)$. This expresses the Radon transform of f(x) as decomposed in Equation 12.4 in terms of the Gegenbauer transform of $g_l(r)$. Thus if a function defined in R^n is expanded into spherical harmonics with coefficients depending on $|x|$, then the Radon transform of that function induces a transform on the coefficients which involves Gegenbauer polynomials. We have followed Deans'[36] derivative here.

IV. THE RADON TRANSFORM: THEOREMS AND PROPERTIES

We now discuss the relation between the function spaces on R^n and P^n induced by the mapping $f \to \check{f}$. The following lemma[33] will be useful.

Lemma 12.1

For each $f(x) \subset \mathcal{L}(R^n)$ the integral

$$\int_R \check{f}(p,\xi)p^k\,dp,\ k \subset Z^+$$

is a k-th degree homogeneous polynomial in $\{\xi_i\}_1^n$.

Proof:

$$\int_R \check{f}(p,\xi)p^k\,dp = \int_R p^k\,dp \int_{x\cdot\xi} f(x)\,dm(x) = \int_{R^n} f(x)(x\cdot\xi)^k\,dx$$

(We will use both forms (12.1) and (12.2) interchangeably.)

Lemma 12.1 has another equivalent formulation which is due to Ludwig.[34] Consider the real spherical harmonics $Y_l(\xi)$ of degree l. These are homogeneous harmonic polynomials in $\{\xi_i\}_1^n$ of degree l over the unit sphere $|\xi| = 1$. These are arbitrary spherical harmonics which can be considered as functions of ξ on the unit sphere. The spherical harmonics Y_{lm} form an orthonormal basis on the function space to which the Laplace functions Y_ls belong. Therefore, $Y_l(\xi)$ can be expressed as a linear combination of Y_{lm}s whose m runs from 0 to 2l.* Now a polynomial of degree k can be expanded in terms of the spherical harmonics of degree $\leq k$ over the sphere. Using these facts, Lemma 12.1 implies that

* For an excellent and highly readable account, see the well-known monograph *Orthogonal Functions* by G. Sansone, Interscience Publishers, New York, 1959.

$$\int_{|\xi|=1}\left[\int_R \check{f}(p,\xi)p^k\,dp\right]Y_l(\xi)\,d\xi = 0, \quad l \neq k \tag{12.10}$$

To prove the contrary, expand the k-th moment given by the integral of Lemma 12.1 in terms of the spherical harmonics. Since all coefficients of harmonics of degrees greater than k must vanish, the moment integral must be of degree $\leq k$.

The necessary conditions for a function $g(p,\xi)$ to be the Radon transform of a function $f \subset \mathcal{L}(R^n)$ have thus been encountered, namely, (1) $g \subset R \times S^{n-1}$, (2) $g(p,\xi) = g(-p,-\xi)$, and the condition of Lemma 12.1 or its equivalent condition given in Equation 12.10. That these conditions are also sufficient can be proved.[34,35] The proof runs thus. We follow Ludwig here. Assuming that $g(p,\xi) = \check{f}(p,\xi)$ of function f(x), use Equation 12.3 to write

$$\tilde{f}(\eta) = (2\pi)^{\frac{1-n}{2}}\,\tilde{g}\left(|\eta|, \frac{\eta}{|\eta|}\right)$$

The evenness of $f(\eta)$ and hence of f(x) follows that of g. The crux of the proof lies in showing that if g satisfies the above three necessary conditions, then $\tilde{f}(\eta) \subset \mathcal{L}(R^n)$ and hence $f \subset \mathcal{L}(R^n)$. It is worthwhile to point out that this defines the range of the Radon transform operator.[35]

Let $g(p,\xi) = \check{f}(p,\xi)$ be the Radon transform of some function f. Expanding g in spherical harmonics and then taking Fourier transform of both sides of this expansion in p yields

$$\tilde{g}(\rho,\xi) = \sum_{l,m} \tilde{a}_{lm}(\rho)Y_{lm}(\xi) \tag{12.11}$$

From the symmetry property of the Radon transform, and using the reflection property of the spherical harmonics,[37] namely, $Y_{lm}(-\xi) = (-)^l Y_{lm}(\xi)$, it follows that

$$\tilde{a}_{lm}(-\rho) = (-)^l\tilde{a}_{lm}(\rho) \tag{12.12}$$

The equivalent condition 12.10 gives

$$\int_R a_{lm}(p)p^k\,dp = 0, \quad \text{if } k < l,\ k \subset Z^+$$

Setting k = 0 one obtains

$$\int_R a_{lm}(p)\,dp = \left[\int_R a_{lm}(p)\exp(-ip\rho)\,dp\right]_{\rho=0} = \tilde{a}_{lm}(\rho = 0) = 0$$

Thus the Fourier transform $\tilde{a}_{lm}(\rho)$ vanishes to order l at $\rho = 0$. From this and Equation 12.12 it follows that

$$\tilde{a}_{lm}(\rho) = \rho^l\tilde{b}_{lm}(\rho) \tag{12.13}$$

and

$$\tilde{b}_{lm}(\rho) = \tilde{c}_{lm}\left(\rho^2\right) \tag{12.14}$$

Using Equation 12.13 in Equation 12.12, substituting $\rho = |\eta|$ and $\xi = \eta/|\eta|$ yields

$$\tilde{g}(\rho,\xi) = \tilde{g}\left(|\eta|, \frac{\eta}{|\eta|}\right) \tag{12.15}$$

Defining $\tilde{f}(\eta) = \tilde{g}(|\eta|,/|\eta|)$ then results in

$$\tilde{g}\left(|\eta|, \frac{\eta}{|\eta|}\right) = \sum_{l,m} \tilde{c}_{lm}\left(\eta^2\right) Y_{lm}(\eta) = \tilde{f}(\eta) \tag{12.16}$$

where we have used the fact that Y_{lm} is a homogeneous polynomial of degree l. That $\tilde{f}(\eta) \subset \mathcal{L}(R \times S^{n-1})$ follows from the hypothesis that $\tilde{g}$ belongs to the same space. Since $\tilde{f} \subset \mathcal{L}(\cdot)$ means that its inverse transform $f \subset \mathcal{L}(\cdot)$, the sufficiency is then established.

This result can be sharpened. The Radon transforms $\check{f}(p,\xi)$ are defined as functions on P^n and are identified with functions $\phi(p,g) \subset R \times S^{n-1}$ such that $\phi(p,\xi) = \phi(-p,-\xi)$. If a space $\mathcal{L}_H(P^n)$ is defined as

$$\mathcal{L}_H(P^n) = \{F \subset \mathcal{L}(P^n)\text{: F satisfying Lemma 12.1}\}$$

then by virtue of Lemma 12.1, the Radon transform $\check{f} \subset \mathcal{L}_H(P^n)$. From Equation 12.11 it follows that $\tilde{a}_{lm}(\rho)$ depends continuously on $\tilde{g}(\rho,\xi)$. By virtue of the relations between $\tilde{a}_{lm}(\rho)$ and $\tilde{c}_{lm}(\rho^2)$ given by Equations 12.13 and 12.14, $\tilde{c}_{lm}(\rho^2)$ also depends continuously on $g(\rho,\xi)$. Hence, from Equation 12.16, $\tilde{f}(\eta)$ depends continuously on $\tilde{g}(|\eta|,\eta/|\eta|)$. This means that the function f(x) depends continuously on the Radon transform $\check{f}(p,\xi)$.

These results can be summarized into the following theorem.

Theorem 12.1

In order that a function $\check{f}(p,\xi)$ be the Radon transform of a function f(x), $x \subset R^n$, it is necessary and sufficient that (1) $\check{f} \subset R \times S^{n-1}$, (2) $\check{f}$ be even in p and ξ, and (3) $\check{f} \subset \mathcal{L}_H(P^n)$. Condition (3) is equivalent to (a) $\int_R \check{f}(p,\xi)p^k\, dp$ a polynomial of degree $\leq k$ in ξ, and (b) $\int_\Omega \int_R \check{f}(p,\xi)p^k Y_l(\xi)\, dp d\xi = 0$, where $Y_l(\xi)$ is a real spherical harmonic of degree l and $k < 1$. Moreover, the mapping $f \to \check{f}$ is continuous from $\mathcal{L}(R^n)$ to $\mathcal{L}_H(P^n)$.

We state two theorems below without proof, which is somewhat elaborate and can be found in Helgason.[33] Condition (3) of Theorem 12.1 has an analogue in Fourier transform theory. Following Helgason, this "Schwartz Theorem" is stated below.

Theorem 12.2: The Schwartz Theorem

The Radon transform $f \to \check{f}$ is a linear one-to-one mapping of $\mathcal{L}(R^n)$ onto $\mathcal{L}_H(P^n)$.

That the mapping is from $\mathcal{L}(R^n) \to \mathcal{L}_H(P^n)$ follows from the second condition of the above theorem. The proof of the onto part is somewhat elaborate.

The second theorem concerns the condition of the map $f \to \check{f}$ being bijective. Let $\mathcal{L}^*(R^n)$ be the space of all $f \subset \mathcal{L}(R^n)$ such that

$$\int_{R^n} f(x)P(x)\,dx = 0 \quad \text{for all polynomials } P(x)$$

Similarly, let $\mathcal{L}^*(P^n)$ be the space of all $\phi \subset \mathcal{L}(P^n)$ satisfying a similar orthogonality condition, namely,

$$\int_R \check{\phi}(p,\xi)p(r)\,dr = 0 \quad \text{for all polynomials } P(r)$$

Then

Theorem 12.3

$$\text{The transform } f \to \check{f} \text{ is a bijection of } L^*\left(R^n\right) \text{ onto } L^*\left(P^n\right)$$

We now extend Theorem 12.1 to functions $\subset \mathcal{D}$. For functions with support in $|x| \le r$, it follows from the definition 12.1 of the Radon transform that $\check{f}(p,\xi)$ also has a support in $|p| \le r$. The converse is, however, not obvious and must be proved. The proof given by Ludwig[34] is by far the simplest and we follow him here. Let $\check{f}(p,\xi)$ have a support in $|p| \le r$. Expand $\check{f}(p,\xi)$ in terms of the spherical harmonics as

$$\check{f}(p,\xi) = \sum_{l,m} a_{lm}(p)Y_{lm}(\xi) = \sum_{l,m} \check{f}_{lm}(p,\xi) \tag{12.17}$$

where $\check{f}_{lm}(p,\xi)$ is the Radon transform of some function $f_{lm}(x)$. If $\check{f}(p,\xi)$ has support in $|p| \le r$, then $a_{lm}(p)$ has support in $|p| \le r$. Let us suppose for the moment that if $a_{lm}(p)Y_{lm}(\xi)$ has support in $|p| \le r$, then the corresponding inverse transform $f_{lm}(x)$ also has support in $|x| \le r$. By Theorem 12.1 it then follows that the sum converges to some function f(x) in $\mathcal{L}$ having a support in $|x| \le r$, that is, to a function f(x) the Radon transform of which is $\mathcal{D}(R^n)$. The onus of the proof then lies in establishing that each element of the sum which we write as $a(p)Y_l(\xi)$ corresponds to some Radon transform that has a support in $|x| \le r$.

Ludwig proves this via the Paley-Wiener theorem for the Fourier transforms in $\mathcal{L}$. It may be recalled that according to the Paley-Wiener theorem a function f in $\mathcal{L}$ has a support in $|x| \le r$ if and only if its Fourier transform $f^\sim(\xi)$ is an entire analytic function of ξ such that for some constants A and r

$$\left|\tilde{f}(\xi)\right| \le A\exp\left(r|\zeta|\right)$$

where $\zeta = \text{Im}\,\xi$, that is, $f^\sim(\xi)$ is an entire function of exponential type r. These were discussed in Chapter 4.

Consider the Radon transform $\check{f}(p,\xi) = a(p)Y_l(\xi)$ and its Fourier transform $\tilde{f}(k,\xi) = \tilde{a}(k)Y_l(\xi)$. Since by hypothesis, $\check{f}(p,\xi) \subset \mathcal{L}$ and has a compact support in $|x| \le r$, then by the Paley-Wiener theorem

$$\left|\tilde{\check{f}}(k,\xi)\right| = \left|\tilde{a}(k)Y_l(\xi)\right| \le A\exp\left(r|\beta|\right) \Rightarrow \left|\tilde{a}(k)\right| \le A\exp\left(r|\beta|\right), \quad \beta = \text{Im}\,k$$

From Equations 12.13 and 12.14, we have

$$\tilde{f}(k,\xi) = k^l\tilde{c}\left(k^2\right)Y_l(\xi) \text{ yielding } \left|\tilde{c}\left(k^2\right)\right| \le \left[\frac{A}{|k|^l}\right]\exp\left(r|\beta|\right) \tag{12.18}$$

Again, as in Equation 12.15,

$$\tilde{f}(\xi) = (2\pi)^{\frac{-(n-1)}{2}} \tilde{f}\left(|\xi|, \frac{\xi}{|\xi|}\right)$$

$$= c(\xi \cdot \xi) Y_l(\xi) \tag{12.19}$$

$f^{\sim}(\xi)$ is, therefore, an entire function of ξ.

It remains to show that $|\tilde{f}(\xi)| \leq B \exp(r|\beta|)$ for some B and some $\zeta = \operatorname{Im} \xi$. Write $\xi = \eta + i\zeta$, η and ζ real. Let ρ be a complex number such that $\rho^2 = \xi \cdot \xi$. It is not difficult to show that

$$(\operatorname{Im} \rho)^2 = \left(\frac{1}{2}\right)\left[|\xi|^2 - |\eta|^2 + \left\{\left(|\eta|^2 - |\zeta|^2\right)^2 + 4(\eta \cdot \zeta)^2\right\}^{\frac{1}{2}}\right]$$

where the square root was chosen to be positive. This then yields

$$\beta^2 \leq \left(\frac{1}{2}\right)\left[|\zeta|^2 - |\eta|^2 + \left\{\left(|\zeta|^2 - |\eta|^2\right) + 4|\eta|^2|\zeta|^2\right\} = |\zeta|^2\right]$$

Then from Equations 12.18 and 12.19 it follows that

$$\left|\check{\tilde{f}}(\xi)\right| \leq \left[\frac{A|Y_l(\xi)|}{|(\xi)|^{-l}}\right] \exp(r|\zeta|) \leq B \exp(r|\zeta|)$$

where

$$B = A \sup_{|\xi|=1} |Y_l(\xi)(\xi)|$$

The Fourier transform $f(\xi)$ is thus an entire function of type r. Its inverse f(x), therefore, must have a support r in x. This then proves that if the Radon transform $\check{f} \subset R \times S^{n-1}$ has a compact support r in p, then the function f(x) must also have a compact support r in x.

Therefore, if f(x) has a compact support in x, then by definition $\check{f}$ has a compact support in p and the above discussion establishes the converse. Combining this with the result of Theorem 12.2 — that the Radon transform maps $\mathcal{L}(R_n)$ onto $\mathcal{L}_H(P^n)$ — it follows that the Radon transform mapping of $\mathcal{D}(R^n)$ must be a bijection onto $\mathcal{D}(P^n)$. Lemma 12.1 can again be used to sharpen the result by replacing $\mathcal{D}(P^n)$ by $\mathcal{D}_H(P^n)$. The bijection is, therefore, between $\mathcal{D}(R^n)$ and $\mathcal{D}_H(P^n)$. Moreover, the mapping is continuous, which is a consequence of Theorem 12.1. Summarizing, the following theorems are obtained.

Theorem 12.4: Support Theorem

If $f \subset \mathcal{L}$, then f(x) has support in $|x| \leq r$ if and only if its Radon transform $\check{f}(p,\xi)$ has support in $|p| \leq r$.

Theorem 12.5: Paley-Wiener

The Radon transform is a bijection of $\mathcal{D}\left(R^n\right)$ onto $\mathcal{D}_H\left(P_n\right)$

The last theorem has been termed the Paley-Wiener theorem because it is analogous to the Radon transform of the Paley-Wiener theorem for the Fourier transform. Note also the difference. Unlike the Fourier transform case, the Radon transform of a space-limited function still remains limited even in the transform domain.

At this point let us introduce some simple, basic properties of the Radon transform and its derivatives. These are standard results and can be found almost in any text on the subject.

1. The Radon transform is an even homogeneous function of degree –1
In order to see this, change $p \to ap$ and $\xi \to a\xi$, $a \subset R\backslash\{0\}$. Then

$$\check{f}(ap, a\xi) = \int f(x)\delta(ap - x \cdot a\xi)\,dx = |a|^{-1}\int f(x)\delta(p - x \cdot \xi)\,dx = |a|^{-1}\check{f}(p,\xi)$$

For $a = -1$, we recover the symmetry property described before.

2. The Radon transform is linear
Given two functions f and g and two constants a and b,

$$\Re(af + bg) = a\Re f + b\Re g$$

3. Shifting Property

$$\Re f(x+t) = \int_{y\cdot\xi=p+t\cdot\xi} f(y)\,dm(y) = \check{f}(p + t\cdot\xi, \xi), \quad x, t \subset R^n \tag{12.20}$$

4. Convolution
Let f(x) be the convolution of g(x) and h(x). Then

$$f(x) = g * h = \int_{R^n} g(y)h(x-y)\,dy$$

Taking the Radon transform

$$\check{f}(p,\xi) = \int_{x\cdot\xi=p}(g*h)\,dm(x) = \int_{-\infty}^{\infty} ds\,\check{h}(p-s,\xi)\int_{y\cdot\xi=s} g(y)\,dm(y)$$

$$= \int_{-\infty}^{\infty} ds\,\check{h}(p-s,\xi)\check{g}(s,\xi) = \left[\check{g} * \check{h}\right]$$

As usual, dm(x) and dm(y) are the Euclidean measures on the hyperplanes p and s, respectively. Thus the Radon transform of the convolution of two functions is the convolution of the Radon transforms of the functions themselves. This is to be contrasted with the convolution theorem either for the Fourier or for the Laplace transform, where the transformed convolution is the *product* of the transformations and not their convolutions.

5. The Radon transforms of the derivatives

Let $f(x) = f(x_1, x_2, \ldots, x_n)$, x a vector in R^n. We wish to determine the Radon transform of the derivative of f with respect to any of its variables. Let us calculate, for example, the derivative $(\partial f/\partial x_k)$. Taking e_k to be a unit vector along the coordinate x_k and Radon transforming yields

$$\Re\left(\frac{\partial f}{\partial x_k}\right) = \Re\left[\lim_{dx\to 0}\left[\frac{\{f(x + dx \cdot \mathbf{e}_k) - f(x)\}}{dx}\right]\right]$$

$$= \lim_{dx\to 0}\left[\frac{\left\{\check{f}(p + dx(\mathbf{e}_k \cdot \xi), \xi) - \check{f}(p, \xi)\right\}}{dx}\right]$$

$$= \lim_{dp\to 0}\left[\frac{\left\{\check{f}(p + dp, \xi) - \check{f}(p, \xi)\right\}}{dp}\right] \cdot \xi_k$$

$$= \xi_k\left[\frac{\partial \check{f}}{\partial p}\right] \tag{12.21}$$

In the above derivation, the shifting property of Equation 12.20 was used. From the linearity property (2) it follows that if ∇ is the gradient operator and a an arbitrary vector in R^n, then

$$\Re\{(a \cdot \nabla f)\}(p, \xi) = (a \cdot \xi)\check{f}'(p, \xi)$$

the prime $'$ denotes differentiation with respect to p.

Equation 12.21 shows that the Radon transform of the derivative of the function f(x) with respect to a coordinate x_k is just the derivative of the Radon transform $\check{f}(p,\xi)$ of f with respect to p times the component ξ_k of the vector ξ along the coordinate x_k. The extension to the cross-derivatives is immediate and yields

$$\Re[\partial_i \partial_j f](p, \xi) = \xi_i \xi_j \check{f}''(p, \xi), \quad \partial_i = \frac{\partial}{\partial x_i} \text{ and } \partial_j = \frac{\partial}{\partial x_j}$$

By induction,

$$\Re\left[\sum_{i=1}^{n}\sum_{j=1}^{n} a_i b_j (\partial_i \partial_j) f\right] = (a \cdot \xi)(b \cdot \xi)\check{f}''(p, \xi) \tag{12.22}$$

These results can be easily generalized to an arbitrary differential operator **L** of the form $\mathbf{L} = \mathbf{L}(\partial_i)_i^n$ with constant coefficients to yield

$$\Re(\mathbf{L}f) = \left[\mathbf{L}\left(\xi_i \frac{\partial}{\partial p}\right)_1^n\right] \check{f}(p,\xi)$$

An important special case is when $a_i b_j = \delta_{ij}$, the Kronecker delta. Equation 12.22 is then Laplace's equation in n dimensions and results in

$$\Re[\Delta f(x)] = |\xi|^2 \check{f}''(p,\xi) = \check{f}''(p,\xi)$$

if $|\xi|^2 = 1$, Δ being the Laplacian in n space dimensions. If we define, after Helgason, an operator $\square = \partial^2/\partial p^2$ for an arbitrary function ψ of the scalar p and the vector ξ, then

$$\square\psi(p,\xi) = \psi''(p,\xi)$$

Identifying $\psi(p,\xi)$ with $\check{f}(p,\xi)$, the Laplace's equation is transformed as $\Re(\Delta f) = \square\check{f}$. Also $\Re^{-1}[\square f] = \Delta\check{f}$. The operators Δ and $\square$ are thus "intertwining" operators.

Finally, if **L** is a differential operator which is a homogeneous polynomial P_m of degree m in its derivatives with constant coefficients, that is, if

$$\mathbf{L} = P_m(D_x)$$

then $\Re[\mathbf{L}f] = P_m(D_p)\check{f}(p,\xi)$. This simple generalization has important implications in the theory of partial differential equations, since it reduces the problem from n dimensions into only one coordinate p. Since the Radon transform operator commutes with the operator $\partial^2/\partial t^2$, this implies that an n-dimensional wave equation $\ddot{f} = \Delta f$ is reduced to a one-dimensional wave equation, namely, $\ddot{\check{f}} = \check{f}''(p,\xi)$.

6. Derivatives of the transform

Here we wish to differentiate the Radon transform $\check{f}(p,\xi)$ of f(x) with respect to the variable ξ. In what follows it will be assumed that ξ is not a unit vector. The results can, of course, be specialized for the case $|\xi| = 1$.

$$\check{f}_{\xi_k} = \int_{R^n} f(x)\delta(p - \xi\cdot x)_{\xi_k}\, dx = -\int_{R^n} f(x)(x\cdot\mathbf{e}_k)\delta'(p - \xi\cdot x)\,dx$$

where the prime is the differentiation with respect to p and we have used the identity

$$\delta_{x_j}(x - x_0) = -\delta_{x_{0j}}(x - x_0)$$

Then

$$\check{f}_{\xi_k} = -\left[\int_{R_n} f(x)(x\cdot\mathbf{e}_k)\delta(p - \xi\cdot x)\,dx\right]' = -\{\Re[x_k f(x)]\}'$$

7. The Radon transform of a linear transformation[36]

Let $\mathbf{y} = \mathbf{A}x$ be a linear transformation in R^n, the matrix of the transformation $\mathbf{A}$ being invertible. Then $x = \mathbf{A}^{-1}y = \mathbf{B}y$.

Consider a function $f = f(\mathbf{A}x)$ and its Radon transform:

$$\Re[f(\mathbf{A}x)](p,\xi) = \int_{x\cdot\xi=p} f(\mathbf{A}x)\,dm(x) = |\Delta\mathbf{B}|\int_{(\mathbf{B}y)\cdot\xi=p} f(y)\,dm(y)$$

$$= |\Delta\mathbf{B}|\int_{(\mathbf{B}^T\xi)\cdot y=p} f(y)\,dm(y)$$

In the above, $|\Delta\mathbf{B}|$ is the determinant of $\mathbf{B}$ and the change of variable from x to $y = \mathbf{A}x$ requires that the resulting expressions be multiplied by the Jacobian of the transformation, which is $|\Delta\mathbf{B}|$. Also, we have used the fact that the scalar product

$$z\cdot(\mathbf{B}y) = (\mathbf{B}^T z)\cdot y$$

Since $\mathbf{A}^{-1} = \mathbf{B}$, we can write the above as

$$\Re[f(\mathbf{A}x)](p,\xi) = \Re\left[f\left(\mathbf{B}^{-1}x\right)\right](p,\xi) = |\Delta\mathbf{B}|\,\check{f}\left(p,\mathbf{B}^T\xi\right)$$

In the event that $\mathbf{B}^T\xi$ is not a unit vector, the transform can be rewritten to put it into the usual form. Let $\mathbf{B}^T\xi = \zeta = (a_1,a_2, .. ,a_n)$, that is, the vector $\mathbf{B}^T\xi$ has components $\{a_i\}_1{}^n$ in R^n. Then

$$|\Delta\mathbf{B}|\,\check{f}\left(p,\mathbf{B}^T\xi\right) = |\Delta\mathbf{B}|\,\check{f}(p,\zeta) = |\Delta\mathbf{B}|\,\check{f}(p,s\rho) = |\Delta\mathbf{B}|s^{-1}\,\check{f}\left(\frac{p}{s},\rho\right)$$

$$= \left\{\frac{\left[\prod_1^n |a_i|\right]}{|\zeta|}\right\}\check{f}\left(\frac{p}{|\zeta|},\frac{\zeta}{|\zeta|}\right)$$

using Property 1 of homogeneity.

Two special cases are important:

1. $\mathbf{B}^{-1} = \mathbf{B}^T = \mathbf{A}$, that is, $\mathbf{B}$ is orthogonal. Then $|\Delta\mathbf{B}| = 1$ yielding

$$\Re[f(\mathbf{A}x)](p,\xi) = \check{f}(p,\mathbf{A}\xi)$$

2. $\mathbf{A} = \lambda\mathbf{I}$, $\lambda \subset R$, and $\mathbf{I}$ is the identity operator. Then

$$\Re[f(\mathbf{A}x)](p,\xi) = \Re[f(\lambda x)](p,\xi) = \lambda^{-n}\,\check{f}\left(p,\frac{\xi}{\lambda}\right) = \lambda^{-(n-1)}\,\check{f}(\lambda p,\xi)$$

The last two identities have been obtained using the homogeneity property (1).

V. THE INVERSE RADON TRANSFORM

Having discussed the direct Radon transform and its properties, we now return to the central theme of this chapter, that of determining a function f(x) from its integrals over the hypersurfaces. The problem is that of inverting the Radon transform. Three derivations will be presented for the inverse Radon transform. These are due to John,[32] Gel'fand, Graev, and Vilenkin,[31] and Ludwig.[34] John's approach is one of partial differential equation and spherical mean of a plane wave function. Their meanings will be discussed shortly. Gel'fand et al. give two different derivations of the inverse Radon transform. Their derivation presented here is based on the plane wave expansion of an n-dimensional delta distribution. Ludwig's treatment is relatively more straightforward and can be considered an approach via Fourier transform. There is also a fourth derivation due to Helgason.[33] Helgason uses the so-called "Riesz potentials" and the fractional power of a Laplacian. However, we will not discuss the Riesz potentials here. The discussion of Radon's problem is incomplete without mentioning P. D. Lax's work.[38] Lax presented Radon's problem in the context of the theory of partial differential equations. The expressions for the inverse Radon transform are different depending on whether the dimension n of the Euclidean space R^n is odd or even. John, Gel'fand et al., and Helgason have given unified formulas of the inverse Radon transform for n odd or even. Before we can present these derivations, some mathematical preliminaries are required regarding plane wave expansion of a function, certain generalized functions, their spherical means, and generalized derivatives. These are discussed next.

A. DERIVATIVES AND SPHERICAL MEANS OF CERTAIN GENERALIZED FUNCTIONS

We begin by introducing the concept of a "plane wave" and the expansion of a function in terms of plane waves. The meaning of "spherical mean" and its properties are discussed next. Certain generalized functions are then introduced and their spherical means and derivatives are calculated. The operation of the Laplacian on these generalized functions is discussed next.

The term "plane wave" is used to imply any function that has parallel hyperplanes as its level surfaces. For a given vector, Radon's transform depends only on the variable p. The transform is constant over any hypersurface perpendicular to ξ and has, therefore, the hypersurfaces as level surfaces. Its argument is the inner product $(\xi \cdot x)$. The Radon transform is thus a plane wave. The most familiar example of plane wave decomposition of a function is, of course, the Fourier transform, where the plane waves are of the exponential type, i.e., of the form $\exp(i\xi \cdot x)$. The exponential character, however, is not essential, and more elementary forms of plane wave decomposition of a function can be used. One such method is to take the spherical mean of the integrals of the function over the hypersurfaces. By a spherical mean is meant the following. Let B(r,x) denote the open ball $\{y: |y - x| < r\}$ in R^n, and A(r) be the area of the sphere $|y - x| = r$. For a continuous function f(x) on the sphere, the mean value

$$(Sf)(r,x) = \left[\frac{1}{A(r)}\right] \int_{|y-x|=r} f(\xi)\, d\xi \tag{12.23}$$

$d\xi$ being the Euclidean measure, is called the "spherical mean" of the function f(x). It is just the average value of the function over the sphere. S stands for the spherical mean and the variables (r,x) imply that the average is taken at a point x and over a sphere of radius r. Gel'fand et al.[31] define the mean in terms of a generalized unit sphere embedded in an n-dimensional Euclidean space. They consider the following equation

$$\int_{R^n} f(x)\, dx = \int_0^\infty dx\, r^{n-1} \int_\Omega f(r\xi)\, d\xi \tag{12.24}$$

where $d\xi$ is the surface element of the spherical surface of radius unity for which Ω_n is the total surface measure. If we identify f with a function $\phi(x)$ where ϕ is either in $\mathcal{L}$ or in the space of $\mathcal{D}$ and define $(S\phi)(r,x)$ as $S_\phi(r)$, then

$$\Omega_n S_\phi(r) = \int_\Omega \phi(r\xi)\, d\xi \tag{12.24a}$$

$S_\phi(r)$ has, therefore, a bounded support since $\phi \subset C_\infty{}^0$. It also follows from Equation 12.24a that for $r > 0$, the spherical mean is infinitely differentiable with r. Even at $r = 0$, the spherical mean possesses derivatives of all orders with the odd order derivatives vanishing there. This can be seen by expanding $\phi(x)$ in a Taylor series around the origin $r = 0$ and then substituting in Equation 12.24a with f replaced by ϕ.

Since we are integrating over a sphere around the origin (a unit sphere in this case), the terms containing an odd number of factors of the x_i's do not contribute to the integral, whereas the terms with 2m even factors of x_i result in contributions of the form $a_m r^{2m}$. This yields, for the spherical mean, an expression of the form

$$S_\phi(r) = \phi(0) + \sum_{i=1}^{\infty} a_i r^{2i} \tag{12.24b}$$

The odd derivatives then clearly vanish at the origin, leaving only the nonzero even derivatives. As written above, the mean $S_\phi(r)$ possesses derivatives of all orders with the odd derivatives vanishing at $r = 0$. The spherical mean $S_\phi(r)$ is thus infinitely differentiable at the origin. From Equation 12.24b the value of the function is recovered at $r = 0$. For a general function $f \subset C_m$, its spherical mean is also C_m in both x and r. Setting $r = 0$ in Equation 12.24a, the value of the function is also recovered. Let us summarize these properties of the spherical mean into the following.

Lemma 12.2

Let $\phi(x)$ be a test function. Then its spherical mean $S_\phi(r)$ over a sphere of radius r is given by

$$\Omega_n S_\phi(r) = \int_\Omega \phi(r\xi)\, d\xi$$

where $d\xi$ is the elemental area of the hypersurface Ω of a generalized unit sphere embedded in an n-dimensional Euclidean space. The mean $S_\phi(r)$ is infinitely differentiable for $r > 0$ and has bounded support. At $r = 0$, the mean $S_\phi(r)$ is also infinitely differentiable with only the even derivatives being nonzero, all odd derivatives vanishing at the origin. Moreover, S is continuous in x and r and satisfies $(Sf)(0,x) = f(x)$. Furthermore, if $f \subset C_m$ then (Sf) also $\subset C_m$.

Next we consider the generalized function r^λ and its plane wave expansion for later use. The quantity r is defined by

$$r = [(x \cdot x)]^{\frac{1}{2}} = \left[\sum_{i=1}^{n} (x_i)^2\right]^{\frac{1}{2}}$$

The generalized function r^λ is defined through the functional

$$\left(r^\lambda, \phi\right) = \int_{R^n} r^\lambda \phi(x)\, dx, \ \ \mathrm{Re}\,\lambda > -n \subset Z^+$$

Recall that a generalized function is a continuous linear functional on $\mathcal{D}$ and that all such regular functions can be defined through the form given for (r^λ,ϕ).

Let w be a point on the unit sphere in R^n and consider the generalized function $s(n,\lambda)$ given by

$$s(n,\lambda) = |w \cdot x|^\lambda = \sum_{i=1}^{n} |w_i x_i|^\lambda$$

Also consider the integral

$$I(n,\lambda) = \int_\Omega s(n,\lambda)\,dw$$

Let λ be > -1. Since the integration is over a sphere, $I(n,\lambda)$ is a spherically symmetric function of x, is homogeneous of degree λ, and equals the generalized function r^λ to within a constant. Thus

$$I(n,\lambda) = C(n,\lambda) r^\lambda \tag{12.25}$$

The constant $C(n,\lambda)$ can be evaluated as follows. Let us consider the vector x such that $x = \{x_i = 0, i = 1,2, \ldots ,n-1; x_n = 1\}$. Then Equation 12.25 is

$$\int_\Omega |w_n|^\lambda\,dw = C(n,\lambda) \tag{12.26}$$

In a spherical coordinate system $\{\theta_i\}_1^{n-1}$ in the n-dimensional hyperspace $w_n = \cos\theta_{n-1}$ and the surface element of the unit sphere is given by

$$dw = \sin^{n-2}\theta_{n-1}\,dw_{n-1}$$

dw_{n-1} is the element of hypersurface area on the unit sphere in $n - 1$ dimensions. The integral of Equation 12.26 is then

$$2\Omega_{n-1}\int_0^{\frac{\pi}{2}} \sin^{n-2}\cos^\lambda\theta\,d\theta \tag{12.26a}$$

where Ω_{n-1} is the total surface measure of the generalized sphere in an $(n - 1)$-dimensional space. This area is known[31] to be

$$\Omega_{n-1} = \frac{2\pi^{\frac{(n-1)}{2}}}{\Gamma\left(\frac{n-1}{2}\right)}$$

Substituting $\sin^2\theta = x$, the integral reduces to

$$\left(\frac{1}{2}\right)\int_0^1 x^{a-1}(1-x)^{b-1}\,dx$$

where $a = (n - 1)/2$ and $b = (\lambda + 1)/2$. The integral is the well-known representation of the beta function $B(a,b)$ and is given by[39]

$$\frac{1}{2}B(a,b)=\frac{\Gamma(a)\Gamma(b)}{\Gamma(a+b)} \tag{12.26b}$$

From Equations 12.26, 12.26a, and 12.26b, the factor C(n,λ) is obtained from Equation 12.25

$$C(n,\lambda)=2\pi^{\frac{(n-1)}{2}}\frac{\Gamma\left(\frac{\lambda+1}{2}\right)}{\Gamma\left(\frac{\lambda+n}{2}\right)} \tag{12.27}$$

Using Equation 12.27, one finally obtains from Equation 12.25 the following expression

$$A\int_{\Omega}|w\cdot x|^{\lambda}\,dw=r^{\lambda} \tag{12.28}$$

where the constant A is given by

$$A=\left(\frac{1}{2}\right)\frac{\Gamma\left(\frac{n+\lambda}{2}\right)}{\pi^{\frac{(n-1)}{2}}\Gamma\left(\frac{\lambda+1}{2}\right)}$$

Equation 12.28 is the "plane wave expansion" of the generalized function r^{λ}. Setting $\lambda = 0$ in Equation 12.28 results in the formula for the area of an n-dimensional unit sphere, namely,

$$\Omega_n=\frac{2\pi^{\frac{n}{2}}}{\Gamma\left(\frac{n}{2}\right)}=C(n,0) \tag{12.29}$$

Differentiating r^{λ} with respect to λ, and comparing with the formula for Ω_{n-1} above, we obtain

$$\left(\frac{\partial}{\partial\lambda}\right)r^{\lambda}=r^{\lambda}\ln r$$

Using this fact and differentiating the plane wave expansion 12.28 yields the plane wave expansion for the generalized function (r^{λ} ln r). This is written as

$$\int_{\Omega}|w\cdot x|^{\lambda}\ln|w\cdot x|\,dw=A^{-1}r^{\lambda}\ln r$$

with a certain numerical constant c_{nk}. We note that these plane wave expansions also give the spherical means of these generalized functions.

These results are summarized into the following Lemma.

Lemma 12.3

Consider the generalized functions r^{λ} and (r^{λ} ln r) in the n-dimensional Euclidean space R^n, $r = (x\cdot x)^{1/2}$. Then the plane wave decomposition of these generalized functions is given by

$$A \int_{\Omega} |w \cdot x|^{\lambda} \, dw = r^{\lambda}$$

and by

$$\int_{\Omega} |w \cdot x|^{\lambda} \ln|w \cdot x| \, dw = A^{-1} r^{\lambda} \ln r$$

respectively, where

$$A = \left(\frac{1}{2}\right) \frac{\Gamma\left(\frac{n+\lambda}{2}\right)}{\pi^{\frac{(n-1)}{2}} \Gamma\left(\frac{\lambda+1}{2}\right)}$$

and c_{nk} is a numerical constant.

Now, the generalized function r^{λ} is defined by the functional

$$\left(r^{\lambda}, \phi\right) = \int_{R^n} r^{\lambda} \phi(x) \, dx$$

Using the spherical coordinate system, this expression becomes (using 12.24a)

$$\left(r^{\lambda}, \phi\right) = \int_0^{\infty} r^{\lambda} \left[\int_{\Omega} \phi(r\xi) \, d\xi\right] r^{n-1} \, dr = \Omega_n \int_0^{\infty} r^{\lambda} S_{\phi}(r) r^{n-1} \, dr$$

$$= \Omega_n \int_0^{\infty} S_{\phi}(r) r^{\lambda+n-1} \, dr = \Omega_n \left(S_{\phi}(r), r^{\lambda+n-1}\right)$$

r^{λ} is a well-defined tempered distribution. From the properties of the spherical mean $S_{\phi}(r)$ summarized in Lemma 12.2, $S_{\phi}(r) \subset C_{\infty}^{0}$ and is of bounded support, that is, $S_{\phi}(r) \subset \mathcal{D}$. Since $\mathcal{D} \subset \mathcal{L}$, it follows that r^{λ} is a tempered distribution. Moreover, it will be seen later that r^{λ} has simple poles at

$$\lambda = -n - 2h, \quad h \subset Z^{+} \tag{12.30}$$

Since r^{λ} is a tempered distribution, its Fourier transform exists and is also a tempered distribution.

The Fourier transform of a distribution T is defined by

$$\left(\tilde{\mathbf{T}}(\phi) = \mathbf{T}(\tilde{\phi}), \quad \phi \subset \mathcal{L}\left(R^n\right)\right) \tag{12.31}$$

Equation 12.31 means that the Fourier transform of a distribution acting on ϕ is identical to the distribution T acting on the Fourier transform of ϕ. The operations are, of course, in the sense of distribution. Let us consider a function $\phi(x) \subset \mathcal{L}(R^n)$ and a distribution $\psi(x) = \exp[-(1/2)t|x|^2]$. Then according to Equation 12.31 we have

$$\int_{R^n} \tilde{\phi}(x) \exp\left[-\left(\frac{1}{2}\right) t|x|^2\right] dx = \left[\frac{2\pi}{t}\right]^{\frac{n}{2}} \int_{R^n} \phi(u) \exp\left[-\left(\frac{1}{2t}\right)|u|^2\right] du$$

The Fourier transform of the distribution $\exp[-(1/2)t|x|^2]$ is just $(2\pi)^{n/2}\exp[-(1/2t)|u|^2]$. For example, see *Generalized Functions: Theory and Technique*, R. P. Kanwal, Academic Press, New York, 1983. Multiplying this equation by $t^{-\alpha/2-1}$ and integrating with respect to t, we obtain

$$\Gamma\left(\frac{-\alpha}{2}\right)2^{\frac{-\alpha}{2}}\int_{R^n}\tilde{\phi}(x)|x|^{\alpha}\,dx=(2\pi)^{\frac{n}{2}}\Gamma\left[\frac{(n+\alpha)}{2}\right]2^{\frac{(n+\alpha)}{2}}\int_{R^n}\tilde{\phi}(u)|u|^{-\alpha-n}\,du$$

We have used the identity

$$\int_0^{\infty}\exp\left[-\left(\frac{1}{2}\right)t|x|^2\right]t^{-1-\frac{\alpha}{2}}\,dt=\Gamma\left(\frac{-\alpha}{2}\right)2^{-\frac{\alpha}{2}}|x|^{\alpha}$$

From this equation and the location of the poles given by Equation 12.30 it follows that

$$\left(r^{\alpha}\right)^{\sim}=2^{n+\alpha}\pi^{\frac{n}{2}}\left\{\frac{\Gamma\left(\frac{n+\alpha}{2}\right)}{\Gamma\left(\frac{-\alpha}{2}\right)}\right\}r^{-\alpha-n}\left(\alpha,-\alpha-n\not\subset 2Z^{+}\right)\qquad(12.32)$$

Lemma 12.4

The Fourier transform of the generalized function r^{λ} is given by

$$\left(r^{\alpha}\right)^{\sim}=2^{n+\alpha}\pi^{\frac{n}{2}}\left\{\frac{\Gamma\left(\frac{n+\alpha}{2}\right)}{\Gamma\left(-\frac{\alpha}{2}\right)}\right\}r^{-\alpha-n},\quad\left(\alpha,-\alpha-n\not\subset 2Z^{+}\right)$$

Finally, we derive the action of the Laplacian Δ on the generalized function r^{λ}. If T be a tempered distribution, then operating on T by the Laplacian Δ and taking the Fourier transform yields $(-\Delta T)^{\sim}=r^2T^{\sim}$. Now identifying T with r^{2-n} and using the identity 12.32, we obtain

$$\left(-\Delta r^{2-n}\right)^{\sim}=\frac{4\pi^{\frac{n}{2}}}{\Gamma\left(\frac{n}{2}-1\right)}=\left(\frac{2\pi^{\frac{n}{2}}}{\Gamma\left(\frac{n}{2}\right)}\right)(n-2)\tilde{\delta}=\Omega_n(n-2)\tilde{\delta}$$

where the expression 12.29 for the area of the unit sphere was used. Inverting the Fourier transform yields

$$\Delta r^{2-n}=(2-n)\Omega_n\delta,\quad(n\neq 2)$$

It is also verified by direct computation that

$$\Delta r^{\alpha}=\alpha(\alpha+n-2)r^{\alpha-2},\quad\left(-\alpha-n+2\not\subset 2Z^{+}\right)\qquad(12.33)$$

Equation 12.33 can be iterated to obtain the action of any power of the Laplacian.

Lemma 12.5

The actions of the Laplacian on the generalized functions r^λ are given by

$$\Delta r^{\alpha} = \alpha(\alpha + n - 2)r^{\alpha-2}, \quad \left(-\alpha - n + 2 \not\subset 2Z^{+}\right)$$

and

$$\Delta r^{2-n} = (2-n)\Omega_n \delta, \quad (n \neq 2)$$

We now present various formal derivations of the inverse Radon transform. The derivation due to John is presented first.

B. DERIVATION OF THE INVERSE RADON TRANSFORM: JOHN[32]

Let f(x) be the function to be recovered from its integrals over the hypersurfaces. The starting point in John's derivation is Poisson's equation $\Delta u(x) = f(x)$. The Green's function for Poisson's equation is given by the equation $\Delta G(x,y) = \delta(x - y)$. From Lemma 12.5, the Green's function is obtained as

$$G(x,y) = \frac{|x-y|^{2-n}}{\Omega_n (2-n)}$$

The solution to Poisson's equation is then

$$u(x) = \int_{R^n} f(y) G(x,y)\, dy = \int_{R_n} f(y) \left[\frac{r^{2-n}}{(2-n)\Omega_n}\right] dy \tag{12.36}$$

We have written r for $|x - y|$. If $f \subset C_m$ then $u \subset C_{m+1}$.

We use the identities of Lemma 12.3 in order to make a plane-wave expansion of G(x,y), that is, calculate the Green's functions in terms of $|y - x|^k$ and $|y - x|^k \log|y - x|$. These Green's functions are then replaced in the solution (12.36) of the Poisson's equation to which the Laplacian is applied once more. Finally, using the plane wave expansions of Lemma 12.5 in the Green's functions the following expressions are obtained on iteration

$$\Delta^m r^\lambda = [\lambda(\lambda-2)\ldots(\lambda-2m+2)]\cdot[(\lambda+n-2)\ldots(\lambda+n-2m)]r^{\lambda-2m}$$

Setting $\lambda = k$ and $m = (n + k - 2)/2$, the above identity reduces to

$$\Delta^{\frac{(n+k-2)}{2}} r^k = [k(k-2)\ldots(4-n)]\cdot[(k+n-2)(k+n-4)\ldots 2]r^{2-n}$$

The above expression can be rewritten as

$$\Delta^{\frac{(n+k-2)}{2}} |y-x|^k = 2^{n+k-1}\left[\frac{\left[\Gamma\left(\frac{k+2}{2}\right)\Gamma\left(\frac{k+n}{2}\right)\Gamma\left(\frac{n}{2}\right)\right]}{(2-n)\pi}\right](-1)^{\frac{(n-1)}{2}}|y-x|^{2-n}$$

For n even and (n + k) even, we use $r^\lambda \ln r$ instead of r^λ and follow the same procedure to derive the following identity, $n > 2$

$$\Delta^{\frac{(n+k-2)}{2}}|y-x|^k\log|y-x| = 2^{n+k-2}\left[\frac{\left[\Gamma\left(\frac{k+2}{2}\right)\Gamma\left(\frac{k+n}{2}\right)\Gamma\left(\frac{n}{2}\right)\right]}{(2-n)\pi}\right](-1)^{\frac{(n-2)}{2}}|y-x|^{2-n}$$

We now replace $|y-x|^{2-n}$ from these identities in the solution of Poisson's equation and then apply the Laplacian once. Using the identity[39]

$$\Gamma(s)\Gamma\left\{s+\frac{1}{2}\right\} = 2^{1-2s}\pi^{\frac{1}{2}}\Gamma(2s)$$

the following expressions are obtained on using Lemma 12.3, namely

$$\Delta^{\frac{(n+k)}{2}}\int\left(\int_\Omega f(y)|(y-x)\cdot\xi|^k\,d\xi\right)dy = 4(2\pi i)^{n-1}k!f(x), \quad k=1,3,5,\ldots \tag{12.37}$$

and

$$\Delta^{\frac{(n+k)}{2}}\int\left(\int_\Omega f(y)|(y-x)\cdot\xi|^k\log|(y-x)\cdot\xi|\,d\xi\right)dy = -(2\pi i)^n k!f(x), \quad k=0,2,4,\ldots \tag{12.38}$$

for n odd and n even, respectively. These two expressions can be formally combined to yield a unified inverse Radon transform for even and odd n, namely,

$$f(x)=\Delta^{\frac{(n+k)}{2}}\Re e\left[\frac{-1}{k!(2\pi i)^n}\int\left(\int_\Omega f(y)|(y-x)\cdot\xi|^k\cdot\log\left[\left(\frac{1}{i}\right)(y-x)\cdot\xi\right]d\xi\right)dy\right]$$

The logarithm in the unified expression is the principal branch with the cut along R^-. Also Δ in all the above expressions act on the variable x.

Let us now take k to be unity and express the plane wave decompositions of Equations 12.37 and 12.38 explicitly in terms of the Radon transforms. The y-integral in Equation 12.37 for n odd can be reduced as follows:

$$\begin{aligned}\int_{R^n} f(y)|\xi\cdot(y-x)|\,dy &= \int_{R^n} dy\, f(y)\int_{-\infty}^{\infty} dp|p|\delta[p-\xi\cdot(y-x)] \\ &= \int_{-\infty}^{\infty} dp|p|\int_{R^n} dz\, f(x+z)\delta(p-\xi\cdot z) \\ &= \int_{-\infty}^{\infty}|p|\check{f}(p+\xi\cdot x,\xi)\,dp\end{aligned}$$

A change of variable $y = x + z$ was made and the shifting property (12.20) for the Radon transforms was used. This result can be reduced further. Changing the variable once more from p to $p = t - \xi\cdot x$, we obtain

$$\begin{aligned}\int_{-\infty}^{\infty}|p|\check{f}(p+\xi\cdot x,\xi)\,dp &= \int_{-\infty}^{\infty}|t-\xi\cdot x|\check{f}(t,\xi)\,dt \\ &= \int_{-\infty}^{\xi\cdot x}|t-\xi\cdot x|\check{f}(t,\xi)\,dt + \int_{\xi\cdot x}^{\infty}(t-\xi\cdot x)\check{f}(t,\xi)\,dt \\ &= \int_{\xi\cdot x}^{\infty}(t-\xi\cdot x)\check{f}(t,\xi)\,dt - \int_{-\infty}^{\xi\cdot x}(t-\xi\cdot x)\check{f}(t,\xi)\,dt\end{aligned} \tag{12.39}$$

The y-integral in Equation 12.38 is also similarly reduced to obtain (k = 0)

$$\int_{R^n} f(y)\log|\xi\cdot(y-x)|\,dy = \int_{R^n} dy\, f(y)\int_{-\infty}^{\infty} dp\,\log|p|\,\delta[p-\xi\cdot(y-x)]$$

$$= \int_{-\infty}^{\infty}\log|p|\,\check{f}(p+\xi\cdot x,\xi)\,dp \qquad (12.40)$$

Let us apply the Laplacian with respect to x once to Equations 12.39 and 12.40. For Equation 12.39 we obtain

$$\Delta\int_{-\infty}^{\infty}|p|\,\check{f}(p+\xi\cdot x,\xi)\,dp = \Delta\int_{\xi\cdot x}^{\infty}(t-\xi\cdot x)\check{f}(t,\xi)\,dt - \Delta\int_{-\infty}^{\xi\cdot x}(t-\xi\cdot x)\check{f}(t,\xi)\,dt$$

$$= 2\xi\cdot\xi\,\check{f}(\xi\cdot x,\xi) = 2\check{f}(\xi\cdot x,\xi), \quad \xi\cdot\xi = 1 \qquad (12.41)$$

For Equation 12.40 we proceed as follows. Applying the Laplacian again in x yields

$$\Delta\check{f}(p+\xi\cdot x,\xi) = \sum_{j=1}^{n}\partial_j\partial_j\,\check{f}(p+\xi\cdot x,\xi), \quad \partial_j = \partial/\partial x_j$$

$$= |\xi|^2\,\ddot{\check{f}}(t,\xi)\Big|_{t=p+\xi\cdot x}$$

$$= \ddot{\check{f}}(t\cdot\xi)\Big|_{t=p+\xi\cdot x}, \quad \xi\cdot\xi = 1 \qquad (12.42)$$

"·" denotes differentiation with respect to the variable t. Using Equation 12.42 and introducing a change of variable, namely, $p = t - \xi\cdot x$, as above, we have

$$\Delta\int_{-\infty}^{\infty}\log|p|\,\check{f}(p+\xi\cdot x,\xi)\,dp = \int_{-\infty}^{\infty}\log|t-\xi\cdot x|\,\ddot{\check{f}}(t,\xi)\,dt$$

$$= -⨍_{-\infty}^{\infty}\left[\frac{\check{f}'(p,\xi)}{(p-\xi\cdot x)}\right]dp \qquad (12.43)$$

The last integral is to be understood as the Cauchy principal value. The prime represents differentiation in the variable p. Equation 12.43 is a Hilbert transform.

Assuming integration over the unit sphere, that is $\xi\cdot\xi = 1$, and substituting Equation 12.41 in Equation 12.37 yields the inverse Radon transform for n odd

$$f(x) = \left(\frac{1}{2}\right)(2\pi i)^{1-n}\,\Delta^{\frac{(n-1)}{2}}\int_{\Omega}\check{f}(\xi\cdot x,\xi)\,d\xi \qquad (12.44)$$

Δ is again in the variable x. Equation 12.44 can be given an alternate expression. Using

Property 6 of the derivative of the Radon transform given in Section IV, we can write, interchanging x_k with ξ_k

$$\Delta^{\frac{(n-1)}{2}} \int_{\Omega} \check{f}(\xi \cdot x, \xi)\, d\xi = \int_{\Omega} \check{f}^{(n-1)}(\xi \cdot x, \xi)\, d\xi$$

where the (n – 1)-th derivative is with respect to p such that

$$\check{f}^{(n-1)}(p, \xi)\Big|_{p=\xi\cdot x} = \check{f}^{(n-1)}(\xi \cdot x, \xi)$$

The change in the order of differentiation from (n – 1)/2 to (n – 1) is because of the fact that the Laplacian is of the form $\partial_i \partial_i$. (For an illustrative and explicit calculations of the various derivatives used here, see Reference 36.)

Hence, the inversion formula for odd n is expressed alternately as

$$f(x) = \left(\frac{1}{2}\right)(2\pi i)^{1-n} \int_{\Omega} \check{f}^{(n-1)}(\xi \cdot x, \xi)\, d\xi \tag{12.44a}$$

Similar substitution of Equation 12.43 in Equation 12.38 yields the inverse Radon transform for even n

$$f(x) = (2\pi i)^{-n} D^{\frac{(n-2)}{2}} \int_{\Omega} \xi \oint_{-\infty}^{\infty} dp \left[\frac{\check{f}'(p, \xi)}{(p - \xi \cdot x)}\right] \tag{12.45}$$

Equation 12.45 can again be reduced further. Applying the Laplacian once to the p-integrand of Equation 12.45 yields

$$\Delta \int_{-\infty}^{\infty} \left[\frac{\check{f}'(p, \xi)}{(p - \xi \cdot x)}\right] dp = \int_{-\infty}^{\infty} \left[\frac{\check{f}'''(p, \xi)}{(p - \xi \cdot x)}\right] dp$$

Replacing this expression in Equation 12.45 yields the alternate expression for the Radon inversion for n even

$$f(x) = (2\pi i)^{-n} \int_{\Omega} d\xi \int_{-\infty}^{\infty} dp \left[\frac{\check{f}^{(n-1)}(p, \xi)}{(p - \xi \cdot x)}\right] \tag{12.45a}$$

Equations 12.43 through 12.45 determine the function f(x) in terms of the spherical mean of its integrals over the hypersurfaces in an n-dimensional Euclidean space R^n.

C. DERIVATION DUE TO GEL'FAND ET AL.[31]: METHOD OF PLANE-WAVE EXPANSION OF DELTA FUNCTIONS

The derivation of the inverse Radon transform by Gel'fand et al. is perhaps the most elegant and conceptually pleasing. It is based on the simple fact that a function f(x) can be expressed in terms of the delta distribution δ(x) as

$$f(x) = \int_{R^n} f(x')\delta(x - x')\, dx'$$

In the present case, f(x′) inside the integral will be a plane wave expansion related to the Radon transform. This suggests that the delta distribution is also expressed in such a manner as to contain plane wave functions as its arguments. Once this is obtained, the inversion will follow at once. This plane wave expansion of the delta distribution is the central point of the derivation by Gel'fand et al.

1. Preliminaries

Consider the generalized function x^{λ}_{+} defined by

$$x_{+}^{\lambda} = 0 \qquad \text{if } x \le 0$$

$$x_{+}^{\lambda} = x^{\lambda} \qquad \text{if } x > 0$$

The function is defined through the functional

$$\left(x_{+}^{\lambda}, \phi\right) = \int_0^{\infty} x^{\lambda} \phi(x)\, dx \tag{12.46}$$

For Re $\lambda > -1$, this functional is regular since x^{λ} is then locally integrable. Equation 12.46 can be continued analytically for $-2 < \text{Re}\ \lambda \le -1$ by "regularizing" the integral. Let us recall the definition of regularization.

Let f(x) have a singularity at some point x_0, and let $\phi(x) \subset \mathcal{D}$. Then the integral 12.46 is generally divergent. However, if $\phi(x)$ is such that it vanishes in a neighborhood of x_0, then the functional is said to be the "regularization" of the divergent integral 12.46. In general, if there exists an integer $m > 0$ such that $f(x) \cdot x^m$ is locally integrable, then the functional 12.46 can be regularized. The standard procedure is to subtract from $\phi(x)$ as many terms of its Taylor series expansion as necessary in order to make the remainder of the order of x^m or greater.

For $\lambda > -(n + 1)$, $\lambda \ne -n$, $n = 1,2,\ldots$, the regularization of the generalized function x^{λ}_{+} is given by:

$$\int_0^{\infty} x^{\lambda} \phi(x)\, dx = \int_0^{1} x^{\lambda} \left[\phi(x) - \phi(0) - x\phi'(0) - \ldots - \left\{ \frac{\left(x^{n-1}\right)}{(n-1)!} \right\} \phi^{(n-1)}(0) \right] dx +$$

$$+ \int_1^{\infty} x^{\lambda} \phi(x)\, dx + \sum_{k=1}^{n} \left[\frac{\phi^{(k-1)}(0)}{(k-1)!(\lambda + k)} \right] \tag{12.47}$$

This defines the generalized function x^{λ}_{+} for $\lambda \ne -n$, $n = 1,2,\ldots$ Equation 12.47 shows that as a function of λ, the generalized function x^{λ}_{+} has simple poles at $\lambda = -n$, $n \subset Z^{+}$. The residue at a pole $\lambda = -k$ is given by

$$\frac{\phi^{(k-1)}(0)}{(k-1)!} = \frac{(-1)^{(k-1)} \left(\delta^{(k-1)}(x), \phi(x)\right)}{(k-1)!} \tag{12.48}$$

This allows us to ascribe the simple pole at $\lambda = -k$ to the generalized function x^{λ}_{+} itself with the residue there

$$\left[\frac{(-1)^{k-1}}{(k-1)!}\right]\delta^{(k-1)}(x), \quad (k = 1, 2, \ldots)$$

Similarly, let us define the generalized function x^{λ}_{-} as

$$x^{\lambda} = |x|^{\lambda} \qquad \text{if } x < 0$$

$$x^{\lambda} = 0 \qquad \text{if } x \geq 0$$

The function is defined through the functional

$$\left(x_{-}^{\lambda}, \phi\right) = \int_{-\infty}^{0} |x|^{\lambda} \phi(x)\, dx$$

All the results for x^{λ}_{+} remain unchanged for x^{λ}_{-} if $\phi(x)$ is replaced by $\phi(-x)$ and $\phi(0)$ by $(-1)^{j}\phi^{(j)}(0)$. x^{λ}_{-} then is analytic in the entire λ plane with simple poles at $\lambda = -n$, $n = 1,2,\ldots$ having the residue at $\lambda = -k$ as

$$\frac{\delta^{(k-1)}(x)}{(k-1)!} \tag{12.49}$$

Note that the factor of $(-1)^{k-1}$ is absent in Equation 12.49.

Finally, we consider a third generalized function $|x|^{\lambda}$ defined by $|x|^{\lambda} = x^{\lambda}_{+} + x^{\lambda}_{-}$. Equations 12.48 and 12.49 imply that the function $|x|^{\lambda}$ has simple poles at $\lambda = -(2m + 1)$ with the residue $2\delta^{(2m)}(x)/(2m)!$, $m \subset Z^{+}$.

It is important to remove the singularities of these generalized functions. This is achieved by constructing functions $x^{\lambda}_{+}/\Gamma(\lambda + 1)$, $x^{\lambda}_{-}/\Gamma(\lambda + 1)$, and $|x|^{\lambda}/\Gamma[(\lambda + 1)/2]$. It can be seen that these are entire functions of λ. For each of these functions, the singularities are common to both the numerator and to the denominator. The values of these generalized functions at the singularities are just the ratios of the residues. We replace the Γ functions by the following integral representations:

$$\left(x_{+}^{\lambda}, \exp(-x)\right) = \int_{0}^{\infty} x^{\lambda} \exp(-x)\, dx = \Gamma(\lambda + 1)$$

$$\left(x_{-}^{\lambda}, \exp(x)\right) = \int_{-\infty}^{0} |x|^{\lambda} \exp(x)\, dx = \int_{0}^{\infty} x^{\lambda} e^{-x}\, dx = \Gamma(\lambda + 1)$$

$$\left(|x|^{\lambda}, \exp\left\{-x^{2}\right\}\right) = \int_{R} |x|^{\lambda} \exp\left(-x^{2}\right) dx = 2\int_{0}^{\infty} x^{\lambda} \exp\left(-x^{2}\right) dx$$

$$= \int_{0}^{\infty} t^{1/2(\lambda-1)} e^{-t}\, dt = \Gamma\left[\frac{(\lambda + 1)}{2}\right] \tag{12.50}$$

The residues are then calculated to yield

$$\left[x_+^\lambda/\Gamma(\lambda+1)\right]\Big|_{\lambda=-n} = \left[\operatorname*{res}_{\lambda=-n} x_+^\lambda \Big/ \operatorname*{res}_{\lambda=-n}\left(x_+^\lambda, e^{(-x)}\right)\right]$$

$$= (-1)^{n-1}\left(\delta^{(n-1)}(x)\right)(n-1)! \Big/ (-1)^{n-1}\left(\delta^{(n-1)}(x), \exp(-x)\right)(n-1)!$$

$$= \delta^{(n-1)}(x);$$

$$\left[x_-^\lambda/\Gamma(\lambda+1)\right]\Big|_{\lambda=-n} = \left[\operatorname*{res}_{\lambda=-n} x_-^\lambda \Big/ \operatorname*{res}_{\lambda=-n}\left(x^\lambda, e^{(x)}\right)\right]$$

$$= \left[\delta^{(n-1)}(x)/(n-1)!\right] \Big/ \left[\left(\delta^{(n-1)}(x), e^{(x)}\right)/(n-1)!\right]$$

$$= (-1)^{n-1}\delta^{(n-1)}(x);$$

$$\left[|x|^\lambda/\Gamma((\lambda+1)/2)\right]\Big|_{\lambda=-2m-1} = \operatorname*{res}_{\lambda=-2m-1} |x|^\lambda \Big/ \operatorname*{res}_{\lambda=-2m-1}\left(|x|^\lambda, \exp\{-x^2\}\right)$$

$$= \left[2\delta^{(2m)}(x)/2m!\right] \Big/ \left[2\left(\delta^{(2m)}(x), \exp\{-x^2\}\right)/(2m)!\right] \tag{12.51}$$

The denominator in the last expression can be further reduced. Note that $\delta^{(2m)}(x)\exp(-x^2) = [(\exp\{-x^2\})^{(2m)}]|_{x=0}$. Expanding the exponential then yields

$$\exp\left(-x^2\right) = 1 - x^2 + \frac{x^4}{2!} - \frac{x^6}{3!} + \ldots + (-1)^m \frac{x^{2m}}{m!} + \ldots = \sum_{m=0}^{\infty} \left(\exp\{-x^2\}\right)^{(2m)}\Big|_{x=0} \frac{x^{2m}}{(2m)!}$$

Therefore,

$$(\exp)\left\{-x^2\right\}^{(2m)}\Big|_{x=0} = \left[\frac{(-1)^m(2m)!}{m!}\right] \text{ yielding } \left[\frac{|x|^\lambda}{\Gamma((\lambda+1)/2)}\right]\Bigg|_{\lambda=-2m-1}$$

$$= \frac{(-1)^m \delta^{(2m)}(x) m!}{(2m)!} \tag{12.52}$$

We still require further computations before we are ready to expand the delta function into plane waves. We need to evaluate the residue of the function $2r^\lambda/\Gamma[(\lambda+n)/2]$ at its singularities $\lambda = -n$. The functional (r^λ, ϕ) is related to its spherical mean through the relation

$$\left(r^\lambda, \phi\right) = \Omega_n \int_0^\infty r^{\lambda+n-1} S_\phi(r)\, dr \tag{12.53}$$

In Lemma 12.2, $S_\phi(r)$ was found to be an even function in r with all the odd derivatives vanishing at the origin r = 0. The mean value function was also found to be infinitely differentiable and of bounded support. These imply that Equation 12.53 is the generalized function $\Omega_n x^\mu_+$ with $\mu = \lambda + n - 1$ with singularities at $\mu = -m$, that is, at $\lambda = -(m + n - 1)$. From Equation 12.48 the residue is

$$(-1)^{m-1}\frac{\left(\delta^{(m-1)}(x), S_\phi(x)\right)}{(m-1)!} = \frac{S_\phi^{(m-1)}(0)}{(m-1)!}$$

at $\mu = -m$. Since the odd derivatives of the spherical mean are zero at the origin, we must have $\lambda = -(n + 2k)$, k = 1, 2, … The residue at $\lambda = -(n + 2k)$ is then given by

$$\frac{\Omega_n\left(\delta^{(2k)}(x), S_\phi(x)\right)}{(2k)!} = \frac{\Omega_n S_\phi^{(2k)}(0)}{(2k)!} \tag{12.54}$$

Particularly at $\lambda = -n$ we have a simple pole with residue

$$\Omega_n S_\phi(0) = \Omega_n \phi(0) \tag{12.54a}$$

since $S_\phi(r) = \phi(0) + \sum a_i r^{2i}$.

Let us now return to Equation 12.33 and iterate 2k times. This yields

$$r^\lambda = \frac{\Delta^k r^{\lambda+2k}}{(\lambda+2)(\lambda+4)\dots(\lambda+2k)(\lambda+n)(\lambda+n+2)\dots(\lambda+n+2k-2)}$$

At $\lambda = -(n + 2k)$, the denominator is regular and the numerator is

$$\left(\Delta^k r^{\lambda+2k}, \phi(x)\right) = \left(r^{\lambda+2k}, \Delta^k\phi(x)\right) = \left(r^\mu, \Delta^k\phi(x)\right), \quad \mu = -n$$

Following the same reasonings as in obtaining Equation 12.54a, the residue of $(r^\mu, \Delta k\ \phi(x))$ at $\mu = -n$ is given by

$$\frac{\Omega_n \Delta^k \phi(0)}{(-2k-n+2)\dots(-n)(-2k)\dots(-2)} = \frac{\Omega_n\left(\Delta^k\delta(x), \phi(x)\right)}{2^k k!\, n(n+2)\dots(n+2k-2)}$$

The residue of the generalized function r^λ is then

$$\frac{\Omega_n \Delta^k \delta(x)}{2^k k!\, n(n+2)\dots(n+2k-2)} \tag{12.55}$$

Normalizing as before, the function $[2r^\lambda/\Omega_n\Gamma(\lambda+n)/2]$ at $\lambda = -(n + 2k)$ is the ratio of the residues of the numerator and the denominator yielding

$$\left(\frac{2}{\Omega_n}\right)\left[\frac{r^\lambda}{\Gamma\left(\frac{\lambda}{2}+\frac{n}{2}\right)}\right]\Bigg|_{\lambda=-n-2k} = \frac{\operatorname*{res}_{\lambda=-n-2k} r^\lambda}{\operatorname*{res}_{\lambda=-n-2k}\left(r^\lambda, \exp\left\{-r^2\right\}\right)}$$

$$= \frac{\left[\dfrac{\Omega_n \Delta^k \delta(x)}{2^k k!\, n(n+2)\ldots(n+2k-2)}\right]}{\left[\dfrac{\left(\Omega_n \delta^{(2k)}(x), \exp\left\{-x^2\right\}\right)}{(2k)!}\right]}$$

$$= (-1)^k \left[\frac{\Delta^k \delta(x)}{2^k n \ldots (n+2k-2)}\right] \tag{12.56}$$

Again we have used the identity

$$\left(r^\lambda, \exp\left\{-r^2\right\}\right) = \int r^\lambda \exp\left(-r^2\right) r^{n-1}\, dw\, dr = \left[\frac{\Omega_n}{2}\right]\left[\frac{\Gamma(\lambda+n)}{2}\right]$$

2. The Derivation

At this point we have all the necessary results required to expand a delta distribution in terms of plane wave functions. We start with the plane wave expansion (12.28). Consider the singularity at $\lambda = -n$. According to Equation 12.56, then

$$\left[\frac{2r^\lambda}{\Gamma\left(\frac{\lambda}{2}+\frac{n}{2}\right)}\right]\Bigg|_{\lambda=-n} = \Omega_n \delta\left(x_1, x_2, \ldots, x_n\right) \tag{12.57}$$

The left-hand side of Equation 12.28 is of the form $|y|^\lambda/\Gamma\{(\lambda+1)/2\}$ and by Equation 12.52

$$\left[\frac{|y|^\lambda}{\Gamma\left(\frac{\lambda}{2}+\frac{1}{2}\right)}\right]\Bigg|_{\lambda=-n} = (-1)^{\frac{(n-1)}{2}}\left[\frac{\left(\frac{n}{2}-\frac{1}{2}\right)!}{(n-1)!}\right]\delta^{(n-1)}(y) \tag{12.58}$$

for odd n. Substituting Equations 12.57 and 12.58 in the plane wave expansion (12.28), we obtain with $y = w \cdot x$

$$\delta\left(x_1, \ldots x_n\right) = \left[\frac{(-1)^{\frac{(n-1)}{2}}\left(\frac{n}{2}-\frac{1}{2}\right)!}{\pi^{\frac{(n-1)}{2}}(n-1)!\Omega^n}\right]\int_\Omega \delta^{(n-1)}\left(w_1 x_1 + \ldots + w_n x_n\right) dw$$

The area of the surface of the unit sphere for odd n is

$$\Omega_n = \frac{2\pi^{\frac{n}{2}}}{\Gamma\left(\frac{n}{2}\right)} = \frac{2(2\pi)^{\frac{(n-1)}{2}}}{1, 3, 5 \ldots (n-2)} \tag{12.59}$$

Replacing Equation 12.59, the final result is

$$\delta(x_1, \ldots, x_n) = \left[\frac{(-1)^{\frac{(n-1)}{2}}}{2(2\pi)^{n-1}}\right] \int_\Omega \delta^{(n-1)}(w_1 x_1 + \ldots + w_n x_n)\, dw \tag{12.60}$$

Equation 12.60 is the plane wave expansion of a delta distribution when n is odd.

For even n, the generalized function

$$|x|^{-n} = |x|^{-2m} = x^{-n} = x^{-2m}, \quad m \subset Z^+$$

This is because at $\lambda = -2m$, $m \subset Z^+$. $|x|^\lambda$ is well defined. This follows from the fact that $|x|^\lambda$ has poles only if $\lambda = -(2m + 1)$, as discussed above. The regularized expression for x^{-2m} is given by

$$\left(x^{-2m}, \phi\right) = \int_0^\infty x^{-2m} \left\{\phi(x) + \phi(-x) - 2\left[\phi(0) + \left(\frac{x^2}{2!}\right)\phi''(0) + \ldots + \left(\frac{x^{2m-2}}{(2m-2)!}\right)\phi^{(2m-2)}(0)\right]\right\} dx \tag{12.61}$$

yielding

$$\left[\frac{|x|^\lambda}{\Gamma\left(\frac{\lambda}{2} + \frac{1}{2}\right)}\right]\Bigg|_{\lambda=-n} = \left[\frac{1}{\Gamma\left(\frac{-n}{2} + \frac{1}{2}\right)}\right] x^{-n} \tag{12.61a}$$

Inserting Equations 12.61a and 12.57 in Equation 12.28 and using Equation 12.59 for even n yields

$$\delta(x_1, \ldots, x_n) = \left[\frac{(-1)^{\frac{n}{2}}(n-1)!}{(2\pi)^n}\right] \int_\Omega (w_1 x_1 + \ldots + w_n x_n)^{-n}\, dw \tag{12.62}$$

Equation 12.62 is the plane wave expansion of the delta distribution for even n.

Let us apply these expansions to a function $\phi(x)$. Then for odd n, we obtain

$$\int \phi(x)\delta(x - x_0)\, dx = \phi(x_0) = A \int_\Omega d\xi \int_{R^n} \delta^{(n-1)}(p - \xi \cdot x)\phi(x)\, dx \tag{12.63}$$

In Equation 12.63, $p = \xi \cdot x_0$ and A is the factor multiplying the integral on the right-hand side of Equation 12.60. The x integral is just the Radon transform of the (n – 1)-th derivative of $\phi(x)$ and, by the results established in Section IV, Property 5, is the (n – 1)-th derivative of the Radon transform of $\phi(x)$ with respect to p. Finally, we then obtain the inversion formula for the Radon transform when n is odd

$$f(x) = \left[\frac{(-1)^{\frac{(n-1)}{2}}}{2(2\pi)^{n-1}} \right] \int_\Omega \check{f}_p^{(n-1)}\big(\xi, (\xi, x)\big)\, d\xi \qquad (12.64)$$

This is the same as Equation 12.44 with $(-1)^{(n-1)/2}$ factor instead of i^{1-n}.

Let us now consider n even and define $\zeta = [p - \xi \cdot x]^{-n}$. As mentioned before, this function is to be understood in its regularization as given by Equation 12.61. Consider the Radon transform

$$\check{\phi}(p,\xi) = \int_{\xi \cdot x = p} \phi(x)\, dm(x)$$

If $\phi(x) \subset \mathcal{D}$, the $\check{\phi}$ also $\subset \mathcal{D}$ and we have from Equation 12.61

$$\phi(x_0) = \left[\frac{(-1)^{\frac{n}{2}}(n-1)!}{(2\pi)^n} \right] \int_\Omega \left(\zeta, \check{\phi} \right) d\xi$$

Replacing x for x_0 yields the Radon inversion formula for even n

$$f(x) = \left[\frac{(-1)^{\frac{n}{2}}(n-1)!}{(2\pi)^n} \right] \int_\Omega \left[\int_R \check{f}(\xi, p) \{p - (\xi, x)\}^{-n}\, dp \right] d\xi$$

where the inner integral is to be regularized as given by Equation 12.61. If the integral is considered in terms of the Cauchy principal value, then only the first term of the regularization expression remains, yielding the more familiar formula for the inverse Radon transform for even n written as

$$f(x) = (2\pi i)^{-n} \int_\Omega d\xi \int_R dp \left[\frac{f^{(n-1)}(p,\xi)}{(p - \xi \cdot x)} \right]$$

The integral is then a Hilbert transform.

D. LUDWIG'S DERIVATION[34]: FOURIER TECHNIQUE

Ludwig's derivation is based on the relation between the Radon and the Fourier transform of a function. In Section III the relation between these two transforms was established. Equation 12.3 expressed the Fourier transform $f(k) = f(k\xi)$ in terms of the Radon transform $\check{f}(p,\xi)$. Inverting the Fourier transform yields

$$f(x) = (2\pi)^{-\frac{n}{2}} \int_{R^n} \tilde{f}(k)\exp(ik \cdot x)dk$$

$$= (2\pi)^{-\frac{n}{2}} \int_{\Omega}\int_0^{\infty} d\xi\, d|k||k|^{n-1}\, \tilde{f}(|k|\xi)\exp(i|k|\xi \cdot x)$$

$$= (2\pi)^{-n} \int_{\Omega} d\xi \int_0^{\infty} d|k||k|^{n-1} \exp(i|k|\xi \cdot x)\int_R dp\, \check{f}(p,\xi)\exp(-i|k|p)$$

$$= (2\pi)^{-n} \int_{\Omega} d\xi\, F(\xi \cdot x, \xi)\, d\xi$$

where the function $F(\xi{\cdot}x,\xi)$ is defined by the above integral and Ω is, as before, the surface of the unit sphere, that is, $|\xi| = 1$. Since we are integrating over a sphere, the odd part of the integral does not contribute and we then work only with the even part. Towards this, we define

$$F_e(t,\xi) = \left(\frac{1}{2}\right)\left[F(t,\xi) + F(-t,-\xi)\right]$$

with $t = \xi{\cdot}x$. Then, the inversion formula is

$$f(x) = (2\pi)^{-n} \int_{\Omega} d\xi\, F_e(t,\xi) = \int_{\Omega} d\xi\, F_e(\xi \cdot x, \xi) \qquad (12.65)$$

Using the symmetry property of the Radon transform, namely, $\check{f}(p,\xi) = \check{f}(-p,-\xi)$, and changing $k \to -k$ in the integrand involving $F(-t,-\xi)$, the integrand in Equation 12.65 can be reduced to

$$F_e(t,\xi) = \left(\frac{1}{2}\right)\int_R dk|k|^{n-1} \int_R dp\, \check{f}(p,\xi)\exp\{ik(t-p)\} \qquad (12.66)$$

Let n be odd. Then the absolute value on k can be dropped and on integrating (n–1) times with respect to the Radon transform variable p, the factor k^{n-1} can be eliminated. The result is

$$F_e(t,\xi) = A' \int_R dk \int_R dp\, \check{f}^{(n-1)}(p,\xi)\exp\{ik(t-p)\} \qquad (12.67)$$

where the constant A' is $i^{(1-n)/2}$, as before. The k-integral can be evaluated to yield the delta function $\delta(p-t)$, which in turn reduces Equation 12.67 to

$$F_e(t,\xi) = A\check{f}^{(n-1)}(t,\xi)\Big|_{t=\xi \cdot x}$$

where $A = \pi i^{1-n}$. Replacing this in Equation 12.65 results in the same inversion formula for the Radon transform as derived in the previous two sections when n is odd.

For n even, the function $|k|$ must be replaced by k(sgn k), where the signum function is defined by

$$\operatorname{sgn} k = 1 \qquad \text{if } k > 0$$

$$= -1 \qquad \text{if } k < 0$$

With this replacement, Equation 12.67 becomes

$$F_e(t,\xi) = A' \int_R dk \operatorname{sgn} k \int_R dp\, \check{f}^{(n-1)}(p,\xi) \exp\{-ik(p-t)\} \tag{12.68}$$

The k-integral is just the Fourier transform of the signum function in the variable (t–p) and is given by $i\hat{H}(1)$. Replacing this in Equation 12.68 yields

$$F_e(t,\xi) = A'\left(\frac{i}{\pi}\right) \oint_R dp \left[\frac{\left\{\check{f}^{(n-1)}(p,\xi)\right\}}{(t-p)}\right]$$

The integral is to be understood as the Cauchy principal value integral. Since the Hilbert transform of a function F(x) is

$$\left(\hat{H}F\right)(x) = \left(\frac{1}{\pi}\right) \oint_R \left[\frac{F(t')}{(t'-t)}\right] dt'$$

the above expression can be rewritten as

$$F_e(t,\xi) = iA'\hat{H}\left[\check{f}^{(n-1)}(p,\xi)\right]$$

Substituting in Equation 12.65 then yields the inversion of the Radon transform in even dimension

$$f(x) = iA \int_\Omega d\xi \left[\hat{H}\left\{\check{f}^{(n-1)}(p,\xi)\right\}\right](\xi \cdot x, \xi)$$

A is as given by Equation (12.70).

The inversion formulas for the Radon transform in R^n can then be combined to give

$$f(x) = \int_\Omega F_e(x \cdot \xi, \xi)\, d\xi = \int_\Omega K \check{f}(p,\xi)\, d\xi \tag{12.69}$$

where

$$K\, g(p,\xi) = \begin{cases} Ag^{(n-1)}(p,\xi), & n \text{ odd} \\ iA\hat{H}g^{(n-1)}(p,\xi), & n \text{ even} \end{cases}$$

$$A = \left[1/2\,(2\pi i)^{n-1}\right] \tag{12.70}$$

and the Hilbert transform is defined as

$$\left(\hat{H}F\right)(x) = \left(\frac{1}{\pi}\right) \oint_R \left[\frac{F(t')}{(t - t')}\right] dt'$$

From Equations 12.3 and 12.66, **K** is an operator that multiplies the Fourier transform of g in p by a factor $A|k|^{n-1}$. The operator **K** is, therefore, positive and symmetric. Some authors such as Helgason define the Hilbert transform with a factor of (i/π) in front of the Cauchy principal value instead of $(1/\pi)$ as has been done here. This introduces only some very minor changes in the form of the inversion formula.

Equation 12.69 is a measure of a spherical mean. In the case of odd dimension, the integration is over all the hyperplanes that pass through a given point x. The inversion formula in this case is local in that the value of the function f(x) at a point x depends only on the integrals of f over the hyperplanes passing through x. In the case of even dimension, the inversion formula is global and one must integrate over all possible hyperplanes before calculating the spherical mean.

VI. THE ADJOINT OPERATOR AND PARSEVAL'S THEOREM FOR RADON TRANSFORM

Equation 12.69 also provides a basis for defining the "adjoint" of the Radon transform. In order to do this, consider the inner product of the complex conjugate of the Radon transform and a function on R^n. Then

$$\int_\Omega \int_R \check{f}^*(p,\xi)\phi(p,\xi)\,dp\,d\xi = \int_\Omega \int_R \int_{R^n} \delta(p - x\cdot\xi) f^*(x)\phi(p,\xi)\,dx\,dp\,d\xi$$

$$= \int_{R^n} f^*(x) \int_\Omega \phi(x\cdot\xi,\xi)\,d\xi\,dx$$

It then follows that

$$\int_\Omega \int_R (\Re f)^*(p,\xi)\phi(p,\xi)\,dp d\xi = \int_{R^n} f^*(x)\left(\Re^*\phi\right)(x)\,d\xi\,dx \qquad (12.71)$$

yielding

$$\left(\Re^*\phi\right)(x) = \int_\Omega \phi(x\cdot\xi,\xi)\,d\xi \qquad (12.72)$$

Replacing $\phi(x\cdot\xi\cdot\xi)$ by $F_e(x\cdot\xi\cdot\xi)$, the following theorem is obtained:

Theorem 12.6

Consider a function $f \subset \mathcal{L}$. Then $f = \Re^* \mathbf{K} \Re f$. $\Re$ is the operation of the Radon transform given by Equation 12.1 or 12.2, **K** is a positive symmetric operator defined in Equation 12.70, and $\Re^*$ is the "adjoint" of the Radon transform operator $\Re$ and is defined in Equation 12.71.

It is noteworthy that the Radon transform and its adjoint form a pair of duals in the sense of integral geometry. The Radon transform operator $\Re$ integrates over all points on a hyperplane. The adjoint operator $\Re^*$ is the reverse. It sums up all the hyperplanes passing through a point. The combined operation of $\Re^*\Re$ is generally called the "backprojection" operator and forms the basis of what is known as "tomography". It is not difficult to show[35] that the operation of $\mathfrak{R}^*\mathfrak{R}$ on a function $f(x) \subset \mathcal{L}$ is that of convolving the function with $\Omega_n|x|^{-1}$. That is, $\mathfrak{R}^*\mathfrak{R} = \Omega_n|x|^{-1}*f$.

The adjoint provides a suitable basis for defining the Radon transforms of distributions. Thus if $f(x) = \delta(x - x_0)$, then

$$\left(\mathfrak{R}^*\delta, \phi\right) = \left(\delta, \mathfrak{R}^*\phi\right) = \mathfrak{R}^*\phi(x_0)$$

It can be shown that if ϕ is even and $\phi \subset \mathcal{L}(R \times S^{n-1})$, then

$$\left(\mathfrak{R}^*\mathbf{K}\phi\right)(x) = f(x) \tag{12.73}$$

implies that $\phi = \check{f}$. The proof makes use of Fredholm's alternative theorem to show that $(\mathfrak{R}^*\mathbf{K}\phi) = 0$ has only a trivial solution, implying that Equation 12.73 has a unique solution. Since by Theorem 12.6, $(\mathfrak{R}^*\mathbf{K}\check{f})(x) = f(x)$, then ϕ must be the Radon transform of f. The proof, which is relatively simple,[34] is omitted. This then establishes the following useful theorem:

Theorem 12.7

If $f \subset \mathcal{L}$, ϕ is even and $\phi \subset (R \times S^{n-1})$, and if $\mathfrak{R}^*\mathbf{K}\phi = f$, then $\phi \equiv \check{f}$.

It is possible to derive a Parseval relation for the Radon transforms. Since the operator **K** defined in Equation 12.70 is positive and symmetric, we can take its square root and define

$$\left(K^{\frac{1}{2}}g\right)(k) = \left[\frac{1}{2}\left(\frac{1}{2\pi}\right)^{n-1}\right]^{\frac{1}{2}} |k|^{\frac{(n-1)}{2}} \tilde{g}(k) \tag{12.74}$$

the Fourier transform being in p. If n is odd, then we have

$$\left(\mathbf{K}^{\frac{1}{2}}g\right)(p) = \left[\frac{1}{2}\left(\frac{1}{2\pi}\right)^{n-1}\right]^{\frac{1}{2}} \left(\frac{d}{idp}\right)^{\frac{(n-1)}{2}} g(p) \tag{12.74a}$$

Consider the integral $\int_{R^n} f^*f\,dx$. In view of the operator $\mathbf{K}^{1/2}$, we have from Equation 12.71

$$\int_{R^n} f^*f\,dx = \int_{R^n} f^*\mathfrak{R}^*\mathbf{K}\mathfrak{R}f\,dx = \iint (\mathfrak{R}f)^*\mathbf{K}\check{f}\,dp\,d\xi = \iint \left|\mathbf{K}^{\frac{1}{2}}\mathfrak{R}f\right|^2 dp\,d\xi$$

This then establishes Parseval's relation given in

Theorem 12.8

$$\text{If } f \subset L, \text{ then } \int_{R^n} |f|^2\,dx = \int_\Omega \int_R \left|\mathbf{K}^{\frac{1}{2}}\mathfrak{R}f\right|^2 dp\,d\xi$$

Let us discuss the mapping $\mathbf{K}^{1/2}\mathfrak{R}$. From Theorem 12.6

$$\mathfrak{R}^*\mathbf{K}\mathfrak{R} = \left(\mathfrak{R}^*\mathbf{K}^{\frac{1}{2}}\right)\left(\mathbf{K}^{\frac{1}{2}}\mathfrak{R}\right) = I \tag{12.75}$$

Furthermore,

$$\mathfrak{R}^*\psi = \mathbf{I}\mathfrak{R}^*\psi = \mathfrak{R}^*\mathbf{K}\mathfrak{R}\mathfrak{R}^*\psi \Rightarrow \mathbf{I} = \mathbf{K}\mathfrak{R}\mathfrak{R}^*$$

yielding

$$\mathbf{K}\mathfrak{R}\mathfrak{R}^* = \mathbf{K}^{\frac{1}{2}}\left(\mathbf{K}^{\frac{1}{2}}\mathfrak{R}\right)\mathfrak{R}^* = I \tag{12.76}$$

Premultiplying both sides of Equation 12.76 by $\mathbf{K}^{1/2}$ postmultiplying by $\mathbf{K}^{1/2}$ yields

$$\left(\mathbf{K}^{\frac{1}{2}}\mathfrak{R}\right)\left(\mathfrak{R}^*\mathbf{K}^{\frac{1}{2}}\right) = I \tag{12.77}$$

From Equations 12.75 and 12.77, we conclude that $\mathbf{K}^{1/2}\mathfrak{R}$ is unitary.

Let us now consider the function $g(p,\xi) \subset L_2(R \times S^{n-1})$ and its one-dimensional Fourier transform $\tilde{g}(|k|,\xi) \subset L_2(R \times S^{n-1})$. Define

$$\tilde{f}(k) = \frac{2^{\frac{1}{2}}\left(|k|, \frac{k}{|k|}\right)}{\left(2\pi|k|^{n-1}\right)^{\frac{1}{2}} g^{\sim}} \tag{12.78}$$

$\tilde{f}(k)$ is then $L_2(R^n) \Rightarrow f(x)$ itself is $L_2(R^n)$. From Equation 12.78

$$\tilde{g}(|k|,\xi) = \left[\frac{(2\pi|k|)^{n-1}}{2}\right]^{\frac{1}{2}} \tilde{f}(|k|,\xi)$$

From Equation 12.74 and the relation 12.3 between the Fourier and the Radon transform, it follows that $g = \mathbf{K}^{1/2}\mathfrak{R}f$. Therefore, if $g \subset L_2(R \times S^{n-1})$, then $g = \mathbf{K}^{1/2}/2\mathfrak{R}f$, where $f \subset L_2(R^n)$. This result then shows that $\mathbf{K}^{1/2}\mathfrak{R}$ is a mapping from $L_2(R^n)$ into $L_2(R \times S^{n-1})$. Since the operator is unitary by Equation 12.77, we have

Theorem 12.9

$\mathbf{K}^{1/2}\mathfrak{R}$ is a unitary mapping of $L_2(R^n)$ onto $L_2(R \times S^{n-1})$.

VII. ILL-POSEDNESS AND REGULARIZATION

The inversion formulas presented thus far determine f uniquely if the Radon transform $\check{f}$ is known over $R \times S^{n-1}$. However, in practice, the data can be obtained only over a subset of this domain and hence is incomplete. The uniqueness of the solution is then problematic and even if a unique solution does exist, the stability problem can be severe. This situation occurs if, for example, $\check{f}$ is given only over a subset of S^{n-1}, the so-called "limited-angle" case, or if $\check{f}$ is given only in the exterior (interior) of some convex, compact ball in R^n, the "exterior" ("interior") problem. The function is, of course, to be determined only in the exterior (interior), and, finally, the integrals of f given only over a certain curve, the so-called "restricted source" problem. The last mentioned case was discussed by Kirillov (see Chapter 2). All these cases share a common difficulty, namely, that of ill-posedness. In view of its importance in practical imaging problems, we will discuss this briefly for the sake of completeness. A thorough discussion of this topic is beyond the scope of this book. We follow Reference 35 in the main.

The nature of ill-posedness was discussed in Chapter 1, where both Hadamard's classical definition and Tikhonov-Arsenin's generalized definitions of ill-posedness were introduced. Given two Hilbert spaces H_1 and H_2 and a linear bounded operator $\mathbf{A}$ from H_1 to H_2, a problem is well-posed in the sense of Hadamard if and only if $\mathbf{A}f = g$, $f \subset H_1$ and $\forall\, g \subset H_2$ and if continuity of f with g exists. Well-posedness thus breaks down if $\mathbf{A}$ does not exist or is not continuous. The problem of continuity is extremely important since it is entirely possible that a solution (even a unique one) may exist, but may differ significantly from g if g varies only infinitesimally. The exterior problem mentioned above is a case in point.[35]

The most well-known solution of the ill-posed problem is that of the "generalized inverse" of Moore-Penrose.[40] This is an operator $\mathbf{A}^+{:}H_2 \to H_1$ such that the solution $f = \mathbf{A}^+g$ satisfies the so-called "normal" equation $\mathbf{A}^+\mathbf{A}f = \mathbf{A}^+g$. However, $\mathbf{A}^+$ is in general not continuous. The restoration of continuity is achieved through what is known as "regularization". It is analogous to the regularization of various generalized functions discussed in connection with the Gel'fand-Graev-Vilenkin derivation of the inverse Radon transform. Briefly, the idea is as follows. One defines a family of linear continuous operators $\mathbf{T}_\gamma{:}H_2 \to H_1$ on all H_2, $\gamma > 0$, parameterized by the "regularization parameter" γ such that limit $\gamma \to 0$, $\mathbf{T}_\gamma g \to \mathbf{A}^+g$ on the domain of $\mathbf{A}^+$. Let $\gamma = \gamma(\varepsilon)$ such that $\gamma(\varepsilon) \to 0$ as $\varepsilon \to 0$, ε being an arbitrarily small positive number such that $\|g^\varepsilon - g\| \le \varepsilon$, g^ε being a close approximation of g in H_2. Then, imposing the condition that $\|\mathbf{T}_{\gamma(\varepsilon)}g\|\varepsilon \to 0$, we have as $\varepsilon \to 0$

$$\left\|\mathbf{T}_{\gamma(\varepsilon)}g - \mathbf{A}^+g\right\| \le \left\|\mathbf{T}_{\gamma(\varepsilon)}\left(g^\varepsilon - g\right)\right\| + \left\|\mathbf{T}_{\gamma(\varepsilon)}g - \mathbf{A}^+g\right\|$$

$$\le \left\|\mathbf{T}_{\gamma(\varepsilon)}\right\|\varepsilon + \left\|\mathbf{T}_{\gamma(\varepsilon)}g - \mathbf{A}^+g\right\| \to 0$$

implying continuity since $\mathbf{T}_{\gamma(\varepsilon)}g^\varepsilon$ approaches $\mathbf{A}^+g$ as $g^\varepsilon \to \varepsilon$.

An interesting example of regularization is an iterative procedure described by $f^{(n+1)} = \mathbf{P}_n f^{(n)} + \mathbf{Q}_n g$, $\mathbf{P}_n$, $\mathbf{Q}_n$ being continuous operators and the superscript n denoting the stage of iteration. Assuming that the iteration converges as $n \to \infty$, we can write the index γ such that $n(\gamma) \to \infty$ as $\gamma \to 0$. Hence the iteration procedure is described by $\mathbf{T}_\gamma f \to f^{n(\gamma)}$. The regularization parameter is simply the number of iterations performed.

The other well known regularization methods are that of singular value decomposition (SVD)[40] and a method due to Tikhonov and Arsenin.[41] In the Tikhonov-Arsenin method the regularization operator is defined by

$$\mathbf{T}_\gamma = \left(\mathbf{A}^+\mathbf{A} + \gamma\mathbf{I}\right)^{-1}\mathbf{A}^+ = \mathbf{A}^+\left(\mathbf{A}^+\mathbf{A} + \gamma\mathbf{I}\right)^{-1}$$

and it is shown that $\mathbf{T}_\gamma g \to \mathbf{A}^+g$ as $\gamma \to 0$. The SVD procedure is the following. If ψ_k and ϕ_k are orthogonal sequences in H_1 and H_2, respectively, then the operator $\mathbf{A}$ is said to have a SVD if it can be written as

$$\mathbf{A} = \sum_{k=1}^{\infty} \sigma_k (f, \psi_k) \phi_k$$

The positive numbers σ_k are called the "singular values" of **A**. If **A** can be represented as above, then it can be shown that

$$\mathbf{T}_\gamma g = \sum_{k=1}^{\infty} F_\gamma(\sigma_k)(g, \phi_k)\psi_k$$

$F_\gamma(\sigma) \to \sigma^{-1}$ as $\sigma \to \infty$ and $\to 0$ as $\sigma \to 0$. The above equation for the regularization is in the limit $\gamma \to 0$. For further details regarding regularization of the Radon transform operator we refer the reader to Natterer's monograph.[35] The most well-known reference on ill-posed problems is Reference 41. A lucid and readable account of ill-posed problems in integral geometric situation is the book by Craig and Brown.[42]

VIII. RADON TRANSFORM OF GENERALIZED FUNCTIONS

We now present a brief account of Radon transforms of generalized functions and follow Gel'fand et al.[31] in the main. We will work with spaces of odd dimensions. Consider condition 3(a) of Theorem 12.1. Integration of the moment integral with respect to p by parts (n – 1) times and Lemma 12.1 show that the integral

$$\int_R \psi(p,\xi)p^k \, dp = \int_R \check{f}^{(n-1)}(p,\xi)p^k \, dp$$

is a homogeneous polynomial in ξ of degree $(k - n + 1)$. For $k < n - 1$, the integral vanishes. $\psi(p,\xi) \equiv \check{f}^{(n-1)}(p,\xi)$. Moreover, if $f \subset \mathcal{D}$, then so does ψ. That is, if f is a test function, then so is ψ.

Now consider functions of the form $p^k a_{-k-1}(\xi)$, where a_{-k-1}'s are arbitrary except for the fact that they are required to satisfy two conditions, namely,

$$a_{-k-1}(\alpha\xi) = \alpha^{-k}|\alpha|^{-1} a_{-k-1}(\xi), \quad \alpha \neq 0, \quad k < n-1$$

$$= \int_\Omega a_{-k-1}(\xi) P_{k-n+1}(\xi) \, d\xi = 0, \quad k \geq n-1$$

for every homogeneous polynomial $P_{k-n+1}(\xi)$ of degree k–n+1. Then

$$\left(p^k a_{-k-1}(\xi), \psi\right) \equiv \int_\Omega \left[\int_R \psi(p,\xi)p^k \, dp\right] a_{-k-1}(\xi) \, d\xi = 0$$

The subspace generated by the functions of the form $p^k a_{-k-1}(\xi)$ is termed the subspace of "unessential" functions by Gel'fand et al. It will be seen that the Radon transform of a generalized function can be defined only up to an unessential function. With this fact in mind, we now turn to defining the Radon transform of generalized functions.

Our starting point is Parseval's theorem discussed in the previous section. Using Equations 12.69, 12.70, and 12.71, Parseval's relation can be generalized to two functions f(x) and g(x) to yield the expression[31]

$$\int_{R^n} f(x) G^*(x) \, dx = \frac{1}{2}\left(\frac{i}{2\pi}\right)^{n-1} \cdot \int_\Omega \left[\int_R \check{f}(p,\xi) \check{G}^{*(n-1)}(p,\xi) \, dp\right] d\xi \qquad (12.79)$$

Let F(x) be a generalized function and $f(x) \subset \mathcal{L}(R^n)$. Then Parseval's theorem for F and f in a space of odd dimension yields

$$\int_{R^n} f(x)F(x)\,dx = \frac{1}{2}\left(\frac{i}{2\pi}\right)^{n-1} \cdot \int_\Omega \left[\int_R \check{F}(p,\xi)\check{f}^{(n-1)}(p,\xi)\,dp\right] d\xi \tag{12.80}$$

Parseval's relation (12.80) can also be written as

$$(F,f) = \left(\check{F}, \psi'\right) \tag{12.81}$$

(,) is, as usual, the inner product and the function ψ' is defined by Equation 12.80. ψ' belongs to a space of functions that satisfy the conditions of Theorem 12.1. Then Equation 12.81 defines the Radon transform of a generalized function. The Radon transform of a generalized function F(x) defined on the test functions $f(x) \subset \mathcal{L}(R^n)$ is a generalized function $\check{F}(p,\xi)$ defined on the space of test functions $\psi(p,\xi)$ satisfying the conditions of Theorem 12.1. Moreover, in view of our earlier discussion about the unessential functions, the transform is undefined only to within an unessential function.

Let g(x) be the "characteristic function" of some bounded region V in a space of odd dimension. This means that g(x) = 1 for $x \subset V$ and 0 otherwise. Then the Radon transform $\check{g}(p,\xi)$ of g(x) is the area of intersection of the hyperplane $\xi \cdot x = p$ with the region V. Denoting this area by $\check{S}(p,\xi)$, we have from Parseval's relation (12.79)

$$\int_{R^n} f(x)g(x)\,dx = \frac{1}{2}\left(\frac{i}{2\pi}\right)^{n-1} \cdot \int_\Omega \left[\int_R \check{S}(p,\xi)\check{f}^{(n-1)}(p,\xi)\,dp\right] d\xi \tag{12.82}$$

or

$$\int_V f(x)\,dx = \frac{1}{2}\left(\frac{i}{2\pi}\right)^{n-1} \cdot \int_\Omega \left[\int_R \check{f}(p,\xi)\check{S}^{(n-1)}(p,\xi)\,dp\right] d\xi \tag{12.82a}$$

Interestingly, Equation 12.82a yields an interpretation of Parseval's relation. It says that the integral of the function f(x) over the bounded region V can be obtained by knowing its integrals over every hyperplane intersecting V. Physically, this is intuitive.

Furthermore, the volume v of the bounded region V can itself be obtained from the areas $\check{S}(p,\xi)$'s. In order to see this, let f(x) itself be a characteristic function. Then Equation 12.82a reduces to the volume v given by

$$v = \left[\left(\frac{1}{2}\right)(2\pi)^{n-1}\right] \int_\Omega \left[\int_R \left\{\check{S}^{(m)}(p,\xi)\right\}^2 \check{S}(p,\xi)\,dp\right] d\xi \tag{12.83}$$

where m = [(n – 1)/2]. The surface integral is over a unit sphere. Letting n = 3, Equation 12.83 yields the familiar fact that the volume of a sphere of radius R in a three-dimensional Euclidean space is just $(4\pi R^3/3)$.

The integral on the left-hand side of (12.82) is the integral of f over the region V. Let us now extend the problem so that the region is unbounded. Equation 12.82 then interprets as : to calculate the integral of f over an unbounded region V if what is known are its integrals over every hyperplane intersecting V. If we find the solution to this problem, then it would generalize the concept of the area of a hyperplane in an unbounded region. Since the characteristic function of an unbounded region is nonintegrable, the solution will also generalize the definition of the Radon transform to nonintegrable functions.

Before leaving this section, we would like to illustrate briefly the concept of the area $S(p,\xi)$ of an unbounded region. Although the area does not in general exist in an unbounded region, the volume can be made to exist by defining a differential volume element as $S(p,\xi)dp$. One considers the unbounded region as the limit of a sequence of bounded regions, that is, limit $h \to \infty$ $V_h = V$. The area $S(p,\xi)$ is then generalized as the $h \to \infty$ sequence of the bounded areas $S_h(p,\xi)$, which converges to some function $S(p,\xi)$.

Two simple examples of Radon transform of generalized functions are given after Reference 31.

IX. APPLICATION TO PARTIAL DIFFERENTIAL EQUATIONS

1. $f(x) = \delta(x)$

Let f(x) be the density with which some mass is distributed in R^n. Let (p,ξ) denote a hyperplane. Then the total mass in the half-space $p' < p$ is given by

$$\check{M}(p,\xi) = \int_{R^n} f(x)H(p - \xi \cdot x)\,dx, \quad p' = \xi \cdot x$$

where H is the Heaviside function. Differentiating with respect to p yields

$$\check{M}'(p,\xi) = \int_{R^n} f(x)\delta(p - \xi \cdot x)\,dx = \check{f}(p,\xi), \quad ' \equiv \frac{\partial}{\partial p}$$

Consequently, the Radon transform of f(x) is $M'(p,\xi)$, which is the total mass in the half-space $p' < p$.

Now let the mass density distribution be a delta distribution, that is, let $f(x) = \delta(x)$. The mass is thus concentrated at the origin. Then M is unity if $p > 0$ and is zero if $p < 0$. The derivative $M'(p,\xi)$ is then a delta function in p, namely, $\delta(p)$. We then have the result that $\delta(x) = \delta(p)$.

2. $f(x) = a(x_1)\delta[x \backslash \{x_1\}]$

The mass is now a line distribution only on the x_1 axis. The distribution is such that the total mass is finite. Therefore, the integral of $|a(x_1)|$ over the real axis is finite. Consider again the half-space $p' < p$. The hyperplane p' crosses the x_1 axis at (p'/ξ_1). If $\xi_1 > 0$, then

$$\check{M}(p,\xi) = \int_{\frac{p}{\xi_1}}^{\infty} a(x)\,dx$$

If $\xi_1 < 0$, we have

$$\check{M}(p,\xi) = \int_{-\infty}^{\frac{p}{\xi_1}} a(x)\,dx$$

Taking the derivative,

$$\check{M}(p,\xi) = |\xi_1|^{-1}\,\check{a}\left(\frac{p}{\xi_1}\right)$$

Thus the Radon transform of $f(x) = a(x_1)\delta[x\backslash\{x_1\}]$ is given by $|\xi_1|^{-1}a(p/\xi_1)$, the total mass remaining finite.

X. APPLICATIONS

One application of Radon transform was already discussed in Chapter 2. The problems of Kirillov discussed there were examples of Radon's transform. Its two- and three-dimensional forms essentially constitute the basis of reconstructing a density distribution by various imaging techniques such as X-radiography. The density function is recovered from the radiographic projections (which are the Radon transforms of the object) via the inversion formulas presented above. Here we discuss another very important application of the Radon transform, namely, its application in the solution of partial differential equations. The central point is that Radon's transform allows a multidimensional problem to be reduced to a problem in one dimension. This section is based on Helgason.[33]

Let us consider the differential equation $\mathbf{D}u = f$, $f \subset \mathcal{L}(R^n)$ and $\mathbf{D}$ a differential operator in R^n given by

$$\mathbf{D} = \sum_{(k)} a_{k_1} \ldots a_{k_n} \partial_i^{k_i}$$

a_i's are constants. We simplify matters by working in a space of odd dimension. As an example, let $\mathbf{D}$ be a differential operator $\mathbf{L}(\partial_x,\partial_y,\partial_z;\partial/\partial t)$ which may be a polynomial of some degree m in the derivatives. Consider the equation $\mathbf{L}(\partial_x,\partial_y,\partial_z;/\partial t)u(x,y,z;t) = 0$. The Radon transformed equation, namely, $\mathbf{L}(\xi_{i\partial p};\partial/\partial t)u(p,\xi;t) = 0$ is clearly a one-dimensional equation in only p as opposed to the original three-dimensional equation, the coefficients ξ_i's acting merely as parameters.

Consider, therefore, the "reduced" equation $\mathbf{D}u_\xi = f_\xi$ and look for a plane wave solution u_ξ. The solution of the multidimensional equation $\mathbf{D}u = f$ is then

$$u(x,t) = A\Delta^{\frac{(n-1)}{2}} \int_\Omega u_\xi(x,t)\,d\xi$$

That this is a solution can be checked as follows. Since $\mathbf{D}$ and Δ commute, we write

$$\mathbf{D}u = A\Delta^{\frac{(n-1)}{2}} \int_\Omega \mathbf{D}u_\xi(x,t)\,d\xi = A\Delta^{\frac{(n-1)}{2}} \int_\Omega f_\xi\,d\xi = f$$

where $f_\xi = f(\xi\cdot x,\xi;t)$. It is assumed that the restriction of D to the space of the plane waves with normal ξ is nontrivial. This means that the reduced equation must possess solutions which exist and vary smoothly with ξ.

Let us consider the following Cauchy problem as an example.

$$\Delta u = \ddot{u} \text{ with the Cauchy data } u(x,0) = 0 \text{ and } \dot{u}(x,0) = f(x), \quad f \subset \mathcal{L}\left(R^n\right) \tag{12.84}$$

If $u_\xi(x,t)$ is a solution of the reduced problem, then

$$v(x,t) = \int_{R^n} f(y)Z(x-y,t)\,dy \tag{12.85}$$

with $Z(x,t) = \int_\Omega u_\xi(x,t)\,d\xi$ and f a continuous function is a solution to the original Cauchy problem (12.84). For more details see John.[32]

Now the function f(x) can be decomposed into the plane wave functions according to Equation 12.37 as follows

$$\Delta^{\left(\frac{n+1}{2}\right)} \int_{R^n} \int_{\Omega} f(y)\left|(x-y),\xi\right| d\xi = 4(2\pi i)^{n-1} f(x) \tag{12.86}$$

Equation 12.86 is the most important relation of this section. The basic idea of solving the Cauchy problem is this. Find an appropriate plane wave function u_ξ which when substituted for $|((x - y),\xi)|$ in Equation 12.86 will result in the solution of the Cauchy problem and is such that $t \to 0$, the time derivative u_ξ will reduce to the quantity $((x - y),\xi)$. Towards this, we take the time derivative of Equation 12.85, set t = 0 and compare with Equation 12.86. This yields

$$\dot{Z}(x,0) = \int_{\Omega} |\xi \cdot x| \, d\xi$$

It is not difficult to verify that u_ξ must be of the form

$$u_\xi(x,t) = \left(\frac{1}{2}\right)\left[(\xi \cdot x + t)^2 \ \mathrm{sgn}(\xi \cdot x + t)\right]$$

The solution to the Cauchy problem is then

$$u(x,t) = \left[\frac{1}{8(2\pi i)^{(n-1)}}\right] \Delta^{\frac{n+1}{2}} \cdot \left[\int_{R^n} f(x-y) \int_{\Omega} \left\{(\xi \cdot y + t)^2 \mathrm{sgn}(\xi \cdot y + t)\right\} d\xi \, dy\right] \tag{12.87}$$

For n = 3, Equation 12.87 can be written as[33]

$$u(x,t) = \frac{-1}{16\pi^2}\left(\frac{\partial}{\partial t}\right) \int_{R^3} f(y)\left[\left(\frac{\partial}{\partial t}\right) \int_{\Omega} \mathrm{sgn}\left\{(x-y,\xi)+t\right\} d\xi\right] dy \tag{12.88}$$

The d_ξ integral can be calculated by replacing the integral over the surface Ω to one in p as follows. Let us consider a sphere of radius r and a vector x in the Euclidean space R^n. The sphere can be thought to be composed of the hyperplanes perpendicular to the vector x. Consider one of these hyperplanes at a distance p from the origin. A plane wave function $g(\xi \cdot x)$ is then simply $g(p|x|)$ on this hyperplane if ξ is on the sphere. This plane wave function is, of course, constant on the hyperplane. The volume of the hyperplane can be written as

$$\left[\frac{\Omega_{n-1}}{(n-1)}\right]\left(r^2 - p^2\right)^{\frac{(n-1)}{2}}$$

Ω_{n-1} is the total surface measure of a sphere in the space R^n. Then integral of g over the volume of the sphere can be written as

$$\int_{|x|<r} g(p,\xi)\, d\xi = \left[\frac{\Omega_{n-1}}{(n-1)}\right] \int_{-r}^{+r} \left(r^2 - p^2\right)^{n-\frac{1}{2}} g(|\xi| p)\, dp$$

Differentiating this expression with respect to r and then setting r = 1 yields the surface integral

$$\int_\Omega g(p,\xi)\,d\xi = \Omega_{n-1}\int_{-1}^{+1}\left(1-p^2\right)^{\frac{(n-3)}{2}} g\left(|\xi|p\right)dp = \Omega_n h\left(|\xi|\right)$$

Specializing this integral relation for the case of $R^n = R^3$, the total surface measure Ω_2 is simply 2π and we obtain

$$\int_\Omega g(p,\xi)\,d\xi = 2\pi\int_{-1}^{+1} g\left(p|\xi|\right)dp, \quad n = 3 \tag{12.89}$$

Using Equation 12.89, the ξ integral of Equation 12.88 reduces to

$$\int_\Omega \text{sgn}\left\{(x-y,\xi)+t\right\}d\xi = 2\pi\int_{-1}^{+1}\text{sgn}\left\{|x-y|p+t\right\}dp$$

The integral is then $4\pi|x-y|^{-1}t$ if $t < |x-y|$ and is simply 4π if $t > |x-y|$. Substituting in Equation 12.88 yields

$$u(x,t) = -\left(\frac{1}{4\pi}\right)\left(\frac{\partial}{\partial t}\right)\int_{|x-y|>t} f(y)|x-y|^{-1}\,dy$$

$$= -\left(\frac{1}{4\pi}\right)\left(\frac{\partial}{\partial t}\right)\int_t^\infty d\rho\int_\Omega \rho^{-1} f(y)\,d\xi$$

$$= t\left[S_f(t)\right]$$

This is an interesting result and shows that in a space of three dimensions, the solution of the wave propagation problem at a point x in space and at a time t is given by the spherical mean of the initial function f(x) over a sphere of radius t around x. In other words, the points on this sphere propagate and then combine to produce the disturbance at x. This is precisely "Huygen's principle"[43] which states that each point on a wave front acts as a point source. The disturbances propagating from these point sources combine to reconstruct the wave front. Huygen's principle is used extensively in optics and electromagnetic wave propagation.

REFERENCES

1. **J. Radon:** Uber die Bestimmung von Funktionen durch ihre Integralwerte langs gewisser Mannigfaltigkeiten, Berichte Sachsische Akadamie der Wissenschaften, *Leipzig Math Phys. Kl.*, 69, 1917.
2. **B. K. Vainshtein:** Finding the structure of objects from projections, *Sov. Phys. Crystallogr.*, 15, 1971.
3. **S. S. Orlov:** Theory of three-dimensional reconstruction, *Sov. Phys. Crystallogr.*, 20, 1975.
4. **A. M. Cormack:** Representation of a function by its line integrals, with some radiological applications. I, *J. Appl. Phys.*, 34, 1963; **A. M. Cormack:** Representation of a function by its line integrals, with some radiological applications. II, *J. Appl. Phys.*, 35, 1964.

5. **A. Yu Sasov:** Microtomography in a scanning electron microscope, *Sov. J. Opt. Technol.*, 53, 1986.
6. **A. M. Mikhailov and B. K. Vainshtein:** Electron microscope determination of the three-dimensional structure of the extended tail of the T6 bacteriophage, *Sov. Phys. Crystallogr.*, 16, 1971.
7. **R. A. Crowther:** Procedures for three-dimensional reconstruction of spherical viruses by Fourier synthesis from electron micrographs, *Phil. Trans. R. Soc. London Ser. B*, 261, 1971.
8. **D. J. DeRosier and A. Klug:** Reconstruction of three-dimensional structures from electron micrographs, *Nature*, 217, 1968.
9. **A. K. Klug:** Applications of image analysis techniques in electron microscopy. Optical diffraction and filtering and three-dimensional reconstruction from electron micrographs, *Phil. Trans. R. Soc. London Ser. B*, 261, 1971.
10. **G. W. Faris and R. L. Byer:** Beam-deflection optical tomography, *Opt. Lett.*, 12, 1987.
11. **R. J. Santoro, H. G. Semerjian, P. J. Emmerman, and R. Goulard:** Optical tomography for flow field diagnostics, *Int. J. Heat Mass Transfer*, 24, 1981.
12. **D. N. Ghosh Roy:** Radon inversion of spectral intensities emitted by an asymmetric and absorbing plasma, *J. Quant. Spectrosc. Radiat. Transfer*, 31, 1984.
13. **M. Zwick and E. Zeitler:** Image reconstruction from projections, *Optik*, 38, 1973.
14. **W. M. Boerner:** Development of physical optics inverse scattering techniques using Radon projection theory, in *Lecture Notes in Physics*, Vol. 130, *Mathematical Methods and Applications of Scattering Theory*, J. A. DeSanto, A. W. Saenz, and W. W. Zachary, Eds., Springer-Verlag, New York, 1980.
15. **C. D. Maldonado and H. N. Olsen:** New method for obtaining emission coefficients from spectral intensities, *J. Opt. Soc. Am.*, 55, 1965.
16. **R. N. Bracewell:** Strip integration in radio astronomy, *Aust. J. Phys.*, 9, 1956.
17. **M. D. Altschuler:** Reconstruction of global-scale three-dimensional solar corona, in *Topics in Applied Physics*, Vol. 32, *Image Reconstruction from Projections*, G. T. Herman, Ed., Springer-Verlag, New York, 1979.
18. **M. R. Kundu and T. E. Gergely, Eds.:** *Radio Physics of the Sun*, IAU Symp. No. 86, Reidel, Dordrecht, Holland, 1980.
19. **C. H. Chapman:** Generalized Radon transform and slant stacks, *Geophys. J. R. Astron. Soc.*, 66, 1981.
20. **R. G. Newton:** Inversion of reflection data for layered media: a review of exact methods, *Geophys. J. R. Astron. Soc.*, 65, 1981.
21. **N. Bleistein and J. K. Cohen:** Inverse methods for reflector mapping and sound speed profiling, in *Topics in Current Physics*, Vol. 8, *Ocean Acoustics*, J. A. Desanto, Ed., Springer-Verlag, Berlin, 1979.
22. **R. G. Romanov:** *Inverse Problems of Mathematical Physics*, 1986.
23. **J. F. Greenleaf:** Computerized transmission tomography in *Methods of Experimental Physics: Ultrasonics*, Vol. 19, P. D. Edmonds, Ed., Academic Press, Orlando, 1981.
24. **G. Wade and A. F. Metherell, Eds.:** Ultrasonic imaging by reconstructed tomography in *Acoustical Imaging*, Vol. 9, Plenum Press, New York, 1980.
25. **P. C. Lauterbur:** Image formation by induced local interactions: examples employing nuclear magnetic resonance, *Nature*, 242, 1973.
26. **L. A. Shepp and J. B. Kruskal:** Computerized tomography: the new medical X-ray technology, *Am. Math. Monthly*, 85, 1978.
27. **K. T. Smith, D. C. Solomon, and S. L. Wagner:** Practical and mathematical aspects of the problems of reconstructing objects from radiographs, *Bull. Am. Math. Soc.*, 83, 1977.
28. **D. C. Solomon:** The X-ray transform, *J. Math. Anal. Appl.*, 56, 1976.
29. **G. T. Herman:** *Image Reconstruction from Projections: The Fundamentals of Computerized Tomography*, Academic Press, New York, 1980.
30. **F. Natterer:** A Sobolev space analysis of picture reconstruction, *SIAM J. Appl. Math.*, 39, 1980.
31. **I. M. Gel'fand, M. I. Graev, and N. Ya. Vilenkin:** *Generalized Functions*, Vol. 5, Academic Press, New York, 1966.
32. **F. John:** *Plane Waves and Spherical Means Applied to Partial Differential Equations*, Interscience, New York, 1955.
33. **S. Helgason:** *The Radon Transform*, Birkhauser, Boston, 1980.
34. **D. Ludwig:** *The Radon transform on Euclidean space, Comm. Pure Appl. Math.*, 19, 1966.
35. **F. Natterer:** *The Mathematics of Computerized Tomography*, John Wiley, New York, 1986.
36. **S. R. Deans:** *The Radon Transform and Some Its Applications*, John Wiley & Sons, New York, 1983.
37. **M. Abramowitz and I. A. Stegun:** *Handbook of Mathematical Functions with Formulas, Graphs, and Mathematical Tables*, NBS Applied Mathematics Series 55, U. S. Government Printing Office, Washington, D.C., 1964.

38. **P. D. Lax:** *Notes on Hyperbolic Partial Differential Equations*, Stanford University, 1963.
39. **I. S. Gradshtein and I. M. Ryzhik:** *Table of Integrals, Series and Products*, Academic Press, New York, 1965.
40. **D. M. Young:** *Iterative Solutions of Large Linear Systems*, Academic Press, New York, 1977.
41. **A. V. Tikhonov and V. Y. Arsenin:** *Solution of Ill-Posed Problems*, Winston & Sons, 1977.
42. **I. J. D. Craig and J. C. Brown:** *Inverse Problems in Astronomy*, Adam Hilger, 1986.
43. **M. Born and E. Wolf:** *Principles of Optics*, Pergamon Press, New York, 1975.

INDEX

A

B

C

D

K

L

M

N

O

P

Q

R

S

T

U

V